SEARCHING FOR MOLECULAR SOLUTIONS

SEARCHING FOR MOLECULAR SOLUTIONS

Empirical Discovery and Its Future

Ian S. Dunn

A JOHN WILEY & SONS, INC., PUBLICATION

Published by John Wiley & Sons, Inc., Hoboken, New Jersey
Published simultaneously in Canada

For general information on our other products and services or for technical support, please contact our Customer Care Department within the United States at (800) 762-2974, outside the United States at (317) 572-3993 or fax (317) 572-4002.

Wiley also publishes its books in a variety of electronic formats. Some content that appears in print may not be available in electronic formats. For more information about Wiley products, visit our web site at www.wiley.com.

Library of Congress Cataloging-in-Publication Data:

Dunn, Ian S.
 Searching for molecular solutions : empirical discovery and its future / Ian
S. Dunn.
     p. ; cm.
 Includes bibliographical references and index.
 ISBN 978-0-470-14682-8 (cloth)
 1. Biochemistry–Research. 2. Drugs–Design. I. Title.
 [DNLM: 1. Drug Discovery–trends. 2. Chemistry, Pharmaceutical–methods.
 3. Combinatorial Chemistry Techniques–methods. 4. Evolution, Molecular. 5.
 Molecular Structure. 6. Quantitative Structure-Activity Relationship.   QV 744
 D923s 2009]
  QP514.2.D86 2009
  612′.015–dc22
                                                     2009019346

Printed in the United States of America

10 9 8 7 6 5 4 3 2 1

CONTENTS

ACKNOWLEDGMENTS

I thank my colleagues for discussions and advice. In particular, I am grateful to Pim Stemmer, Val Burland, David Stockwell, Allan Sturgess, Peter Evans, Michael Yarus, and Ken Zahn. Special thanks are due to Penny Dunn for finding mistakes tirelessly, but as always any remaining errors at any level are my own responsibility. I also thank Jim Kurnick for general support.

I would also like to thank all the scientists whose creative work, always methodical and often brilliant, I am privileged to describe in these pages. On that note, I apologize to the many workers whose additional worthy and relevant contributions could not be discussed and cited due to space limitations, but more can be found in the accompanying website (ftp://ftp.wiley.com/public/sci_tech_med/molecular_solutions).

Finally, I thank my editors for their patience, and in particular Darla Henderson for taking interest in this project.

INTRODUCTION

Most of the ordinary matter in the universe is not complex. The great bulk of it is hydrogen and helium, with the former consisting of a simple diatomic molecule except under conditions of extreme temperatures and pressures. Helium, the most inert of the "noble" gases, is a chemical loner and never enters into bonding partnership with other atoms. Yet this picture of the universe is utterly contrary to our everyday experience, where highly complex molecules and molecular systems are the very stuff of existence. In a very real sense, highly organized and staggeringly complex molecular systems are what we are. Matter in the form of molecules matters, and matters immensely to us. Finding the right molecule can be a matter of life and death.

Humans made use of molecules from their environments long before they had the slightest inkling of what a molecule actually is. Indeed, for most of recorded history, many cherished supernatural beliefs actively impeded understanding of the nature of physical reality. But there are levels and levels of understanding. Long before systematic scientific investigation, people observed their worlds and made connections between different phenomena. They could make a link, for example, between a red sky at night and mild weather to follow, or they might find that eating some plants reduced some sicknesses. Consider the latter a little more. If a plant really helped an ill human being, by what means could this occur? Traditional lore might have claimed that a god just made it that way, and leave it at that, or even propose that a god was somehow resident in the plant itself. Either way, the effects of the plant and its usefulness would remain the same. But knowledge can enable powerful means for improving this state of affairs. After a long and convoluted pathway leading to modern physics, chemistry, and biology, we can take such a plant and find the actual essence of its usefulness. We now understand a great deal about the atomic basis of reality, how different atoms bond together to form molecules, and how molecules interact with each other. In quite recent times historically, we have also come to perceive that living things themselves are also composed of molecules, in highly (but not infinitely) complex systems. So we could take the plant in question, break it up into its components, and eventually find the specific molecule responsible for its medicinal properties. "Finding" the molecule also means understanding its structure, how its atomic constituents are precisely bonded together. Further on, it may be synthesized in the laboratory, thus removing a direct requirement for the original plant itself to act as a source of the molecule. Though fundamentally important, these are still only the first steps. The identified molecule can be changed in various ways to search for improved versions of it, and the mechanism of its activity in the human body can be defined. At this level of biological and chemical understanding, it may be possible to design molecules far better than the original, and which are unrelated to the original both in their structures and their modes of activity.

This sequence of progress would be recognized by many pharmacologists and molecular biologists. It is a seemingly logical series of steps from early chance discoveries (serendipity) to pragmatic observations of cause and effect (empirical discovery), through to the modern era of "rational" drug design. And indeed, this interpretation has much in favor of it. What could be more reasonable than to assume that we are at last moving beyond the primitive time when the discovery of drugs, or any molecule of use to human beings, was a hit-and-miss affair? Is this not simply an extension of "the more you know, the more you can do" principle, which leads to increasing mastery over the physical world? If we are entering a future where a molecule can be designed to fit virtually any need, is not rational design the only pathway currently worth the expenditure of time and money?

In fact, the aim of this book is to present the case that finding useful molecules is far from as simple as this and that empirical strategies will complement and even outperform rational approaches in many areas for the foreseeable future. To continue with this theme, let us step back and consider some of the issues in more detail. Although molecular discovery is applicable to a large range of human needs, the biomedical area offers a rich field of historical and ongoing studies from which to choose.

The discovery of penicillin by Alexander Fleming is often noted as a classic example of a chance-based finding of major subsequent importance. The familiar story goes that Fleming noticed zones of inhibition of bacterial growth on culture plates accidentally contaminated with the *Penicillium* fungus. Thus, serendipity played a major role in this discovery, combined with the good luck to have this observation followed up by very competent chemists. Over the span of history, your chances of recovery from a deadly bacterial infection are strongly influenced by whether or not you were born before or after the advent of antibiotics. True, the effectiveness of antibiotics has been blunted in recent times by bacterial resistance, but historically the number of lives saved has been massive. Most people would optimistically assume that ongoing progress also applies to advanced cancers and other terminal illnesses and infections that still resist cures today. In other words, we in the early twenty-first century are still awaiting the arrival of treatments that will consign such dire maladies into the category of bad memories. On a timeline reaching back into the past and extending outward, the moving dot (or moving finger, if you prefer) representing our "now" is at some undetermined point behind the future dates when these wonderful innovations will be achieved.

Then from where will these effective treatments emerge? Is there a "penicillin" for various advanced metastatic tumors, a drug with ideal properties of killing the invasive tumor cells while completely sparing normal ones? If such a drug exists, will it inevitably emerge from rational design, or will empirical approaches remain important? If a drug for any such condition is to match the appeal of penicillin, it must excel in activity and specificity. Paul Ehrlich, the German bacteriologist who pioneered antimicrobial chemotherapy, coined the term "magic bullet" for a treatment that would zero in on a target disease-causing agent while selectively sparing all normal components of the body. He and his assistants exhaustively screened a large number of compounds before finding the first effective drug treatment for syphilis with the arsenical compound salvarsan 606. A numerical tag was appended to compounds as they were successively evaluated; thus, it was not until the 606th test that something useful emerged. Yet salvarsan fell far short of being a real magic bullet. While it was more toxic for the causative agent of syphilis (the spirochaete bacterium *Treponema pallidum*) than human cells, it was a fine line between eradicating the bacterial parasite and eradicating the human host. Nevertheless, the magic bullet concept was and remains the ideal to strive toward in any drug-related therapy. Magic or not, such a "bullet" is a tool with special properties, where the keyword is "recognition". The tool must recognize its target, and only its target. The ideal tool must act like a unique key seeking a very specific lock among all other possible locks. At this point, our tool-key is but an imprecise speculation, but such conjectures can be useful and in many cases have been realized by modern science and technology.

Thus far, we have been thinking in terms of therapies for diseases, and indeed a great many molecular applications fall into this class. But many other categories of molecules of use to humans can easily be brought to mind. The diagnostic, food, chemical, and cosmetic industries routinely use both natural and artificial molecules in widely diverse fields, and improvements in many such applications would be extremely beneficial. This observation also raises a very general issue regarding the size of molecules of practical utility. To many individuals, the word "drug" conjures up an impression of a small molecule, yet both macromolecular proteins and nucleic acids can function as molecular solutions in their own right. Antibodies, for example, are an important category of modern therapeutic and diagnostic proteins. These are sizable molecules in comparison to conventional drugs, even allowing for modern "minimized" antibody versions. Protein enzymes are of immense value in a vast range of diagnostic, research, and industrial settings. Nucleic acids can also perform many protein-like functions, and we will consider these topics in more detail in Chapters 6 and 7. For the moment, it is sufficient to note that the best molecular solution to a given task may not necessarily be small. Beyond this, and even beyond biology as such, for solving problems on a molecular scale in many cases, it is necessary to think in terms of entire interactive molecular systems. But once again, "recognition" at some level or another applies to most molecular functions, therapeutic or not.

But now let us return to the three ways of metaphorically reaching into vast sets of molecules and pulling out the right one for the job (or at least a prototype), which were noted above: serendipity, empirical observation and deduction, and rational design. In the first and oldest way, people obtained natural products as therapeutics, or for other purposes. Through a hit-and-miss process clouded with superstition and lacking mechanistic understanding, many useful and often potent concoctions were derived from plant, animal, or fungal materials. This kind of knowledge must often have accrued from the chance observations or lucky accidents often referred to as serendipity, but at least in principle a second way was possible even in early times. For example, if it was accepted that small herb-like plants as a group were a potential source of useful treatments, then a systematic screening program for remedies for specific illnesses could be instituted. Of course, there would be numerous obstacles preventing early humans from doing this truly effectively, including placebo effects, toxicities, variation in active constituent levels, and lack of knowledge of statistics. The point is, such empirically based approaches need not in principle be high tech in their implementation.

Only within recent times has the third way become possible, which is usually termed "rational design" or "rational drug discovery." This is essentially a knowledge-based pathway, and this knowledge resource has to encompass several levels for there to be much prospect of success. Chapter 1 introduces this theme, and Chapter 9 is devoted to an overview of the current state of play in rational molecular design from a broad perspective. In a volume of this limited size, such a survey of rational design is necessarily brief and primarily included for contrasting with the preceding chapters where more empirical strategies predominate.

It is certain that the ability to implement rational design is of profound importance and that such design strategies will continue to grow in power and significance. Yet systematic empirical methods will continue to offer viable solutions in many areas for the foreseeable future. Though much older than rational design, efficient implementation of empirical approaches is also reliant on technological advances. Empiricism *per se* does not depend on high technology, but it can massively benefit from it. To explain this, imagine that one could reach into a universe of molecules and evaluate (for a desired activity or function) a very large number of randomly picked arbitrary compounds, such that functionally suitable molecules could be identified. Having fished such candidates out from the ocean of molecular forms, one could then identify their natures for further development. Given these abilities, in principle you would find something useful, and such is the potential power of the empirical approach. But in the real world, it is only technological advances that permit this kind of "fishing," and the subsequent molecular characterizations. Actual collections of molecules for this kind of evaluation are termed "libraries," and these can range from small chemical compounds to large proteins. Another crucially important point is that such library evaluations (or screenings) can be done successively, by taking the first result and using it as the starting point for further variation (a secondary library) and new functional evaluation. When this is done in repeated rounds, it becomes a case of applied artificial evolution.

Even so, in its actual implementation, the molecular "structure space" that is sampled with technologies currently available is naturally a tiny subset of the universe of all possible structures. Despite this, very useful findings can emerge from such theoretically limited (but still very large) libraries of molecular forms. But at this point, we come to an intersection between the rational and the purely empirical. The above randomly selected libraries can be reduced in size, or focused in a predetermined manner, if some knowledge as to the nature of the solution is available. For example, if it is clear that only a specific type of molecule can offer a potential solution, the spotlight of searching can be cast entirely in that direction, with a commensurate increase in the likelihood of success. So, in fact, the opposite poles of the purely empirical and the entirely rational are spanned by a spectrum of hybrid strategies with varying degrees of interplay, as we will see. Defining "rational design" itself depends upon where one sets the limiting boundaries, and this too we will consider in the final chapters. Empirical design can be described in generalities, but this rapidly becomes unsatisfactory. So between Chapter 1, which provides an overview, and the final chapter lies a body of detail that takes us from natural precedents to several major areas of modern research, development, and practical applications of molecular sciences. All of the fields that are included are vast and impossible to do justice to in a single volume, but attention is given to some of the major themes of interest.

There are a great many examples of biologically oriented writers pointing to the "nature did it first" principle. These authors chalk up a sizable list of human institutions and inventions (and accompanying hubris) that are allegedly humbled by comparison with phenomena in the natural world which pre-date the experience of *Homo sapiens* by millions of years. So we have jet propulsion (squids), sonar (bats; whales and their kindred), gliding (marsupial gliders and numerous others), slavery, armies, warfare (ants), dams (beavers), and on and on. More facetiously, the list is sometimes extended to powered flight (birds, including parrots), talking (parrots), and drunkenness (parrots). Since these comparisons are invariably made after the fact, following the advent of the corresponding human innovation, I have never been particularly convinced that this is of generalizable significance. (Although I will certainly concede that protohumans may have learned much from watching parrots feeding on fermented fruit, and ironically some opinion holds that observation of animals has indeed been a factor in the origins of tribal drug lore.)

On the other hand, I am not suggesting that there are no real examples at all of learning from nature; a relatively recent and well-known case is the origin of that modern wonder velcro, the inspiration for which is attributed to Georges de Mestral's alpine observations of burdock seeds, whose microscopic hooks allow adherence to fibrous material such as clothing or hair. More significantly, in the context of empirical screening and selection for useful molecules, there are examples in nature that are so pervasive and instructive that it would be foolish to ignore them. The most important is the process of evolution by natural selection, the principles of which are actively applied artificially in mutagenesis, screening, and selection processes today. Following this is the marvelous vertebrate immune system, and how it copes with the vast range of pathogens that assail higher organisms. It is appropriate for this book to include these natural precedents for empirical design, and how artificial "directed evolution" is used for analogous ends. As a final note to this "contributions from nature" theme, perhaps the most fundamental message of all, wrapped up in the secret of life itself, is the importance and power of combinatorial molecular alphabets and macromolecular folding in achieving a vast array of functional molecular possibilities. This point is implicit in several areas and will be explicitly amplified in the final chapter also.

It is not at all a trivial matter to generate a very large library of variant molecules in itself, but coming up with a successful screening process is another thing again. For the screening technology to be useful, it must allow a user to wade through the myriads of alternative forms and pull out the best one for the target at hand. Much effort in recent times has been devoted toward making screening as efficient as possible, to permit truly high-throughput processing. Here each component of the library is evaluated by an assay, which is amenable to performing on a mass scale, to maximize the number of library members that can be evaluated in a given time. This is the "big pharma" avenue, where a very substantial investment in the mass-screening infrastructure must be made such that the chemical libraries available can be processed in a cost-effective time period. Where biologically based diverse libraries are concerned (peptides or proteins), screening can be transformed into a selection process by linking the information that encodes each variant structure to the physical structure itself. This enables, in both theory and practice, a single candidate for a desired functional feature at the molecular level to be amplified sufficiently for routine analysis. Since library generation, screening, and selection are fundamental to advanced empirical molecular discovery, this book devotes a substantial portion toward their description.

From some sections of the scientific community, there has been in the past a negative attitude toward empirically based discovery processes, and to some extent this still continues despite the effective application of such approaches in many fields. In part, this may originate from the desire of many individuals to gain the satisfaction of rationally constructing solutions from first principles. For example, some pure mathematicians dislike mathematical strategies that invoke randomization (such as Monte Carlo methods), on the grounds that these lack the elegance of stringent mathematical theorems. But applied mathematics seeks practical solutions for real problems that might otherwise prove intractable, and pragmatic molecular discovery is the same. Some people have referred to molecular discovery with empirical underpinnings as "irrational design," but true irrationality, if any there be, may be more likely located in the stances of such individuals than the discovery approach itself, if it truly is the most efficient available alternative. It is unfortunate that the word "empirical" has taken on the connotations of randomness, chance, and blindly stumbling onto something useful, which is more accurately applied to "serendipity." A primary dictionary definition of "empirical" is simply that which is derived from experimental observation, rather than from theoretical considerations. Possibly, the listed subsidiary meaning of "pertaining to medical quackery" is one factor that has led to its disfavor among hard-nosed researchers. So we need to bypass such derogatory meanings and focus on the primary attribution of "empirical," as simply a very practical approach toward a problem. The central issue becomes: if a solution to such a problem is the primary goal, the strategy that gets you there first, or at least points you in the right direction, is the most rational. If such a solution is defined as "empirical," then so be it. The "problem" may be a search for a therapeutically useful drug or simply an interesting scientific question that has resisted conventional assaults.

The essential aim of this book is then to compare different modes of molecular discovery, with an emphasis on the historical and ongoing importance of empirical strategies. At the same time, it is most certainly not the intent to take sides for empirical versus rational methods, for that would be absurd. If a clear-cut rational approach exists for designing any functional molecule with a high probability of success, no one in their right minds would opt for anything else. Yet if one examines rational molecular design closely, its historical development is inevitably based on an underpinning of empirical knowledge acquisition. Moreover, the interplay between empirical and rational design approaches continues in many cases to the present day, where empirical results allow the bootstrapping of a subsequent rational strategy. But increasingly, areas of molecular discovery that were purely in the empirical realm move into the light of productive rationality. Given the pace of scientific and technological advancement, does this process continue to a logical end point where empirical (and serendipitous) discovery is relegated to the history books? In other words, the questions ultimately become, "Will rational design entirely supplant empirical methods for all molecular solutions?" and "If so, when and how long will this take to occur?" In the final chapter, the future of rational design will be considered in the broad context of future molecular discovery in general.

In grappling with these questions, the chapters of this book themselves serve as a self-referential metaphor for *modularity*, a property of great importance in biology. This is so owing to their functioning as independent topics that can be separately digested, but can also be joined together in different combinations. In turn, this implies that numerous interconnections exist between these subthemes, whose existence boils down ultimately to the entangled nature of biological informational and effector molecules and the complex connections between biological pathways. The term "interactome" has been coined within modern genomics in reference to the components of interconnected biosystems, and as a metaphor for this, the "chapter interactome" for this book is portrayed in Fig. I.1, in order to help communicate its content and message. Given the scope of the fields visited in these chapters, there is much material that could not be included through limitations of space, and some further details are provided at the ftp site.

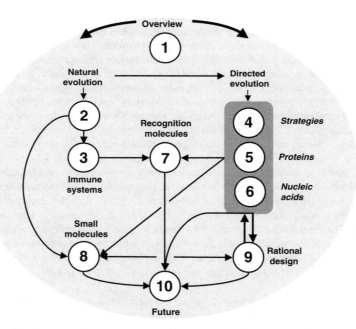

FIGURE I.1   The *Searching for Molecular Solutions* "interactome." Circled numbers refer to chapter numbers, and the major themes within each chapter are as indicated. Chapter 1 provides an overview of ways of molecular discovery, Chapters 2 and 3 provide background information on natural processes relevant to empirical molecular design, Chapters 4–8 cover molecular discovery with a largely empirical basis at different levels, Chapter 9 gives a perspective on rational design, and Chapter 10 looks at rational design limits and the general future of molecular discovery.

In the opinion of some people, the acceleration of technological innovations, particularly with respect to computing power, will enable routine rational molecular design for all manner of purposes in the future, be they concerning drugs or nanotechnological molecular devices. And not just at some vague and remote time, it is also often predicted that the exponential curve of technological progress will result in this happy state of affairs being realized within 20 years or so. Apart from remembering a profound aphorism attributed to a number of sages including Yogi Berra ("prediction is hard, especially about the future"), there are several points to make concerning these optimistic assertions. It is unquestionably the case that improvements in computational modeling and more rapid acquisition of macromolecular structural data will greatly benefit current efforts in the field of molecular design. This will also undoubtedly include the enabling of many new pathways (not yet feasible) for successful *de novo* design of both large and small effector molecules. It is also important to note that as our knowledge base increases, more and more it will be the case that "molecular solutions" concern not just single molecules but entire sets of molecules cooperating in highly complex supramolecular systems. This too can be addressed by both empirical and rational strategies, or more usually a combination of both.

Nevertheless, until such time as these advances are available (even with the most optimistic technological growth curve), it is still a matter of choosing the fastest and most cost-effective approach on hand, which will often be empirical. If the full power of empirical technologies is implemented, the historical timeline for drug discovery is potentially greatly accelerated. While rational design will eventually solve many current problems, it is not clear that it can deal with all complexities of multivariate interacting molecular systems, at least for the foreseeable future. "Blind" natural evolutionary processes have generated biological systems of staggering complexity in the first place, and artificial empirical approaches based on evolutionary principles can be an effective way of modifying them or generating novelty in its own right. While rational molecular design will certainly greatly progress in years to come, so will the technologies that currently limit applied empirical means of obtaining functional molecules. In some areas, this might lead to an undeclared "arms race" between the two broad strategies, but most likely their mutual development will complement each other, as in the present time.

By shining the light of all molecular discovery through a prism spanning the empirical–rational spectrum, we basically see it from a different perspective to a view that is narrowly focused on experimental outcomes alone. The empirical–rational principles that underpin experimental advances help us understand how these successes have developed, what more they can realistically achieve, and where they are going. And the continual feedback and interplay between these two superficially diametrically opposed approaches gives a powerful metaphor for the pragmatic nobility of human striving.

The ftp site for this book is: ftp://ftp.wiley.com/public/sci_tech_med/molecular_ solutions

1 If It Works, It Works: Pragmatic Molecular Discovery

OVERVIEW

In the Introduction, three distinct pathways toward finding a useful molecule were introduced. To start this chapter, more detail on the nature of serendipitous and empirical discovery is provided. Then, we survey how the marriage of empirical approaches and modern technology can be highly productive, and how this overlaps and leads to rational design. And there is more to rational design itself than may at first meet the eye.

THREE WAYS OF DISCOVERY

The Three Wise (*Serendipitous*) Men and Other Stories

Scientific progress is often viewed as an orderly path of advancement based on systematic testing of reasoned hypotheses. Certainly, this process is the underpinning of the scientific method, which has had a spectacular track record in unraveling and interpreting nature's mysteries. And when a scientific paper is prepared, it is necessary for reasons of economy and clarity to present the results in a developmentally logical manner. Yet in reality, the process of discovery is frequently far from such a straightforward, linear progression. This is especially so when a major advance is achieved, where initial observations are hard to reconcile with pre-existing conceptions. An experiment designed to answer a specifically posed question may yield totally unexpected results, which direct the worker into a new and perhaps revolutionary field.

This kind of process, fueled by a sizable element of chance, has come to be termed "serendipity," based on an ancient name for Sri Lanka in a Persian fairy tale "The Three Princes of Serendip." The protagonists in this story constantly make useful but accidental discoveries, prompting the eighteenth-century English earl Horace Walpole to coin the term as a result of a serendipitous event in his own life, and his chance familiarity with the Persian fable. In the latter half of the twentieth century, serendipity and its historical importance have gained a higher profile, although this may not always be acknowledged as a significant factor in scientific research. A quick search of abstracts (article summaries) on PubMed (the free online U.S. National Library of Medicine repository of published biomedical information) with "serendipity" or "serendipitous" as keywords reveals a total of around 1120 hits, a small number indeed, considering the size of database (over 18 million published articles). Throwing in the related word "fortuitous" yields an increased total score (around 2800), although still a

Searching for Molecular Solutions: Empirical Discovery and Its Future. By Ian S. Dunn
Copyright © 2010 John Wiley & Sons, Inc.

tiny portion of the total (these specific citation figures will change with time, but the proportionate use of these terms is unlikely to vary greatly). Part of the reason for this, of course, is the way scientific papers are typically prepared, as mentioned above. In reporting a series of linked findings that constitute a research article, citation of a chance event as having a major influence on the study will often appear inelegant, almost an embarrassment. So, while it is difficult to quantify, the number of acknowledged instances where chance has significantly influenced research progress is likely to be only the tip of the iceberg. To be sure, a fortuitous observation or event can only be useful if the researcher can correctly interpret it in the first place, and then has the capacity and will to follow it up. One is reminded of the famous quote from the great French chemist and microbiologist Louis Pasteur, "In the field of observation, chance favors the prepared mind."

It is hard to imagine that serendipity will ever cease to play a role in scientific and technological advancement. But before the conscious application of the scientific method, most human knowledge (such as it was) was obtained by hit-and-miss trials where chance was a major partner. For all human history until very recent times, the use and development of any sort of natural pharmaceutical has largely been a serendipitous process. Certainly, many tribal and traditional medicines do indeed contain potent and therapeutically useful drugs, but this information can be developed further in useful ways with modern technologies. These processes can have aspects of both empirical testing and rational design, so before proceeding further we should first draw some contrasts between the meanings of serendipitous, empirical, and rationally directed discoveries. The following section revolves around this theme and introduces a whimsical fellow traveler in the field of molecular discovery.

An Empirical Fable

A human progenitor (not necessarily *Homo sapiens*) becomes ill with an intestinal parasite. We can call her Lucy if you like, and consider her at least as an honorary Australopithecine, even if her cognitive endowment is a little more advanced. She wanders off from her band, most likely to succumb to the infection. At this point, for the sake of argument, we will assume that the other members of her group are not acquainted with the very concept of herbal remedies. While staggering through the scrub, Lucy notices an unusual leafy plant not previously familiar to her. For reasons we can only surmise (perhaps related to the effect of her illness itself on her better judgment), she eats about a dozen leaves. A short while later, she undergoes violent purging, but subsequently feels markedly improved. Making the cause-and-effect connection with the new plant, Lucy returns to her band and passes on her *serendipitous* observation (speaking Proto-World?).

Soon an epidemic of stomach troubles strikes the band, and thus having a number of patients and little to worry about from ethics committees, Lucy decides to test each sufferer with a different number of leaves from her plant. She is a very gifted individual and does this systematically, and on more than one occasion. The carefully interpreted results of her study show that on average, five leaves cure the affliction with the least number of side effects. Her *empirically* determined dosage is used by the band from that point on, but Lucy is a perfectionist and still feels that matters could be improved. She has no conception of how consuming a plant could cure an intestinal ailment, nor indeed of the existence of parasites (at least those not visible to the naked eye), or that different parasites might require different therapies. But in a flash of insight, she thinks of trying a large number of different plant leaves for their abilities to treat bowel illnesses. By so performing this *empirical* screening, she may discover promising new candidate plants, but her sample size of both new plants and patients is small. If she is extremely lucky, during her screen for treatments of a specific class of illness, she may also discover *serendipitously* some useful agents for entirely different medical problems. Having a need to make sense of the universe, she may later invent explanations for the effects of her pharmacopoeia involving magic or spirits. These ad hoc stories become accepted wisdom and are passed on into the folklore of the band—but do not change the degree of efficacy of her treatments. Life is improved, but still far from perfect.

One night Lucy has a very strange dream, involving creatures similar to herself, yet physically different in subtle ways. One of them speaks to her, "This is how we can help you. We'll take samples of your plant and identify the chemical constituent with the potent anti-parasitic activity. With its structure in hand, we'll be able to test hundreds of analogs to characterize structure/function relationships, and in the end come up with a compound with greatly enhanced activity!" Another of the creatures pipes up, "We can do better than that! We'll focus on the parasite itself, and define which organism-specific proteins represent the best targets for therapies. With the protein crystal structures in hand, we'll use virtual docking software to design low molecular weight compounds as specific inhibitors!" Of course, Lucy understands not a word of any of this, and her memory of it is as fleeting as dreams usually are. All she remembers for a time is one of the creatures saying, "*Rationally* designed compounds will solve your problem . . ."

Productively Applying Empiricism

These distinctions between alternative avenues for discovery of new therapeutic agents are important in the context of what is presented in this book. It would certainly appear at first glance that the rational approach (so inaccessible to Lucy's people but often within reach by us) is "the way to go." There is much to say about this, and most of this falls within the territory of Chapter 9. By its very nature, pure serendipity has a chance or "wild-card" quality, which suggests that it will essentially be quite unpredictable as to when or how often it will rear its pretty head during the conduct of research. Also, serendipity is an interactive process in the sense that it requires both a fortunate observation and the correct interpretation of the data by the observer. This itself can span a spectrum from an extremely rare occurrence acting as a "lucky break," which most capable observers would seize upon (our fabled Lucy is favored by the Fates and also talented), to mundane events that serendipitously set off a chain reaction in the minds of only a small gifted set of individuals. What better example of the latter could one cite than the (possibly apocryphal) falling apple that very indirectly planted the Law of Universal Gravitation into Isaac Newton's head? This requirement for good fortune in both experimental results and the receptivity of experimenters themselves naturally renders serendipitous discovery impossible in principle to foresee. But an empirical approach, as we took pains to note in Lucy's Fable, can be applied in a systematic manner.

Finding useful drugs from the environment is a very old human activity, which (as we have seen) can occur either purely serendipitously or by a directed empirical process. But the empirical approach can be harnessed in the laboratory to maximize its effectiveness. Rather than relying on what nature can provide in the environment, the essential technological innovation here is to produce an artificial large collection of variant molecular forms, commonly referred to as a "library." Again referring back to the fable of Lucy, the health problems of her band of protohumans stem from an intestinal parasite, which is therefore the central target for therapeutic intervention, whether or not empirical experimentalists are aware of the parasite's presence and effects. Identifying the parasite as the disease-causing agent is thus a very good start, but this will often require technologies not available until relatively recently, especially if the pathogen is invisible to the naked eye. (The existence of micro-organisms has been known since the time of Antonie van Leeuwenhoek, only a little over

300 years ago. This is a short span even compared with the duration of human recorded history, let alone the time since the arising of *Homo sapiens.*) If it is possible to grow the relevant parasitic organisms in the laboratory, it then becomes feasible to systematically screen as many drug candidates as possible to test which ones can kill the parasites or halt their growth. This empirical approach is a great advance over testing potential treatments directly on sick individuals, but it is still fairly cumbersome to the extent that each test involves a separate growth of the parasites treated with one candidate drug. To be sure, modern technologies have greatly streamlined this kind of process (whether searching for agents effecting parasite killing, or some quite distinct biological target) leading to "high-throughput" screening strategies.

Screening Versus Selection   A short detour into semantics will be useful at this point. The words "selection" and "screening" occur throughout this volume, and although they have operational relationships, they are not at all synonymous. Some precise definitions are thereby in order, especially with respect to empirical molecular identification strategies. "Screening" involves a systematic evaluation of a (usually large) series of alternatives,[*] at a variety of possible levels ranging from molecules to cells to whole organisms. As such, a screen can be conducted with any type of testing mechanism, provided the assay that is used is informative toward the desired end property. Also, it is clearly important that specific positively-scored members from the available set of alternatives (a library of some kind) can be identified, isolated, and characterized. A direct selection process, on the other hand, allows a desired alternative to "pop out" from a large background without the need to plow through the evaluation of each alternative possibility. Biological selection exerted by natural processes is a fundamental aspect of natural evolution, and often defined as "differential survival." We will consider this further in the next chapter, but for the present purposes, an example taken from simple molecular cloning can help distinguish the selection/screening dichotomy.

Extrachromosomal loops of DNA with the ability to replicate are frequently found in bacteria, and these "plasmids" have been extensively used as vehicles for DNA cloning. Insertion of a foreign segment into a plasmid allows the replication of the novel DNA along with the rest of the plasmid *vector*. But how can you distinguish between the recombinant plasmid bearing the desired foreign sequence and the original vector alone, or plasmids that have recombined with some other spurious sequence? Consider if the desired extraneous DNA segment happened to encode and permit the expression of an enzyme that enables the bacterial host of the plasmid to escape killing by an antibiotic. One such enzyme is β-lactamase, which breaks down penicillins and thus allows bacteria producing it to survive in the presence of penicillin and other β-lactam antibiotics. Now, if β-lactamase itself was the target, one could laboriously screen numerous clones of bacteria for its expression by some assay that identified the appropriate enzyme activity (*in vitro* assays determining the rate of breakdown (hydrolysis) of appropriate β-lactam antibiotics). Though certainly possible, this would be rather foolish, since a vastly better approach is to use antibiotic resistance itself to "pull out" the clone of interest (Fig. 1.1). It is fairly obvious that only bacterial cells that possess the antibiotic resistance "marker" can grow in the presence of the specific antibiotic. Therefore, if the mixed population of bacterial clones is propagated along with the antibiotic, only those bearing the desired resistance gene can form colonies.

---

[*]While on the topic of semantics, it may be noted that instead of "alternatives," the word "candidates" could have been reasonably substituted in its ordinary usage. This was avoided, though, since "candidates" is often used to refer to a relatively small subset of possibilities identified through early rounds of library screening, rather than the whole library itself. A candidate molecule is thus on a molecular short list.

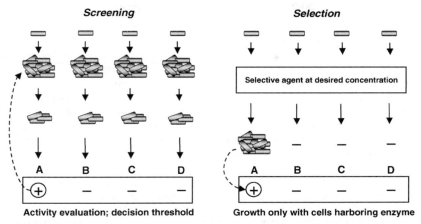

Screening

Selection

Selective agent at desired concentration

A   B   C   D

⊕ — — —

A   B   C   D

⊕ — — —

Activity evaluation; decision threshold   **Growth only with cells harboring enzyme**

FIGURE 1.1   Screening versus selection. For screening, individual bacterial cells (represented by rods) grow into macroscopic colonies, from which samples can be taken (and repropagated if necessary) to allow evaluation of an activity. If measured activity surpasses a decided threshold, these data identify the corresponding colony bearing the desired genetic information encoding the enzyme (or other protein) of interest. If a selection process is applicable, it is exerted at the initial single cell level, and only cells expressing an appropriate enzyme or other protein (enabling growth in the presence of the selective agent) will survive and form colonies. For screening, the information from the assays allows colony identification, whereas the selection itself provides evidence that the desired gene product is present (dotted line arrows).

Of course, this is a very special case, and most cloned segments will not be so readily selected. A very common strategy is to ensue that the plasmid vector itself bears an antibiotic resistance marker, so that cells that have taken up a plasmid (whether the original parent or its derivative bearing a foreign DNA insert) can be readily selected from background of cells with no such plasmid. A foreign segment that is not directly selectable can then be identified through some screening process (often nucleic acid hybridization). In such cases, *select* for the plasmid, *screen* for the insert. This process is as equally applicable to eukaryotic cells as it is to bacteria.

These examples help to demonstrate the differences between selection and screening, but perhaps have still not quite pinpointed the essential distinction. We can make a better definition of selection as a process applicable *at the level of individual replicators* of any description, which allows specific replicators[*] within a population to be directly isolated, amplified, and identified as unitary entities. Let us explain this further by considering the above model of β-lactamase enzymes and antibiotics in the context of Fig. 1.1. In the screening model, we have started with a population of individual bacterial cells (individual replicators for our present purposes) and allowed them to grow into visible colonies, some of which have a plasmid encoding and expressing β-lactamase activity, and some of which do not. Assaying samples of each colony for β-lactamase levels will allow identification of the specific colony that is

---

[*]A "replicator" here is defined as a supramolecular unitary entity that carries both effector molecules and informational molecules, which enable its self-replication. We will see in Chapters 3–6 that selection is mediated at the *phenotypic* level, which is usually (but not exclusively) comprised of different molecules from the informational molecules carried by replicators. The phenotype is accordingly encoded by the latter informational molecules.

producing sufficient enzyme to exceed a prechosen threshold. In contrast, the selection model acts directly on bacterial cells at the outset, by only permitting the growth of cells expressing a high enough level of β-lactamase. Hence, in the selection model, individual replicators are "chosen" and amplified, which holds true for any biological selection process (including natural selection[*]).

An additional point flows from this: humans can in effect act as the "choosers" for selection, and this can potentially cause confusion between the levels where screening and selection operate. "Selective breeding" is a familiar term to most people, which conjures up images of dogs and other domestic animals, or many domesticated plants. We will touch upon this again at the beginning of Chapter 4, but let us examine the "selective" connotations of this a little more closely for the present purposes, with another metaphorical tale:

> An evolutionarily minded farmer with a herd of cows decides for obscure reasons to breed for a dark-coat color in his bovine charges. He systematically evaluates each animal and chooses the darkest subset of them for further use and correctly concludes that this undertaking is classifiable as a screening process. Then, he rather pointlessly announces to his cows, "Now, consider that in effect an environmental change has occurred, such that a pallid coat has become a definite low-fitness phenotype. *I* am in fact the relevant change in these circumstances!" He then sacrifices all cows except for his chosen dark-coat subset, and enables them to breed. Through this differential propagation, he concludes that he also has acted as the selective agent discriminating the "fittest" cows from the remainder.

> Selection can thus be an entirely natural process, but screening is a human activity that systematically evaluates a large number of alternatives for a desirable property. Artificial selection can be superimposed on such a screening process through its enabling of a human-based choice for differential replication. Nevertheless, the distinction between the processes is important, and here is a way to remember it:

State clearly just what you mean
(Because misuse is obscene)
Is direct selection
Your correct direction?
Or is your intent to screen?

How to be a Librarian   Screening can be carried out on a small scale, or taken up to "high-throughput" levels (Fig. 1.2). But another means of empirical molecular identification exists, in the form of directly selecting a candidate from a library through a specific molecular binding interaction. Before looking more closely at this, we should first think about the nature of targets and probes, in relation to molecules of interest isolated from a library. It should be noted that a "target" molecule in this case is simply the "starting point" defined molecular structure for which an appropriate functionally interactive compound is sought. Again there are important semantic issues to take note of. In this terminology, a "probe" is essentially synonymous with a target, as in a statement of the type, "the target protein was used to probe the library for molecules which would bind to it." The pharmaceutical industry routinely refers to the choice of "drug targets," and the search for new ones. Choosing consistent word usage in this area is not trivial, since loose terminology may be confusing.[†]

---

[*]An exception would be selection putatively operating on groups of replicators rather than individuals, as with selection at the group or species level. This often controversial topic will also be noted in Chapter 2.

[†]By an alternative viewpoint, library "hits" (primary active candidates) might be seen as the targets for the screening process itself, but this is not the standard meaning of the word "target" in this context.

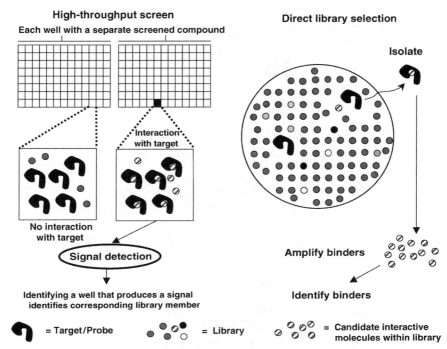

High-throughput screen

Each well with a separate screened compound

Interaction with target

No interaction with target

Signal detection

Identifying a well that produces a signal identifies corresponding library member

Direct library selection

Isolate

Amplify binders

Identify binders

= Target/Probe   = Library   = Candidate interactive molecules within library

FIGURE 1.2 Depiction of empirical screening of large collections of candidate molecules (libraries) for interaction with a specific target molecule (or probe). In high-throughput screening, a large parallel series of tests are performed, physically separated in discrete wells each with the same target (which can be a single molecule or a complex system such as a mammalian cell). Each well is tested with a defined separate drug candidate from the library. Some form of positive read-out signal is necessary to recognize potential positive candidates, such that specific positive wells identify the corresponding candidate drug. In a "direct library selection," the library is treated with the target molecule. It is necessary to have the capability of separating bound candidate molecules, and then amplifying the bound fraction in order to obtain sufficient material for identification purposes.

"Target" for our purposes will therefore be defined in general as a specific molecule or system that serves as an objective, toward which one seeks another molecule that will modify the properties of the objective molecule or system in a desired manner. This usage is consistent with the original wording of Paul Ehrlich referred to in the Introduction, where a "magic bullet" (drug) is one's holy grail against disease. And bullets naturally must be aimed at a target, pathogenic microorganisms in Ehrlich's case. Successful "hits" within a library must by definition interact with the target system, and where the target is a defined molecule, the interactive library molecules may themselves be termed "ligands."

But for molecular discovery in general, both large and small molecules can be of interest as functional mediators, and either can act as targets or library members. For example, libraries of mutant variants of a single protein can be screened for gains in thermostability. The target in this case can be considered the original "wild-type" parent, variants of which are sought to exhibit improvements over the parental properties. But the library of mutants itself does not contain the original target, although obviously this parental protein is used as the yardstick by which

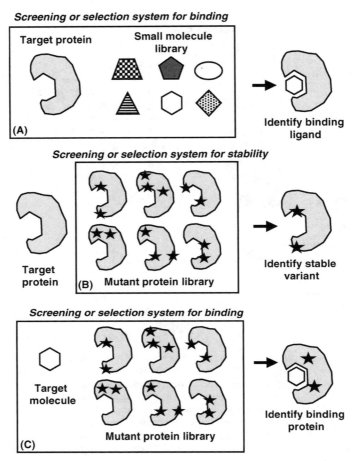

FIGURE 1.3   Invariant targets and diverse libraries, where the screening/selection systems A–C are boxed. (A) With the depicted small molecule library, the designated target is physically part of the system for identifying a binding ligand. (B) When a library of mutant variants of the target protein is screened or selected for stability, the target is not physically part of the library itself, but may be used as a reference for measurement purposes. (C) Searching for a binding protein in a library of mutants for a specific target molecule ligand. This is conceptually the same as seeking a specific antibody.

improvements are gauged. A library then is always a diversified collection of molecules, while the target is invariant. (Note, though, that invariance does not mean that the target must necessarily be a single molecular entity; whole cells or even whole simple organisms can be screened with molecular libraries.) Different types of target and screening arrangements are depicted in Fig. 1.3.

The more information available concerning the biological system that one wishes to modify, the more favorable the chances of defining the ultimate functional target molecule(s), and in turn, the better the chances for designing an optimal screening process for candidate drugs. Accordingly, if one or more specific proteins of a parasite that are essential to its functioning are known, they can be used as targets for drug development. This can in principle be either through rational design, as in Lucy's dream, or by "applied empiricism" where such target proteins are used to screen a molecular library in the laboratory. Very broadly, this can be done either by a screening assay with maximal possible speed and processing efficiency (hence *high-throughput*) or by a selection process, subject in both cases to the nature of the library itself. In Fig. 1.2, the principles of high-throughput screening versus direct library selection are contrasted. The basic difference between the two approaches concerns the means for identifying the specific candidate binding molecules. In the case of high-throughput screening, evaluation of each member of a library is done as a separate test, which requires devising some measurable assay for a positive response, whether this is killing of a parasite or a tumor cell, changing expression of a specific gene, or a huge range of other biological responses. Thus, the separate screening reactions for each library member can be performed in minute wells of special plates, and set up and assayed collectively by robotic mechanisms. The library chemical members (of whatever nature) are added to each plate as a pre-arrayed grid, such that a positive assay signal from a specific well automatically provides the grid position and identity of the library member.* Because each library member is separately assayed, the target system can be indefinitely complex provided an unambiguous read-out assay for the desired effect can be devised.

And what of the selection-based alternative? In such a process, a molecular target is mixed with the combined collection of molecules within the library, under conditions where specific molecular forms (if represented in the library in the first place) can interact with the target/ probe. The underlying premise here is that a molecule interacting with the target will bind to it with significant affinity. (The functional consequences of such binding are another matter, but it is the binding itself that enables one to "pull out" candidate library molecules of interest.) For such a library-based selection to work effectively, some other fundamental requirements must also be met. The bound complexes between the target and candidate library molecules must be purified away from all other irrelevant library members. Then, the interactive library molecules must be identified, but this is generally not possible directly. Since library collections of molecules are large, any specific molecule constitutes a tiny fraction of the total range, and the amount that binds to the target is commensurately small. So an additional *amplification* step† is needed, where the bound library molecules are increased in number until such a stage where they are amenable to characterization. The necessary isolation of complexes between target and bound library molecules, and subsequent amplification of bound molecular candidates is also depicted in Fig. 1.2. Since this complete operation applies at the level of individual replicators that are amplified as unitary entities, we are entitled to indeed refer to it as a "selection process" by the earlier definition we have arrived at. In practice, the first pass of the target through the library will often yield a set of molecules highly enriched for true target-binding candidates but not yet free from extraneous library members. A second or third pass of the target through increasingly enriched libraries may accordingly be needed before useful candidate molecules can be evaluated.

Another way of looking at both of the library screening processes of Fig. 1.2 is to see them as the implementation of search algorithms for finding members of the library of interest that fulfill preset search criteria. A flowchart of the algorithm (Fig. 1.4) refers to a sequential evaluation operation, which would mirror a laborious one-by-one screen of a set of *N* compounds for useful activity. Such a process is similar to the pioneering chemotherapeutic experiments of Paul

---

*As we will see further in Chapter 8, this is really a special case of encoding the library content (by *spatial positioning*), but other means for library encoding also exist.
†Biological nucleic acids can be amplified by replication, and proteins indirectly amplified through encoding them with replicable nucleic acids. This is detailed in Chapters 5–7 and revisited in the final chapter.

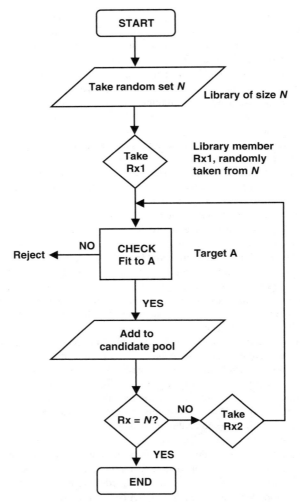

FIGURE 1.4 Empirical library screening as a search algorithm. "Check fit" refers to the process for assigning a potential "hit" (often binding of a library member to target A, but complex screening processes are possible). The algorithm is nondeterministic since the library of size N is taken randomly from a much larger molecular space.

Ehrlich referred to in the Introduction. Both the high-throughput screening and library selections of Fig. 1.2 side step this problem by engaging in extensive *parallel processing*. In the former case, each library member is screened separately, but as components of a very large array such that each library member is identifiable through some encoding process (spatially as in the plate grid example of Fig. 1.2). For direct selection from a library, the parallel processing is done in a single mixture, with the unbound "rejects" physically separated from those bound to the target (and thus satisfying the primary search criterion).

But this is still not the end of the story. Such a search process (by any strategy) really only constitutes a single round of evaluation, and in practice a workable solution is unlikely to emerge directly. First, because there is always "noise" in the experimental operation, the first-round pool of candidates would need to be rescreened to confirm their correct status. Beyond this, the primary candidates typically serve as frameworks to generate secondary variant libraries based on the demonstrably useful first-pass molecules. Repeated rounds of screening or selection and identification of improved candidates (often under increasingly stringent conditions) is a way for cumulative beneficial changes to accrue, and is the essence of evolution, of which there is much more to say below and in later chapters.

Having considered these points, let us now think about libraries from a somewhat different stance . . . .

Demon-strating the Power of Empirical Screening and Selection   The great nineteenth-century physicist James Clark Maxwell, famed for demonstrating the unity of electromagnetic phenomena, once imagined tiny "demons" that could sort atoms or molecules by virtue of their temperatures, and thereby reverse entropy. Although later physicists have shown the impossibility of this process even in principle, a looser version of Maxwell's demons can be used as a metaphor of sorts for a device or structure that is capable of performing some useful function on a nanoscale level. While reversing entropy is indeed a tall order, a molecule-sized demon could be proposed to perform a vast number of more modest but highly useful tasks. If a "task" is stripped down to "recognize a specific target molecule, and no other molecule, and bind to it in a specific way," then a demon becomes nothing more than a tool-key, a magic bullet, or an idealized drug molecule. But in order to help remind us of the "demonic meaning" in the context of this specific metaphor, perhaps an acronym for DEMON (Discovered Empirically, Molecules Of Note) would be useful. At the same time, another acronym (Don't Ever Molest Other Names) cautions us not to get carried away with this sort of thing.

To clarify the different approaches to empirical screening, in a brief interlude, metaphorical "demonic" models for molecular libraries can illustrate the process of extracting molecules of interest from them. First, as a metaphor for the type of chemical library involved with high-throughput screening, think of a very large number of boxes, each with the same target sitting inside. The target is the object or group of objects that you want to modify by means of a molecular interaction, and this target may be a highly complex system in its own right. Let us visualize it as a number of balls, where each ball has a hole with a different specific shape. Previously, you have shown that if a hole in a ball is filled with a closely matching "key," then the ball changes in some way, but it is very hard to predict what the overall effect will be (especially since a changed ball can in turn influence other balls within the same box). Remember that the target itself is comprised of all such balls collectively in the same box (there might be thousands of balls per box defining one specific target), and you have many, many copies of these target-containing boxes. You are looking for a way to change this target in a particular manner; perhaps to make the balls jump up and down in unison, for example. This specific change in the target is actually the *signal* that will enable you to identify an agent that produces the effect that you are seeking. So you have your target and your aim; now you want to find something that will produce the results you want, and a very large and obliging library of demons is ready to help.

Somewhere within this vast demonic collection is the right one for the job, meaning that you have to figure out how to screen the library to find it. You can picture the demons any way you like, but each demonic individual carries a tool with a unique shape, which will be tested for its ability to fit into any hole that the demon finds. It is your hope, then, that a particular demon within the library will hold a tool that will fit a *specific* target ball, which in turn will cause *all* of the balls in the target box to jump up and down. (Other demons may have tools that fit different target balls, but without eliciting any trace of the desired effect. Also, although the vast majority of demons will be irrelevant to your needs, it may be the case that more than one specific demon can trigger the same result that you are seeking.) So you arrange a huge number of your target boxes, and put one specific demon next to each box. Every demon has a unique number as well as its unique tool, so you place the demons in a set pattern alongside the boxes. When the demons jump into their target boxes on your command, if you know which box has the right response you will then also know which demon ("it is box 300 from the left and 2000 from the top . . . that means it is demon no. 60,000"). But there is a catch. The demons are not so obliging that they will yell out and tell you when the objective is achieved. They are not too bright, actually, since all they will do for you is try out their tools for a fit in a hole (although this they do very diligently). To make matters worse, they insist on closing the lids of the boxes when they jump in. So you are stymied unless you can come up with a way of telling independently when one of the demons has been successful. You realize that you can measure the jumping of the balls by the sound they create inside the box, and the more the balls within each box jump up and down, the greater the sound.

Inspired by this, you arrange a monitor on each box that will automatically measure the sound after the demons jump in and send the information back to you. In practice, you may not need absolutely 100% of the balls to move, perhaps 80% would be satisfactory. In any case, you may find a range of sound levels, where the vast majority of demons cause no sound to issue from their respective boxes whatsoever, but a small number produce varying sound levels. You could simply pick the demons that produced the loudest results and study them further. Even better, it might be possible to set up preliminary tests where you independently make the balls jump, and measure the loudness in order to calibrate the sound signals (from the target boxes after demon entry) with the numbers of jumping balls. But in any case, you have empirically "fished out" some candidate demons for the desired effect. You may not know how changing one ball directly could affect the majority of balls, but you surmise that there may be "master balls" that respond to an exact matching "key" to their holes (provided by specific demons) with a cascade of actions that ultimately affect most or all of the other balls. This you can study further; perhaps, it will lead to other ways of rationally changing the actions of the balls within your total target.

There may be additional properties of this target (or other unrelated targets) toward which you might want to search for useful modifying demons, but you find that there are practical limitations on how many target boxes you can use in your screening of a demon library. This in turn limits how many demons you can check, and the more the demons you can screen, the greater the chances of success. A friend makes an interesting suggestion to you. "Why screen only one demon per target box? You could make pools of say 10 random demons and have 10 of them jump into a box at once. From this you could identify promising pools of demons in the usual manner, and then split these pools up into individuals for rescreening to find the one that's really active. So if you used the same maximal number of target boxes as before, you could increase the number of screened demons 10-fold! This procedure is called sib-selection and it does work in some circumstances." You investigate this further, but find a potential problem. Some demons within the library can knock out the ability of the "master balls" to respond to the very demons you are searching for (they do this indirectly by binding to other balls that in turn modulate the "master balls"). So you might miss a positive signal from some demon pools by this kind of interference effect. You realize, though, that this problem would not apply for a very simple target box, namely one with only a single ball within.

This leads us to the second metaphor for the process of direct selection from a library. In this case the target is simpler, and the aim initially is to find demons that will bind tightly to it. (Binding in itself is required as a prerequisite for any functional changes to the target, and any such alterations can be investigated after binding demons are found.) This time instead of having the target in separate boxes, you can simply imagine an enormous number of copies of a one-ball target (with the same kind of hole that can receive a specific key) bobbing around in a vast swimming pool. Each ball is free to move, but is firmly attached to the bottom of the pool by a pegged rope. You also have a demon library again with each demon armed with a unique tool, but with important differences. These demons are not individually numbered, but have the ability to multiply to form exact copies of themselves if they receive the right stimulus. This is very important because another strange property of these demons is that by themselves (as individual demons) they are invisible, but large quantities of perfect copies of specific demons can be collectively seen and identified. So you take this type of demon library and throw it onto the pool with the one-ball targets. They cannot be seen, but the demons energetically swim around trying out their tools for matches with the single type of target holes. Almost all the demons fail to find matches, but a tiny set of them (from the multitude within the library) are successful. Now, despite the size of the pool, you have the ability to rapidly drain all the water away, and anything not tied down is also drained away. You do this, and the only things remaining are the attached targets and any demons that may have found a good and strong hole match. Snap your fingers (or whatever stimulus is needed) and the demons multiply until you can see and identify them, and you have your candidate binders of the target.

In making the library comparisons as in Fig. 1.2, it should be noted that there are certain variations on these two overall themes that will be described in later chapters. If the biggest advantage of high-throughput screening is its ability to use high-complexity targets and complex screening assays, the great benefit of direct selection from a library is the sheer size of the collection of variant molecules that can be practically evaluated. Whether biologically active agents are screened from environmental sources or artificial libraries, and regardless of the screening process itself, one of the most fundamental issues is the size and diversity of the total pool of molecules that is available. If an appropriate molecule for a given target is not represented within a molecular collection, then it is clear that no amount of sophisticated selection or screening will produce the desired molecular solution. What is an appropriate way of visualizing the nature of molecular libraries in general, in order to gain insight into both their strengths (diversity, size) and weaknesses (constraints on diversity, exclusion of useful compounds)? One way is to arrange molecules of various classes into mathematically defined multidimensional spaces. For the present purposes a metaphorical space in which all stable molecules are found can be used to illustrate some principles involved with real collections of molecules (natural or artificial libraries) at later points.

Dreaming of Pandemonium: A Universal Molecular Space

In the above "demonic library" explanations, the well-known metaphor of Maxwell's demons was extended to include useful chemical agents because in reality specific modification of molecular systems is best done with other molecules. And finding the correct molecular tool for a task may seem demonically difficult, even if one accepts that in principle such a tool exists. So any given "demon" in this sense is a specific grouping of atoms held together by well-understood chemical bonds, with sufficient stability under normal conditions as to be practically useful. Having defined such entities, can we imagine all *possible* molecular "demons," large or small, as a single vast set? If so, it would be hard to resist calling this pandemonium, following Milton's coinage for a place of all demons. Demons and pandemonium in general have been popular themes in various contexts. For example, the human mind has been modeled as the outcome of interactions between mental "demons" corresponding to various sensory and cognitive functions,[1] whose interplay in the arena of mind constitutes a pandemonic synthesis.

So our demons, as demons will, all reside in a vast pandemonium, which may invoke infernal associations. But being able to exploit this molecular sea-of-all-demons at will would be anything but hellish, since it would be an immense force for human progress. (Although evil applications of it could also be found given malevolent intent.) Then it is justifiable to think about this in more detail, to ensure that this molecular pandemonium, if not the original term, as a useful metaphor.

In many circumstances people need to think about large sets of objects that vary over a wide range, but with specific rules for the smallest discrete changes that can be applied to change object A into another object. This second object must by definition be closely related to the original object A, since it only differs by a basic "quantum" unit of change. Around the original A object, a "cloud" or neighborhood of such closely related entities thus exist.[2] If the rules determining transition from one object to another can be applied successively, such "near-A" objects could be modified in turn, continuing *ad infinitum*. Now, if it were only possible to modify each object in one discrete way at a time, there would be a linear transition from objects A → B → C → D, and so on. But B, C, D, and onwards in turn can be part of a branching transition series, and so likewise can each new object arising from the B, C, D pathway, and so on. Clearly a one-dimensional depiction of these branching transitional series will not suffice, but even before we decide how many dimensions are required, it can be seen that the series of objects are being arrayed into a "space" based on their relationships with each other. A fundamental aspect of this depiction is that each point or "node" represents a specific unique object in the artificial space, and the same object cannot be found at any other point in this space. And then what if multiple ways of discrete modification of the objects are allowed rather than just one? New transition series vector lines would have to originate from our original starting point of object A, radiating out and continuously branching into this theoretical N-dimensional space from each new object node.

Chemists have given many terms to theoretical spaces where the "objects" are organic molecules, including "chemical space," "design space," "diversity space," "structure space," "topological structure space," and "chemogenomic knowledge space."[3] Considerable effort has been devoted to defining these and allied spatial constructs with mathematical precision, although they may still fall short of the precision of abstract mathematical vector spaces.[4] "Shape space" has been a useful theoretical construct for the modeling of antigen–antibody interactions.[5] These studies aim to give practical guides as to the minimal number of compounds required to cover a maximal amount of diversity, or theoretical space volume.

As an explanatory aid for this book, it will be useful to invoke a universal, all-inclusive molecular space, although this is certainly not (nor could ever be) a precise mathematical construct. It is intended rather as an instructive metaphor to illustrate some real features of large molecular collections (libraries) that enable screening or selection for specific molecules with desired properties. The tag "universal" is simply meant to distinguish this mental conception from other molecular spaces with defined restricted ranges. For example, "structure space" has been used specifically in a protein folding context[6–8] as opposed to protein or DNA sequence spaces. Let us term our particular construct "OMspace," for Overarching Molecular space. Om (or Aum) in Hindu tradition denotes the ultimate reality (which includes, but is not limited to, the observable physical universe), so a universal molecular space should fit within "Om" through the rather broad mandate of the latter.

Spatial models of varying mathematical precision have been frequently used in a number of fields, especially those related to chemical sciences. Given this widespread usage, perhaps is it justifiable to refer to this "spatial categorization" as a special "meme," or mental construct that tends to replicate itself by "infectiously" spreading from mind to mind.[9] Indeed the originator of memes has himself used "animal space" as a convenient way of depicting evolutionary transitions in the animal kingdom.[10] All computer networks constituting the Internet are very often seen as a spatial array (cyberspace), but another information-based and potent meme has some relevance to molecular spaces. In this I refer to the tale of the Universal Library of Jorge Luis Borges[11] (the "Library of Babel"), where not only all books that have ever been published are represented, but also all *possible* books including all imperfect versions of real literature and oceans of gibberish are present as a single copy.* So the complete works of William Shakespeare are there, along with all the mistake-ridden variants completed by those tireless hypothetical monkeys locked in a room and typing away at random until the final works emerge. By relevance to OMspace, I do not mean that the ideal version of *Searching for Molecular Solutions* is contained within the Borges Library, although what you are reading surely falls far short of the perfect edition somewhere in the Library's metaphorical vastness. Of course, this could be said about any written work; perhaps even Shakespeare's plays could be improved by the judicious insertion of a word or two here and there. This precise issue, in fact, has been leveled as a criticism of the Library—how could you ever know when you had pulled out the perfect edition of anything; by what standards could you judge it? For the Library by definition contains not only all possible accurate knowledge, but also all possible falsehoods and red herrings.

We can nonetheless compare OMspace and the Borges Library if we consider them both as universal repositories of information. In the case of the former (or in a real physical molecular library), the information is revealed through the means for "interrogating" the library, as with a target/probe molecule that is used to try to identify an interactive binding molecule. (Thus, in effect, we are seeking the information specifying which library molecules will act as efficient ligands with the target/probe.) A potent message of the Library of Babel is indeed that all knowledge can be distilled into a problem of selection,[12] if one can but ask the right question, or deploy the appropriate "probe." A thought experiment could accordingly be devised where the Borges Library is searched for a technical book precisely specifying (in words) the composition and structure of the ideal molecule that interacts with a desired target. However, again there would be a huge (infinite?) number of books with suboptimal solutions to the binding problem, and indeed a vast range of books with wrong "solutions" (those specifying molecules that do not complete the desired task of high-affinity interaction with the target probe). The "right questions" to ask in this instance would remain obscure. The crucial difference between pulling a candidate molecule from OMspace and picking a structure described within the Borges library lies in the availability of an objective standard for assigning high-level function (if not perfection) for the former. In other words, in the case of OMspace, the "right question" is embodied by the target molecule used to "interrogate" the universal library.†

This process is simply the pragmatic evaluation of the performance of the candidate molecule for its ability to bind and modify the properties of the target/probe. In the real world, a molecule isolated from a physical library (a minute subsection of OMspace) would very rarely be the ideal form; much more probably it would constitute a useful *lead* for further refinement. Optimizing such lead molecules can be approached by systematic chemical modifications, or by reiteratively probing a specialized library whose members vary around a central theme based on the information obtained via the original compound. Thus, reaching into a molecular space and extracting a reasonable approximation to the desired solution is a major achievement. The take-home message is that while all knowledge may reduce to a selection process as a matter of

---

*Within Borges' story, the size of each book is set within a specific boundary. Although thus not infinite, the number of Library volumes is nonetheless of such a vast magnitude as to make the label "hyperastronomical" meaningless. For the purposes of this rumination, we can consider the Library contents as effectively unlimited.
†Of course, as an imaginary universal set, OMspace must also encompass the target molecule itself, and all other molecules that interact with it.

theoretical principle, it is often hard to empirically apply this dictum to real-world problems. But when seeking a functional molecule, empirical strategies, embodied by real molecular libraries, are both possible, logical, and potent.

MODES OF MOLECULAR DISCOVERY

A process that results from "blind chance" without any predictive basis is hard to rely on, and serendipity, as we have seen, falls into this category. Empirical discovery is also often thought of as a chance-based process, but the reality is far more complex than this. The major differences between serendipitous and empirical ways of finding useful molecules are shown in Table 1.1, also contrasted with rational approaches. In some cases, systematic empirical screening will almost always yield useful molecules, although by definition their exact nature is not known at the outset. The determining factors are the size of the molecular library to be screened, the desired properties of the sought-after molecule, and the nature of the screening process. An analogy (pursued in more detail in Chapter 3) may be made with natural selection, where the raw material for selection (genetic variation) may be random, but the process of cumulative selection itself is certainly not. Cumulative repetitive selection for molecular function by laboratory "directed evolution" is an analogous and parallel process where the selective pressure is determined by the experimenter. Since artificial molecular library size is

TABLE 1.1   Comparison of the Three Pathways to Molecular Discovery

Feature	Molecular Discovery Mode		
	Serendipitous	Empirical	Rational
Principle of discovery	Chance	Experimentation	Knowledge
Operational algorithms	None	Nondeterministic	Deterministic/ nondeterministic
Raw material for molecular discovery	Local environment/ unspecified	Specified by experimenter[a]	Precisely defined
Amenability of the *discovery process* to optimization and development	None	Highly optimizable	Optimization inherent in the design process
Prior knowledge of *target molecule*	Not required	Not required[b]	Required, at detailed structural and/or system level
Prior knowledge of chemical nature of *discovered molecule*	None	Limited to class of molecules screened or selected[c]	Predicted in advance
Specific structure of *discovered molecule*	Not known in advance	Not known in advance	Predicted/designed in advance of synthesis or expression[d]

[a] The experimentalist determines whether a complex system or a single-molecule target is to be screened.
[b] Not required for empirical screening *per se* (e.g., complex systems such as whole cells can be empirically screened for compounds affecting their viabilities or morphologies). However, a defined single-molecule target can also be effectively subjected to many forms of empirical screening methods.
[c] For example, if a peptide library is screened, a successful ligand found within this library is obviously a peptide.
[d] Although generally as a lead molecule requiring rounds of optimization.

a definable entity, it constitutes one factor contributing to the probability of success of empirical screening.

How then do we contrast this kind of molecular discovery with rational design? In essence, this hinges on foreknowledge of a target molecule or system and the technological ability to use this knowledge in order to make structural predictions. The key point distinguishing true rational design from an empirical strategy can thus be highlighted with an operational definition: Successful rational design can *specify a lead molecule with desired functional properties in advance of its actual physical realization, through processing of relevant structural information*, where the "physical realization" refers to synthesis or genetic expression. No matter what the power of an empirical strategy, its ability to predict the outcome of an experimental search is limited (Table 1.1). There is more to be said about this definition for rational design, which we will come to shortly.

At first glance, the categorization within Table 1.1 would suggest that molecular discovery can be cleanly rendered into a discrete tripartite arrangement. But although these discovery modes are quite distinguishable in their broad characteristics as shown, between the empirical and rational domains lies a gray area of detail, and this we should enter before moving on. This will be the opportunity to further note some of the features that distinguish these modes of identifying useful molecules.

The Borderland of Rational Design

Unlike serendipity, purely empirical discovery can be viewed as a nondeterministic search algorithm[*] (Fig. 1.4), since an arbitrary physical library (either from natural or from artificial sources) can be regarded as a very small random subsection of a vastly larger set of all possible molecules (ultimately OMspace). Nondeterminism in this context indicates simply that at certain decision points multiple different ways of continuing are possible, and a different answer will be delivered if the entire process is repeated. In contrast, with deterministic algorithms, the decisions and outcomes are precisely determined, such that processing of the same input data will always deliver the same output. Note that an empirical screening process performed on a defined library should produce an unequivocal result (each library member can be reproducibly evaluated as useful or not); it is the arbitrary sampling from a much larger set that results in the formal label of nondeterminism. This remains the case when the larger set itself is very far from a random sampling from the universal OMspace of molecules, as is certainly the case within the world of natural bioproducts. (Some of the factors influencing the molding of this "natural molecular space" are raised in Chapter 8.)

An empirical process can also involve rediversification and reiteration of the screening or selection rounds, in which it becomes an adaptive or evolutionary process, though still nondeterministic (Fig. 1.5). But here it is necessary to make a point that will occur repeatedly: an empirical procedure is in itself a rational pathway to molecular discovery, based on knowledge available at the time of action. In "pure" empirical sampling of natural molecular space, a completely arbitrary sample picking is made, but in the modern world there are many rational factors that channel such choices in very directed ways. An empirical approach can thus proceed logically without any pre-existing information (as the systematic screen instigated by

---

[*]To use Lucy's earlier screening project as an example, it is assumed that she randomly picked leaves from a variety of plants until she had a prechosen number. But this sample itself is taken randomly from a much larger set, so although the evaluation of the chosen sample is definitive, the outcome from the entire search would vary if she repeated the entire process, hence its nondeterministic nature.

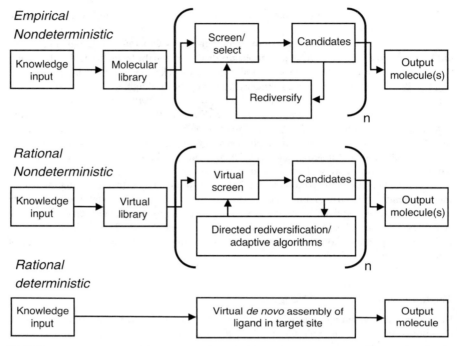

FIGURE 1.5 Schematics for empirical nondeterministic, rational nondeterministic, and rational deterministic molecular discovery processes. The *n* reiterations refer to evolutionary processes. A "pure" empirical process would have no knowledge input into the constitution of the molecular library, but the process becomes "semirational" as the pre-existing knowledge fund steadily increases. For computational rational nondeterministic processes, the rediversification itself can be undirected or knowledge based, or via an adaptive optimization algorithm. In such cases the nondeterministic status owes to alternative (non-predetermined) program steps, rather than the nature of the outcome. The idealized rational deterministic process as shown directly refers to small molecules, but can also apply in principle to *de novo* design of functional proteins or other macromolecules. This single-step representation represents an ideal for this design class; in practice, multiple assessments would be required.

Lucy above), or it can be progressively guided by increasingly refined models of the molecular problem that requires a specific molecular solution.

There are many examples of this that can be illustrative. Consider a situation where it is recognized that blocking the action of a natural hormone or mediator would be therapeutically beneficial. In this scenario, while the structure of the hormone is familiar, the mechanism by which it exerts its effects (its receptor and downstream signaling) is quite obscure. Given this starting point background information, a rational pathway is to use the structure of the known hormone to construct a series of chemical analogs (a small chemical library) one or more of which might act as an *antagonist* of the natural hormone's receptor and signaling. The assembled members of this chemical library could be screened serially or *en masse* by high-throughput methods as above: the efficiency is greatly different between such alternatives, but not the general principle. Yet the pathway followed is not as simple as indicated so far, since initial screenings may well provide continuing information that can be used to further guide the screening process. For example, a particular substituent at one site of the molecule of the above hypothetical hormone might have low but measurable activity as an antagonist. This could in turn direct a focus upon this region, or on the specific substituent involved. Since the initial chemical analog library in such cases is far from a random collection, and candidate compounds in turn are modified for improvement in a directed fashion, the entire procedure is clearly not purely empirical, but neither is it strictly rational according to the definition given above. This gray area could be referred to as "guided empiricism," or (as most commonly seen) "semirational," perhaps an instance of glass-half-empty versus glass-half-full stances. (From another point of view, semirational might seem comparable to "semipregnant," but we will not be so pedantic.) Regardless of semantics, semirational strategies can greatly shorten the pathway toward identifying a useful drug, and in consequence there is very often constant feedback between available information and rational decisions made for molecular discovery approaches.[*]

When enough information is available, "true" rational design becomes a possibility. Making a prediction for molecular solution is inherent in a rational design scheme, where a target molecule's structure, combined with high-level chemical understanding, allows a specific molecule to be designed from "scratch," which will fulfill the desired functional properties. But there are levels of prediction, too, ranging from broad generalities to the precise and rock hard. Prediction levels span the type of molecular discovery involved, where a purely empirical process can only make a trivial specification based on the type of molecules screened (Table 1.1). In between, there are cases such as the above hormone scenario, where of course the prediction is that the sought-after compound will be a chemical analog of the natural active molecule. Still, an important issue here is that the latter prediction is restricted in chemical space, but still imprecise and potentially requiring evaluation of a very large number of possible analogs with chemically diverse substituents. A rigorous ideal for rational design would demand the specification of a final desired molecule in advance of its actual physical realization, but in the real world this strict definition is a little too demanding. What if the design process came up with less than 10 possibilities, one of which proved ideal after testing each in turn? It might seem unfair and indeed "irrational" to insist that this process should still be labeled empirical discovery, or even just "semirational." A better compromise might be then to consider rational design as the knowledge-based prediction of a desired molecular solution within a narrow range of alternatives, but this only shifts the focus onto a definition of "narrow," a likely can of worms in this context. (Where would one draw the line? Obviously not 10^7, but 200? 20 or less?) But a simple concept from conventional drug discovery can be brought in here—that of the "lead" compound noted earlier, which *leads* toward a structurally related final optimal objective. (There are parallels between this and Darwinian evolution, which we will come across later.) The essential concept is that in any technology, it is virtually impossible to move in a single jump to an ideal form. Progression from a prototype is the only practical way in reality. In a likewise fashion, optimal drugs for human use rarely come ready-made, and require considerable "tuning."[†] So, while it is unreasonable to insist that rational design should in one swoop be capable of pulling out an *ultimate* molecular solution like a rabbit out of a hat, if it is truly rational it should be able to point directly to a useful lead compound for further function-based optimization.

More and more we find an increasing impact of computational approaches on not only rational design (not surprisingly) but also the design of optimal libraries for various (otherwise empirical) screening requirements. This leads to an interesting scenario that also bears upon the

---

[*]The example of the development of the histamine H2 receptor antagonist cimetidine is a case in point here, noted in Chapter 9.
[†]Some drugs originating as natural products have entered the market with little of no modification from their initial status as leads, but in such cases the "optimization" has been achieved by natural evolution over very long time periods.

distinctions between empirical and rational molecular discovery. Computer-aided design can be used for candidate drug modeling, and *virtual screening* or "docking" software can evaluate and help distinguish between likely positive design contenders. But consider taking this to a high level where both a single target macromolecule (such as a protein, with a specific binding pocket) and a very large library of compounds are screened virtually for interactions of suitable energetics. In such circumstances, the spatial and structural characteristics of each library member are rapidly modeled for their binding to the target protein without any preconceptions. In effect, this hypothetical "blind" virtual screening is transferring a physical library screen with real proteins and candidate chemical ligands into an *in silico* surrogate. If the computational screen (all done without anyone venturing into a laboratory) yields a candidate subsequently proven to be the correct choice, then is this a rational design? After all, it has entirely flown from the results of human knowledge, ingenuity, and technological sophistication. Yet does it not in essence remain the same process, if it was indeed performed in this manner? Before processing the entered "virtual library," no prediction of a lead compound can be made, so is this not "electronic empiricism?" In one sense, perhaps, but we should not forget that the key factor enabling rational design in the first place is information, and accurate spatial and structural computer modeling of library and target requires a very large amount of information indeed. And most notably, this informational requirement completely distinguishes the allegedly "empirical" computational screening from its real-world counterpart, where even rudimentary knowledge of the target protein structure is not necessary in principle for fully empirical evaluation of compound libraries. Possession of the required information, and the means for processing it in order to accomplish the electronic library screen, therefore places it into the domain of rational design.

An ideal for rational molecular discovery could be viewed as a deterministic process: data are acquired, sophisticated processing by previously designed algorithms is instituted, and a set of lead molecules is generated. The processing step here could involve virtual library screening, if the library members are chosen solely based on rational criteria, and is deterministic if reiteration of each step in the program will result in the same final output. For small molecule discovery, the virtual library can be rationally designed from the characteristics of the target protein binding pocket itself, and then used for virtual screening. Real implementation of virtual library screening also involves prefiltering of compound libraries based on "drug-likeness" or other relevant criteria, which we will look in more detail in Chapter 8. In comparison with the considerations of empirical discovery at the beginning of this section, as semirational design converges toward the fully rational, the sampling of universal OMspace becomes increasingly ordered and nonrandom, until an ultimate end point of molecular definition is attained and the search algorithm becomes deterministic.

But in practice, much current use of computational evolutionary algorithms and other "adaptive" optimization strategies also qualify as rational design, since this approach requires sufficient information to enable its practical implementation. Genetic algorithms[13] and genetic programs[14] are strategies that exploit evolutionary principles for optimizing computational problem solving and have application in the area of molecular discovery.[14,15] (These and other "adaptive algorithms" are considered in a little more detail in Chapter 4.) Rational design itself can then be viewed as the implementation of either deterministic or nondeterministic algorithms (Table 1.1; Fig. 1.5). Note, though, that pursuing either class of rational algorithms may return a design solution, but this is not necessarily the same thing as finding an ideal global optimum for the problem of interest. This point we will pursue further in Chapter 9.

As noted above, rational design is enabled only if sufficient knowledge regarding the target molecule or molecular system has been accumulated, but a few more words on this are in order. "Relevant structural information" will usually refer to the target(s), but consider the following scenario as an extension of the above hormone receptor antagonist semirational design: If a large range of ligands for a receptor were described, along with their precise biological effects, it may become possible to rationally model an antagonist even without the target structure available. In effect, the database of relationships between ligand structure and their activities provides an indirect surrogate model for the receptor binding pocket, and application of this ligand-based principle is recognized as a *de novo* ("new," or first-principles) design approach (Chapter 9). But how does this stack up as truly rational design, given that the target in this scenario has not been structurally defined? This is where we can return to the "relevant" part of the above definition. If the ligand information is detailed enough, the model for the binding pocket could approach perfection, but it cannot provide such detail for the rest of the receptor, which might be a very large and multifunctional structure. Therefore, it cannot in principle be as strong a level of rational design as when the entire receptor structure is accurately defined. While we have seen that the transition between purely empirical and rational design includes a span of semirational strategies, even when rational design is attained (by our earlier definition), it too can be seen as a range of achievements, rather than as a single edifice. In considering the "strength" of rational design, for most purposes one should not stop at the level of a single target molecule, but continue into the entire complex milieu in which the target is located, which we will pursue further in Chapter 9.

The utility of nondeterministic computational design, though designated as "rational" through its knowledge-based implementation (Fig. 1.5), nonetheless mirrors empirical design carried out in an analogous manner with real molecules and (possibly undefined) targets. This is not to suggest that empirical screening has always been carried out in such a structured and logical manner, but even Lucy's plant screening project we noted earlier could indeed fit the empirical process chart if we substituted "samples" for molecules and "confirm/refine" for "rediversify" in Fig. 1.5. It should be noted, though, that while early empirical screens could be conducted logically and systematically, they lacked the ability to reiteratively accumulate change. Consequently, though organized and systematic, such early approaches could not be termed evolutionary, which the empirical process of Fig. 1.5 essentially is. Modern molecular discovery methods routinely exploit evolutionary principles, and although using advanced technologies, the output of these pathways is not determined in advance. As a consequence, they fall within the empirical side of the spectrum of empirical–rational design.

Before we return to rational design in Chapter 9, we will examine many areas of molecular discovery that can be labeled as empirical by such criteria, even though (as we have seen) there is much overlap between these two areas. Pragmatically speaking, as suggested by this chapter's title, if a promising molecular candidate for a given task or problem is found initially through a chance-based event, then obviously it will be pursued enthusiastically regardless of its origins. But for the time being, let us focus on empirical strategies in more detail. To start with, it is hard to go past the incredible range and diversity of natural biomolecules and biosystems as inspirations for "blind" molecular discovery, as the next chapter will pursue.

2 Empirical Miracles: Nature's Precedents for Empirical Solutions

OVERVIEW

Before thinking about modern approaches to empirical selection for functional molecules, it is highly instructive to consider some fundamental biological processes whose action over time has seen the generation of staggeringly complex systems. All of these involve selection from an array of alternatives where a predetermined solution to the existing "design problem" does not exist, which can therefore be considered an empirical process. The tag "miracles" in this chapter title is not, of course, to be taken in any literal sense, but is intended only to convey the sense of amazement that often accompanies an appreciation of the incredible levels of organization in so many biological systems. Certainly this was the case in pre-Darwinian times, when the seemingly special nature of complex living organisms could only be conceived as having occurred through divine intervention. We now have an explanation for the nature and origins of living organisms in all their richness and diversity, through the agency of evolutionary natural selection. This cumulative process over time has in itself led to the development of biological systems, which themselves exploit processes of diversification and selection to obtain "design solutions." The best understood examples of this are adaptive immune systems, which will be examined in some detail in the next chapter. But before considering these phenomena, we should first look at the basic process that allowed these complex biological systems to arise from relatively simple prebiotic components.

EVOLUTION ITSELF

No contemporary biologist of any repute seriously disputes that all biological organisms, with all their attendant complexity and apparent "design," have arisen through evolutionary processes. This, of course, includes the most complex of biological systems, the human brain, and all its activities. The engine for biological evolutionary change is biological variation combined with natural selection, a concept which even most fundamentalists will associate with Charles Darwin, although Darwin's nineteenth century contemporary Alfred Russel Wallace independently conceived of the general idea.

Natural selection in principle is a simple process, but is frequently misunderstood, even by those who should know better. The "dangerous idea"[16] of Darwin still sits uneasily in many minds today, although most who reject evolution and natural selection have a demonstrably rudimentary understanding of it (if even that!). A particular misconception seems to be the "improbability argument." A familiar refrain goes that the probability of complex living things

Searching for Molecular Solutions: Empirical Discovery and Its Future. By Ian S. Dunn
Copyright © 2010 John Wiley & Sons, Inc.

emerging on earth is similar to the chances of a tornado passing through a junkyard managing to reconstitute the late Sir Fred Hoyle or some such miracle. But this simply stems from a basic misunderstanding of natural selection and its "power to tame probability."[17] A frog does not suddenly spring from mud through the agency of natural selection (or via any other agency, for that matter), for that is not the way natural selection operates. And, as we will see in later chapters, neither does the artificial laboratory counterpart of natural selection.

Evolution is not often explicitly termed an "empirical" process, but for the present purposes it can be viewed that way, in the same manner that "design" is often used to refer to the products of evolution (of which more shortly). Natural evolutionary processes are empirical in the sense that the solutions to "design problems" are not themselves designed in advance—they are progressively and cumulatively selected from the available range of diversity among an organism's population (Fig. 2.1). Certainly, there is no rational designer; the point of under-standing biological evolution is that it is a natural process governed only by physical laws operating throughout the universe. (Hence, the famous label of natural selection coined as "the blind watchmaker."[10]) What is the relevance of natural evolutionary processes to obtaining molecules by empirical means? In a nutshell, natural selection demonstrates that solutions to highly complex molecular design problems are achievable given a mechanism for generating diversity in the genetic material, a selection process, reproduction, and time. (Remember it as the "GODSTAR" principle—generation of diversity, selection, time, and reproduction (or replication).)

The generation of diversity (GOD) itself refers to processes that result in variation occurring within a population of replicators, or entities that can make copies of themselves. Reproduction of living organisms is controlled by their genetic material (specifically nucleic acids, DNA or RNA), but the accuracy of replication of these informational macromolecules is not perfect. And a pool of variant organisms resulting from such replicative flaws provides the diversity (whether naturally or artificially induced), which is the raw material upon which selection can potentially be exerted. (A hypothetical absolutely invariant population of replicators could offer no such selective opportunities, since by definition all the individual replicators would be identical.) But where does selection actually operate in natural settings? This is generally not directly at the level of replicable informational molecules themselves (individually, genes; for an organism, collectively its *genome*), but almost always at the level of the functional proteins or RNA molecules that are encoded and expressed by genes, termed a *phenotype*.* A specific phenotype of an organism, which confers a survival (fitness) benefit thereby allows its differential reproduction over alternative competitors. The gene(s) that encode the successful phenotype are consequently favored by natural selection, and a phenotype may accordingly be viewed as the vehicle by which replicating molecules (genes) secure their differential survival in a competitive world.[†,9]

Figure 2.1 depicts the natural selection process, which can alternatively be written as a simple reiterated algorithm.[18] At the same time, application of this process could equally well apply to an artificial evolutionary strategy, where the experimenter ensures the diversity of his or her starting population (be they are individual molecules or units of higher level organization), and supplies whatever selective pressure is desired. The "reproduction" of the selected species

---

*As we will see in later chapters, in some cases, a replicable molecule (RNA) is itself functional and selectable, or its own phenotype. In a majority of cases for complex organisms, a functional phenotype will be manifested by multiple proteins (sometimes including RNA molecules) acting as parts of a higher order system.

†In living organisms, the replication of genes and the proteins they encode is "entangled." Thus, some genes of an organism encode polymerase enzymes that replicate genomic DNA or RNA, including the genes that specify such enzymes themselves. The ultimate origin of the "entanglement" between informational nucleic acids and their encoded proteins is a conundrum for elucidating the origin of life, as we will see later in this chapter.

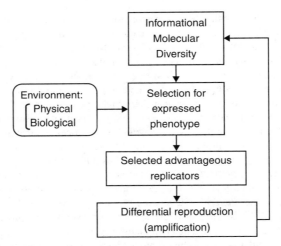

FIGURE 2.1   Process of natural selection from the raw material of natural variation, with environmental selective pressure resulting in the differential reproduction of advantageous forms, which in turn diversify and are subject to continuing selection. This process is directly applicable to *in vitro* artificial molecular or cellular systems, where the selection criteria are determined by the experimenter, and the "reproduction" step is amplification of the selected molecular or cellular species.

is more usually viewed as an amplification step, which may be *in vitro* itself or within a host organism exploited for such purposes. It is the ability of the experimenter to apply any appropriate selection toward obtaining a desired outcome that distinguishes artificial evolutionary processes from our visually challenged natural watchmaker. Even so, they both remain empirical processes.

To account for all observed natural phenomena, the "selection" level in the evolutionary process has considerable complexities that have generated long-standing controversies. The major question has been the "units" upon which selection can operate, be they are at the gene, cell, individual organism, species, or even higher levels.[19] The likelihood of real-world operation of natural selection at higher levels (groups or species) has been discounted by many evolutionists, although this notion has re-emerged in "multilevel" models in relatively recent times.[*,20,21] Many of the proposed selection complications have arisen to accommodate social or altruistic behaviors in organisms, which at face value seem to contradict the "selfish" dictates of natural selection. Thus, *kin selection* and other models have been invoked to explain the origin of altruistic behavior where it can be shown that natural selection should favor cooperation between close relatives.[20]

Another evolutionary bone of contention has been whether all aspects of organismal diversity are explicable entirely through adaptive mechanisms. In other words, this question asks whether all biological genotypic and phenotypic features have actively arisen through natural selection alone or by selection operating in concert with other processes. An interesting and important issue in this respect is the effect of the size of the populations of organisms for the potential nonadaptive fixation of genomic configurations.[22] Population genetic analyses can rationalize certain differences in genomic complexities between high-population organisms (typical prokaryotes) and those with relatively low populations (multicellular eukaryotes), and

---

[*]In any case, when human beings themselves are recapitulating evolutionary processes in the laboratory, selective pressure can be applied at will in a precisely defined manner, and (as outlined in Chapter 1) this is based at the level of replicators.

suggest that these distinctions do not necessarily arrive through natural selection.[22] Since our concern is using natural evolution as a guide and model for analogous artificial processes for molecular design, much of this is beyond our scope, although there is a little more to say in related areas, which we will pursue later in this chapter. For the time being, the diversification mechanisms (GOD) and the versatility of the evolutionary process itself are more immediate issues from a molecular design perspective. Before looking at the (hopefully less than mysterious) ways of GOD in an evolutionary context, as an introduction, let us first briefly compare natural and artificial evolutionary processes.

Evolution: Natural and Artificial

Consider a biologist who is seeking an enzyme with improved catalytic activity, say, toward the breakdown of a specific sugar against which it normally has very low activity. This diligent researcher might produce a population of randomly mutated enzymes (focusing either on a known active site of the enzyme or globally for the whole protein molecule) and then devise a system for identifying specific variants with improved catalytic performances. Let us leave aside for the time being how such selection/screening processes are instituted. (We have briefly considered this in Chapter 1, and we will also revisit this topic in succeeding chapters.) We can also leave aside until later the important issue of whether such an improved enzyme can be obtained by a rational design strategy.

Repeated rounds of this type of experiment successfully allow the isolation of enzymes that function better on the chosen substrate, and the researcher then identifies the specific amino acid residue changes that cause the desired effects. Now contrast this with a bacterial population whose natural biological environment has changed such that it now provides a source of the same sugar that had not been previously used for bacterial metabolic or cellular structural purposes. Bacteria expressing variant enzymes with an ability to use the sugar as a substrate (even if rudimentary) might gain a selective advantage over their normal counterparts; recursive application of this process (as in Fig. 2.1) could lead to high-efficiency utilization of the sugar. So in both of these hypothetical cases, the end result is the same in the form of an improved enzyme for a specific task.* In the case of the experimentalist, the process is not "blind" in that the novel substrate has been consciously predetermined, but in the natural parallel situation, the groping hands of the unseeing craftsman have yielded comparable results. Yet the real similarity between these two scenarios is that in both cases, the improved enzyme was not designed in advance—the experimentalist only finds out which specific residue changes contribute to the superior molecular performance after the fact, and the bacterium merely continues to hydrolyze its new energy source without paying much attention to the triumph of molecular design that it may be heir to. It is this feature of "finding the best solution from a large range of alternatives" that allows us to classify both natural and artificial evolution as empirical processes, in contrast to rational molecular design where a proposed molecule is specified in advance of functional testing.

So it is important and instructive to review the process of natural evolutionary change as a powerful inspiration and natural exemplar of empirical molecular discovery. We then need to consider several features of natural evolution, which are of special interest in this regard. But first, a short word about a short word.

---

*It should be noted that this is a functional definition of enzyme similarity only; the starting point enzymes may have been different between the artificial versus natural scenarios. However, if the catalytic properties (rate constants, equilibrium constants, etc.) of the artificial and natural enzymes were comparable, it is possible that similarities at the respective active sites might be observed—a case of "convergent evolution."

To Opine on Design   Much has been written through history about the order of nature implying a designer, but modern evolutionary science shows that complex order can arise from natural processes. (And these processes can indeed be productively harnessed in the laboratory.) It might be noted that "design" has been used above with respect to the products of evolution, and it is important to make a defense for the convenience of this usage. (These points are also applicable to the above use of "empirical" in reference to natural evolution, as opposed to the interpretation that empirical processes must have conscious instigators.) Obviously, if "design" is strictly interpreted to mean "produced by a conscious agent" (a *designer*), then it is inappropriate in this context. Yet when speaking or writing of the products of evolution, "design" is a compellingly useful word, but, of course (if one uses the strict definition as above), it really means *as if* consciously designed, *as if* produced by a pragmatic experimentalist, not literally so. (And the last thing one would wish to do would be to give aid or comfort to creationists.) With this understanding, "design" is often found in the scientific literature with respect to evolutionary products, whether molecules, molecular systems, or cells. Certainly, it has been used in this sense with respect to artificial directed evolution.[23,24] One could insist that "design" in these circumstances should be continually embraced with inverted commas, but eventually these tend to get omitted, and the meaning of the word broadens. Short of inventing a new term,[*] this is probably inevitable.

In the beginning was the word, and the word was design. Designerless design, that is, the great discovery of Darwin.[25]

NATURAL EVOLUTION AT WORK

The philosophy for including the following sections concerning evolution and natural selection is based on the premise that to artificially exploit evolution for molecular discovery, one should have as comprehensive an understanding as possible of the original biological process. So a certain level of detail follows, but it is still, of course, but an overview of a vast field of study. This is broken down into sections on the raw materials of diversity upon which natural selection can operate, and then certain features of the process itself and other factors of interest.

The Raw Materials: Diversity Enough and Time

GOD from Mutation   From the earliest of times, it was an obvious fact that the individual members of any type of organisms were not created equal—that is, individuals vary. Heritable diversity provides the starting point for natural selection and incremental evolutionary changes to biological organisms. Accordingly, this is the "GOD" (generation of diversity) part of the GODSTAR principle noted above, and the classical way toward GOD for biological evolution is genetic mutation. As we will see below, there are other avenues toward genetic diversity and certain additional complicating factors, but given the importance of mutation and its relevance to artificial empirical pathways for improving biomolecules, we should consider it further. Mutations are often single base "point" changes, but more complex multibase insertions or deletions ("indels") can also occur, and can be biologically significant.

---

[*]One could use "e-design" for "evolutionary design," but this has obvious potential for confusion with electronic terminology. "Endesign" could be coined for "evolutionary natural design," but would exclude directed evolution. The simplicity of "design" itself is likely to prevail.

An understanding of mutation at the molecular level requires knowledge of both the structure of nucleic acids and the means for their propagation. DNA and both structural and functional RNAs provide templates for their own replication (via base-pairing complementarity), but cannot replicate without catalytic assistance from polymerase enzymes. A sufficient level of accuracy in template copying by DNA polymerases is necessary to ensure genomic integrity, without which successive compounding of errors would rapidly lead to a breakdown in the intergenerational transmission of genetic information. (Indeed, certain theories for the origin of life have been faulted for failing to solve the problem of such "error catastrophe" in early replicators before accurate polymerase functions could have evolved.) There are a wide variety of polymerases for nucleic acids in nature, but those with the highest replicative fidelity come with specific enzymatic proofreading mechanisms. A "module" (or separate domain) within such polymerases has a specific separate enzymatic activity (as a 3' to 5' exonuclease; ftp site), which can remove a mispaired base and thereby allow the polymerase function to continue with the incorporation of the correct base.

So fidelity is a good thing, but a striking conundrum is that literally perfect genomic hi-fi would forever freeze an organism into a set genomic configuration and disable subsequent evolutionary development. One can imagine an organism approaching an asymptote of replicative perfect accuracy, although never exactly reaching this limit as is its evolutionary rate continuously declined. But there are a number of compelling reasons why this has not happened, and indeed could never happen. Since most mutations are deleterious, there is a selective advantage to minimize them, and thereby improving the reproductive fitness of progeny.[*] But achieving maximal levels of DNA replicative fidelity does not come "for free"; higher and higher levels of polymerase accuracy require more elaborate mechanisms and energy consumption that becomes a counterinfluence toward its selection in the first place (sometimes termed the "cost of fidelity"[26]). So these competing trends tend to result in organisms with replicative editing functions "set" within fairly tight limits.

Until this point we have been treating mutation within complex genomes as though it were set at a fixed rate for each organism, but mutational rates at different sites within the same genome can vary widely. This has long been noted from observing the occurrence of mutations at different genetic loci. With the advent of molecular biology, it became clear that mutation rates can depend on the local sequence context as well as more long-range effects on a number of scales up to the chromosomal level.[27] Since many mutational events occur during DNA replication, it has long been recognized that male germline cells (sperm cells undergoing many cell divisions) have a mutational bias compared to ova.[28] After the identification of eukaryotic introns, it was soon noted that the rate of mutation in many intronic (and other noncoding) regions and silent codon positions was higher than that for nonsynonymous positions.[†,29] This was interpreted as strong support for the neutral theory of evolution largely developed by Motoo Kimura, which proposes that the majority of mutant base substitutions occurring through time are selectively neutral and randomly fixed into populations.[29] Truly

---

[*]Note that this refers to germline transmission of mutations to offspring, which directly affect their phenotype; somatic mutations arising during development or maintenance of an organism also become significant if the organism suffers impaired fitness through a very high somatic mutational load. Such somatic mutations themselves are influenced by the fidelity of nucleic acid replication encoded by the organism's genotype. Cancers resulting from somatic mutations tend to occur at a higher frequency in postreproductive individuals, against which no evolutionary selective counterpressure can then be applied.

[†]The genetic code is degenerate, such that of the 64 possible triplet base codons, many encode more than one amino acid. The third "wobble" position of codons is where this degeneracy is located. Thus, for example, the codons CCA, CCC, CCG, and CCT all encode (at the DNA level) the amino acid proline. Therefore, a mutation that substitutes any base at the third position of a proline codon will not alter the resulting expressed protein. Such mutations are hence termed "silent" or "synonymous."

neutral mutations in themselves could not serve as direct substrates for Darwinian natural selection, since by definition they are nonadaptive and rest as an invisible background against the panorama of adaptively selectable mutational change. But as we will see shortly, neutral (or near-neutral) mutations can become evolutionarily significant by offering a different platform for the effects of secondary mutations.

Even synonymous mutations within protein-coding sequences may be nonneutral in an absolute sense. Some synonymous codons are not translated with equal efficiency owing to the variation in the levels of isoacceptor tRNAs that recognize each distinct codon type.[*] Recent studies have concluded that numerous synonymous base substitutions can alter mammalian mRNA stability or pre-mRNA processing involving intron splicing.[30] Surprisingly, synonymous codon changes can alter gene product formation in more direct ways. Rare alternative synonymous codons can cause pausing of translation into protein, which in turn affects the folding pathway taken by the nascent protein itself. At least in some cases, this can have functional ramifications for the nature of the final translated product.[31] While many synonymous codon alterations are therefore clearly nonneutral, not all can be so categorized. But since the original assignment of such "neutral" mutations was based on inadequate knowledge of the nuances of genetic transcription and translation, further information in this area may, in fact, allow recognition of additional "neutral" substitutions as being subject to selective pressures.[30]

Even if a protein mutation is truly neutral in a specific background, it is not necessarily so if the remainder of the protein changes (through secondary mutations) or its cellular environment alters. A marginally deleterious nonsynonymous mutation in a protein-coding sequence might also be rendered neutral by a second mutation that overcomes the negative effects of the first. In principle, the second mutation in this "compensatory neutral evolution" effect might involve another interacting protein, but would most likely occur within the same protein-coding sequence.[29] These "compensated" doubly modified proteins could then provide a different starting point for subsequent evolutionary change (a third substitution in these proteins might have quite different effects than when present (as a single mutation) within the original parental wild-type sequence). But the same point applies to any set of genetic alternatives even if truly neutral with respect to each other, as the downstream consequences of a new mutation may be dependent on the specific background in which it occurs. Thus, synonymous (silent) substitutions can influence the subsequent pathway of coding mutations. As an example, consider a synonymous mutation exchanging the threonine codon ACA to the alternative threonine codon ACG. A mutation in the second codon position takes the original codon to isoleucine (ATA), but the synonymous mutation is converted into a unique methionine codon (ATG); therefore, the nature of the previous silent mutation is not irrelevant. This kind of effect has been proposed to be of particular significance in retroviral populations, potentially influencing mutational pathways leading to drug resistance in HIV.[32] Neutral mutations in a functional macromolecule can be depicted as a network, which provides an evolutionary pathway to acquisition of a novel function. (A productive model of this notion is seen with functional RNA molecules, to be considered in Chapter 6.)

A truly neutral mutation elicits no fitness change by virtue of its inherent properties. Yet there is an additional interesting possibility, where mutations are effectively rendered neutral (or nearly so) by the agency of another gene product. Mutation or chemical blocking of the

---

[*]tRNA molecules function as adaptors during protein synthesis, and isoacceptor tRNA molecules are charged with the same amino acid by aminoacyl tRNA synthetase enzymes. But two degenerate codons for the same amino acid may require distinct isoacceptor tRNAs (with different anticodons), expressed at different levels. Codon usage frequencies and translational efficiencies vary between organisms. Thus, in the bacterium *E. coli*, a mammalian protein coding sequence needs to have its synonymous codons adjusted to *E. coli* codon preferences for optimal expression levels. Further details on tRNA molecules and aminoacyl tRNA synthetases are provided in Chapter 5.

heat-shock protein Hsp90 in fruit flies has been shown to enable the phenotypic expression of (otherwise silent) mutations in signaling pathways relevant to development. The result is diverse phenotypic morphological variation affecting a wide range of structures.[33] Moreover, since Hsp90 function may be abrogated during the periods of stress, it was proposed that Hsp90 acts as a "capacitor" for morphological adaptive evolution, by enabling the release of constraint on a pool of otherwise neutral mutation in response to environmental changes.[33] Subsequent positive selection could then allow changes endowing a fitness advantage to become permanently expressed in that lineage. The ability of Hsp90 to mask or buffer the effects of mutation can be considered as a special case of *epistasis* (where one gene can modulate the phenotypic expression of another). At the same time, the ability of Hsp90 to buffer mutations has some important differences to more conventional instances of epistasis, especially with respect to the role of stress in the unmasking of the Hsp90 effect. Also, Hsp90 buffering tends to favor specific signaling pathways that normally result in invariant phenotypes with a correspondingly low ability to evolve. As such, Hsp90 has been viewed as a mediator of *evolvability*,[34] which is defined as the capacity of an organism to undergo evolutionary change.[*]

One important difference between natural evolution and its artificial laboratory counterpart is that while both processes are empirical (the final results are not predicted in advance), the human evolutionary "watchmaker" can choose the starting material using criteria beyond the initial fitness of a target molecule. As such, directed evolution can indeed begin with a protein bearing nominally neutral or even deleterious amino acid substitutions, as means for probing the functional outcomes of alternative mutational pathways. The general subject of mutations has other implications for artificial empirical molecular discovery, in particular, the optimal mutational load to place on a target molecule for screening purposes. (The higher the number of engineered mutations per target, the greater the number of combinatorial possibilities, which eventually collides with the upper limit of screening.) Thus, while screening multiple mutations (whether scattered through a sequence or in a linked "cassette") simultaneously for a single (relatively small) protein may be a viable strategy, it will tend to become progressively less useful as the complexity of the system increases. These topics will be extended and developed in Chapters 4–6.

In this short voyage through the sea of mutationally significant themes, we have only toured a number of islands of knowledge judged to be of interest and relevance, out of a large archipelago. But within the larger ocean of natural genetic diversification, there lies in wait another important region to visit, where we should now proceed.

GOD from Recombination and Other Things   Mutation *per se* is not the only pathway to GOD. The interchange of gene segments, usually referred to as recombination, is an ancient way of generating diversity through reassortment and shuffling of alleles. The enzymatic machinery that promotes recombination is associated with DNA repair mechanisms and is virtually universal (only certain viruses lack such coding capacity, and then rely on their host cells to provide this function). In fact, recombination is a fundamental biological process. Just how fundamental is still contentious, but a very good case can be made for the "extreme" antiquity of recombinational processes. This viewpoint asserts that recombination has gone hand in hand with the origin of life itself.[35] We noted briefly toward the beginning of the previous section that many theories of abiogenesis have foundered on the notorious error catastrophe principle, where the compounding of replicative errors in the absence of adequate proofreading mechanism serves as an entropic full stop to a struggling protoreplicator. An early

---

[*]There will be more to say on this theme at later points within this chapter. It is important to note that evolvability is not a "forward-looking" effect, but concerned only with the ability of a system to undergo evolutionary change.

recombination model offers a potential escape from this conundrum[35] through the shuffling of informational macromolecular segments, and as we will see, sexual recombination itself has often been proposed as a kind of mutational cleansing mechanism. Interestingly, it appears that mutation and recombination *per se* are not unlinked modifiers of the genome, as recombinational hot spots often coincide with regions of high mutation.[36]

Recombination has been compared to "fusing the back half of an airplane to the front part of a motorcycle."[37] This operation might be presumed to result in an "aircycle" or a "motorplane," neither of which would have good prospects for functionality, to put it mildly. In fact, this is clearly a metaphor for not just any type of recombination, but for recombining nucleic acids with a low degree of sequence relatedness, or homology. We will see later that such processes of *nonhomologous* (or "illegitimate") recombination are significant in evolutionary terms, despite the apparent poor prospects for improving function from these kinds of rearrangements. This can immediately be contrasted with genetic exchange occurring between segments of DNA which indeed have significant sequence similarities (homologies), and such processes are accordingly referred to as *homologous* recombinational events. To extend the above metaphor, homologous recombination might involve two airplanes of similar but nonidentical design, one of which had a superior cockpit configuration, and the other that possessed a better tail fuselage section. "Crossing over" the two planes by "homologous recombination" could result in the best of both worlds for aviation design in this metaphorical world.

But this metaphor illustrates a fundamentally important aspect of recombination, the power to cross evolutionary "valleys." A common device for modeling selection and adaptation is the *fitness landscape*, where a continuously variable genotype of an organism is represented in one dimension, and local fitness optima as peaks in a second dimension (or additional dimensions as required). In this parlance, an adaptive progression may see an organism "marooned" on a local (suboptimal) fitness peak, where any further improvement might require such a radical redesign of existing systems that incremental mutational changes are uniformly deleterious. In such cases, it becomes impossible by mutation to jump the gap to a novel superior form, inaccessible across a deep valley void of any stepping-stones. Many biological examples of this effect have been ascribed, and among the most fundamental may be the use of the existing four bases in nucleic acids.[38] A widely used metaphor for this "historical fixation" or "frozen relic" effect is the "QWERTY" alphanumeric keyboard, which is suboptimal in configuration[*] but so entrenched in usage that all efforts toward mass implementation of design improvements have run aground. In biological circumstances, though, recombination offers a potential escape route from a local fitness peak by (in effect) mixing separate peaks together, and allowing selection to pick a product that constitutes a new and previously unscalable height. (But this too has limits, as suggested by the very existence of biological cases of putative "frozen" phenotypes.)

In the real world, homologous recombination begins with single or double-stranded breaks in DNA. Specialized proteins promote the base pairing between single stranded regions of DNA resulting from strand scission events, and corresponding homologous regions on another DNA duplex. One such recombination-catalyzing protein that has been extensively studied is the recA protein of *Escherichia coli*,[42] but proteins with comparable functions are known in eukaryotes from yeast upward, including mammals. In addition to proteins mediating homologous strand pairing in recombination, to complete the recombinational process a

---

[*]The details of this are controversial. It is commonly believed that the original typewriter QWERTY keyboard was designed to deliberately slow down typists to avoid typebar jamming,[39] but this, and the relative efficiency of supposedly optimzed rivals such as the Dvorak keyboard, has been disputed.[40] But although important at one level, we should not get too bogged down with such considerations; the point, of course, is the utility of the QWERTY system as a metaphor for frozen design solutions. Another example of an artificial "locked-in phenotype" is the computer millenium bug,[41] albeit one that was forced to change due to an implacable "selective pressure."

considerable number of additional proteins are required, many of which are evolutionarily conserved and widely expressed. Homologous recombination can result in the reciprocal exchange of DNA segments between participating DNA duplexes, or nonreciprocal exchange, where the sequence of one DNA segment is replaced by that of another homologous sequence (a process known as gene conversion). But as noted above, not all recombinational events involve homologies between participating DNA segments. Nonhomologous recombination takes place through double-stranded DNA breaks and the process of nonhomologous end joining.[43] A relevant pathway whereby nonhomologous recombination is of significance for the generation of genetic diversity is via exon shuffling,[44] where exons from different genes may be exchanged or exons within the same gene deleted or duplicated.[*]

Recombination is intimately linked with DNA repair mechanisms since both processes are characterized by responses to DNA strand scission events, and share many enzymatic components. An alternative to the "recombination-first" theory[35] is that the machinery for direct DNA repair (to deal with DNA nicks and double-stranded breaks) evolved initially, and recombinational mechanisms arose by subsequent evolutionary piggybacking upon these initial enzymatic tools.[45] In fact, a variety of existing DNA transactional processes use these biological tools in multiple roles, such that a single protein may be associated with alternative combinatorial protein complexes assembled for different DNA repair or recombinational functions. While most recombinationally based systems share certain ubiquitous DNA repair enzymes, some also require the input of specific proteins with dedicated activities. Examples of these fundamentally important recombination mechanisms, as we will see, occur within the immune system and mammalian sexual reproduction.

A Sexual Interlude   Sex in its broadest sense is relevant to molecular discovery, and I am not referring to the chemical structure of viagra. This link, hopefully, will become explicable in Chapter 4, but nowhere else is a discussion of sex more appropriate than in a section dealing with the generation of diversity from recombination. Recombination during the generation of eggs and sperm cells (by the specialized form of cell division termed meiosis) is the major reason why human siblings (excepting monozygotic twins) are nonidentical.[†] And this inevitably leads to ruminations about recombination and sexual processes, which are fundamentally nothing more than the organized exchange of genetic information. The problem of what sex is really for has preoccupied many cellular, molecular, and evolutionary biologists, although lay people are generally untroubled by the question. (The laity are far more likely to wonder what cellular, molecular, and evolutionary biologists are really for, thereby showing the need for "sex" education.) But sex is indeed popular, biologically speaking. It is ubiquitous across the domains of life, and yet not universal. The problem of understanding the origin and purpose of sexual reproduction, as has often been noted,[46] lies in its apparent high costs. Sex can bring together beneficial combinations of alleles, but it can just as readily break them up again. Also, it seems genetically disadvantageous for sexually reproducing females who will transmit only half their genes compared to corresponding asexuals. In contrast, males transmit half their genes with relatively little reproductive burden.

A perennially proposed solution to the existence and spread of sex is that it counters the negative effects of the accumulation of deleterious mutations, a process known as "Muller's ratchet." It was originally shown by (who would guess?) Muller[47] that since the great majority

---

[*]Exon shuffling via retrotransposition (involving reverse transcription of transcribed sequences) may be the dominant mechanism whereby exon shuffling occurs.[44] The reintegration step of retrotransposition may itself involve nonhomologous recombination events.

[†]Epigenetic factors may also play a role in phenotypes of multicellular organisms, as considered later in this chapter.

of mutational changes are deleterious (reducing fitness), for an asexual organism a statistically inexorable increase in this mutational load would occur, ratcheting up through each generation with steadily decreasing population fitness. (Mutations that reverse the bad changes are too rare to compensate for this downward trend.) Sexually mediated genetic exchange and recombination can in principle reverse or greatly reduce this mutational burden, and experimental support for this notion have been obtained in bacteriophages,[48] mammalian viruses[49] and microcrustaceans.[50] Alternatively, the promotion of genetic variability by sex may improve an organism's adaptive responses over time. For an asexual organism, selective adaptation can occur only through beneficial mutations, and the limits of adaptation are set by mutation rate and population size.[51] But sexual reproduction with its associated shuffling of alleles may provide an advantage by increasing genetic variance and thus improving the power of selection, which modeling shows applies to realistic population ranges.[52]

It is commonly held that while adopting an asexual mode of reproduction may confer transient selective advantages for an organism, inevitably Muller's ratchet will catch up with it. Or if that fails to do the job, the reduced ability of such an asexual renegade to adapt to changing environments will see it off rapidly in evolutionary time. Or one of other explanations for the necessity of sex would become manifested.[53] Therefore, so this logic goes, one might indeed expect to find relatively recent asexual lineages, but antique asexuals ought never to occur. Compelling as this argument may seem, an inconvenient finding for its veracity is that ancient asexuals do, in fact, appear to exist. This has been termed an evolutionary "scandal" (in evolutionary terms, it is the lack of sex, not the occurrence of it, which is likely to be considered scandalous). So, it is important in the first place to ascertain with confidence that an organism is, in fact, asexual. As cases in point, the supposedly "asexual" protist organism *Giardia* (of infamy in intestinal infections) has been found to possess genes associated with meiotic (sexual) recombination,[54] and "asexual" species of the minute crustacean ostracod *Darwinula* have recently yielded males.[55] Such findings raise valid questions as to whether many organisms classified as asexuals are not in actuality furtive "cryptosexuals."

In favor of the true antiquity of some asexual lineages, fossil information has been proffered, admissible where male and female organisms can be readily distinguished.[56] But unconventional means of gene transfer and exchange, which could not be resolved by examining the fossil record alone, have to be considered before assigning an organism as completely asexual. (We will discuss "lateral" or "horizontal" genetic transfer in the next section.) So it is molecular evidence for ancient asexuality that is most convincing, particularly in a group of organisms known as bdelloid rotifers. Rotifers in general are a group of tiny multicellular animals, ranging in size from the microscopic to just visible to the naked eye. Possessing a number of unique features in their body plans, they are sufficiently divergent from other animals to be entitled to a phylum all their own (termed Rotifera, logically enough). Not all rotifers are asexual, but the bdelloid class of these organisms* has no need for males, and in all likelihood have maintained their asexuality for a very long time. Sexually mediated genetic exchange will normally set measurable limits on the allelic diversity within an individual. (In other words, mutational variation between cognate homologous chromosomes will be tempered by the sexual processes of recombination and allelic segregation.) This is not the case in a diploid asexual organism and chromosomal pairs would be predicted to become highly divergent over time, the longer the organism persists in an asexual state. Indeed, the extent of variation within former alleles in individual bdelloid rotifers is consistent with many millions of years without sexual influences (the "Meselson effect"[57]). Another interesting and useful yardstick here is the prediction

---

*"Bdelloid" literally means "leechlike," and the name "rotifer" itself refers to the ciliated organs of these creatures, which create the illusion of a rotating wheel.

that asexual organisms will tend to lose "selfish" transposable genetic elements, which is borne out by the absence of retrotransposons in bdelloid rotifers.[58] Recent analysis has found that transposable elements can be expected to be eliminated from asexual populations provided such elements are excisable.[59] Thus, the combined molecular evidence, and also the fossil record,[56] comes down strongly in favor of the ancient asexual reproduction in this class of rotifers. Despite evolutionary theory, these organisms seem to go on and on. . . .

So there is a message for us in the existence of ancient asexuals that is very likely to have general ramifications for the origins and functions of sex,[56] but as yet we may not be hearing it correctly. The possibility that asexuals have particularly efficient DNA repair systems has been raised.[60] Evidence (albeit inconclusive) has been produced suggesting that nonreciprocal recombination events may create allelic diversity in asexuals.[61] A convincing mechanism whereby asexuals undergo sufficient recombination to allow adaptive evolution would, it is felt, enable them to escape their "scandalous" reputation.[61]

Recombination and sex are thus intimately linked, but interestingly, recombination has been co-opted in many organisms for the formation of germ cells themselves. During meiotic cell division, which enables the formation of ova (eggs) and sperm, the normal complement of chromosomes is halved. Meiosis begins with a round of DNA replication and doubling of chromosome number, during which replicated chromosomes (sister chromatids) remain physically linked, and also form close association with the corresponding homologous chromosomes (maternal/paternal pairs). This is followed by a cell division (Meiosis I) that separates homologous chromosomes. To complete the process, another division ensues which separates pairs of sister chromatids (Meiosis II) such that cells with half the original numbers of chromosomes result.[62] In many organisms, including placental mammals, the process of association of homologous chromosomes (formation of a "synaptonemal complex") is accompanied by extensive double-stranded DNA breaks and homologous recombination (crossing over of chromosomal DNA strands[62]). This is important for the physical segregation of chromosomes, and thus recombination in such organisms has been recruited for a fundamental purpose in the mechanics of sexual cell division.

By its activity in "shuffling the deck," and then reforming a new zygote by the fusion of two haploid germ cells, meiosis is clearly a driver of genomic diversity, whatever the adaptive implications of this may be. Nevertheless, illegitimate recombination has been proposed to be the main recombinationally relevant evolutionary driver, through its role in mediating non-homologous gene fusions and gene duplication.[63] (An important component of this is retro-transposition, the indiscriminate insertion of reverse-transcribed RNA molecules such as retrotransposons.[64]) By this view, homologous recombination shuffles alleles within homologous genes, whereas illegitimate recombination has the potential to create greater diversity and novelty through nonhomologous gene transpositions.[*] At the same time, it should be noted that such "illegitimate" processes would be more likely than homologous recombination shuffling to generate a deleterious genetic event than one with a beneficial phenotype. Thus, nonhomologous recombination would carry greater risks, but greater potential payoff in evolutionary terms.

Having made a brief foray into the always-fascinating topic of sex, we have touched upon some areas of interest and relevance. Although there are still many question marks surrounding sex and its processes, and much to be learned,[46] we trust that this particular sweet mystery of life

---

[*]To continue with a metaphor of playing cards, while homologous recombination can shuffle the deck, illegitimate recombination might fuse the queen of hearts and the jack of spades into a single card. A newly invented card game in which such a card had a special role would be analogous to an illegitimate gene fusion acquiring a new function. In time, the rules of the game with the novel card might be amended repeatedly to improve its entertainment value, as for a modified gene undergoing successive rounds of selective optimization for function.

will eventually yield up its remaining secrets. But let us first consider some other recombinationally related mechanisms for the generation of diversity.

Doubling Up and Reassorting   If you were in the business of trying to design an organism with a competitive advantage over its rivals, a problem likely to be encountered early on is that many genes will be quite sensitive to tinkering. That is to say, a large number of proteins expressed from many essential genes will not be able to tolerate radical changes in their sequences. A few alterations here and there may be acceptable, and maybe some selectable improvements on their original functions can be engineered, but no major shift in function is likely to be possible because (practically by definition) the original essential function and/or shape will be left wanting by a large move in sequence space.

It has long been considered, though, that a way around this conundrum is via duplication of a gene, which liberates one copy from heavy selective pressure against significant change. In such an arrangement, the second copy is then in theory free to accumulate mutations and transform into a potentially highly divergent functional molecule, provided the successive mutations in the "tinkered" gene copy do not significantly reduce fitness. ("Purifying" selection will act to remove such mutations[65] if fitness is compromised.) Genes of an organism with divergent functions, which are derived from duplication of a common ancestral gene, are termed paralogs.* There is no doubt that gene duplication *per se* is a frequent and functionally significant biological event. In fact, it is only in the recent era of whole-genome sequencing that the full extent of gene duplications has been realized. For example, it has been estimated that the duplication events are more important than point mutations in accounting for global genomic differences between humans and chimpanzees.[66] In many cases, it is not just individual genes, but large-scale genomic segments up to the level of entire genomes that have undergone duplication. Whole genome duplication has been documented in plants[67] and is believed to have occurred at the beginning of the evolutionary lineage leading to bony fishes.[68] Genome sizes themselves cover a very wide range, which bears little correlation with protein-coding capacity. Thus, it has been noted that nuclear genomes (as opposed to organelles) vary in size over a range of 300,000-fold, while the size of the transcriptome (the portion of the genome transcribed into RNA) varies by only about 17-fold.[69] Macroduplication events are believed to be heavily involved in driving these genome size effects, which in turn impact on cellular size. The flip side of genomic expansion is genomic miniaturization, which is a frequent observation in both eukaryotic and prokaryotic intracellular parasites.[69]

The prevalence of gene duplications notwithstanding, it is still necessary to account for the processes that enable their occurrence in the first place. As noted briefly above, illegitimate nonhomologous recombination generating tandem repeats is a driver of duplication,[44,70] as is homologous recombination involved with the generation of segmental duplications by "unequal crossing over" events.[70] Reverse transcription of mRNAs and subsequent random genomic integration of the DNA copies (retrotransposition) can generate an active gene copy, or a defective *pseudogene*, both of which lack intron sequences as a consequence of their origins.[64,70] Pseudogenes in general are defective gene copies which cannot be normally expressed, and are also a frequent fate of duplicate genes after recombinational duplication events (as well as retrotransposition[71]). Yet pseudogenes are not necessarily utterly useless in the genome, as their sequences can be an informational resource for diversification of active genes via the recombinational process of gene conversion, which is a very significant process in the vertebrate adaptive immune system (Chapter 3).

---

*A paralog should be distinguished from an "ortholog," which is a gene from another species with a similar function and sequence believed to be derived from a common ancestor.

What then befalls each gene copy after a duplication event? The early stages of gene duplication may involve positive selection, possibly linked to beneficial effects of an increased gene dosage,[72] although altered levels of gene expression *per se* may also have negative effect on fitness.[*,73] The rate of gene duplications and their fixation into populations has been noted (from studies of recent genomic data) to be very high,[74] higher, in fact, than expected from population genetic models for even beneficial mutant alleles.[70] An early view was that one copy of a duplicate pair would tend to degenerate by deleterious mutations and suffer negative ("purifying") selection, or transform by mutation into a nonfunctional and nonexpressed pseudogene. Certainly, there is evidence that sequence divergence in duplicated genes does not occur symmetrically between both copies.[75] In rare cases (as we introduced this section), one copy might acquire mutation(s) such that a positively selectable beneficial phenotype was conferred (that is, a novel function or *neofunctionalization*). This has been framed as a race between "entropic decay" of a gene copy and acquisition of a beneficial (and preserving) mutation.[76] Yet to account for the observed high duplication rate, a different model termed duplication–degeneration–complementation[76] was proposed, where both gene copies tend to suffer degenerative mutations, but functional complementation between each other acts to prevent loss of the duplication as such.

Allied to the concept of functionally coupled mutations in both copies of a duplicated gene is the principal of *subfunctionalization*.[76] (Here, the meaning of "sub" as "under" is used only in a hierarchical sense, not as a reflection of the quality of a function *per se*. Thus, it should not be confused with "subfunctional.") Basically, this effect describes the splitting of functions (formally relegated to one protein) into two. Some gene products wear many figurative hats; they perform separate functions in different cell lineages or in different combinatorial configurations with other proteins (as we have already noted for proteins involved with recombination). Such functional versatility clearly has advantages in terms of genomic informational efficiency, but it is hard to be many separate things at once and still perform at an absolutely optimal level in each case. (In other words, some compromises may have to be made for natural selection to find a protein that best fits multiple roles. The result may be slightly suboptimal in at least one of these functions, bringing to mind the old adage celebrating the virtues of specialization, "A jack of all trades but a master of none.")

The fact that a selective advantage may be obtainable from a process of gene duplication followed by subfunctionalization can be taken as evidence that at least in some cases protein "multifunctionalization" carries a certain cost. But by the same token, one must consider what selective pressures were involved in molding the parental gene product toward assuming multiple roles in the first place, in the progenitor organism. (It is rather unlikely that two distinct roles arose in a given protein simultaneously.) The evolution of multifunctionality within a single protein in itself may suggest that a novel protein function can evolve without necessarily involving gene duplication,[65] but the importance of gene duplication for protein functional evolution is widely accepted.[77–79] A major determinant of protein function is protein shape (topology),[†] dictated by the folding of the primary amino acid sequence. Since the diversity of distinguishable protein folds is much less than the total numbers of natural proteins,[81,82] the origin of novel protein folds has been a focus of interest. Gene duplication, combined with gene fusions and truncations, is at least one pathway to novel protein topologies,[79] which has been successfully modeled in the laboratory as a multistep

---

[*]Deleterious effects engendered by effective doubling of gene expression also bring to mind that human disorders linked to inherited defects in genomic imprinting raised briefly the later section on epigenetics.

[†]Folding determines the specificity of ligand binding, but enzyme specificity is also determined by a limited number of key amino acid residues for catalysis.[80]

process.[83] Thus, neofunctionalization of members of a duplicated gene pair can be a driver of protein diversity at a fundamental level.

The existence of gene paralogs with subtly differing functions raises the general issue of real (or apparent) gene redundancy as a consequence of duplication.[84] While genetic redundancy has been studied in many organisms by a number of approaches, experimental evidence of such apparent backup systems in mammals has been derived through "gene targeting." This powerful approach uses homologous recombination to make in principle any change to a genomic sequence, but its earliest use in mice in the late 1980s was mainly to study the consequences of ablating specific gene function on the biology of the whole organism. Many of these studies showed surprisingly little phenotypic effects of certain gene "knockouts," suggesting functional gene redundancy in such cases. True functional redundancy is nevertheless unproven by mere viability and an apparent null phenotype of knockouts, since laboratory organisms may not be exposed to the full range of environmental factors that can exert subtle (or not so subtle) selective pressures which favor overlapping, yet nonidentical functions between "redundant" sets of genes. In any case, the phenotype of specific gene knockouts in mice can vary greatly in different murine strain backgrounds,[85] indicating the importance of "background" genes and their interactions with the targeted gene. Also, similar but nonidentical proteins may participate in different combinatorial settings in processes requiring complex interactions. The partially redundant, overlapping activities of cytokines (protein mediators of the immune system) is a good case in point.[86] Apparent redundancy itself is certainly far from universal in any case, since there are indeed essential genes whose "knockout" also knocks out viability (causing death during embryonic development[87] and many others where phenotypes are nonlethal but detectable and definable. The ability to ablate a gene with minimal effect on phenotype (often termed "genetic robustness") may result not only from the existence of a duplicate copy but also from the redundancy of an entire gene network or metabolic pathway, unrelated directly in sequence homologies. In yeast, the significance of duplicated genes in the robustness phenomenon has been supported,[88] but unrelated alternative gene networks have been ascribed as actually having a greater contribution.[89] In fact, if all wild-type environments (as opposed to laboratory conditions) are considered, true functional redundancy may exist rarely if at all.[90]

And now, another diversification mechanism from the sidelines. . .

Lateral Thinking   The traditional view of the operational pathway of natural selection has been of gradual change over very long periods of time, by means of selection for favorable variants under environmental pressures (which includes the competing local segment of the biosphere as well as any relevant physical factors). One challenge to this account of the process, developed by Stephen Jay Gould and his colleagues, has claimed that rather than a more or less constant "flow" of accumulated change, evolution is better characterized as proceeding in relatively sudden bursts followed by long periods of comparative stasis. This proposal, termed "punctuated equilibrium" or "saltation,"[91] was framed as a refinement to Darwinism and was certainly not considered a challenge to the fundamental process of natural selection itself. Nevertheless, this debate seems to arouse passions. Some other evolutionary thinkers have felt that this theory of relative quiescence preceding rapid biological change and speciation is most accurately termed "evolution by jerks." It is said that supporters of punctuated equilibrium responded by rephrasing the more conventional view of gradual and progressive change as "evolution by creeps." Yet instances of "punctuated" evolution arising from combinations of rare beneficial mutations in laboratory bacterial experiments have indeed been documented, in the form of sudden morphological changes sweeping through bacterial populations.[92] The ins and outs of this are beyond our scope, although "punctuated" change is not irrelevant for directed evolution, as we will see in Chapter 4. But we can consider by what mechanisms a sudden large change in an organism's structure or underlying functions could possibly come about.

Mutations that cause large and obvious changes in a multicellular organism's development have been long known, and have been more recently characterized at the molecular level. These "homeotic" mutations (within the *hox* genes) have been much studied in the fruit fly *Drosophila melanogaster*, and have provided a wealth of information about the mechanisms of development. Moreover, despite the seeming huge gap between fly and "higher animals" (including *Homo sapiens*), these homeotic fly genes have clear-cut equivalent forms (homologs) in mammals. But can these developmental mutations ever result in a beneficial effect, a significantly improved phenotype that would allow a sudden jump in evolutionary terms (or a rapid move within "animal space" if you so prefer) and "punctuate" a period of stasis? Certainly, the homeotic mutations seen in *Drosophila* are not encouraging in this regard, generally eliciting monstrosities. For example, the well-known *antennapedia* gene (as its name implies) results in the formation of fly leg structures in the place normally reserved for antenna development.[93] These observations themselves, of course, could not rule out the possibility that rare homeotic mutations (sometimes referred to as "macromutations"[*]) might confer an advantageous phenotype. The possibility that sudden dramatic shifts in phenotype might be capable of driving speciation events harks back to the colorful and evocative term of "hopeful monsters," coined by Richard Goldschmidt in 1933.[94] Most observers, it should be noted, have not been at all hopeful that such phenotypic shifts could actually confer a fitness advantage, but some possible examples in arthropods have been cited.[95]

One proposal for circumventing some of the difficulties associated with macromutations has been the activities of mobile genetic elements. As the name implies, these sequence elements have the ability to insert themselves (transpose) into different genomic sites. They constitute a significant fraction of eukaryotic genomes and are evolutionarily relevant.[96] Such elements may be independent segments of "selfish DNA," which carry the information that enables their mobility and transposition from site to site (hence the name "transposons"[†]). They may also be associated with retroviruses or retrotransposons that use reverse transcriptase enzymes to convert their RNA genomes to corresponding DNA copies. Many sequence elements in the human genome are thought to have originated by such a process of retrotransposition; indeed, such elements compose more than 10% of mammalian genomes and may drive evolutionary processes through their roles in promoting illegitimate recombination.[64] In the context of macromutations, the key theoretical advantage of viral spread of a mobile genetic element is the fixation time, or the period required for the new genetic (and phenotypic) feature to become stably established in an intrafertile breeding population.[97] This kind of "lateral" (or "horizontal") transfer with "infectious mutations" has thus been proposed as a hopeful pathway for hopeful monsters.[97] The sudden acquisition of novel genetic material, as exemplified by lateral transfer, has also been termed "genetic catastrophism" (usually by those less charitably inclined toward accepting its biological significance[98]).

But is there evidence that this type of genetic transfer has occurred to a significant extent, whether via retroviruses or other means? Lateral transfer between bacterial populations is a well-established phenomenon, to the point where it has been seen in some circles as a threat to the whole concept of species in prokaryotes, although this is probably something of an alarmist

---

[*]Note that the prefix "macro" here refers to the consequences of the mutation, not the extent of the mutation in terms of DNA sequence changes. Thus, changes in expression of a *hox* gene might have profound developmental consequences yet involve relatively small sequences changes in genetic control regions.

[†]Transposons have been described in many organisms from bacteria upward. Barbara McClintock was awarded Nobel Prize in 1983 for her pioneering work in genetic transposition in maize.

view.[99] But it is important to keep the concept and definition of species in mind in this context, since lateral or "horizontal" transfer between species would seem to violate the principle that speciation represents a *restriction* on genetic exchange (at least between species that cannot normally produce viable offspring, of which more below). Indeed, a working definition of a sexual species is an interbreeding population within which new allelic combinations can form, while excluding allelic variation from different species.[100] The existence of interspecies lateral transfer thus indicates that this definition could not be an absolutism, and in turn that certain "backdoor" means of genetic exchange must exist. It has been postulated that during the early development of life before species were "crystallized" by restrictions on genetic exchange, lateral transfer was rampant.[101] For multicellular organisms, a successful lateral transfer event must clearly establish itself in the germline as opposed to somatic cells (the "horizontal" event can then be stably passaged "vertically" through the germ cells). Usually retroviruses or mobile genetic elements are the postulated agents of lateral transfer, but for some organisms more direct routes may apply. (For example, in bacteria, conjugation (a means of physical transfer of genetic information) enables crossing of species barriers.[102]) Perhaps, the most significant "horizontal" event of all has been the acquisition by eukaryotic cells of functional organelles (principally chloroplasts and mitochondria) that were in the remote past free-living prokaryotes.[103]

The initial sequencing of the human genome revealed a number of genes that seemed to be clear-cut examples of transferred genetic information from bacteria.[104] Yet caution is clearly needed in the interpretation of such results, since contamination of the original material with bacterial DNAs has been demonstrated to be a real problem and source of artefact.[105] Also, some of the putative vertebrate genes of microbial origin were later shown to have clear antecedents in earlier eukaryotic organisms.[106] Candidates for horizontal gene transfer remaining after such screening require systematic evaluation, and while the overall incidence of this type of genetic mobility may not be large,[107] specific instances of it may be significant. The acquisition of genetic sequences laterally from other organisms has been referred to as genetic "plagiarism,"[104] which is perhaps an unfortunate turn of phrase given its underhanded connotations. All's fair in love, war, and the acquisition of useful genetic information. One person might see lateral gene transfer as a dirty trick; another might see it as further evidence for the essential unity of life on earth.[*]

It is also important to note that significant lateral genetic transfer need not be associated with a rapid alteration in an organism's phenotype. Lateral transfer can be (and may most often be) a means for acquiring novel raw material for incorporation into an organism's tool kit for evolutionary change. This would logically apply more often when there is a large phylogenetic gap between the donor and recipient organisms in the lateral gene transfer event, as where a prokaryotic gene requires "tinkering" before it can efficiently be useful in a eukaryotic context. Somewhat analogously to a gene duplication event for endogenous eukaryotic genes (as discussed above), a laterally introduced gene (lacking an essential cellular role in its new context) can undergo successive rounds of mutation until its altered descendents assume roles that confer a selective benefit to the host organism. An interesting example of this is the probable origin of the recombination-activating genes (RAG) of the mammalian immune system[109] because they are also a fundamental part of a system that itself uses an evolutionary strategy for deriving functional proteins. But this is putting the cart before the horse, and we must defer this until the Chapter 3, which gives an overview of the immune system itself and its relevance in this context.

---

[*]A Barbara McClintock quotation ("Basically, everything is one. There is no way in which you draw a line between things. What we do is make these subdivisions, but they are not real."[108]) suggests she would be in sympathy with the latter point of view.

After considering ways for the generation of genetic diversity that constitute the raw material for natural selection, we are now led to consider the process of natural selection itself. In doing so, we will consider an old discredited idea, which certain recent findings have brought back into mainstream consciousness, at least from certain limited viewpoints.

The Process: Old Ideas and Newer Wrinkles

The idea of natural selection was first put forward before the founding of modern genetics (or at least the widespread recognition of Gregor Mendel's contributions to this field) and certainly prior to any comprehension of the physical nature of the genetic material. In other words, an understanding of genetics *per se* was not an essential requirement for development of evolutionary (and revolutionary) proposals in the line of Darwin and Wallace. Nevertheless, natural selection as an explanation for biological diversity and evolution was a relative latecomer in the field of human thought. It has been noted that it often seems perversely difficult for Darwinian ideas to take root in human minds,[10] and perhaps Lamarckian notions (of the transmission of acquired characteristics across generations) are somehow easier for humans to conceive and believe. As an example, an Australian aboriginal legend traces the dreamtime origin of the kangaroo's tail to one of its tailless precursors receiving a spear in the rump from tribesmen, which subsequently gave rise to the tails we are familiar with today[110] (This particular account would seem to combine lateral transfer of sorts with a Lamarckian sequel.) Legends of this nature are common in many diverse cultures, but stories of natural selection are very thin on the ground until Darwin and Wallace.

Jean-Baptiste Lamarck (1774–1829) is widely noted (and frequently ridiculed) as the chief originator of the proposal that an organism could transmit to the next generation useful characteristics acquired during its lifetime. (Probably the most widely used example is that of the giraffe having acquired its long neck through successive generations straining to reach higher and higher tree leaves.) Certainly, such ideas in general have had a long folk history predating Lamarck, and some biologists of his time and later have shared or at least been broadly sympathetic toward these notions. But the opprobrium usually directed at Lamarckism by the majority of biologists from the late nineteenth century onward, of course, stems from the fact that there is no evidence for Lamarckian principles contributing to biological evolution, notwithstanding certain very interesting and quite recent data we will consider shortly. An often-cited case in point is the failure of many generations of circumcision in certain ethnic groups to render the circumciser's role unnecessary. In a more controlled setting, many generations of mice suffered successive laboratory tail removals to prove that stumpy-tailed mice could not be bred in a Lamarckian manner (the devil is in the de-tails, at least from a mouse's point of view). This work was done by August Weismann (1834–1914), who is also renowned for establishing the distinction between the immortal germline versus mortal soma (somatic cells) of an organism. "Weismann's doctrine" states that traits acquired by the soma during an individual organism's lifetime cannot be transmitted back to its germline cells through which the species is maintained. In fairness to Lamarck, it appears he did not regard mutilations acquired during an organism's lifetime as candidates for Lamarckian transmission, as opposed to traits "needed" by the organism, such as increased neck length in the giraffe. But considered as a competing idea, the much greater "fitness" of the Darwinian viewpoint trumped that of Lamarckism and was strongly "selected for."

The Marks of Lamarck? In relatively recent time, evidence has emerged that challenges the dogma that an organism's inheritance is solely determined by its genetic composition, or more specifically by its gene sequences. There is considerable evidence that many nonclassical inheritance phenomena are mediated by physical changes to genomes that do not alter the original genetic-coding sequence, or *epigenetic* changes. One such physical process is the methylation of DNA at cytosine residues (to form 5-methylcytosine), mediated by methyltransferase enzymes. Epigenetic changes can also result from alterations in the proteins involved in the packaging of genes into the nuclear supramolecular protein–DNA complex called chromatin. It has been known for a considerable time that either genetic DNA methylation or the specific nature of genetic packaging into chromatin can strongly influence gene expression. Cytosine methylation in transcriptionally critical sites (such as promoter regions) is a means for repressing gene function and maintaining expression control.[111] "Epigenetics" is usually defined as chemically superimposed changes on the primary genetic material which modify the expressed phenotype, but a somewhat broader definition is any process that elicits an observable phenotypic variation in the absence of direct genetic change.

The "epigenome" refers to the subset of the conventional genome which is epigenetically modified, and it is well documented that epigenetic changes are mitotically preserved (that is, the pattern of epigenetic changes are preserved during normal somatic cell division). It was originally assumed that the somatic epigenome would be erased from the germline during the process of meiosis, that the germ cells would begin afresh with a clean slate, as it were, and certainly the epigenome is "reprogrammed" during the germ cell maturation.[112] If this were not the case, epigenetic information (contributing toward a specific phenotype) that was not erased from the germline could in principle be transmitted to the next generation and thereby propagate the phenotype by a nongenetic mechanism. Yet considerable experimental evidence in plant systems indicates that certain parts of the epigenome are preserved in the germline and transmissible.[113,114] There is also documentation for specific cases of inheritance of somatic epigenetic modifications in mammals.[115] It is accepted that a specific component of the methylated epigenome is normally reconstituted during development (or more accurately, specific genomic segments are re-established with the appropriate pattern of epigenetic methylation "marks" corresponding to their parental chromosomal origins[112]). This epigenetically "imprinted" part of the genome is known to be essential for normal mammalian development[116,117] and imprinting defects have been linked to certain human developmental disorders.[118]

So processes with a Lamarckian flavor have been shown to occur, but is it time to rehabilitate poor derided Jean-Baptiste? Let us consider first whether a taste of Lamarckism is the same as the real thing. It must be stressed that conclusive proof of epigenetic inheritance *per se* (whether RNA-mediated or via methylation "marks") does not at all immediately vindicate true Lamarckism. As opposed to epigenetic inheritance itself, "true Lamarckism" can be defined as the directed transfer to the germline of the information specifying *beneficial* (adaptive) somatically acquired characteristics. The stress on the word "beneficial" is intended to highlight the general problem of how a true Larmackian mechanism could discriminate useful phenotypes from deleterious ones and feed them back into the germline. In contrast, natural selection for fitness-promoting phenotypes is at the heart of Darwinian evolution.

Leaving aside the issue of true Lamarckism, we can still consider how important the general area of epigenetic inheritance is with respect to evolution and its mechanism of natural selection. We may concede that it is a real effect, but is it a significant evolutionary mechanism, and does it contribute to the formation of new species? One point to consider initially in this context is that (as far as known) heritable transmission of epigenetic information occurs at the level of gene regulation rather than structural genes (epigenetic variation still comes within the "diversity" box of Fig. 2.1, but the diversity in this case lies within the expression levels of genes rather than within gene products themselves). But changes in gene regulation (which can affect combinatorial gene expression and expressed protein interaction networks in very complex ways) can indeed be critically important for speciation events. A good case in point in this regard can be made by considering the high degree of genetic similarities between humans and chimpanzees in their structural genes. (This has been greatly facilitated by the sequencing of the genomes of both *Homo sapiens*[119,120] and *Pan troglodytes*, or chimpanzee.[121]) The human–chimp similarity has become something of a cliché, finding its way even into the popular literature, but it remains essentially true nonetheless (~99% at the nucleotide level, excluding insertions, deletions, and rearrangements[121]). Key differences in their respective regulatory systems (as understood at present) may be critical in determining the important distinctions between these species[122], especially at the neurological level. It is significant that the expression of the chimp and human genomes at the RNA level appears to be significantly more divergent than their respective genome sequences *per se*.[123] The take-home message here is that changes in genetic regulation can be potent evolutionary drivers.[*]

Of course, to make this point, the above human–chimp comparisons refer to changes in key DNA sequences affecting gene regulation, rather than epigenetics. But since the epigenome can also modulate an organism's phenotype through changes in gene expression, advantageous epigenetic alterations should be selectable as for conventional genetic mutations, *if* relevant epigenetic "imprints" are stably transmissible through the germline. Also, if such changes to the epigenome are propagated across generations, it would afford a rapid means for selectable phenotypic change. The machinery for generating epigenetic changes is itself encoded by the conventional genome (for example, with methyltransferase enzymes). Yet a considerable problem with the role of epigenetics in evolutionary processes is the issue of epigenomic stability. Loss of a selected epigenetic methylation "imprint," for example, will also result in a reversion or alterations in phenotype. This can occur somatically in tumor cells (a selectable advantage for the tumor, perhaps, but a rather negative event for the host organism), but also has been observed in germline transmission in certain human developmental pathological conditions.[118] Epigenetic changes may therefore require more secure "fixation" into the line of inheritance to significantly impinge upon evolutionary speciation. It has been suggested that chromatin modifications might affect susceptibility of separate genomic loci to mutation.[115] (This effect would not be true directed mutation but rather a biasing of mutation toward previously epigenetically "marked" sites.) Epigenetically methylated cytosines (5-methylcytosine) can also undergo spontaneous deamination to thymine[115] and thereby act as potential mutagens.

So at the present time, we might conclude that epigenetic inheritance is a reality as such, but remains an interesting adjunct to the main game of Darwinian natural selection. All of the above neo-Lamarckian possibilities do not pose any threat to the fundamental mechanism of evolutionary change, although they do raise the level of complexity of the totality of processes that may influence the genetic raw material of natural selection. It seems almost a general rule that as understanding of biological systems increases, an appreciation of the inherent complexities also increases. This has to be coped with if biosystems are to be described, understood and ultimately mastered. This is not to say, though, that certain underlying simplifying rules exist, and the importance of Darwinian natural selection itself as the prime mover of evolutionary processes is a good case in point.

---

[*]One way for gene regulation to affect evolutionary development may be through its effects on gene duplications. The duplication–degeneration–complementation model of postduplication gene fate predicts that certain mutations in regulatory regions increase the likelihood of the preservation of duplicate genes[76] (see the Recombination section earlier in this chapter).

Evolution by Order? Yet as we noted earlier, some biologists have long questioned whether natural selection itself is the only process determining biological evolution. While such workers acknowledge the importance of natural selection, the suggestion has in effect been that the metaphorical "watchmaker" has been given a given a "leg up" in some significant areas (maybe being directly handed some premade watch components, rather than having to make everything from scratch). Since an important theme of this chapter is the utility of natural selection as a model for artificial evolutionary methods for molecular discovery, it is pertinent to consider such ideas. Of these, one of the most important and best-defined is molecular self-organization, a subset of the relatively new sciences of complexity and self-organizing systems.

The study of these phenomena can find applications in many fields ranging from pure chemistry to economics, but it is obviously highly pertinent to biological issues. Of these, the origins and subsequent evolutionary development of life are at the forefront. Many important insights have been obtained from work with computer-generated "artificial life" models, where networks of regular cells obeying certain simple sets of rules dictating their existence or propagation can (quite counterintuitively) show the emergence of orderly patterns. By tuning some of the rules for such cellular automata, states with either a frozen stable configuration or disordered chaotic behavior can be generated. At a point in between, a "phase transition" occurs where ordered patterns may be observed, leading to the coining of the evocative phrase "order at the edge of chaos."[18,124] The supposition is then that life itself, and natural selection, must operate in this productive zone between chaos and rigidity, and indeed it has been further postulated that natural selection will inevitably take biological systems there. An allied proposal deriving from analyses of cellular automata, Boolean networks, and other studies is that much order in complex systems will emerge spontaneously, or as "order for free."[18,124]

Biological order is readily apparent in the formation of many regular patterns (for example, the orderly presentation of pigmented stripes or spots) and the development of body forms, or morphogenesis. It was Alan Turing, of fame as a pioneer of the theory of computation and the Turing test measure of artificial intelligence,[125] who first provided a mathematical description to account for the formation of biological patterns and morphogenesis by the application a reaction–diffusion mechanism. In his model, diffusion and reaction between chemical morphogens results in the development of specific shapes in three dimensions or pattern formation in two. "Turing patterns" have been demonstrated in chemical systems.[126] Recent evidence has served to confirm these principles in biology, using the maturation and skin surface distribution of hair follicles in mice as a convenient system. The WNT signaling pathway is involved in follicle development, and is regulated by proteins expressed from *Dkk* genes, which are inhibitors of WNT protein. Artificial manipulation of *Dkk* genes in transgenic systems, allowing variation in *Dkk* expression and consequent WNT inhibition, produced changes in follicle distribution that corresponded to computer models and strongly argued for WNT and *Dkk* participation in pattern formation via a reaction–diffusion mechanism.[127] Other pathways are also involved in follicle development, which have been associated with self-organizational phenomena as well, modeled with simple cellular automata.[128]

Biological systems may indeed be major beneficiaries of universal principles that empower spontaneous order, and indeed for those interested in artificial molecular systems and molecular discovery in general, it is well to pay attention to developments in this field. (We will briefly revisit allied themes in Chapter 9 in the Systems Biology section.) But what does self-organization have to say about biological evolution through natural selection? Some opinion has depicted evolution as a kind of team effort, where self-organization does some bits, while natural selection completes others. An analogy along these lines holds that natural selection is part of the motor of evolution, while "order for free" constitutes another part.[18,124] But a simpler view is that spontaneous order increases the size and versatility of the toolbox upon which natural selection can act, where the tools are self-organized autocatalytic molecular sets acting as supramolecular building blocks. To return to the motor analogy above, natural selection is still the motor, but spontaneous order is one of the processes supplying the nuts and bolts, and maybe even the rotors and solenoids. In other words, there are many physical constraints on what the results of evolutionary change can be, and one of these constraints is molecular self-organization.[16] By the same token, it is reasonable to propose that natural selection will itself favor systems with the ability to self-organize if they confer a reproductive advantage. It may be, of course, that the propensity of certain molecular systems to organize themselves without outside intervention is more of a boon to natural selection than a constraint, by supplying "bits and pieces" that enable the selection of more complex processes or structures. Thus, the arrangement of complex genetic circuits, up to the level of entire genomes, may result in part from the self-organizational properties of complex systems.[18,124]

It is well established that the application of a simple set of rules in model systems can generate complex patterns,[124,129] and this can be extended to encompass the view that the process of natural selection is equivalent to the running of an algorithm (as noted earlier). If evolution cannot be described in any more compressed terms than this, then its operation is "algorithmically incompressible." That is, the process represented by the evolutionary algorithm is its own shortest description of itself. In turn, this means that no simplifying law can be found to describe evolutionary outcomes in a more succinct fashion than to let the original algorithm run through its operation. It seems highly likely that this is the case, although as has been pointed out, it cannot be formally proven that a shorter algorithm might exist.[18]

The model of evolution as a "tinker," originally attributed to the Nobel Prize winning molecular biologist François Jacob, tends to be rejected by proponents of the importance of self-organization. The argument against "tinkering" (where random small changes are introduced by mutation, combined with selection for variants with improved fitness) is that in a complex highly interrelated system (such as an intricate genetic circuit), almost all changes will be deleterious. While random mutation (and/or recombination) and natural selection alone, by this view, can only throw a wrench into the works, the power of self-organization can still come to the rescue. But in the context of "tinkering," we should recall the discussion of gene duplication and related effects earlier in this chapter. A duplicate gene copy can provide a pathway for much tinkering without endangering viability, as would occur with a singleton essential gene.[130] In any case, by the same argument that molecular self-organization must be of significance to natural selection by providing "raw material" (molecular self-organized subsystems) for biological evolution, self-organizing phenomena will undoubtedly be relevant to empirical discovery by directed molecular evolution. Both the blind watchmaker and the sighted experimentalist can thereby benefit from the automatic operation of self-organizing molecular phenomena.

Orderly Beginnings   The above considerations aside, the notion that natural selection acts on pre-existing order must be true in at least one aspect: the point of origin. Darwinian natural selection requires a link between phenotype and genotype, where a successful phenotype allows amplification (reproduction) of the genotype that encodes it. But in the prebiotic world, this kind of selection by definition could not have existed. Therefore, by some process, the abiotic world achieved a level of organization that allowed conventional natural selection to begin operations. How this occurred remains enigmatic, although there is certainly no shortage of ideas. Those who study molecular self-organization may sometimes tend to brush over the question of abiogenesis,* taking it as a natural consequence of the physical properties of this universe. Central to the role of self-organization in abiogenesis is the concept of an autocatalytic set,[124]

---

*Literally the origin of life from nonliving matter. An alternative term is *biopoesis,* or literally "making of life".

where at a certain level of complexity an extended series of catalytic reactions can loop back on itself, such that a downstream product can catalyze the formation of the initial catalytic chemical species itself. When this occurs, the network of catalytic reactions becomes autocatalytic, and is self-perpetuating provided appropriate input molecules continue to be present.

Autocatalytic self-replication of simple oligonucleotides[131] and peptides[132] has been achieved in laboratory systems. Simple artificial autocatalytic networks have also been established.[132] And yet the origin of life, as we know it, is still a fiendishly difficult problem to account for. A major conundrum is the origin of "entanglement" between functional and structural proteins and informational nucleic acids, very different types of molecules from a chemical point of view. When the first ribozymes (RNA enzymes) were described, it was a logical leap to consider an "RNA world" with RNA molecules fulfilling both functional and information-transmitting roles.[133] An appealing prospect, this proposal is widely, but not universally, accepted.[134,135] One observation in particular, though, has considerably increased the attraction of the RNA world hypothesis. It has been found that the heart of the ribosome, the universal biological protein synthesis machine, is in fact a ribozyme that catalyzes the formation of peptide bonds (described in more detail in Chapters 5 and 6). Logically, RNA catalysis of peptides and proteins then preceded the involvement of proteins themselves in the same process. This in turn avoids one of the apparent chicken-and-egg paradoxes of the origin of life ("if you need proteins to make proteins, how did the first protein ever arise?") Darwinian selection itself may have arisen very early in the development of the RNA world.[136] But then, of course, one has to account for the origin of the RNA world itself. . . .

A majority of proposals regarding the mechanism of abiogenesis assume the requirement for primacy of self-replicating informational macromolecules. Theoreticians of molecular self-assembly have proposed that autocatalytic molecular loops might enable a "metabolic" origin of life, where self-sustaining autocatalytic molecular systems precede their entanglement with informational molecules which encode them.[18,124,134,135] This concept has, however, been criticized on a number of logical grounds.[137] The quandary of the origin of life has led to some interesting, if radical, ideas concerning the possible role of quantum processes.[138] Quantum effects are certainly relevant to living systems at least in the trivial sense that progressive reductionism of all physicochemical systems will ultimately move into the domain where quantum mechanics becomes significant. (For a comprehensive understanding of the electronic covalent bonding and noncovalent weak interactions critical for the existence of life, quantum mechanical analyses are required.) Spontaneous mutations (themselves obviously relevant to evolutionary processes) have been interpreted as the outcome of quantum decoherence of alternative chemical forms of hydrogen-bonding bases in DNA.[138] These speculations aside, no convincing evidence for a nontrivial role of quantum processes in biological systems has yet been demonstrated.[139]

The Arrow of Complexity?

As we noted previously, natural and artificial evolution can both be viewed as empirical processes, but differ in a key aspect. Evolution conducted in a laboratory setting usually has "directed" appended to it, which is not an unreasonable description given that the experimenter can arrange both the starting materials and the goals. (Of course, the solution to the molecular goals is not known at the outset.) In contrast, "blind" natural selection has no predetermined goal, being shaped only by the natural and biological environment, and the genetic raw materials that we have spent some time thinking about. Yet there is a common perception that evolution has a goal in the broad sense of an increasing drive toward complexity. As entropy is the arrow of time, complexity then is the arrow of evolution, according to this view. It cannot be disputed that some groups of organisms appearing relatively late in the evolutionary sequence are more complex (by any standard criteria) than any early forms, but what does this signify? Is a trend toward complexity in natural evolution relevant to artificial evolutionary strategies for molecular design? This question is amplified by the inherent distinction between the two processes that we restated at the beginning of this paragraph.

The conception that evolution is purposeful, with a predetermined direction and end point, by definition likewise conceives of evolution as a *teleological* process. Such a notion is implicit in the numerous cartoons which show a linear chain of organisms starting from the primal ooze, moving to dry land, and eventually culminating in *Homo sapiens*. (As has been often noted, this pinnacle of evolution is always, of course, depicted as a good-looking white male[*].) Claims that evolution has an explicit teleological focus are anathema to most evolutionary biologists. A champion of contingency in evolution, Stephen Jay Gould, has used the metaphor of rerunning the tape of biological change over evolutionary time, whereupon the re-emergence of humans would be highly improbable.[140] (An obvious source of contingency in this regard is unpredicted environmental change, ranging from climate variation to impacts of extraterrestrial objects.)

So the correct interpretation of the natural selection process (as in Fig. 2.1) is constant selection-based movement through time under environmental pressures, without any mysterious guiding hand. In the language of fitness landscapes, an organism may ascend local peaks of fitness, but can never take a "bird's eye view"[18] of where it is all globally going. If at some point the development of increased intelligence happens to provide a selective advantage and is a feasible "move," as evidently occurred on this planet within the past ten million years or so, then it will occur. This is not to suggest, of course, that by performing the imaginary tape rerunning experiment that absolutely anything could happen. There are clearly many constraints that will necessarily divert evolutionary change into certain broad channels of design, one of which is likely to be the self-organizational propensities of molecular systems referred to above. Mathematical chemical models have found that certain patterns of self-propagating replicators, and their tendency to assemble into higher order systems, would be expected to be repeated upon "replaying the tape."[141]

The role of contingency in evolution has also been investigated in the laboratory with bacteria, of great utility in this regard owing to their short generation times. In a classic study, the effects of chance and prior history were judged to be of low importance for the evolution (more than 1000 *E. coli* generations) of traits subject to strong selective pressure (for example, the ability to metabolize a specific type of sugar as the carbon source in a minimal growth medium), but significant for less important traits.[142] Nevertheless, it was acknowledged that historical effects might become evident only after very long periods of time. More recent work with greatly extended numbers of *E. coli* generations has found parallel evolution of specific sets of genes, although changes were not identical at the nucleotide level.[143] Bacterial phenotypic diversity may, in fact, exist when overt measurements suggest approximately equivalent fitness,[144] suggesting multiple solutions toward approaching a fitness peak. Beyond bacteria, physical limitations on the morphology of multicellular organisms certainly exist, such as a practical requirement for bilateral symmetry in mobile organisms. Within these broad dictates, there is likely to be much contingency in the details[140] and the origin of human intelligence is (in all probability) one such detail.

---

[*]These sterling representatives of our species might also be expected to be heterosexual, since how could gay individuals transmit their genes? Yet, "gay genes" (still undefined) would be expected to confer some fitness advantage in heterozygotes or they would not persist in the general population.

Drivers of Complexity   If we should avoid evolutionary teleology at all costs, what does this say about the observation noted at the beginning of this section, that more complex organisms are found in some lineages as the evolutionary progression moves through time? The term "in some lineages" serves as a reminder that there is no observable general principle of automatic increasing complexity in any organisms. Bacteria, for example, (very complex in their own right, but simpler than eukaryotes and multicellular organisms) are highly diversified, but with their basic features unchanged for billions of years. It is one thing to point to commonsense differences between (for example) a bacterium and a chimpanzee as evidence for changes in complexity, it is another to have a well-formulated definition of complexity (or a reliable means for measuring it) as the basis for investigating the dynamics of evolutionary complexity[*]. In fact, defining biological complexity, be it structural, organizational, hierarchical, or genomic, has been a long-standing problem.[145] For example, for some time, it has been ap- preciated that genome size *per se* bears no relationship to the complexity of an organism[†] (the so-called C-value paradox[147]). So how to quantitate complexity? Entropy, a fundamental component of informa- tion theory,[148] has also been applied toward deriving formulas that measure complexity in biological organisms, using nucleotide sequence[145,149] or metabolic networks[150] as yardsticks. In simplistic terms, in such calculations, there is an inverse relationship between entropy and degree of complexity.

There is no generally accepted comprehensive theoretical reason why biological complexity *per se* should necessarily rise during evolutionary time,[146] but computational artificial life evolutionary models have been to used to make the case that complexity must increase under some circumstances.[149] In an unchanging artificial life environment, only mutations decreasing entropy (and thus increasing complexity, by the above measures) can be "selected for" and increase fitness. This process was compared with Maxwell's demon (discussed in Chapter 1), since it amounts to the directed selection of favored entities with a resulting fall in entropy, and this analogy was furthermore related to the generation of complexity in the natural world.[149] In both the real and the virtual worlds, however, changing environments can certainly result in at least a temporary drop in complexity. Classic examples are cave-dwelling organisms (living in darkness) whose eyes, and the attendant unrequired complexity, are rapidly shed.

A global factor mediating increased complexity in biological systems is its general irreversibility,[146] since acquisition of a complex phenotype in response to selective pressures will usually involve a series of changes whose reversal may be a practical impossibility. Different processes contributing to the advancement of complexity have been noted, including gene duplication, differential gene regulation, and symbiosis.[146] The symbiotic partnership between ancient eukaryotes and prokaryotically derived organelles such as mitochondria and chloroplasts was alluded to earlier. These are also a pertinent example of the irreversibility of increased complexity, as many organelle genes over time became relegated to the host nucleus. Duplicated genes (which we have considered previously in this chapter) have been found to be more complex on average than single-copy genes, attributed to the sub/neofunctionalization process.[151]

Signaling pathways are a fundamental aspect of cellular regulation, often functioning through the sequential transmission of a post-translational modification (such as phosphoryla- tion) to a series of enzymes through activation of their abilities to act as kinases (catalyzing the specific phosphorylation of proteins). It has been proposed (and supported by modeling) that

---

[*]This relates to the issue of whether *maximal* or *average* complexity should be the most relevant factor for measuring temporal changes in complexity through evolution, which has also been a contentious issue.[145]

[†]Many genomes have large amounts of repetitive DNAs, of which much is due to transposable elements. Coding sequence is a better correlate of complexity[146] although still imperfect.

such pathways will have a tendency to increase in complexity even in the absence of specific selective pressures,[152] owing to the predominance of mutations that add new interactions or protein participants (increasing complexity) over those which decrease such effects. The intricacy of pathways may also be increased through "natural restoration,"[153] where pathways that have been damaged (reduced in fitness to a degree) are rescued through a secondary mutation that acts by engendering an additional interaction or protein component, again ratcheting up the net level of complexity. The new component could increase the versatility of the signaling pathway, or allow crosstalk with other pathways, at least after further evolutionary change. The "secondary mutation" could arise from one gene copy of a duplication event, as we have previously considered.

Apart from the factors that push molecular systems toward increasingly complex arrangements, the complexity of organisms is also driven by their biological environment or their ecological competitors. The interdependent evolution of ecologically linked species that affect each other's adaptive histories is termed coevolution, and this effect is not an inevitable driver of organismal complexity *per se*. (Here, again, we must be wary of definitions. Coevolution might result in phenotypic changes in a species without an inherent increase in the underlying complexities of the relevant molecular systems controlling expression of that phenotype.) But increasing complexity of a competing ecosystem of organisms may well favor a net increase in the corresponding complexity of their component molecular subsystems in some circumstances. Evolutionary "arms races" between predator/prey or parasite/host pairs[10] are a well-noted example of such an ecological effect, where increasingly complex systems are a common result. The development of the exquisite complexity of the vertebrate immune system, to be considered in more detail in Chapter 3, is a case in point.

If evolvability (the ability of organisms to undergo evolutionary change, which we previously discussed in GOD from Mutation section) itself evolves, then the rate of acquisition of increasing complexity can likewise change with time. It has been suggested[154] that flexible, robust processes tend to be good for both individual fitness and evolvability itself (such that the latter is a side effect of primary selection for the former). Regulatory pathways themselves in eukaryotes are characterized by *modularity*, where individual components can be utilized in many different permutations, another feature promoting the advancement of evolvable systems.[154] Evolvability is dependent on the rate of genetic change (as the raw material for selection), and thus evolvability logically correlates with population diversity. Selection that is based on differential effectiveness of DNA replication and fidelity or recombination might theoretically enhance evolvability itself, in response to varying environmental pressures. The prediction of selectable evolvability has been confirmed in model simulations, and postulated to be an important factor in many real-world situations, including the adaptive immune system and rapidly evolving pathogens such as HIV.[155]

Complexity and Molecular Discovery   So the broad picture that emerges is that there are certainly factors whose operation over time can promote increased evolutionary complexity, and that natural selection is probably routed into a limited number of solutions for many design problems at the molecular level. Now it is time to return to the question posed at the beginning of this section, whether a trend toward complexity in natural evolution is relevant to artificial evolutionary strategies for molecular design.

One issue raised in this section is that of the "directedness" of evolution, which in turn relates to the inherent numbers of options potentially available for any molecular design problem, and the constraints that may limit the range of alternative solutions. Note the latter phrase, innocent although it might sound, carries considerable hidden implications. To go back to the useful idea of fitness landscapes, there might exist many alternative local fitness peaks that correspond to alternatives for a design problem, all of which fall short of a global fitness peak. Where different species show similar evolutionary solutions (at either the gross morphological or the molecular levels), this "convergence" in evolution[*] may suggest corresponding convergence toward a design optimum (even if the ultimate optimal global peak is an undefined abstraction). Gene regulatory pathways involving multiple proteins (genetic circuits) have been found to have significant similarities between *E. coli* and the yeast *Saccharomyces cerevisiae*, ascribed to convergent evolution.[157]*In vitro* evolution of viral populations has shown high degrees of convergence at the nucleotide mutational level.[158,159] In contrast, while laboratory experiments with bacterial cell evolution show that a limited number of key genes relevant to a specific adaptive evolutionary response will recur and thus be identifiable, the specific mutations involved are rarely congruent.[143] In other circumstances, a varied number of genetic changes may result in approximately equivalent levels of phenotypic bacterial fitness.[144]

Alternate adaptive pathways with similar probability of occurrence may exist toward meeting an environmental challenge,[144] but each solution is not necessarily equally efficient. (And it is assumed that only one global efficiency optimum will exist in any case, meaning that only one "alternative" can correspond a theoretical ideal.) This can be interpreted to suggest that there will often be "room for improvement" in natural biosystems for the aspiring biotechnologist, and that global optima remained to be scaled. For example, natural proteins can indeed be often improved by directed evolution or changed in directions away from their natural functions,[23,160,161] a theme to be expanded in Chapters 4 and 5.

We have seen that complexity may emerge at least in part in natural evolution through the practical irreversibility of certain design solutions, which force compounding levels of intricacy in response to subsequent environmental challenges. For human directed evolution, at least the potential exists to circumvent "forced design complexity" by rationally adjusting the initial components of systems prior to *in vitro* diversification and selection. This interplay between rational input and evolutionary methods is an important theme, which we will encounter repeatedly. A practical implication for molecular discovery and the existence of local optima (local fitness peaks) is "knowing when to stop." One would like to have the ability to distinguish local from global peaks, or be able to rationally decide when an acceptable trade-off is attained between performance and the costs of further sequence exploration. At least in the case of single enzymes, such a global peak is in principle definable through the concept of "catalytic perfection," although even here there are caveats involving enzymatic "superefficiency," which we will consider later in this chapter and Chapter 5. Defining a global optimum for a highly complex system involving many proteins (and varied additional molecules both small and large) is another matter entirely. Even at the morphological level of organisms, where the "best" design solution might initially seem self-evident, proving superiority over alternatives may not be a simple prospect. Four-legged locomotion might appear ideal for herbivores that need rapid motion to escape carnivores, but the existence of kangaroos could give this opinion pause for thought.

MOLECULES OF EVOLUTION AND DESIGN BOUNDARIES

Since the overarching theme of this book is the role of empirical approaches to molecular design, it is worth contemplating the range of products that natural evolution has delivered, and

---

[*]Although true convergent evolution is taken correctly as an indicator that an optimal peak of design exists that has been independently approached, not all cases of "convergence" previously described are likely to be correct (at least in an absolute sense). Ancient genes shared by many lineages may be used toward the same design ends in organisms, which have long diverged in other ways.[156]

also whether there are inherent limits on this process. What can "designerless design" deliver at the molecular and supramolecular system levels? And more to the point, perhaps, what *cannot* it produce, if anything? We can begin by re-emphasizing that the principle engine of observed biomolecular diversity is natural selection. As we have seen, certain aspects of biological complexity *per se* may come from self-organizational phenomena at the molecular level, and some genomic features of complex multicellular organisms may arise from nonadaptive processes, but these do not detract from the power of natural selection itself. Likewise, the exploitation of *in vitro* evolution is based on diversification and selection, laboratory surrogates of the natural counterpart process. The products of natural selection will thus serve to demonstrate the inherent power of this process, and allow us to consider some of its limitations.

The question of the limits of biological design and complexity is relevant at two viewpoints: both at an absolute level (that is, does natural selection, through inherent or practical reasons, have limitations in certain areas of molecular design) and the more practical level of how to alter proteins and complex interactive protein systems to maximize performance and attain desirable functional ends by artificial intervention, especially directed evolution. The latter question we will leave until later chapters, but the former warrants our attention here. First, let us take molecular design from complex systems back down to the protein level, and its constituent amino acid building blocks. Then we can look at the process of protein evolution itself.

Evolving Proteins

The genetic code specifies 20 amino acids (with a couple of additional special ones, to be addressed later[*]) that fall into different groups based primarily on hydrophobicity, hydrophilicity, charge, and size. It has been proposed that the genetic code and usage of amino acids have themselves expanded to this level during early biological development, such that some amino acids were acquired early and some relatively recently.[162] Furthermore, phylogenetic studies have suggested that the usage of the "recent" set of amino acids has become more common, continuing to the present time,[163] but the interpretation of these patterns has been controversial.[164,165]

Most large biological polypeptides with various combinations of incorporated amino acids (residues) *fold* into functional proteins with an astonishingly diverse range of activities. Since proteins can range from having less than 100 amino acid residues (many small regulatory proteins) to gargantuan molecules of more than 35,000 residues (as with the titin proteins of muscle[166]), the number of potential residue permutations in "sequence space" is more than astronomical. (A modest 100-residue protein has a specific sequence, one of the 20^{100} (about 10^{130}) separate possible sequences for a polypeptide of this length.) So a tiny subset of these sequences has been "chosen" by natural selection, although a vast number of the total would be excluded through being incapable of assuming a stable (and useful) folding pattern.[†] But even after conceding that all actual biological protein sequences are but a drop in an ocean of theoretical possibilities, the global structural diversity of proteins is relatively much less again. In relatively recent times (largely as a consequence of the accelerating wealth of genomic sequence information) it has become apparent that the sequence diversity

---

[*]As well these natural "extensions" of the 20 amino acid "alphabet," the code has been extended by artificial engineering (further detailed in Chapter 5).

[†]An interesting qualification here is that some proteins appear to function in normal cellular environments in an inherently unstructured state without a definable fold, and we will consider these a little further in Chapter 5. Nevertheless, these "intrinsically unstructured" protein sequences have specific properties of their own, and are still very "special" when considered as a minute fraction of the totality of protein sequence space.

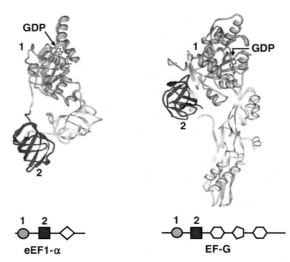

FIGURE 2.2   Modularity of protein domains, using as an example factors required for protein synthesis (elongation factor 1-alpha (eEF1-α) and elongation factor G (EF-G), both in complex with guanosine diphosphate (GDP)).[1] The N-terminal domains of both (the GDP-binding domain 1 and the "translational protein" domain 2, as shown) are in structurally homologous families, but their C-terminal domain structure is divergent (indicated also with domain schematics below their corresponding structures). *Sources*: Protein Data Bank[2] (http://www.pdb.org); eEF1-α: 1JNY[3]; EF-G: 1DAR.[4] Images generated with Protein Workshop.[5]

of proteins is much greater than the structural, or folding diversity.[8,81] In other words, the observed global variety of structural protein folds is much less diverse (and much more evolutionarily conserved) than corresponding sequence diversity itself. This can be represented by mapping protein sequence space to protein structure space, which is inherently asymmetric since many sequences can assume the same type of structural fold.[8] Before thinking broadly about the evolution of protein sequences, let us consider some general features of protein sequence structural organization.

A fundamental concept of protein structure is the *domain*, which can be defined as a relatively small structural and functional unit (usually in the range of 100–250 residues, although exceptions are known[77]) that can often fold independently of the remainder of the polypeptide sequence in which it is contained. By virtue of this modularity (Fig. 2.2), domains can be shuffled around through recombinational processes while preserving their key properties, and thus are viewed as important units of protein evolution.[81] While small proteins will usually be limited to a single domain, most larger proteins will be multidomain in organization, and a large fraction of eukaryotic proteins are in the latter category.[77,81] The very existence of classifiable protein families is due to their evolutionary relationships, and their corresponding origins from the processes of gene duplication and recombination.[167] Protein structural organization also operates at a higher level of modularity termed supradomains, since some domain pairs occur together at much higher rates than could be explained by chance.[168]

Domains can be categorized according to their structures, and upward of 3000 domain families have been currently identified from structural criteria,[169] although overlap between certain folding motifs sometimes makes domain classification difficult.[81] Some domains recur frequently, perhaps owing to their associations with basic life functions rather than deriving from an intrinsic energetic advantage of specific folds themselves.[81] The general distribution of domain families throughout all organisms is nonetheless very uneven, with only about 200 common to all living kingdoms. In the overall universe of natural proteins, it has been furthermore estimated that $\sim$5000 domain structural families exist.[81] Given the evolutionary modularity of domains, using this number to combinatorially produce 2-domain proteins should allow 5000^2 (2.5×10^7) possibilities, and for 3-domain proteins, 5000^3 (1.25×10^{11}). Yet the number of protein families (classified by domain composition) has been gauged at only $\sim$50,000.[81] While this may be a considerable underestimate from incomplete sampling of the biosphere,[170,171] even if increased by 10-fold it is clearly far less than the full potential domain family combinatorial diversity.

Thus, proteins use far fewer domain arrangements than the total possible number, and in turn (to return to the point made above) the minute subset of potential protein sequences that are actually employed biologically is special indeed. Functional proteins are thus highly non-randomly distributed in sequence space, and evolutionarily speaking, this is not a trivial point.[172] Access to altered protein functions can be obtained by evolution through sequences closely clustered in sequence space or by the shuffling of a limited number of protein domains. Were it necessary to plumb remote regions of sequence space every time a novel function was biologically "required," there would be little for natural selection to work with.

A number of other factors have been proposed to influence the rate of protein sequence and structural evolution. All these parameters are susceptible to increasingly refined analyses through high-level cross-species genomic data and high-throughput experimental approaches.[173] First, let us look back to our earlier thoughts on neutral mutations, since this issue overlaps with some additional qualities of protein evolution. We saw earlier that neutral changes can be significant in providing different backgrounds in which subsequent rounds of mutation can operate. But what renders a mutation as nominally "neutral" in the first place is the ability of a protein to tolerate a sequence change without significant perturbation of its folding, and in turn its function. This quality has been termed the *robustness* of a protein. Since robustness reflects the ability of a protein to tolerate mutations (to accept them as neutral or near neutral), this quality would be predicted to correlate with the evolvability of a protein. Evolution toward robustness (and thus increased evolvability) itself can be indirect consequence of selection for protein stability, since stabilized proteins are more accommodating of sequence variation than their less stable precursors.[174] (Mutations that can lead to novel activities will often be destabilizing, and tolerable only in a background where protein scaffold stability has already been increased.) Allied to the concept to robustness is *designability*, or the number of variant protein sequences that converge toward a single folded structure with the lowest energy state.[8,175]

Another proposed factor in protein evolution, *dispensability*, also has some relationships with neutral mutation theory. The rationale goes that a protein that is of secondary importance for an organism's fitness (that is, to say, one that is more likely to be dispensable) is also more likely to be capable of accumulating mutations whose effects are not globally detrimental. In other words, selection against mutations in dispensable proteins should on average be less potent than for highly essential proteins, and in turn the evolvability of dispensable proteins would be predicted to be higher.[*] With modern technology, this is amenable to experimental analysis in model organisms. Systematic study of mutant genes or gene "knockouts" (see Recombination section) allows assignment of dispensability values on a global basis, by assessing the effects of gene

---

[*]By way of a metaphor, one would have a much better chance of preserving the essential functions of a motor vehicle by making random changes to the seating decor or dashboard arrangements than similar arbitrary tinkering with the engine.

inactivation on growth rate[*] (for a completely essential gene, no growth can occur; while a gene with complete dispensibility would not show significant growth retardation). In bacteria, it has been concluded that the essential genes evolve more slowly than the nonessential (more dispensable) genes,[178] and in yeast a comparable significant effect has been found,[179,180] although the generalizability of this has been disputed.[181]

Functional density of a protein, defined as the proportion of sites (out of all accessible residues) that are functionally important,[182] has been proposed to be evolutionarily significant if a high level of functional constraints (with accompanying constraints on structural features) put a brake on the subsequent rate of protein evolution. Support for this has come from analysis of the relationship between the proportion of solvent-exposed (surface) to buried protein residues (where the latter are inferred to dictate constraints on protein structure) and evolutionary rates, which suggested that a high proportion of structural constraints correlate with slow evolutionary change.[183] *Expression levels* of proteins also appear to be a factor in evolvability. Proteins vary widely in their expression levels, and the highly expressed subset within an organism's total protein complement (the proteome) shows decreased rates of evolutionary change than those that are expressed at low levels. This is attributed to the acquisition by highly expressed proteins of a phenotype termed "translational robustness" or resistance to errors introduced during protein expression itself. Once such proteins evolve this useful property, any subsequent mutations have a high probability of reversing this type of robustness phenotype and lowering fitness, which thereby decreases the rate of ongoing evolutionary change.[184]

The remaining proposed protein evolutionary factor is modularity, which we have discussed previously in the general context of evolvability. Modularity refers to systems where certain components can have interchangeable roles, and in biological pathways this correlates with the extent of a protein's interactions with other proteins (alternatively termed the *connectivity* of a protein). It might logically be predicted that a protein with high connectivity would have more constraints on evolutionary change (and thus would evolve at a slower rate) than a protein with low levels of interactions, but this has not been corroborated in *E. coli* and only weakly so in yeast.[185] In the latter type of organisms, the evolutionary effect was only found to be significant for a highly interactive subset, the so-called hubs of the cellular interaction networks.[186] Core proteins in bacterial photosynthetic pathways show extensive evolutionary conservation linked to their high levels of connectivity (defined in a broad sense to include cofactors (see below) as well as protein–protein interactions). In this example, the high connectivity factor is believed to limit further evolutionary improvements in photosynthetic efficiency.[187] In the context of connectivity/evolutionary analyses, it has been pointed out that the nature of protein–protein interactions is an important factor. "Obligate" interactions, as occur in stable multiprotein complexes, have different selective pressures in comparison with transient short-lived interactions such as those associated with protein signaling pathways. Residues at the contact interfaces of obligate complexes accordingly have relatively slower evolutionary rates.[188]

To put all these parameters together, if one found (or engineered) a protein with low levels of expression, low functional density, and low connectivity, at the same time maintaining high dispensability and stability, then it might seem a safe bet that it would be able to set speed records for subsequent evolutionary change. Perhaps these constitute an improbable combination,

---

[*]Assessment of normal (mitotic) growth is a good first approximation for dispensability, but would exclude detection of genes purely required for some other contingent environmental need or meiosis in a sexual organism. Therefore, assessing dispensibility in this way requires analysis of a sizable number of proteins before cross-species comparisons can be made.[176] This also emphasizes that dispensibility is not an absolute property, but is conditional with environmental change. For example, detoxification genes dispensible in some environments may become essential in others. Problems with defining "essential" genes in bacteria in evolutionary terms have been noted.[177]

but at least some of these factors are logically interrelated. (For example, the functional density of a protein is correlated with the extent of its connectivity.) On the other hand, the effects of expression levels, functional density, dispensability, and modularity appear to operate as independent evolutionary variables, and a better understanding of their relative importance and relationships should lead to a more comprehensive picture of protein evolution.[173]

Enzymatic Bounty

Now we should take a brief look at the evolution and function of a special and fundamentally important class of proteins, concerned with the catalysis of chemical reactions. These biocatalysts, known of course as enzymes, can mediate a huge range of reactions, but these can broadly be grouped into six major categories.[*] Enzyme evolutionary history is evident in the ability to group enzymes in superfamilies with certain shared core fundamental chemistries, such as the formation of a specific type of enzyme–substrate intermediate through a conserved enzymatic residue.[190] Enzymes of different origins may have dissimilar structures but catalyze the same reaction.[191] Yet even when extensive differences exist in overall protein structures between enzymes with related activities, similar properties of their active sites can sometimes be discerned, suggestive of convergent evolution toward an optimal function.[†] On the other hand, similar structural scaffolds (protein "catalytic folds") have been exploited by evolution for divergent catalytic tasks.[192]

It has long been recognized that enzymes mediate catalysis by stabilizing a chemical transition state between the substrate (the molecular target of the enzyme) and the final product by binding the transition state with high affinity. The ability of the enzyme to do this lowers the activation energy of the reaction such that the rate of the reaction is dramatically increased. The enzyme may undergo covalent interactions with the substrate during the catalytic process, but, as a catalyst, the enzyme emerges unchanged (and capable of a new cycle of substrate recognition) after product generation. To accomplish catalysis, a region of an enzyme molecule must exist which can physically accommodate the shape of the transition state, and this is referred to as the active site. The ability of an enzyme to recognize its substrate and stabilize the corresponding molecular transition state determines enzyme specificity, which is generally very high. Yet in some cases, enzymes can recognize and catalytically act on multiple substrates, even if the latter show considerable divergence chemically in some cases ("catalytic promiscuity"). Recognition of a substrate per se is necessary but not sufficient for catalysis; the enzyme has to be able to overcome the activation energy barrier leading to the transition state itself, even though this is far less than the activation energy required in the absence of enzyme. And this point leads us back to the things that natural selection can do with proteins, and specifically enzymes, composed of 20 amino acid building blocks. Or to look at it from a negative viewpoint, which chemical manipulations (if any) would stymie any natural enzyme, or only allow inefficient performance?

---

[*]The enzyme groups and their associated reactions are: oxidoreductases (oxidation–reduction reactions), transferases (chemical group transfer reactions), hydrolases (chemical cleavage reactions involving the participation and cleavage of a water molecule), lyases (nonhydrolytic cleavage reactions often characterized by the formation of a double bond), isomerases (reactions promoting the formation of chemical isomers), and ligases (reactions joining two molecules and consuming an energy source, usually adenosine triphosphate).[189] It is very important to note that these categories refer to enzyme specific functional activity and not directly to the enzymes themselves. Thus, two enzymes from different organisms can catalyze identical reactions and have the same "EC" (enzyme commission) code numbers (which denote types of catalytic activities) and yet have very different structures (see also Chapter 5).

[†]Caution should be noted here, however, since distinguishing true convergence of evolutionarily unrelated genes from extreme divergence of an original ancestral gene is not trivial.[81]

First, we should consider again what an enzyme needs for activity, remembering that substrate recognition alone is not enough. Many enzymes are indeed catalytically functional solely as folded polypeptide chains, but many more are not. Yet even if it should prove that a particular function is difficult (if not impossible) to perform with a folded polypeptide *alone*, that is not at all the end of the story. Let us first make these points with an imaginary scenario, and co-opt Lucy from Chapter 1 as the agent.

Combinatorial Catalysis and Its Helpers   Lucy dreams. She has been provided with a series of 20 different building blocks that she can string together in any order she likes, to make products ranging from short (<100) to very long (>10,000) polymeric sequences. The right combinations of these fold up into defined configurations, and of this "foldable" polymer subset, she is happy to find that a significant number have many useful activities. (This functional group is only a miniscule fraction of the total number of possibilities, but in a dream "high-throughput screening" can take on staggering dimensions.)

Yet try as she might, she finds there are some desirable activities that stubbornly resist solution, and some activities for which only inefficient and unsatisfying partial answers can be found. This seems like an insurmountable hurdle, but then she has an idea. Although she cannot string together anything other than combinations of the 20 available building blocks, additional "blocks" could certainly extend the function of her polymers. Other materials in her working environment are not directly suitable, but she finds metals can bind certain polymers and assist their activities. But still, many other needs remain unmet. Then, with an additional inspiration, she realizes that a "bootstrapping" approach should work. First, make one polymer for the dedicated task of creating a specific and useful "cofactor" block, which then enables *another* polymer to take on a new activity. By this means, the range of activities of her polymer strings would be greatly extended. She can foresee a potential catch, though. If polymer A needed cofactor X to make cofactor Y, then how does she make cofactor X? This might require polymer R and cofactor S, and in turn polymer U and cofactor V are required to make cofactor S, and so on, into infinite regress? Then how can she ever start the ball rolling? She is hopeful, though, that this will not be a problem in reality, since if one or more cofactors can be made with a "naked" (unsupplemented) polymer, or if direct environmental factors such as metals can assist in at least one cofactor synthesis, then the bootstrapping could take place. Maybe, she thinks, one could construct a catalytic loop of sorts, where a chain of polymers, cofactor requirements, and syntheses feed back to the beginning of the set. There is much that can be devised and tested. . . .

And so it is in the real world of natural selection, evolution, and biocatalysts, with the obvious caveat that there is no foresight involved. From a vast set of combinatorial alternatives, the correct sequence of certain biopolymers yields efficient catalysis, although cofactors are often involved. We are thinking about protein enzymes in the present context, but thoughts of cofactors go further than this, back to the hypothetical RNA world noted earlier. But the principle remains the same: the extension of catalytic capabilities beyond the limitations of a polymeric "alphabet" by co-opting other molecules, whether they can be obtained from the environment or synthesized by cooperating polymers. Many organic enzyme cofactors[*] have a structure derived from a ribonucleotide core, leading to the long-standing proposal that extant enzyme cofactors are survivors from the remote RNA world when they were used by RNA enzymes (ribozymes).[193] Proteins in general have a greater chemical diversity than nucleic acids, which is very advantageous from a catalytic point of view.[194] For any specific enzyme,

---

[*]This is a general term for any chemical assistance to enzyme catalysis, including metal ions. Small molecule organic cofactors are usually termed coenzymes.

a limited number of residues at the active site directly mediate catalysis. Of the usual 20 amino acids, 11 polar and charged residues participate in enzymatic catalysis in general, and of these, histidine is the star player in terms of frequency.[195] But an important additional point to note is that the relevant chemical properties of catalytic amino acid side chains (such as ionization state) cannot be simply judged in isolation. Interactions between residues in folded polypeptide chains can significantly modulate such properties and this is often demonstrably critical for specific catalytic mechanisms.[*,195]

Nevertheless, cofactors are still widely used by protein enzymes. A variety of metal ions[†] (and other nonmetallic elements) are extremely widespread as mediators of enzymatic activity. Metals and other inorganic elements (notably sulfur) also occur in some enzymes as atomic clusters with specific configurations (especially those involved in oxidation–reduction reactions).[196,197] Organic coenzymes (many of which are derived from vitamins) are synthesized by the activities of other enzymes.[††] Coenzymes may be tightly noncovalently associated with enzymes or specifically covalently linked. It is clear that some reactions catalyzed in nature by enzymes with cofactors do not require a cofactor in an absolute sense. (In other words, although a natural cofactor-associated enzyme will be inactive in the absence of the cofactor, another enzyme or even a variant of the original enzyme may be able to perform the required reaction independently of the cofactor.) This is evident both from the natural examples of enzymes from different organisms performing the same type of catalytic task with or without a cofactor requirement (as with oxygenases[198]) and from the ability to engineer some proteins to a cofactor-free state without loss of catalytic activity.[199,200] Natural selection cannot systematically scan the whole of protein sequence space to reduce cofactor requirements. The solution found by nature for a catalytic design will depend on evolutionary history, environmental access to potential cofactors, and other effects; the resulting biocatalyst will not necessarily be the only possible general design. (On this theme, it can be noted that precedent exists for distinct enzymes performing the same catalytic task with different cofactors. Thus, *E. coli* has two methionine synthase enzymes, one of which uses cobalamin (vitamin B_{12}) and the other uses zinc ions.[201]) The evolution of enzymes is also channeled by pre-existing coenzyme-binding domains. As an example, the "Rossmann fold" (a structural fold that binds certain nucleotide coenzymes) is phylogenetically conserved and believed to be evolutionarily ancient.[202]

Another "cofactor" of sorts that we should note in passing is the solvent itself—water, in other words. We can note that in the above Rossmann-fold motif (at least in its "classic" variety) a conserved water molecule is involved with coenzyme binding.[203] But beyond a structural role in enzyme complexes, water molecules are directly important in many enzymatic catalytic mechanisms. Hydrolytic reactions involve the chemical breakdown of the H_2O molecule, to assist in the cleavage of another molecule. Thus, hydrolysis of the familiar drug aspirin (acetylsalicylic acid) is predominantly carried out in humans by a particular esterase. Human carboxylesterase 2 hydrolytically cleaves the substrate acetylsalicylic acid[204] to yield acetic

---

*This has also been demonstrated to be the case for ribozymes, which we will pursue further in Chapter 6.

†Magnesium, calcium, and zinc ions (Mg^{2+}, Ca^{2+}, and Zn^{2+}, respectively) are commonly exploited, but many other metals including iron, nickel, copper, manganese, molybdenum, vanadium, and tungsten are used by at least some organisms. It should also be noted that metal ions have other roles in protein function beyond direct participation in catalysis, such as global protein stabilization or stabilization of specific protein structures.

††But not necessarily from the same organism. Bacteria such as *E. coli* make everything they need, but many mammalian species (including humans) have lost the ability to synthesize many vital coenzymes. When coenzymes (as vitamins) can readily be obtained from the environment (in food), selection favors loss of the metabolically complex and energetically expensive coenzyme synthesis pathways. The inevitable drawback of this, however, is the onset of deficiency diseases if for any reason the environmental coenzyme source is withdrawn.

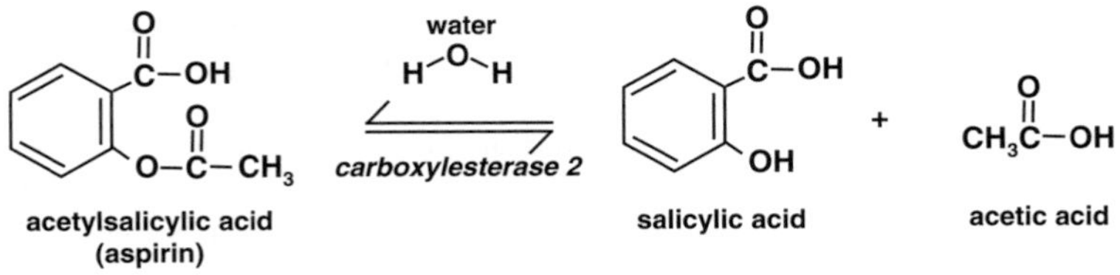

FIGURE 2.3   Catalytic action of carboxylesterase 2 on aspirin.

acid and salicylic acid (the reverse reaction eliminates a water molecule from the two reactants) (Fig. 2.3).*

Enzymology is a vast field, barely scratched here. But we have approached the whole area from a fundamental molecular design point of view, asking what are the basic requirements for protein enzyme catalysis as a "parts list." Given that natural evolution can follow many different pathways in divergent lineages, simple answers are not forthcoming. Evolutionary convergence for some catalytic tasks may suggest design optima with the biologically available protein/ cofactor tool kit, but in other cases, divergent structures can accommodate the same catalytic goal. Some catalytic reactions can be performed with alternate cofactors, or in a cofactor-free state. Different biological enzymatic alternatives for the same reaction can be rationally assessed for their relative catalytic efficiencies, and the ultimate winner would demonstrate formal catalytic perfection, limited only by the diffusion rate of substrate.† But if the existing alternatives fell short of this yardstick, the best case example would not exclude the possibility that a radically different polypeptide/cofactor design (even if not in natural existence) might indeed reach the theoretical ultimate. But enzymatic functional diversity itself can offer encouragement for artificial design, by showing that many catalytic problems may offer multiple (if not globally optimal) solutions. When a catalyst must be integrated into a more complex system, this kind of flexibility will logically become a useful feature.

Enzymes to the Limit   The above observations tell us that a requirement for a cofactor for a specific enzyme (catalyzing a particular reaction) cannot be taken as proof that no polypeptide exists in protein space that could do the same job in a cofactor-free state. It is clear, though, that cofactors extend the range of reactions catalyzable by polypeptides, and for at least some reactions it is likely that no cofactor-free polypeptide configuration is available which can be catalytically efficient. Then, to return to the main question of interest in this section, do cofactors or any covalent post-translational modifications†† allow any conceivable chemical reaction to proceed by enzymatic means, or are there inherent physical limits? The answer to this might seem obvious, by simply pointing to chemical reactions that require conditions way outside those tolerated by living organisms, even bacteria living in extreme environments.**

---

*Of course, human carboxylesterase 2 did not evolve expressly for the purpose of processing aspirin. This enzyme hydrolyzes many esters, and thus exhibits a degree of catalytic promiscuity, which we have encountered earlier. Nevertheless, this enzyme shows a specific pattern of molecular substrate preferences in its activity.[204,205]

†Even so, the diffusion limit is not an absolute end point for enzyme catalytic efficiency, as "super-efficient" enzymes possess mechanisms for accelerating access to substrate, as continued in Chapter 5.

††Any modification to a protein by an enzymatic process following protein synthesis itself (hence post-translational). These include glycosylation (attachment of carbohydrates), phosphorylation, lipidation, and derivatization with certain small regulatory proteins such as ubiquitin.

**Another caveat here is naturally that both an enzyme's substrate and its product have to be stable in an aqueous environment. So in this context, it is interesting to note that some enzymes can function in organic solvents under anhydrous conditions and elevated temperatures (with altered reactivities[206]), but these properties are clearly not selected for directly in evolutionary terms.

But consider the "fixation" of nitrogen from the atmosphere. Diatomic nitrogen (N_2) is very stable and unreactive, and its commercial reaction with hydrogen in the Haber–Bosch process to form ammonia (NH_3) requires an optimal 200 atmospheres pressure, 450°C, and suitable inorganic catalysts. If one was ignorant of biology, it might then be thought unlikely that biological enzymatic processes could efficiently fix nitrogen from the air. And yet they can, and do, and at room temperature to boot. Biological nitrogen fixation, an activity found only among prokaryotes, is a highly complex process that overcomes a formidable activation energy barrier to utilize atmospheric nitrogen to form ammonia, which is then rapidly converted into aspartate and glutamate for continued assimilation into cellular metabolism. For the electron flow that results in reduction of nitrogen, key cofactors are essential, in the form of inorganic clusters of iron, molybdenum, and sulfur (vanadium in some organisms[207,208]). Despite much progress in recent years, the complete details of bacterial nitrogen fixation have still to be unraveled.

So natural selection, time, and cumulative change can deliver impressive design results for enzymes, and indeed for proteins in general. In the context of possible limitations, however, we have to bear in mind the "selection" feature as a paramount determinant of design results. Irrespective of its physical feasibility, no biocatalytic mechanism will emerge without its conferral of a fitness advantage onto its host organism. Despite its considerable complexity, nitrogen fixation is not "irreducible"; even low-efficiency fixation of nitrogen would endow a primordial prokaryote with a considerable competitive boost. Judging the limits of protein-based catalysis from the global array of natural proteins is thus itself subject to the inherent bias determined by selective environmental pressures over eons of time. In other words, "holes" in the natural catalytic repertoire will exist if there is no selective "need" for a particular reaction in existing biosystems. But aside from the issue of the channeling of enzyme evolution by selective pressures, a hole in a repertoire (absence of evidence) does not necessarily mean that a novel catalytic function could never be attained with biological molecules (evidence of absence). An "absent function" might exist if all available protein scaffolds offer poor starting points for evolving toward a novel functional protein conformation remote in sequence space from extant biological polypeptides. In other words, this interpretation would hold that the entire existing protein world is committed to a sizable but nevertheless limited set of scaffold forms, and cannot mutationally or recombinationally traverse a very wide gap into entirely novel forms that might offer new functional perspectives.

An alternative, of course, is that absent functions are truly unfeasible for biological systems in an absolute sense. With these thoughts in mind, let us take a broader look at the issue of "holes" in natural evolutionary design, at a higher level than single enzyme molecules.

Absent by Irreducibility, or Lack of an Access Route?

A common strategy of "intelligent design"* supporters is to point to supposed cases of "irreducible complexity," where certain complex biological systems requiring multiple components are judged to constitute unitary entities that could not come into being in a single step, and which could not have viable intermediate precursors. Among the favorite examples of "irreducible complexity" are bacterial flagella, the blood-clotting cascade, and the immune system. The champions of intelligent design are correct in believing that these systems did not appear in a single bound, but quite erroneous in assuming that precursors, or other applications

---

*"Intelligent design," the pseudointellectual recycling of basic creationism, has been comprehensively refuted.[209,210]

of complex system components, do not exist. The relationship between bacterial flagella and Type III secretion systems is a well-characterized case in point.[211] Furthermore, artificial life computational models have demonstrated that complex features (in this case, those requiring the interaction of many encoded instructions) can accrue during evolutionary development.[212,213] These arrangements resulted from progressive building upon successive precursor genotypes, each of which was selectively favored. A general principle of natural selection and evolutionary development, consistent with the "tinkering" metaphor, is the pre-emption (or "exploitation"[*]) of gene products for a novel function (often proceeding by gene duplications, as we have seen). This is a means for generating complex interdependent structures whose evolutionary origins would be otherwise difficult to account for. There is definitely a place for intelligent design in the context of molecular discovery—but (as we will see in succeeding chapters) that is in the laboratory, practiced by intelligent human beings. But is there such an entity as a truly "irreducible" complex structure that thwarts evolution and is thus absent from all biological organisms?

If it should be possible to identify any inherent limits on natural molecular design, or the properties of complex natural interactive systems, that would be useful information with respect to artificial molecular design. (We could then consider what these constraints might signify for artificial molecular evolutionary processes.) Here, I am referring to limits that might be imposed by the intrinsic nature of a proposed hypothetical function for which no practical solution exists within protein space, or a complex system that cannot be produced by successive accumulated changes (that is, a system that is, in fact, truly irreducible). These kinds of limits are thus clearly distinguishable from failure to find representation of a protein through lack of any selective pressure (as noted above for enzymes).

Limits imposed on biological systems attaining a certain threshold of complexity have been proposed.[18,124] In such theoretical "complexity catastrophes" for any organized chemical system, at a certain level of molecular diversity competing undesirable side reactions create havoc and reduce fitness. Yet the astounding complexity of cellular organisms suggests that, however compelling this argument in theory, in practice in the real world there is plenty of room to move. The complexity catastrophe scenario then becomes just another physical constraint, among many others, on the directions where natural selection can take biological systems. Constraints on biological diversity can also be observed in ecological systems at the higher level of species and families.[214]

To consider the issue of natural selection design limits from a different viewpoint, if we could conceive of a function for which no biological precedent exists,[†] or a supramolecular system that is likewise biologically absent, it might in principle be due to their fundamental irreducibility. An evolutionary biologist might say, "Fine, if you have truly dreamed up an irreducible system, then by definition it will be conspicuously absent from the biosphere. No further problem; we are interested in understanding what is out there in the real world, not flights of fancy" But the crucial point is that *in vitro* evolution directed by humans need not suffer from the natural restrictions of selective pressure (we can introduce selection for a desired goal, "unnatural" or not), or from the starting material (we need not be marooned on a solitary peak in a fitness landscape, but can move the starting point to a new site). That is to say, even a truly irreducible problem for biological natural selection (which remains only hypothetical at present) may yield to human intervention provided it is physically possible. In the area of

---

[*]This effect has also been termed "pre-adaption" (by Darwin) or "exaptation."

[†]Insofar as we are aware. Sampling of the biosphere has been very limited to date, mainly through rudimentary knowledge of the (often uncultivatable) prokaryotic world. "Metagenomic" projects in recent times seek to redress this deficiency (further details are provided in ftp site).

molecular design by *in vitro* evolution, humans can "go one better" than natural selection by setting the right selection criteria (which might never arise in nature), or ensuring that the starting molecule is likely to be compatible with the desired final function (insofar as that is possible).[*]

Food for Evolutionary Thought   Let us consider use of an argument from the vantage point of "natural design," which will also serve to show the limits of this kind of theorizing. Although famed as a science-fiction writer, the late Isaac Asimov was anything but soft-headed when it came to alleged "paranormal" phenomena, and he used the natural world as an argument against the existence of the unconventional mind-to-mind information transfer popularly known as telepathy.[215] (Before going any further, I must stress that there is zero evidence that there is any such thing as unexplained mental communication transcending distances, whether in humans or animals.) He made the argument that it would be a strong selective advantage for an organism to possess telepathic communication; therefore, its absence argues that its existence is highly unlikely. The specific nature of a telepathic facility would determine its fitness benefit, as (for example) whether it could operate only between members of the same species, and whether it could involve involuntary "mind reading" or only work cooperatively. Certainly, a "telepathic" function that gave long-distance warning of an approaching predator would unquestionably provide a survival advantage. But this is a fruitless speculation for an utterly hypothetical property where one could propose any number of different factors that could allow its selection.

While the benefits that telepathy might confer are thus highly dependent on exactly what one means by the "extrasensory perception" telepathic effect, this is not the point I am wanting to make. Here, rather than a physical structure that is uncontroversial for human artifice but absent or limited in nature, we are considering an effect that in itself is hypothetical. I would argue nonetheless that the observed natural nonexistence of telepathy in itself is not a strong argument for its physical impossibility *per se*. There are two ways of viewing Asimov's proposal. If telepathy was an evolvable sense like light perception (which has evolved independently multiple times), then we would indeed expect that its existence should be obvious throughout animals with nervous systems. Its palpable absence, then, certainly indicates that telepathy of this order is a nonstarter. But again, since we cannot really define what a completely hypothetical sense would require for its functional operation in the first place, this argument does not go too far.

A second way to interpret Asimov's argument (which admittedly is a little unfair to Asimov) is to see it carrying an implicit message that if something is physically possible and potentially very useful, then natural selection will see that it is done. Many counter arguments are possible here, but in the context of our hypothetical telepathy rumination, we can think of three possibilities: (1) telepathy is physically impossible; (2) telepathy is physically possible in a biological context but the "design solution" which enables its realization has not (yet?) been reached on this planet, possibly through required levels of neural complexity which are in principle attainable; (3) telepathy is physically possible but not with the molecular and supramolecular tools that natural selection has at its disposal. The third possibility is thus proposing in effect that there are some things natural biosystems cannot do, even if they are physically possible in themselves. Proposal (3) thus further holds that telepathy may be indeed be impossible for conventional neural systems, but some kind of extension of the molecular

---

[*]This is not to forget advances in rational molecular design, which are not limited in principle by anything other than the laws of physics and chemistry, potentially enabling exploration of designs not seen in nature. This topic is contained within Chapter 9.

repertoire available for natural selection could enable its realization. Consider that radio telecommunications do not tend to raise eyebrows in either technical or lay circles, but biological radio transmitters and receivers for communication purposes are definitely in short supply.[*]

While the first possibility cited above for telepathy (its intrinsic impossibility) is by far the most likely based on all evidence, one must look to physics to be confident of this, rather than terrestrial design by natural selection. Again, absence of evidence is not evidence of absence . . . The take-home message for attempts to define the absolute design limits of biology is caution. But theoretical absolute design limits and practical limits imposed by the pathways taken by natural evolution do not necessarily coincide. As we will see later, extensions of biological systems by human ingenuity have the prospect of overcoming limitations imposed by the natural biological tool kit, and can consciously bypass the difficulties of extant natural biology in overcoming "design space jump" limitations imposed upon natural selection. Before we reach this point, there is another major category of products of natural selection at the system level, which we have not yet considered in detail. The systems I refer to are of such complexity, and yet so instructive for empirical molecular design, that they warrant a chapter all its own, which thus follows.

---

[*]Also consider an implicit assumption of the Search for Extraterrestrial Intelligence (SETI) project: If any *bona fide* coded radio signals are received from space, they will originate from intelligent life, not from an extraterrestrial nonsentient biological communication system. Again, absence of "bioradio" on earth in itself does not prove that it is universally impossible, but it may well be.

3 Evolution's Gift: Natural Somatic Diversification and Molecular Design

OVERVIEW

It is a remarkable observation that certain products of evolution that operate within individual organisms can recapitulate the essential Darwinian principles of evolution itself. These products, immune systems, are no mere optional extras but essential requirements for existence in a ceaselessly competitive world. Higher level adaptive immunity of vertebrates has such extraordinary abilities, combined with such exquisite complexity, that stepping back and seeing the "big picture" all at once can be a challenge. Indeed, it is probably quite fair to judge the adaptive immune system as the second most complex system in the universe, as far as we know. For our present purposes, we particularly want to focus on the aspects of immunity in general that are relevant to somatic diversity, selection, and molecular design. But as with any efficient complex system, the bits and pieces of immune systems do not exist in isolation, but are intricately intertwined. In order to provide a coherent presentation of the place of the various subsystems in immunity in general, a summary of some important features of immune systems (especially adaptive immunity) is needed. Apart from its didactic role as the premier case in point for rapid natural evolution, the immune system has historically been of major importance as a source of recognition molecules, and its ongoing significance in this area will be introduced in this chapter and continued in a more general context in Chapter 7.

If adaptive immunity is the runner-up in the complexity stakes, it begs the question of the actual winner. The answer, of course, is often stated to be the higher order neural systems of the human brain. Even though this notion has arisen within human brains themselves, a good case can be made for it. In the final section of this chapter, we take a look at the brain and its somatic diversity, to ask if it will render any further lessons for molecular design. For these purposes, it will also be very useful to consider an important extension of the brain, chemosensory recognition.

IMMUNE SYSTEMS: A BIOLOGICAL FUNDAMENTAL

All organisms must maintain their functional integrities (their homeostasis) in the face of the environment, and a major part of this is the competing biological environment. An organism with no defenses would rapidly succumb to the assaults of any of the wide range of macroparasites and microparasites, ranging from worms to viruses. Protecting the host biosystem from such external threats is the duty of the immune system, but not its only task.

Searching for Molecular Solutions: Empirical Discovery and Its Future. By Ian S. Dunn
Copyright © 2010 John Wiley & Sons, Inc.

Multicellular organisms without exception depend on coordination of cell growth and cooperation between their constituent cell types. A cell that grows independently without restraint (and thereby escapes the controls mandating its normal role as a cog in a much larger machine) behaves essentially as a selfish parasite and may endanger the survival of its host organism. The control of such transformed tumor cells has been termed "immunosurveillance" and has been the subject of much controversy, but evidence for its operation has accumulated.[*] Of course, in targeting either parasitic organisms or parasite-like tumors the immune system can fail, or more often, immunity can be thwarted by sophisticated parasite countermeasures. In contrast to such visible failures (often tragic for human beings), the constant successes of the immune system are frequently hidden, and the power of its protection only obvious when extensive immune failure occurs.

The Selfish Scheme

It is highly relevant to consider the immune system in the context of empirical molecular design, and indeed as an inspiration for this type of design strategy. But first let us clarify what we are referring to. An "immune system" can be used in a very general sense to describe any organized biological defensive system, and as such, it can and does operate at the level of single cells. Before thinking further about such processes, though, we should consider a very basic point: a defense mechanism cannot be effective (and indeed, might be counter-effective) unless an invader of any description is clearly demarcated from the biosystem to be protected. In other words, there is a need to tell the difference between molecules and structures of the host biosystem from everything else.

Yet the need to preserve "a sense of self" in fact operates at the fundamental level of the maintenance of genetic integrity and the desirable promotion of sexual genetic interchange. This can be observed in the discrimination and pairing of homologous chromosomes in meiosis,[62] or in specific mechanisms which ensure that self-fertilization in hermaphroditic species is avoided. Plants have systems of self-incompatibility for pollen fertilization, such that cross-pollination is promoted. Animals that form as cellular colonies (notably certain ascidians or sea squirts) have a potential problem of fusion with genetically foreign individuals, and to avoid this they have developed mechanisms for recognition of such unwanted cellular intruders of the same species, or *allorecognition*.[216] Fungi can undergo a cell fusion process (anastomosis) associated with genetic exchange, and in fact fungal species have developed mechanisms whereby they can distinguish themselves from others. (Recognition of cell fusion between self and nonself fungal partner cells results in programmed cell death.) To recognize genetically distinct members of the same species, a *polymorphic* structure is required; one that has many alternative genetic alleles. (It stands to reason that an invariant gene product is of little use in this regard.)

Self/nonself recognition is thus widespread, essential, and not recent in evolutionary terms. But as well as featuring in mechanisms for defining genetic boundaries, self-definition, as we have seen, is a first-principle requirement for any form of immune system. Environmental threats can come in many forms, but in molecular terms we can break them down into chemical or biological agents. These broad areas are actually interlinked because many environmental chemical threats to an organism in fact originate from other organisms. The key distinguishing point is that biological threats can replicate and undergo adaptive responses, while chemical dangers cannot do so directly. First let us look at how the latter kind of molecular danger is dealt with.

---

[*]A host antitumor response can act as a potent selection agent for tumor "escape" variants that have lost antigens through which they are recognized by the immune system, and as such, the process has been referred to as "immune editing" more than surveillance *per se*.

The Xenobiotic "Immune System" and Beyond   Humans sampling natural molecular space in search of useful products by traditional methods run a risk of being exposed to harmful compounds. But irrespective of conscious intent, it is impossible for any organism to avoid continuously encountering potentially toxic substances in the environment. These may be ingested through foods, water, or even in an air-borne state in some circumstances. Normal bacterial flora and metabolic processes can also produce toxic by-products.[217] In the context of exposure to the natural molecular environment (whether deliberate or not), it is worth briefly noting the means by which vertebrates recognize and neutralize potentially damaging compounds.

A large family of proteins, the nuclear receptors, has long been known to contain members involved in both binding certain hormones or vitamins, and subsequently activating specific genes responsive to these mediators.[218–221] Although these receptors collectively recognize diverse small molecules, certain sequence features in the nuclear receptor family are conserved. Beginning in the 1990s, this observation was exploited by means of sequence homology probing and then computer-based genomic approaches to isolate the remaining complement of these receptors. Since the ligands for newly identified nuclear receptor family members were initially unidentified, these were termed "orphan nuclear receptors."[222] More recently, some of these so-called orphans have been shown to bind diverse ligands and subsequently activate a detoxification pathway involving enzymes of the cytochrome P450* family.[224–226] These receptors and enzymes collectively constitute a recognition and processing system for both foreign chemicals ("xenobiotics") and potentially toxic products of endogenous metabolism. Primary sensing of such possible chemical danger falls on a specific nuclear receptor class, but both the relevant receptors and processing enzymes for xenobiotics must be capable of recognizing and binding the appropriate potentially noxious molecules.

The xenobiotic receptors, which have certainly gained freedom from real orphan status, are principally the constitutive androstane receptor (CAR) and pregnane X receptor (PXR; termed SXR in humans[217,227–229]). The large binding site or "pocket" of PXR has a structurally flexible, spherical configuration lined with hydrophobic residues, probably related to its promiscuous ability to bind hydrophobic ligands of diverse shapes.[227,230,231] Although the ability of PXR to bind diverse molecules is wide, it is not random.[232] In accord with this, different species show divergences in residues lining the binding pocket, possibly associated with specific environmental detoxification needs as directed by natural selection.[226,233]

Consistent with the view that xenobiotic receptors signal the presence of environmental toxins, mice whose PXR genes have been artificially inactivated have a normal phenotype but respond abnormally to challenge with xenobiotic chemicals.[227] As might be predicted, sensing of exogenous compounds by xenobiotic receptors is important in drug pharmacology[224,230,234] and can partially determine the effects that an environmental molecule will have *in vivo* (in other words, how such a molecule will behave as a potential drug). In this vein, it is of interest that polymorphisms (genetic variants within the same species) in xenobiotic receptor genes may be associated with natural variation in drug responses.[217,235] It is also worth noting briefly that other systems for limiting xenobiotic damage exist. One such strategy is to physically remove offending compounds from cells, and these natural "efflux pumps" can be subverted by tumor cells for preventing cytotoxic drugs from completing their tumoricidal task.[236]

Although many poisonous environmental chemicals have extraneous biological sources, active invasion by a biological parasite can present at least as serious a challenge to an

---

*In common with hemoglobin, these enzymes bear the heme molecule as a cofactor. Heme binds carbon monoxide (CO), and in cytochrome P450 enzymes the CO-bound form has an absorbance at 450 nanometers, hence the name.[223]

organism's integrity. Accordingly, there is a need for a large evolutionary investment in defense against biological threats as well as resistance to chemicals. An ancient response to this imperative is based on specific self-modification, where we can frame the general principles of this strategy as: (1) pick a common target molecule that is shared by yourself and a broad range of potential threats; (2) make a weapon that attacks this target; (3) before deploying this weapon, modify your own target molecules such that they are resistant to attack by it. Having achieved this, in theory anything that invades is destroyed and you continue to thrive, and you do not have to waste time worrying about nonself, since the latter is anything that has not been specifically modified. A particular example of this can be found with "restriction–modification" systems in bacteria, where the basic strategy is to modify the host genome (by DNA methylation) such that it is resistant to cleavage by restriction enzymes also produced within the same bacterial organisms. Invading DNAs, principally bacterial viruses, are then cleaved and destroyed by the host restriction enzymes. An analogous situation occurs where certain prokaryotes produce toxic metabolites acting on molecular targets shared by competing organisms. To avoid self-destruction, such organisms modify their own target molecules such that they can function normally in the presence of the produced metabolites.[*]

But self-modification may be complex and energetically costly, and competing organisms inevitably "catch up" by modifying themselves in turn or devising effective countermeasures. If external target molecules are sufficiently unrelated to self, then the self-modification pathway may be unnecessary. In such cases, it would seem logical to pick a target that was as widespread among one's biological enemies as possible, and (even better) one that your competitors and parasites were unable to change at a fast rate (such an inability would also presuppose that such molecular targets were important for pathogen fitness). In the previous chapter we spent some time thinking about factors controlling protein evolution, and from this it might be proposed that (among other possible determinants) a region of a protein with low dispensability and high functional density would be slow to evolve and would also be a good candidate target. The same reasoning applies to other pathogen-associated molecules that are the products of protein (enzymatic) activity, although in a more indirect fashion. If sufficient evolutionarily conserved targets exist, it would be a worthwhile strategy to invest in producing one's own proteins (or other biomolecules) that could bind to such markers of pathogenic nonself and either directly inactivate them or act as a warning siren. And essentially, this is what natural selection has delivered, in the form of *innate* immune systems.

Innately Selfish   Innate recognition of pathogenic organisms is the ancient approach used by virtually all eukaryotic species in the biosphere. Invertebrate animals, fungi, and plants use innate mechanisms exclusively, and vertebrates use innate immune systems *as well* as adaptive processes, which we will consider shortly. RNA mediators feature in many innate defense systems, in the form of the relatively recently discovered phenomenon of RNA interference or RNAi. This system uses specific proteins to process double-stranded RNA molecules into short segments, which ultimately recognize foreign target RNAs through base complementarity and thereby program their destruction. Hence, while this "interference" effect involves both RNA mediators and targets,[†] cellular proteins are crucial for its operation. Antiviral mechanisms

---

[*]A good example of this is biosynthesis of the antibiotic erythromycin in the bacterial organism *Saccharopolyspora erythrea*, which involves the concerted activity of multiple enzymes.[237] In order to avoid self-toxicity, this organism also has a gene that confers resistance to erythromycin itself, through production of a protein that specifically modifies the organism's ribosomal RNA such that it is not susceptible to the antibiotic.

[†]In natural biodefenses, the targets for RNAi are foreign RNA strands, but the endogenous RNAi protein machinery can be artificially co-opted to generate an RNA interference effect against any cellular mRNA. This approach is consequently a powerful (and very widely used) means to "knockdown" gene expression.

mediated via RNAi have been identified in plants and insects, and RNAi has also been implicated as a mammalian frontline defense against viruses. A distinct subclass of RNA interference is specifically associated with innate defense against "selfish" replicating elements (*transposons* of varying types) that act as genomic parasites. Defense of the genome against such molecular parasitic elements takes defensive immunity to its most fundamental level, and RNAi indeed has ancient roots in eukaryotes.[238]

Another innate mechanism directly targeting foreign nucleic acids is the process of editing, where host-directed alteration of DNA or RNA sequences through modification of nucleobases ("editing") can act as an intracellular "poison" against viral genomes (generally by altering adenines to inosines or cytidines to uridines[239]). When applied against retroviruses such as the human immunodeficiency virus (HIV), the editing itself is directed at reverse transcripts of HIV viral RNA (via cytidine deamination), and thus operates in this case at the DNA level.[*] An intriguing point is that while heavy editing of retroviral genomes is inhibitory, a low level may "backfire" as far as the host is concerned through the resulting increase in viral diversity.[239] As defense mechanisms both targeting RNA, it is of interest to consider the interactions between RNAi and RNA editing. In many cases these systems appear to be antagonistic (the operation of one negatively affecting the other), but editing may also be a means for increasing the diversity of target recognition for at least some forms of RNA interference.[241] Elegant as these processes are, countermeasures by the pathogenic target are often all too effective at circumventing the innate host response. In the case of HIV, the relatively recently recognized inhibition of host-protective viral editing by the *vif* protein is a pertinent case in point.[242] Suppression of host RNAi as an escape mechanism for invading pathogenic viruses is likely to be a widespread phenomenon associated with virulence.

Sensing of molecules or specific regions of molecules derived from pathogens (often termed pathogen-associated molecular patterns, or PAMPs) is accomplished by a variety of innate recognition systems, including lectins that bind specific types of sugars.[243] In an evolutionary diverse group of organisms (ranging from fruit flies to humans) sensing of pathogens involves a number of different recognition proteins, and a family known as toll-like receptors (TLRs) are particularly important. Some of these innate receptor molecules are expressed on the surface of cells, while others act as intracellular sensors. Specific members of these receptors bind nucleic acid signals characteristic of pathogens (including double-stranded RNA), but other members of this receptor family recognize a variety of other molecular structures associated with pathogenic organisms,[243] including bacterial lipopolysaccharide and cell wall peptidoglycan. We noted earlier that in order to be very useful, effective targets for an innate immune system should be slow to undergo evolutionary change. One example of a protein that is recognized as exclusively pathogen-associated is bacterial flagellin (a fundamental component of bacterial locomotory flagella; recognized by a specific TLR5) that contains highly constrained and conserved regions associated with filament assembly. In plants, it has been shown that recognition of flagellin is important for innate antibacterial immunity.[244] Even so, certain successful bacterial plant pathogens evade plant defenses by elaborate countermeasures.[245,246] Yet the observation that bacterial pathogens must resort to such lengths rather than using the simpler strategy of mutation of flagellin (or other innate targets) is consistent with the supposition that pathogen-associated structures recognized by the innate immune system tend to be constrained for further evolutionary diversification.

---

[*]Host editing enzymes operating directly on RNA (adenosine deaminases acting on RNA, or ADARs) are also known.[240] Hyperediting of double-stranded RNAs by ADARs can induce RNA cleavage that may be an antiviral defense, but viruses appear to frequently also co-opt ADAR pathways to their own advantage.[240]

Innate immunity has served a wide variety of organisms over vast stretches of time and has been tuned by natural selection to be a very capable discriminator of self from nonself. And yet it is clear that innate immune mechanisms have limitations. As we have seen, pathogens have evolved ways to circumvent innate immunity, usually by sequestering some component of an innate immune system signaling pathway. In such circumstances, if no backup system exists, the host organism under attack may be literally defenseless. Vertebrate organisms have developed a highly sophisticated immune response system whereby in principle *any* foreign structure* can be recognized by specific binding proteins; this we term as the adaptive immune system. It would nevertheless be quite mistaken to say that the adaptive response has replaced innate responses, for the former has evolved as an overlay on the latter, and they are closely coupled in many ways, some aspects of which we will visit shortly.

But there is one key feature of innate immunity which adaptive immune responses can never match, which is speed. Innate defense mechanisms are activatable virtually the instant that a pathogen-associated molecular structure is engaged by an appropriate receptor of the innate system and downstream signaling initiated. Immediate activation of innate immunity buys time while the processes of the adaptive immune responses unfold, and consequently the in-built inherited repertoires of the innate immune system remain the "front line" of defenses for vertebrates, including us. Although some features of the innate system (such as phagocytosis) have been long recognized, a full appreciation of its importance has only emerged since the 1990s.[243] The function of innate immunity as the first sensor of the intrusion of potentially dangerous pathogens also confers upon this system a central role in self/nonself discrimination, a topic that we will return to later. Having thus paid our respects to the varieties of innate protective responses, let us now think about adaptive immunity, since this is the major focus of interest in the context of evolutionary methods for molecular design. An essential feature of adaptive immunity is somatic diversification, but as we will see, GOD is necessary but not sufficient for the implementation of an adaptive immune system. And innate immune systems themselves are not necessarily a GOD-free zone.

How to Shoot Moving Targets

It is logical that the best targets for the innate immune system should be those that are slow to undergo evolutionary change. But at the same time, we have also noted that pathogen-associated countermeasures can override the innate immune system, and thus a pathogenic enemy can in effect present as a moving target over time. Also, a diversity of both pathogens and innate escape mechanisms will mean that investing in a new innate system against a specific microbial defense mechanism will be of limited effectiveness. The radically different approach of an adaptive system, introduced above, has the potential to cut through this problem. The adaptive immune system is itself of course a product of evolution, but here evolution has selected for processes that rely on evolutionary principles themselves.

The unlimited potential of a fully adaptive system for target molecule recognition would seem to offer a potent survival advantage to organisms that are equipped with it. Certainly adaptive immunity has persisted and diversified in vertebrate lineages for a long time.[247] But are things really this simple? Why then do only vertebrates possess adaptive immunity? If innate

---

*More specifically, depending on their size, foreign macromolecules are recognized through specific regions termed *epitopes*. Thus, an antibody against a foreign large protein does not recognize its global structure but rather a relatively small peptide epitope of the target protein that is compatible with the size of the antibody combining site and bound by it. It should be noted that while an antibody epitope may be a short contiguous string of amino acid residues, it is not necessarily so, since an epitope can be formed from residues brought into spatial proximity by protein folding. Antibodies against a "discontinuous epitope" in a protein characteristically will fail to recognize the protein if it is denatured (unfolded).

immune systems are "good enough" for fungi, plants, and invertebrate animals, then why not for vertebrates? While this is a compelling question, the observable importance of vertebrate adaptive immunity is not in doubt. It can be demonstrated in mice by selective knockout of genes crucial to the immune system's development and also with certain natural mutations. Such animals inevitably succumb to "opportunistic" infections from organisms normally easily kept at bay. In our own species, the human immunodeficiency virus and its triggering of AIDS is a stunning and terrible example of what crippling the adaptive immune system can do.

But by the same token, if adaptive immune responses are so essential for vertebrates, how do invertebrates survive in their absence? The latter question is commonly answered by noting the small sizes and relatively short lives of many invertebrates, but this certainly does not apply across the board. Some of them are very large (consider giant squid) and some very long lived (such as certain molluscs and echinoderms). It is also certain that invertebrates do not enjoy any less exposure to potential pathogens than vertebrates; bacteria and a whole gamut of parasites are only too happy to exploit invertebrates if they are able to. So it is necessary to seek other features of vertebrate biology that may have provided strong selective pressures in favor of the evolutionary development of adaptive immunity, and the associated highly complex processes that enable its deployment. It has been suggested that the evolution of adaptive immune systems has been driven by the much more complex coevolved microbiotas of vertebrates over invertebrates. By this conjecture, the need to "manage" complex commensal or symbiotic microbial populations (such as gut microbes) has determined the acquisition of adaptive immune recognition.[248]

Before leaving this unresolved issue, it should be noted that although there is no evidence for fully functioning invertebrate adaptive immune systems, many invertebrates have considerably greater diversity than previously believed in the range of immune detector and effector molecules that they can produce.[249] Given that innate immune systems have worked successfully over a vast period of time in a huge range of organisms, it perhaps should be expected that such "primitive" immunity would have a few tricks up its proverbial sleeve. We have already seen that genetic polymorphism (significant within-species diversity at specific gene alleles) allows allorecognition to occur. Polymorphism within genes of the innate immune system can also aid in enabling some members of a species to stay ahead of rapidly evolving parasites. While this type of germline GOD in invertebrates is not unprecedented, as our knowledge base increases so do the numbers of examples.[*]

We could compare natural somatic GOD with an artificially devised diverse molecular library. As noted in Chapter 1, the nature of such a library and the means for its generation are not trivial exercises. Getting to this point, however, is only the first step, since a means for selecting the desired molecule and amplifying it up to the point where it can be identified are also essential. By this analogy, to make a fully adaptive immune system, natural GOD must be accompanied by a means for selection and amplification of the useful immune effectors. The vertebrate immune system can do this, but this has not been shown among invertebrates. It must also be recalled that somatic diversification in immunity is a double-edged sword: good for thwarting parasites but liable to run the risk of generating deleterious self-reactivity. Much of the elegant sophistication of the vertebrate adaptive immune system is devoted to solving this conundrum. If invertebrates indulge in extensive somatic GOD, they inevitably have to deal with the same problem, in ways that are not yet clear. At least in insects, a certain level of self-damage accompanying innate antipathogen responses appears to be an acceptable evolutionary compromise[251]; perhaps the same applies to an "acceptable" level of self-recognition. A degree of "autoimmunity" in such organisms may be part of the evolutionary package.

---

[*]Certain highly polymorphic immunoglobulin-like genes in the protochordate Amphioxus[250] are a case in point.

At this point we have considered the origins and general utility of the adaptive immune system, but what are the basic requirements for the establishment of an adaptive immune system in the first place?

The Empirical Basis of the Adaptive Immune System

In a close parallel with the mechanism of natural selection itself, an adaptive immune system requires generators of diversity and selection, which were the first two components of the "GODSTAR" principle noted in Chapter 2. Time is also required; we noted earlier that innate immune systems have an in-built advantage over adaptive systems in terms of rapid deployment. Of course, to be in any way useful as a weapon against foreign invaders, an adaptive response must be marshaled in a reasonably short period with respect to the kinetics of parasitic invasion and the lifetime of the host organism. Nevertheless, a significant time interval (often on the order of weeks) is needed to "work-up" effective adaptive immunity. Finally, for the adaptive immune system, the analogy with the "Reproduction" part of the GODSTAR acronym for natural selection lies in the essential need for clonal expansion of the immune cells required for a specific adaptive response. Thus, adaptive immunity, a product of natural selection, encapsulates the principles of natural selection itself. Let us look at each major feature of the adaptive immune system in more detail.

The origins of the GOD of the adaptive immune system are of special interest in the context of molecular design, and we will focus on this in the next section. If we leave this aside for the time being, we can still gain an impression of the overall organization of vertebrate adaptive immunity, which has traditionally been divided into "humoral" immunity (mediated by circulating factors in the blood, namely antibodies), and cellular immunity (mediated, as its name implies, at the cellular level). But all adaptive immunity is "cellular" at the levels of selection and expansion. A fundamental tenet of the clonal selection theory of immunity[*] is that the immune system has a high level of diversity in cells expressing unique recognition receptors on their surfaces (that is, one cell is limited to one specific receptor type) and that after encounter with specific nonself antigen, expansion of the appropriate immune cell clones occurs. Cells expressing antibodies with the ability to combine with a foreign molecule are thus expanded, and later undergo further differentiation into cells actively secreting soluble antibody with the same combining specificity. We have already seen the importance of self/nonself discrimination; the clonal selection theory requires elimination (or suppression) of clones that can recognize self.

Such are the complexities of the adaptive immune system, a full description of its known features is neither possible nor appropriate for our purposes here. But an appreciation of some of its major functional aspects is important for an understanding of how its diversity-generation and selection mechanisms are applied. The so-called humoral and cellular arms of adaptive immunity are interlinked, and innate immune recognition processes are also of fundamental importance to the adaptive immune system by providing key activation signals. The cellular units of this intricate immune network are lymphocytes and cells of the myeloid lineage (all arising from hematopoetic stem cells in the bone marrow). Lymphocytes that express unique antibody molecules anchored to their surfaces are termed B cells (Bone marrow-derived), and lymphocytes that control antibody generation and many other facets of cellular immunity are referred to as T cells (Thymus-derived). The latter differentiate and are "educated" in the thymus, an important lymphoid organ that is crucial for controlling self/nonself discrimination (as we will consider further below). Already we have heeded the fact that B cells express

---

[*]Put forward by the Australian immunologist Sir Macfarlane Burnett, a Nobel-prize winning insight (1960).

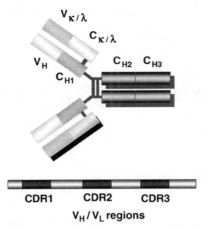

V_{H}/V_{L} **regions**

FIGURE 3.1   Schematic of general antibody (immunoglobulin) structure. V_H, heavy-chain variable region; C_{H1}–C_{H3}, constant region (domains 1–3); $V_{\kappa/\lambda}$, κ (kappa) or λ (lambda) light-chain variable regions; $C_{\kappa/\lambda}$, κ or λ light-chain constant regions. Chains are joined by disulfide (S—S) bonds. CDR = complementarity-determining regions for heavy and light chains. Chapter 7 (Fig. 7.1) provides more structural details for antibodies.

antibodies capable of binding a vast range of different molecular shapes, and a specific antibody with a unique combining site is expressed on each mature B cell and functions (along with certain other associated proteins) as a receptor for a complementary antigen molecule. Antibodies (immunoglobulins) consist of heavy and light protein chains, the N-termini of which are highly variable in sequence at specific regions (complementarity-determining regions, or CDRs) and which mediate the binding of antigen (Fig. 3.1). The C-terminal regions of antibodies are invariant in sequence for each of the several different antibody classes, and these "constant regions" mediate other functions unrelated to antigen recognition.

T cells also express receptors with high levels of diversity that share certain features of their diversity-generating mechanisms with B cells. Compared to antibodies, the T cell receptor was a relatively elusive entity to track down and define, only yielding to molecular biological approaches in the early 1980s. This receptor consists of disulfide-linked alpha and beta chains,[*] which are structurally within the same superfamily of proteins as immunoglobulins, and which also have the same pattern of variable N-terminal V regions and constant C-terminal regions.

But well before the T cell receptor's structure was understood, it was known that the way T cells recognize antigen is quite distinct from antigen recognition by B cells (with surface antibody) or free soluble immunoglobulin. While the latter bind directly to an antigen (or more accurately, to a specific target region known as a determinant or epitope), T cells recognize antigen only in the context of other proteins encoded by the major histocompatibility complex (MHC), so known by its original genetic definition in controlling the acceptance of tissue grafts. Cells presenting antigen to T cells process foreign protein antigens and display (on their surface membranes) specific peptide fragments of the antigens bound within a "groove" present within the structure of MHC proteins. The most effective "professionals" of such antigen-presenting cells are termed dendritic cells (or DC; Fig. 3.2), which also are pivotal in the integration of the innate and adaptive immune systems. Activation of dendritic cells

[*]Separate gamma/delta T cell receptors also exist as a relatively minor component of the total.

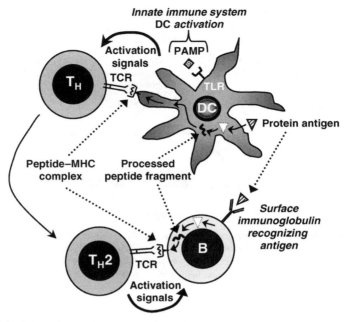

FIGURE 3.2   Schematic showing some of the important primary interactions required for B cell activation in the adaptive immune system. Abbreviations: DC, dendritic cell; T_H, naïve T helper cell; T_{H2}, T helper cell Type 2 (specific subclass of thymus-derived lymphocyte); B, B cell (bone-marrow derived lymphocyte); Ig, immunoglobulin; PAMP, pathogen-associated molecular pattern; TCR, T cell receptor; TLR, toll-like receptor; MHC, major histocompatibility complex protein. Innate molecular sensing can occur through either surface or intracellular recognition molecules (exemplified by TLRs here, but additional molecules also exist[6]).

through signals transduced through receptors of the innate immune system is required for full T cell activation.[243] In consequence, at this level innate immunity has a role in discriminating self from nonself, and innate and adaptive immune systems appear to cooperate in self/nonself recognition.[252]

B and T cell responses are also in fact closely interlinked by the requirement of the former to receive "T cell help" for the generation of most antibody responses.[*] B cells with surface immunoglobulin binding a specific antigenic protein can internalize the bound target protein and display specific fragments of it in the context of MHC molecules on their surfaces. T cells with receptors recognizing the *same* specific MHC–peptide complex thus also recognize B cells that express the correct antibody specificity. It should be emphasized that although the linkage for T–B cell help is thus corecognition of the same antigen, the *epitopes* of the antigen recognized in each case can be quite distinct (as depicted in Fig. 3.2). This T cell recognition therefore provides the necessary help that enables the appropriate B cell clone to proliferate and thereby to be amplified, a fundamental requirement of the adaptive immune response.

---

[*]The requirement for T cell help can be waived for certain bacterial polymeric and carbohydrate "thymus-independent" antigens, which can directly activate B cells.

Three essential features of biological adaptive immune systems can therefore be identified: to produce a vast range of receptor variants without foreknowledge of the required binding specificity, to "pick-out" a cell with the appropriate recognition of antigen as required, and to signal the amplification of such a cell. These attributes render the mechanisms of such immune systems clearly Darwinian and not at all Lamarckian, and (as stated earlier) an encapsulation of the process of evolution by natural selection itself. The adaptive immune system serves as a potent natural demonstration of the great power of empirical selective processes in functional molecular design, and it is thus an inspiration and guide for *in vitro* evolutionary approaches. But in order to fully benefit from a study of the immune system, we must understand where its ability to generate diversity comes from.

The Evolved GOD of the Adaptive Immune System

In thinking about biological evolution, we examined pathways leading to the generation of diversity (GOD) at the genetic level, as the raw material for natural selection. With the development of the vertebrate immune system, natural selection has favored a system that needs to "deliberately" generate diversity in order to implement a somatic selection process within individual organisms. In turn, acquisition of adaptive immunity provides a fitness benefit; vertebrates that are impaired in this regard are at a severe survival disadvantage (although as we have also seen, it is not clear how most invertebrates can dispense with adaptive immunity). It became clear from the early days of immunology that antibodies could combine with a very large range of proteins and other molecules. Moreover, antibodies can be prepared against chemical structures that are entirely artificial and that could never have been naturally encountered or naturally selected for.* This would clearly be hard to account for if one antibody corresponded to one gene, but the conundrum of the scope of antibody diversity remained until evidence for somatic diversification mechanisms was first obtained.

If the power to generate proteins that can recognize and bind virtually any foreign structure provided a strong selective advantage, how could it be best implemented? It is quite unfeasible to encode in the genome a sufficiently huge battery of proteins, each with a different recognition specificity, such that all potential targets could be covered. The elegant solution to this problem is to arrange for a system where the diversity is generated somatically within each individual, rather than rigidly encoded in the germline, and this evolution has done. There are several ways that this somatic GOD can be achieved. Consider a gene encoding a region of an immune receptor concerned with antigen recognition. If such a gene is divided into sets of segments, each of which has diverse sequences, and then randomly combinatorially recombined to reconstitute a full antigen-binding region, then high diversity can be generated with far greater economy than encoding each region separately. This is depicted in Fig. 3.3 for a gene with three sets of diverse segments, where the combinatorial diversity is simply the product of the number of recombinable genes for each segment set. Combinatorial recombinational events of this type, involving deletion of the intervening DNA sequences between recombining segments, are observed for both the immunoglobulin heavy and light chain genes, and also the genes for the T cell receptor alpha and beta chains. In both mice and humans, the variable regions of immunoglobulin heavy chain genes are encoded in three recombinable sets: a sizable number of "V_H" segments (<150), a lesser number of "D" segments (<50), and additional "J_H" segments

---

*Many pioneering immunological studies were performed by Karl Landsteiner, who was awarded the 1930 Nobel Prize for Physiology and Medicine for his discovery and classification of blood groups. Antibodies against small nonnatural chemical groups (termed *haptens* in an immunological context) can be prepared if they are conjugated to a macromolecular carrier such as a protein or polysaccharide.

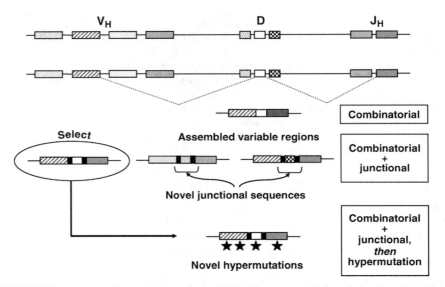

FIGURE 3.3   Processes of somatic immunological GOD. Three genomic loci are depicted, each of which is composed of multiple copies of related but variant genes. (These are denoted V_H, D, and J_H as for immunoglobulin heavy chain genes, but the numbers of each does not reflect the actual gene numbers in each case.) Random combinatorial recombination between such segments results in a range of diverse products (24 ($4 \times 3 \times 2$) in this example); this is further increased if mechanisms promoting junctional diversity (black bars) between the recombined segments are included. Following selection by specific antigen and amplification, one such specificity can be further "tuned" by somatic hypermutation as depicted.

(4–6; as shown schematically in Fig. 3.3). A similar arrangement exists for immunoglobulin light chain genes, except only V and J segments are involved.[*] The T cell receptors are analogous again, with the beta chain having V, D, and J segments, and the alpha chain V and J only.

Another level of diversity is introduced at the junctions between recombining segments (Fig. 3.3). One type of this junctional diversity (albeit probably a relatively small contributor) is termed P-nucleotide addition,[253] which results from the nature of the recombinational process. Removal of nucleotides at recombinational junctions by exonuclease activity is another diversifying mechanism,[253] and nucleotides can be added by yet another process termed N-nucleotide (or N-region) addition, which is active on immunoglobulin heavy chain genes and genes for both T cell receptor chains. For many years, it was suspected that these nongenomic N-regions were the result of the activity of the enzyme terminal deoxynucleotide transferase, which can add nucleotides to the 3′ end of DNA strands without the aid of a template (a very unusual feature for a DNA-polymerizing enzyme). The role of this enzyme was elegantly proven by showing that mice with the terminal deoxynucleotide transferase gene artificially "knocked out"[†] had no such N sequences in their assembled variable region genes, with accompanying changes to their immunological repertoires.[254]

---

[*]Two classes of immunoglobulin light chains exist, of which the major class are termed kappa light chains, but alternative lambda light chain genes also exist. Both have variable and constant regions as depicted in Fig. 3.1, and the variable regions for both are encoded by V and J segments, although their genomic arrangements differ.
[†]This refers to gene ablation in whole animals through specific gene targeting in embryonal stem cells by homologous recombination, as previously mentioned in Chapter 2.

These mechanisms of GOD in separate immune receptor chains, occurring independently in each case, are further extended by assortment between the chains to form fully assembled proteins. Thus, the diversity of N_H immunoglobulin heavy chain genes and N_L light chain genes (kappa or lambda) becomes $N_H N_L$ total diversity when functional immunoglobulin is expressed (as in Fig. 3.1). This likewise applies to the alpha/beta or gamma/delta chains of the T cell receptor. In the context of immunoglobulin heavy and light chain assembly, it is important to note the operation of the principle of allelic exclusion, which determines that only one assembled and functional antibody heavy and light chain is generated per cell. (Once one of the chromosomal alleles is rearranged, the other is blocked from further recombination and remains nonfunctional.) This ensures that each mature B cell expresses only one antibody combining specificity on its surface.[*] For the T cell receptor, however, it appears that allelic exclusion is less rigidly enforced, with up to 30% of T cell bearing the alpha/beta receptor expressing both alleles of the beta chain.[256] The role of this apparent relaxation of the "one cell/one immune receptor" rule in the normal immune system is not certain, but it may allow rescue of certain T cell specificities from deletion during screening in the thymus (as explored further below).

Another important layer of immune diversification exists. In humans and mice, rearranged immunoglobulin genes within B cells that have been activated by antigen-binding and T cell help (Fig. 3.2) undergo diversifying mutations within the variable regions of both heavy and light chains.[257] These nongermline mutations occur at a rate approximately one million-fold that of the spontaneous mutation rate; hence, the process is termed somatic hypermutation (depicted schematically in Fig. 3.3). Although it is generally believed that the hypermutational targeting is essentially limited to immunoglobulin variable regions, it does not occur randomly within them. Hypermutation shows sequence preferences for certain motifs, including the complementarity-determining regions encoding antigen-binding domains. As well as point mutations, insertions and deletions of sequences can ensue from the same hypermutational processes.

These novel changes, occurring within immunoglobulin genes that have undergone a primary antigen-based selection, in effect constitute a second round of diversification. The result of this second tier of GOD is the "affinity maturation" of antibodies, evident in the classical observation that the affinities[†] of antibodies steadily increase up to a maximum after the commencement of immunization with a foreign antigen.[258] As well as acting to maximize antibody affinity toward specific antigens, it has been proposed that secondary somatic hypermutation in general allows rapid and effective antibody responses toward rapidly mutating molecular targets of pathogens.[259] Mutations that result in higher binding affinities toward the original antigen are positively selected by B cell activation, while mutations ablating antigen binding (or reducing binding affinity below a threshold level) are in general selected against by programmed cell death (apoptosis).[258] The changes to antibodies engendered by somatic hypermutation are thus quintessentially Darwinian in nature, in that the somatic mutational patterns are not in any way directed by the nature of the original antigen, and improved variants result from selectional processes. At the same time, this is not to say that variable regions cannot evolve such that somatic mutations occurring within their sequences are more likely to be productive. It has thus been noted that the biases in codon usage in

---

[*]Relatively recently, a minor subset of B cells expressing two separate immunoglobulin specificities has been reported as an apparent violation of the allelic exclusion rule.[255]

[†]The *affinity* of an antibody is a measurement of the strength of one of its combining sites toward antigen (monovalent binding). This should be contrasted with the *avidity*, which is the net binding strength of an entire antibody. Natural antibodies (as in Fig. 3.1) are at least bivalent. IgM antibodies consist of a complex of five joined bivalent antibodies. Thus, while an IgM antibody may have low affinity, the net avidity of an IgM pentamer may be quite high.

immunoglobulin complementarity-determining regions are such that somatic hypermutational events are more likely to be nonsynonymous.[*,260] Aside from this issue of coding sequence bias in sections of the variable region germline sequences that increase their likelihood of hypermutational targeting, it must be recalled that the selection process of antigen binding is of paramount importance in the outcome of this secondary immune diversification process. It has been suggested that the observed average patterns of selected amino acid replacements in complementary-determining regions result from selection for residues with a greater average ability to enhance antigen-binding.[261]

But another aspect of antibody maturation is very important. Primary germline and affinity-matured immunoglobulin molecules are distinct not only in their affinities for antigen, but in a more general qualitative sense as well. Many structural studies have been performed comparing germline antibodies assembled from genomic coding sequences and their somatically matured counterparts, both in isolation and in complex with antigen. A take-home conclusion has been that the former primary antibodies have higher conformational diversity and flexibility,[262,263] which enables their binding sites to adopt significantly different conformations. As a result, the binding of antigen by germline antibodies frequently involves immunoglobulin conformational changes, which can be viewed as an "induced fit" mechanism. In contrast, following affinity maturation, antigen binding by antibodies can often be modeled along the lines of classic "lock and key" recognition, and matured antibodies themselves tend to show higher levels of rigidity and reduced flexibility accompanying their enhanced affinity for target antigen.[†,265] This primary/mature antibody distinction is ideal for optimal design of an adaptive immune system. Not only are there mechanisms in place for generating a large primary repertoire (combinatorial recombination and junctional diversity), but primary antibodies themselves are also flexible agents of molecular recognition. The conformational plasticity of germline antibodies has the effect of greatly increasing the effective range of antigen binding sites, maximizing the chances that foreign nonself will be recognized and immunity triggered.[††] This comes at a cost in terms of the affinity of antigen binding, but this is solved in the second tier of the adaptive antibody response through somatic hypermutation.

This consideration of somatic hypermutation in the context of immune diversification has referred only to immunoglobulin genes, since it has not been detected in the rearranging genes of the T cell receptor.[258] The MHC–peptide recognition repertoire of T cells resulting from their differentiation and selection in the thymus is subject to positive and negative checks to guard against autoimmunity (as we will note in a little more detail below). If T cell receptor somatic hypermutation took place subsequent to this, it might result in the generation of self-recognizing clones and threaten orderly immune regulation.[258] It is also notable that although

---

[*]"Productive" alterations in this context for single-base mutations are those that alter encoded protein sequence in the hypermutated tract. Given that somatic hypermutation is not random, and affects some sequence motifs more frequently than others, certain choices of synonymous codons (alternative codons specifying the same amino acid) will have a higher propensity for change to different amino acid codons than others. (Changes arising from directed somatic hypermutation in a coding sequence with this type of bias are therefore more likely to be nonsynonymous.) Selection for complementarity-determining regions with such properties consequently enhances their "evolvability" under the auspices of the somatic hypermutation process.

[†]But note that biology rarely provides tidy dichotomies. There is also substantial evidence for conformational flexibility and induced-fit conformational changes upon antigen binding by some fully matured antibodies,[264] which we will return to at later points.

[††]One might note that high flexibility of primary antibodies might also tend to create dangers of self-binding and autoimmunity, but this is subject to strict control processes. The T cell repertoire and its role in the activation of antibody-producing B cells (as in Fig. 3.2) is one such measure. A low affinity cross-reactivity for a self antigen in a primary antibody would be lost as the antibody undergoes maturation toward high-affinity recognition of the nonself target.

we have referred to somatic hypermutation as a second-level GOD in humans and mice, this is not necessarily the case in all vertebrates. Some species appear to have relatively limited numbers of germline V genes and achieve most primary diversification after immune segment rearrangements. In this regard, the sheep immune system is a case in point, where antigen-independent hypermutation is an important contributor to primary antibody diversity.

Some other species also have an alternate means for achieving immune receptor diversity. In the discussion of homologous recombination in Chapter 2, the nonreciprocal process of gene conversion was referred to. This type of recombinational process is a major pathway to GOD in the immune system in some vertebrates, including chickens and rabbits. In such cases, the germline diversity of functional variable region genes is low, but sequences from nonfunctional V gene copies (pseudogenes) can be transferred to assembled variable region segments by gene conversion processes. These observations show that while adaptive immune systems are a feature of vertebrate evolution, there is divergence in the specific mechanisms of achieving immune receptor diversity. Recombinational, junctional, and mutational mechanisms feature in varying degrees, but all are used to some extent. Many details of the enzymatic processes that enable somatic rearrangements and hypermutation have been discovered, although some gaps in knowledge still remain. For our present purposes, though, let us now take a look at the other major side of adaptive immunity.

The T Cell Repertoire and Its Special GOD   Thus far we have mainly focused on antibody immune diversity, but the central role of T cells and their molecular recognition systems in immune regulation needs to be emphasized. A major reason for this is that T cells are gatekeepers of most B cell activation events involving engagement with antigen (Fig. 3.2), and thus have control over the expression of the B cell repertoire. Also, T cells are central guardians for ensuring proper discrimination of self from nonself.[*] The nature of the biological imperatives imposed on T cells is reflected in the mechanisms for GOD in the formation of the primary T cell receptor repertoire, and the winnowed repertoire of mature T cells. The power of combinational and junctional mechanisms can generate an enormous primary potential T cell receptor diversity (alpha/beta chain; approximately 10^{15} potential variants[266]). But there is a need for this high starting-point diversity, given the bottlenecks that the T cell repertoire must pass through in the process of gaining biological utility.[267]

In order to understand the T cell repertoire, we must first consider the special nature of T cell molecular recognition. It must be recalled at this point that the T cell repertoire is "tuned" to recognize antigenic peptides in the context of proteins encoded by genes of the MHC locus. Two major classes of heterodimeric MHC molecules exist (Class I and II), which have similar overall molecular architectures but differ in their biological roles, expression patterns in different tissues,[†] and certain specific features of their association with peptide antigens. To gain an understanding of how T cell receptor and peptide–MHC recognition occurs, many structures of both T cell receptors and peptide–MHC in isolation, and their specific complexes, have been determined. Specific binding of T cell receptors to peptide–MHC involves contacts between the three complementarity-determining regions (CDRs) within both of the alpha and beta chains of the receptor, and both MHC itself and MHC-bound peptide residues[††] (Fig. 3.4). The CDR1 and CDR2 segments are of limited germline diversity (encoded in $V\alpha$ and $V\beta$ germline segments),

[*]Self/nonself discrimination also occurs during B cell maturation; strongly self-reactive B cells are eliminated during their development.[258]
[†]Class I MHC are virtually ubiquitously expressed; Class II are restricted to the thymic epithelium, antigen-presenting cells and certain other immune system cells, including B cells.
[††]While this can be demonstrated with structural analyses of complexes *in vitro*, additional associated coreceptor molecules such as the CD3 molecule are crucial for the full functioning of T cell receptor signaling.

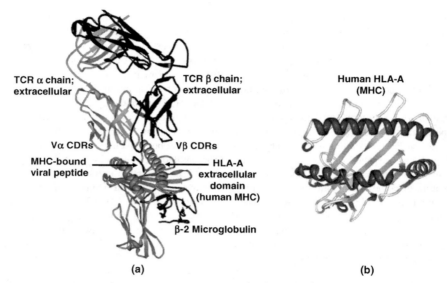

(a)   **(b)**

FIGURE 3.4   Structures of T cell receptor and MHC. (A) Structure of a specific T cell receptor (TCR; α/β) complexed with Class I MHC (human HLA-A 0201) with bound viral peptide (LLFGTPVYV; tax peptide fragment from human T lymphotropic virus Type 1). β-2 microglobulin is associated with the heavy chain of HLA molecules. Only extracellular domains present. (B) Structure of Class I MHC (Human HLA-A2) heavy chain viewing the N-terminal peptide-binding groove (minus peptide) formed from a platform of antiparallel β-strands and enclosed by α-helices (dark segments). In Class I MHC, the peptide-binding groove is formed solely from the heavy chain (as in B); in Class II MHC, it is formed by the association of both the separate α and β chains (not shown[7]). *Source*: Protein Data Bank[2] (http://www.pdb.org). Panel A: 1BD2[8]; Panel B: 2GUO.[9] Images generated with Protein Workshop.[5] (Further details in ftp site.)

and mainly mediate contacts with MHC residues.[268] Most of the great T cell receptor diversity created by combinatorial V(D)J rearrangements and associated junctional alterations is localized to the CDR3 segments, and these mediate the majority of contacts with MHC-bound peptide. It is striking that although immunoglobulins are not constrained in their recognition spectrum in the same manner as are T cell receptors, diversity within CDR3 regions has also been identified as the major contributor to immunoglobulin heavy chain binding specificities.[269]

The T cell receptor then has a unique and seemingly onerous task. It is obliged to efficiently recognize processed peptide fragments from the universe of foreign antigens, but always in a specific context requiring corecognition of MHC molecules at the same time. In order to think about how this is done, first it will be useful to consider the matter from the MHC viewpoint. An MHC protein, required to present a great multiplicity of peptides on cell surfaces for immune recognition, would seem to have almost an opposite molecular recognition role to the specificity shown by T cell and other immune receptors. Though peptides bound by MHC are not long (usually nine amino acids for Class I MHC; somewhat longer for Class II), this still encompasses billions of possible peptide sequences. Moreover, the association of peptides in the MHC groove (Fig. 3.4b) must be reasonably tight, such that the surface peptide–MHC complexes are sufficiently stable.

An essential point is that any one MHC molecule can bind a large but limited number of peptides. The association between peptides and MHC is determined by specific "anchor

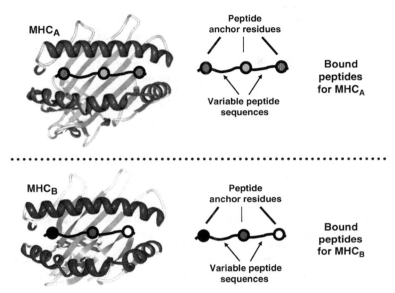

FIGURE 3.5 Depiction of MHC peptide binding. Two MHC Class I alleles (A and B, structures shown with superimposed schematic peptides in binding grooves) each can bind a large but restricted set of peptides, restricted by defined anchor residues (circles). Remaining peptide residues (lines) are less constrained. *Source*: Protein Data Bank[2] (http://www.pdb.org). MHC$_A$: HLA-A2, 2GUO as for Fig. 3.4; MHC$_B$: HLA-B*2705, 3B6S.[10] Images generated with Protein Workshop.[5]

residues" (Fig. 3.5), which set constraints on the characteristics of bindable peptides by any particular MHC molecule.[*] It is estimated that an individual MHC protein can bind approximately 10^6 chemically distinct peptides.[266] Although this is far below the number of possible nonamer peptides ($\sim 5 \times 10^{11}$), a number of other factors come into play in this regard. Display of a foreign peptide–MHC complex on a cell acts as a nonself tag, which can ensure either killing of the cell by a special type of T cells (for peptide and Class I MHC complexes) or activation of B cells recognizing the same antigen (as in Fig. 3.2; for peptide bound to Class II MHC). As such, it is not necessary to display an entire foreign protein; only segments of the protein that are compatible with MHC binding need be considered. (Identifying regions of proteins that are corresponding T cell "epitopes" has kept many immunologists busy and has been useful for vaccine design.) Peptides are continuously generated from cellular proteolytic turnover, and specific mechanisms exist for processing and transport of peptide–MHC, although by different pathways for Class I and Class II.[258] MHC molecules in fact require bound peptide for their surface stability, so in one sense the peptide is an obligate but variable "subunit" of the peptide–MHC complex. Surface expression levels of peptide–MHC may be on the order of 10^5 copies per cell,[271] and if foreign peptides are not present, MHC will be "filled"

---

[*]For example, for a nonamer peptide to bind to a specific MHC protein, it might be necessary to have a tyrosine residue at position 2, certain other preferences at three other positions, and unspecified residues at the remaining positions. This restricts the total number of peptides that are capable of binding in this specific instance. A complicating factor here is the phenomenon of "register shifting," where a single peptide can shift its "frame" within the MHC groove (using alternative anchor residues), and thus expose different residues for T cell recognition (requiring a different T cell specificity for recognition). This has mainly been observed in Class II MHC where longer peptides are bound.[270]

with peptide fragments of self proteins. (Only nonself peptide–MHC will be recognized by T cells, and thus cause T cell activation, in a normally functioning immune system.)

Even so, an individual MHC molecule would be limited in its ability to present peptides from a wide range of pathogens. The solution to this problem is diversity again at two different levels: multiple genes and multiple alleles. In humans, there are three MHC Class I genes (HLA-A, -B, and -C), and three sets (isotypes) of α and β Class II genes (HLA-DR, -DP, and -DQ, with an extra β gene for -DQ).[258] In the case of Class II MHC, further diversity can result from assortment between different combinations of α and β chain alleles (for example, for each separate DR α and β allele on both chromosomes, four heterodimers could be formed, as α1β1, α1β2, α2β1, α2β2), or even between Class II isotypes. MHC genes are highly polymorphic, meaning that a very large number of alternative alleles for each gene exist in the population (quite unlike the vast majority of other genes). In turn, this means that in most cases an individual will have different alleles on each chromosome for each MHC gene (for example, two different alleles for HLA-A, etc.), thus doubling the overall MHC and peptide-presentation diversity.[*] This type of diversity is therefore promoted by MHC polymorphism, and the higher the range of MHC-presentable peptides, the higher the "coverage" of potential pathogens (both at the individual level and for the species as a whole). In consequence, selective pressures have favored the high diversity of MHC alleles in populations. Also, MHC diversity reduces the effectiveness of inevitable counterattacks from pathogens through their generation of specific interfering proteins.[258]

So, the above material represents a bird's eye view of the world according to MHC, a case study of multigene and multiallelic germline diversity, in contradistinction to the somatic diversification mechanisms of T cell receptors. Let us now return to the continuing (but MHC-related) question of the mature T cell repertoire that we began with. It can be seen that T cell receptors act as "constrained variable recognition molecules," in that they always bind under the imposition of MHC recognition, with highly diverse peptide recognition layered upon that requirement. The MHC component of T cell receptor molecular interaction is itself quite specific. This means that out of the very large number of T cell α/β receptors, some will recognize MHC Class I of a specific allele (and specific peptide) only,[†] and others will interact only with similarly specific MHC Class II and associated peptide.[258] Also, recognition of self peptides must be avoided (at least the same type of recognition as seen against nonself, which leads to immunological activation and its attendant consequences). It is evident from these observations that the mature T cell repertoire is very far from a random collection of binding site specificities, and this leads directly to the question of the origin of this bias. Much experimental data have demonstrated the crucial role of the thymus in eliciting the required characteristics of the T cell repertoire. In this organ, differentiating T cells must go through rounds of both positive and negative selection processes in order to qualify as fully functional (in the somatic periphery of the organism), and here the major essential safeguards against recognition of self are instituted.

A large body of experimental results suggest that a single type of process is insufficient for achieving a T cell repertoire with the necessary self/nonself discrimination. Picking out receptors with desirable qualities, actively culling those with potentially bad effects, and patrolling for self-reactivity are all apparently required for an optimally functioning adaptive immune system. While the general arrangements within this complex multitiered biology have

---

[*]MHC polymorphisms cluster at sites affecting both T cell recognition and peptide binding, and are believed to be generated by point mutations in combination with recombinational mechanisms including gene conversion.

[†]This "genetic restriction" on T cell activity (in cytotoxic T cells recognizing a viral epitope) was one of the first indications of the true nature of MHC and T cell recognition, for which its discoverers (Peter Doherty and Rolf Zinkernagel) were awarded a Nobel Prize in 1996.

emerged into view, many details remain to be filled in. One such issue that is not fully resolved is the question of the origin of the T cell receptor's "obsession" with MHC, and this is particularly interesting from the viewpoint of molecular design.

One proposal is that the initial T cell repertoire is entirely random, and it is only the fine-tuned selection processes of the thymus that pull out a mature repertoire with the required MHC–peptide recognition qualities. The very large numerical potential of the recombinatorial and junctional T cell receptor diversification processes is consistent with a need for high initial diversity,[267] as would be predicted by the "entirely random initial repertoire" school of thought. Yet only a very small fraction of a total "universal" repertoire could be expected to have a bias toward any one specific molecular framework, let alone both MHC classes and their many allelic variants. (It must be also kept in mind that a theoretical diversity level is limited temporally by the maximal number of lymphocytes circulating in an organism. So while 10^{15} T cell receptor variants may exist in the potential repertoire,[266] only about 10^{12} human T cells exist at any one time.) Apart from such theoretical qualms as to the practical utility of selection only from a random receptor pool, experimental evidence favors the existence of an inherent presetting of the T cell repertoire.[268] T cell receptors appear to have an "innate programming" for MHC recognition associated with CDR1 and CDR2 of the receptor gene variable regions that primarily encode the regions of the receptor which make contact with MHC residues rather than bound peptide (as noted earlier). These CDRs therefore appear to carry germline information for directing T cell receptors toward an MHC bias. While the structures of many T cell receptor/peptide–MHC complexes have been solved, it has proved difficult to reduce proposed "intrinsic" skewing of recognition toward MHC into general structural rules.[272]

It is clear, though, that whatever the explicit molecular basis of the built-in germline predisposition to favor MHC binding, the thymic selection processes are still essential for mature repertoire establishment. Indeed, the initial repertoire size is reduced by as much as 100-fold by thymic intervention.[266] It follows that the bulk of T cell receptors in the raw repertoire are deemed inadequate, and only those with the correct immunological binding properties emerge from the thymic bottleneck. The intrinsic germline targeting of T cell receptors thus serves to skew an otherwise random collection of variable binding domains in the biologically desirable direction and (presumably) greatly increases the probability of finding suitable molecules in the initial "raw material" of the primary receptor diversity.

So while the GOD of the primary T cell receptor repertoire has many similarities to immunoglobulins, the T cell-specific selection processes shape a very different kind of mature repertoire. Special indeed, and with special lessons for students of molecular design by empirical selection. Principally, the T cell receptor is a natural demonstration (and "proof-of-principle") of a variable molecule whose recognition properties are highly diverse yet bounded in a very defined manner, by a specific molecular target structural framework. As noted above, the complete "rules of engagement" for turning the T cell receptor repertoire toward a specific (MHC) context remain to be fully understood, although its association with the receptor's CDR1 and CDR2 regions appears likely. Another lesson driven home by study of T cell biology is the variety of downstream effects that appear to result from changes in T cell receptor binding kinetics, some of which seem relatively subtle, yet cause radically different signaling processes (at the most extreme, death versus proliferation).

Despite the intricate sophistication of their evolutionary design and implementation, it now appears that antibodies and T cell receptors are not the only solution to the higher level design problem of adaptive immunity itself. In very recent times, it has become clear that the jawless agnathan fish (lampreys and hagfish) possess an entirely different recognition system that qualifies as an adaptive immune system in its own right.[273] But to focus on the more familiar mammalian systems, we have seen that adaptive immunity can deliver proteins capable of recognizing a vast range of targets. Since our primary interest in looking at these natural systems is in the overarching context of molecular design, let us think in more detail about the range and potential limits of immune binding proteins.

The Range and Utility of Adaptive Immunity

At this point we have considered mechanisms for immune receptor diversification, but not yet what the potential range of such diversity entails. First, how many different kinds of antibodies can be made? From the immunoglobulin heavy chain variable region segments noted above (V_H, D, J_H), the combinatorial diversity alone would theoretically be simply the product of the numbers of each genomic segment ($N(V_H) \times N(D) \times N(J_H)$). The actual outcome is not so simple, however, owing to the nonfunctionality of a sizable number of these gene segments. Such defective pseudogenes cannot result in productive proteins through the presence of stop codons or frameshift mutations. Additional ostensibly intact variable region segment genes may be nonfunctional through defective promoters. For example, in the human immunoglobulin heavy chain locus 123 V_H genes exist, but 79 of these are pseudogenes and only 40 are represented in the expressed repertoire.[274] Combined with the number of functional D and J_H segments (25 and 6, respectively), the total combinatorial diversity for the heavy chain locus is calculated at $\sim$6000.[*,274] Much potential diversity is thus lost to the specific immune recombination system through the high frequency of defective coding sequences or expression, which also occurs at light chain loci. Moreover, nonfunctional versions of variable region segments also exist at other genomic sites.[†] Relics of this type are common in vertebrate genomes, but variable region segment pseudogenes are not necessarily entirely useless, since parts of the sequences can still contribute to immune receptor diversity through gene conversion (and indeed, as we have seen earlier, this is a major pathway for GOD in some species).

When immunoglobulin combinatorial heavy chain diversity is assorted with the corresponding kappa or lambda light chain range of variants, the primary human antibody diversity from combinatorial processes *only* has been calculated to about 3.5×10^6.[258] This is likely to be something of an overestimate from the number of functional recombining segments involved, but an approximation to 10^6 is appropriate. In any system diversifying by a random purely combinatorial process, the number of possible combinations is in principle fully predictable if exact information about the numbers of each set of functional recombining segments is available. But consider the other diversification processes of the adaptive immune system of higher vertebrates. When these are superimposed upon the primary combinatorial diversity, precision in calculating the resulting number of possibilities becomes much harder. First, "unconventional" recombinational events may extend the initial estimate of combinatorial diversity.[††] Then, it is necessary to estimate the number and frequency of junctional nucleotide alterations (P- and N-nucleotides and junctional deletions), the influence of gene conversion events, and the range of somatic hypermutation. Though the latter is nonrandom and includes certain biases for point mutations (see above), base insertions and deletions must also be

---

[*]The approximation results from allelic variation in the numbers of functional recombining heavy chain segments. Thus, while a precise figure for the combinatorial diversity can be given for an *individual* of exactly defined genotype, the overall diversity for the species is specifiable only within the bounds of the range of functional V segment allelic variants.

[†]V-region pseudogenes may tend to accumulate through evolution since selective pressures for the functional preservation of any single V segment will usually be weak,[258] and V gene losses may be offset by other diversification processes such as gene conversion, as noted previously.

[††]Rearrangements resulting in D–D fusions, or inversions, have been reported, although the significance of this in the repertoire is controversial. Direct V_H–J_H recombination can also occur in some circumstances.

factored, resulting in a huge number of potential variants. Taking these factors into account, maximal antibody diversities ranging from 10^{11} to as high as 10^{16} have been estimated.[258,275] With this kind of spread, one is tempted to modestly state that the theoretical diversity is very high, but with a perceivable upper boundary. In fact, these figures have not paid heed to the critical requirement to avoid self-recognition, and thus the usable number of specificities is well below the raw number based on diversification mechanisms alone. (We will return to this issue later in this section.)

But these types of comparisons beg the question as to how much diversity is "needed" in the first place to cover the entire universe of potential antigenic molecules, and whether this is theoretically possible with antibodies or any other alternative combining molecules. But even though potential immune receptor repertoires are very large, could there be in principle certain structures (in the hypothetical all-encompassing OMspace of Chapter 1) that cannot be efficiently recognized, at least at high affinity? That is to say, are there structures for which there are inherent "holes" in any natural immune repertoire?[*] This question is also of general interest in the context of the optimal design of molecular recognition structures.

Antibodies and Molecular Recognition in General   Relevant to the intrinsic breadth of the antibody recognition range is the concept of "shape space," where structures involved in binding to antibodies have been mathematically modeled as generalized shapes in hypothetical spaces.[5,276] The notion of shape complementarity between antibodies and antigens is based upon the types of noncovalent molecular forces involved, which are weak short-range interactions (including van der Waals forces, hydrogen bonds and electrostatic forces). To become effective, these require proximity between interactive groups best achieved with surfaces of complementary shapes. Theoretical models have predicted that a finite number of molecular shapes (approximately 10^8) will cover the entire spatial construct.[5,124,276] Moving from point to point in shape space can then be defined in terms of changes in the affinity of binding toward a specific antibody. Also implicit in such calculations is the view that different molecular chemistries and configurations can converge toward the same shape, and in turn that a point in shape space does not necessarily represent a unique molecule. Most antibodies, by this proposal, would therefore cross- react with some other molecular structure approximating to the shape parameters of the "normal" antigen to which the antibody is directed. Numerous examples of cross-reactions of this general nature have been documented, which can also have implications for both the normal immune response[277] and the breaking of tolerance to self (autoimmunity). In much theoretical modeling, the shape of the antibody-combining site is assumed to be fixed, but this is not always the case. Relatively recent work has shown that some mature antibodies can assume distinct conformations (or isomeric forms) associated with binding different antigens.[277] (And recall the general distinction made earlier between germline and mature antibodies in their conformational flexibilities.) Relating this effect to the shape space concept, some antibodies could thus potentially bind not only two chemically different but shape-related molecular determinants, but also (by isomerization) an entirely different set of shapes at a different point in shape space.

The above estimates for the extent of vertebrate adaptive immune diversities would appear to easily accommodate 10^8 specificities, even after subtracting self-reactive receptors. If that could truly "cover" the universe of molecular shapes, then it is implied that most immune repertoires (by whatever mechanisms of diversification) will be capable of recognizing virtually

---

[*]Note that this question is asking whether antibodies *in general* have limits of recognition toward certain structures. This is quite distinguishable from a repertoire hole resulting from a genetic defect in an individual (as further discussed below).

all possible molecular targets. Categorizing antigen–antibody recognition by the theoretical criteria of shape space, however, often failed to match real binding affinities.[278] An additional complication is the phenomenon of "induced fit," (as noted earlier with germline antibodies) where the final interaction between an antigen and an antibody only attains high affinity after conformational changes induced by the initial binding interaction.[*],[264]

With immunoglobulin combining sites, natural selection has come up with a family of proteins whose features permit high variability at specific sites (complementarity-determining regions) while maintaining a stable overall framework. A frequent recognition mode for antibodies is a "concave" state involving a defined recognition pocket, although (as we will see further in Chapter 7), "convex" recognition is also possible, especially for some small antibody variants. Recognition "grooves" or extended contact surfaces for antigens are also known within the highly versatile antibody design.[258] The secret of this success is the "beta-sandwich" immunoglobulin fold[279] with framework antiparallel beta sheets linked by complementarity-determining variable region loops that determine the conformation of the antigen-binding site[264] (see Chapter 7 for additional details). As such, antibodies are a superb compromise between the need for generalizability combined with high-affinity recognition. Binding antigens tightly is a requirement for immunologically effective antibodies, since foreign structures must be recognized in the solution phase (in the circulatory system) as well as via B cell surfaces.

In discussing molecular binding, it has to be kept in mind that the effectiveness of such recognition needs to be precisely defined, and specificity and affinity are useful measures.[†] At the same time, it should be noted, while very high affinity is "good" for antibodies, this is not necessarily the case for other natural recognition systems. Any population of receptor–ligand molecules in an isolated system will reach an equilibrium between bound and unbound forms, the concentrations of which will depend on the inherent binding affinity. As the affinity approaches extremely high levels, a receptor–ligand interaction approximates toward irreversibility (the dissociation rate becomes correspondingly low). Many biological interactions, including a variety of cell surface receptors, have been selected by evolution for triggering when ligand levels (and in turn the levels of receptors engaged with the ligand) reach a threshold concentration, and excessive binding affinities may be detrimental to this process.[††] A consequence of this is that the optimal binding affinity of a protein (or protein complex) in natural circumstances is not necessarily the maximal level that can be inherently attained. Even antibody recognition has been proposed to have an upper boundary on optimal affinities under biological conditions (an "affinity ceiling" where the K_d is around 10^{-10} M)[280]. Although higher affinities than this have been reported in some *in vivo* circumstances, the antibody affinity ceiling can be well and truly smashed through artificial manipulation, into the femtomolar (10^{-15} M) range.[281] Piercing this "ceiling" thus strongly suggests that naturally formed antibodies do not represent the ultimate potential binding power of the immunoglobulin fold in terms of affinity toward target.

The existence of biological constraints on the properties of molecular recognition underscores the difficulties in distinguishing a limitation in any particular immune receptor repertoire from absolute limits to recognition specificities dictated by the specific molecular structures

---

[*]It has been pointed out that apparent "induced fit" can be explained by antigen-mediated selection of specific conformational isomers (conformers) within a conformationally dynamic antibody population, which is analogous to the isomeric explanation for antibody multispecificity noted above.[277]

[†]A measure of affinity is the dissociation constant, or K_d, expressed as molarity (M). The lower the K_d value, the higher the affinity.

[††]An example of a biological process where differential affinities (or strictly speaking, avidities) are crucial is T cell receptor/peptide-MHC recognition, upon which the maturation of T cells is obligatorily dependent, as noted earlier.

involved. A case in point in this regard is "heteroclicity" in antibody responses. A *heteroclitic* immune response occurs when immunoglobulins (or T cell receptors) are generated with higher affinity toward another antigen than against the immunizing antigen itself (although usually both immunizing and heterologous antigens are structurally related). This could result from a limitation in the primary immune response repertoire such that the "best available" receptor is an imperfect match for the immunogen, and may better fit a structural analog of it. But as we have seen, the second round of diversification and selection in adaptive immunity (principally somatic hypermutation) can "tune" an antibody's affinity away from heteroclicity and toward specific recognition. Thus, the second tier of adaptive antibody immunity can (in principle at least) "fill-in" any pre-existing holes in a repertoire. In any case, repertoire holes resulting from genetic deficiencies or the agency of pathogens do not of course represent inherent limitations in immunoglobulin recognition potential.

On the other hand, it would be most surprising if the immunoglobulin fold were the ideal solution for all instances of protein-based molecular recognition. Certainly the antibody beta-sandwich structure is not ubiquitous in all binding proteins! Returning then to the question of the maximal scope of antibody recognition, we could compare examples of "generalist" antibodies binding a specific target, with other natural proteins that are "dedicated" to binding the same target molecule. Since molecular recognition in general is a fundamental biological requirement, there are a huge range of dedicated recognition systems to choose from. A few major classes of these are carbohydrate-binding proteins (natural lectins), ligand-binding signal transducing receptors, enzyme–substrate interactions, and sequence-specific DNA-binding proteins (such as transcription factors). Artificial nucleic acid recognition molecules (aptamers) are another case in point for a class of molecular interactions that in some cases may rival antibodies (as we will see in more detail in Chapter 6). Where interacting proteins within an organism have evolved toward optimal recognition, it might be expected that such committed partners would also display optimized complementarity. Indeed, a "shape correlation statistic" designed to measure the closeness of packing of protein–protein interfaces has indicated better "fitting" between surfaces in oligomeric proteins and other protein–protein interactions than for antigen–antibody interfaces.[282] To continue with one of the above examples of natural molecular recognition, let us compare dedicated sequence-specific DNA-binding proteins with anti-DNA antibodies.

Certain natural proteins have functions that require sequence-specific DNA interactions, including the direction of transcription, control of recombination,* DNA modification by methylation, and site-specific DNA cleavage. A variety of structural domains in such DNA-binding proteins can recognize specific duplex DNA sequences from 4 to >12 bp (target binding sequences of up to 40 bp are known in some cases but have certain degeneracies of recognition†). Only a minority of DNA-binding proteins in fact have DNA recognition domains that share some structural similarities to the immunoglobulin beta-sandwich, where specific protein-DNA contacts are mediated by loops connecting regions of secondary structure.[285] Certain antibodies are known that can recognize DNA, prominently noted in the autoimmune syndrome systemic lupus erythematosus, but DNA itself has low immunogenicity, and anti-DNA antibodies are usually not sequence specific. Nevertheless, immunization with a protein–DNA complex (the E2 transcription factor of a human papillomavirus bound to its specific target site) has been shown to result in antibodies with demonstrable sequence

---

*An example is V(D)J and T cell receptor recombination in the immune systems, which requires specific site-specific recognition proteins as well as a number of other proteins.[283]

†That is, the recognition sequence is not completely specific; at some defined positions within the sequence alternate base pairs can be tolerated. The largest sequence motifs recognized are from the "homing endonucleases" that initiate transfer of intron sequences in some organisms.[284]

specificity.[286] The natural E2 factor (interacting with the same target sequence as the latter antibody) is one of the small subset of DNA-binding proteins whose recognition domain has similarities to an immunoglobulin fold. Yet the mechanism of the antibody–DNA recognition in this case appears divergent from that seen with E2, although not fully structurally defined. While the antibody binding in this case was of reasonable affinity, it had a slow rate of association with its specific target sequence and certain other unusual features.[286] Hence, in spite of the observation that many (if not all) short DNA sequence motifs can be recognized by multiple types of specific protein folds,[285] immunoglobulins are not tailored exclusively for DNA binding, and the best anti-DNA antibodies to date fail to match natural DNA interactive proteins in performance.

This is not to say that further artificial manipulation of antibody variable regions could not result in better sequence-specific binding, especially given the intriguing fact that a subset of different natural DNA-interactive protein folds are immunoglobulin-like. From the point of view of antigen binding in general, it has already been shown that the use of an expanded genetic code and the incorporation of unnatural amino acids (explored in more detail in Chapter 5) provide an advantage in the selection of antibodies against a protein target.[287] Nevertheless, specific DNA sequence targets (especially longer or bipartite recognition sites) are likely to fall outside the ambit of truly effective binding by immunoglobulins.* The take-home message from this rumination is that while antibodies are wonderfully capable in their ability to bind diverse structures, they have limitations toward certain types of recognition challenges, which can be met with other distinct molecular binding solutions. It also seems probable that this could be generalized for any single type of alternative framework to antibodies. In other words, any single type of molecular framework is unlikely to be capable of providing an optimal solution to all molecular design problems involving recognition and binding.

Having briefly looked at the notion of antibody repertoire holes, let us now return briefly to certain fundamental pillars of immune repertoires in general, to achieve something approaching a holistic view of adaptive immunity itself. This again takes us back to the fundamental self/nonself distinction. . .

Selfishness Revisited

There is a curious consequence of the generation of unique immune receptors against foreign antigens that have been long understood in principle. The variable parts of selected receptor antigen-binding regions are assembled by combinatorial, junctional, or mutational processes that in the end produce protein sequences not directly encoded in the genome (as described earlier in this chapter). Such immunological sequences and their associated structures, arising somatically within a mature organism, will thus often be novel with reference to the entire background of molecular forms that constitute the "self." Or in simpler terms, in order to recognize and bind novelty (foreign antigen), you also may need novelty in your weapon (immune-binding protein). As a consequence of this, a successful immune response itself can generate what is in effect "nonself," in an immunological recognition sense. (In other words, in defining "self" by either central thymic selection or peripheral regulatory mechanisms, the immune system cannot possibly anticipate every novel structure on an immune-binding protein that itself has been somatically selected from a very large set of variants.) These novel structures on immune receptors therefore can elicit immune responses against themselves.

---

*The ability to design proteins at will to recognize any desired DNA sequence tract has been a long sought-after goal toward which much progress has been made (ftp site). However, these strategies have not used antibodies for this purpose.

Immunogenic structures unique to the variable regions of immune receptors are termed *idiotypes*.* and a single recognition element (or epitope) within an idiotype is therefore an idiotope. By far the greatest number of studies have been performed on immunoglobulin idiotypes, but T cell receptor idiotypes have been defined as well, which likewise inevitably arise from somatic immune diversification. This general issue may cause us to think about the concept of "self" in a little more detail, at least to avoid confusion. We might think of "self" as any molecular structure directly encoded by an organism's normal genome or synthesized by its genomically encoded enzymes. But this is unsatisfactory, since it is the normal immune system that mediates the distinction of self from nonself. A refined operational definition of vertebrate "self" could be then phrased as "molecular structures encoded by an individual organism toward which its developing adaptive immune system specifically learns to avoid activating responses, or from which its immune system is functionally shielded".† In this view, "self" is actively defined rather than simply a global default position for any molecule that is synthesized by cells of an organism. It also follows from this same definition that certain normal processes (somatic immune receptor diversification and idiotypes) and aberrant events (such as novel tumor-specific antigens) occurring within an organism will fall outside the self-boundary.

Having undertaken an idiotypically prompted rethink of selfish matters, perhaps now we can consider a few more implications of immune idiotypes themselves. If idiotypic regions on immunoglobulins (or immune receptors in general) can stimulate an immune response, an interesting situation arises. In the case of antibody responses, anti-idiotype antibody can in turn generate an anti–anti-idiotype antibody, and (in principle) so on indefinitely.†† The result of this, at least theoretically, is a network of idiotype/anti-idiotype interactions (Fig. 3.6), which was proposed and conceptually developed by the Nobel Prize winning immunologist Niels Jerne.[289] Idiotypic network theory has been amenable to mathematical approaches, including the modeling of nets of interactions in shape space.[290] One implication flowing from analyses of idiotype networks is that second-generation anti-idiotypic antibodies (Ab_2 of Fig. 3.6) would recapitulate the shape of the antigen to which the original antibody was directed. Such anti-idiotypic antibodies would then act as "internal images" of the immunizing antigen, and in turn third-generation antibodies (Ab_3 of Fig. 3.6) would bind the antigen itself in a comparable manner to the first antibody (Fig. 3.6, right panel).

Numerous instances of "molecular mimicry" by anti-idiotype antibodies have been reported that are consistent with the internal image concept. Anti-idiotypic "images" of nonprotein antigens have also been described. In this area, the acquisition of hard structural data cut through much speculation (as it usually does) and provided an explanation for this kind of

---

*This uses the Greek stem *idio-*, meaning "private" or "own," as in idiopathic or idiom.
†The term "activating responses" is used rather than "recognition" *per se* to exclude low-affinity self-MHC/T cell receptor interactions in the periphery. "Learns to avoid" (irrespective of whether via thymic central clonal deletion or active suppression mechanisms) reflects the observation that discrimination of self from nonself is not predefined but largely arises during T cell development, dependent on thymic influences. Thymic expression of self-antigens aimed at elimination of self-reactive clones is thus dependent on the gamut of self-antigens themselves, which can be experimentally manipulated (as with transgenic mice). Another complication to the immunological definition of self requires the caveat "functionally shielded." This refers to the phenomenon of antigen "sequestration," where antiself responses are not provoked owing to physical compartmentalization of certain self-antigens from exposure to immune cells. Sequestration can be temporal as well as spatial, as where certain self-proteins are only expressed during embryonal development before immune systems come physiologically "online." Some of the latter antigens are reactivated in transformed tumors and function as novel tumor antigens.
††It must also be noted that for an antibody response against an immunoglobulin idiotope, a peptide segment of the protein must also function as an MHC Class II binding element recognizable by T helper cells (as in Fig. 3.2). The function of such T cell idiotopes has been demonstrated.[288] Such T cell elements will often differ from the segments (within the same total idiotype) recognized by antibody, but can coincide on some occasions.

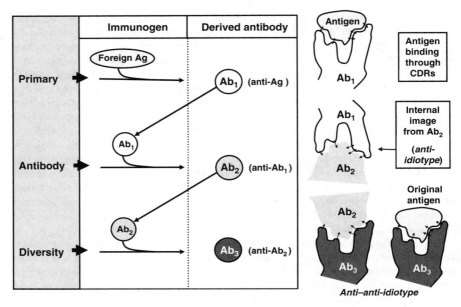

FIGURE 3.6   Depiction of antibody (Ab) anti-idiotypic networks and internal images. Exposure to foreign antigen (Ag) leads to an antibody response (Ab$_1$), which (through unique idiotypes on Ab$_1$) in turn leads to Ab$_2$, and so on. CDR (complementarity-determining region)-mediated contacts (represented by double-headed arrows) of Ab$_2$ with Ab$_1$ (shown in stylized form in right panel) result in an "internal image" of original antigen. As such, Ab$_3$ against Ab$_2$ can also bind the original antigen itself.

mimicry at the molecular level. When crystal structures of idiotype–anti-idiotype antibody complexes were solved, it was found that an internal antigen image was functionally produced through anti-idiotype antibody CDR loops making similar binding interactions to the initial antibody as for the original antigen itself.[291] The molecular "mimicry" involved here is functional, in that binding interactions between original antigen-Ab$_1$ and Ab$_1$–Ab$_2$ (as in Fig. 3.6) are homologous; the mimicry does not extend to close topological matching.[291] Indeed, full topological matching would be impossible between non-α-helical CDR loops and an α-helical epitope, or even more so between CDRs and nonprotein structures.[291]

We can understand the limits of mimicry in a simple thought experiment. Consider antibodies produced against penicillin or ascorbic acid, when these two molecules are presented as immunogenic haptens.[*] Anti-idiotypic antibodies for each would bind the primary antibody and act as antigen mimics. But could they act as functional mimics for the original antigens as well? Could a penicillin-mimic anti-idiotype antibody kill bacteria in the same way as the drug itself? Could an ascorbate-mimic anti-idiotype antibody allow humans to avoid scurvy even without Vitamin C? In both cases, the answer is a firm "No." The anti-idiotypes are molecular shape mimics, but with the normal complement of natural amino acids they could not reproduce the chemistries needed to affect the functions of these two small-molecule examples.

---

[*]Haptens are any small molecules recognized by immune systems, but haptens are only able to stimulate immune responses (that is, to show immunogenicity) if they are coupled to a macromolecular carrier (usually a protein) or sometimes when polymerized. Haptens are important in the generation of catalytic antibodies, which we will see in more detail in Chapter 7.

This could be viewed as a molecular demonstration of the aphorism, "the map is not the territory." In any case, the closeness of match between antigen and anti-idiotype is dependent on complex variables such that the fidelity of mimicry is not constant in different cases, and indeed sometimes is not apparent at all.[292]

While anti-idiotypic interactions and internal images have been confirmed and characterized at the molecular level, the significance of anti-idiotypic networks in normal immune regulation is another matter again. A role for idiotype/anti-idiotype networks in controlling potentially self-reactive clones has long been proposed, and some experimental evidence for the existence of such networks *in vivo* has been proffered.[293] Yet studies on central and peripheral cell-based control of self-reactivity have tended to disfavor a major role for idiotypic regulation, although in recent times a renewal of interest in network theory and applications has arisen. Wherever the eventual placement of idiotypic networks in the pantheon of immune regulation, the phenomena involved with idiotype/anti–idiotype interactions provide important lessons in the nature of molecular recognition and are intimately linked with the heart of the concept of self and nonself.

Anti-idiotypic effects can also provide useful applications. An anti-idiotypic antibody recapitulating an original antigen as an "internal image" can in principle be used for vaccine purposes, by acting as an antigen surrogate. Antibodies against the idiotype of such a vaccine antibody should then in turn recognize the original target antigen itself (as in Fig. 3.6). But what could be the advantage of using an anti-idiotypic antibody "antigen image" over the real thing? There are several answers to this. Anti-idiotypes can create protein mimics of molecules with initial poor immunogenicity, or which are difficult or costly to obtain. (In contrast to some antigens, monoclonal anti-idiotypic antibodies can be made *in vitro* in large amounts.) In particular, with anti-idiotypic approaches, one can make peptide mimics of nonpeptide structures. Earlier in this chapter, it was noted that some specific types of antigens can activate B cells in the absence of T cell help, and these thymus-independent antigens include certain polymeric and carbohydrate (polysaccharide) molecules. When an anti-idiotype bears an internal image of such a polysaccharide epitope, in effect a thymus-independent antigen is converted into thymus-dependence, since the immune response to immunoglobulin idiotopes requires T cell-mediated help. This may allow higher affinity secondary antibody production toward the anti-idiotype internal image (and thereby indirectly against the polysaccharide also) through affinity maturation processes* than would be the case where the original polysaccharide itself is used to generate an immune response.

Sometimes anti-idiotype molecular interactions that mimic the original antigen (Ab$_2$ of Fig. 3.6) are predominantly within a contiguous segment of CDR3.[294] If so, the relevant anti-idiotype interactions themselves may be localized to a contiguous peptide tract, which can be recapitulated with an appropriate synthetic peptide. By this process peptide mimics of carbohydrate and phosphorylcholine antigens have been derived. The exploitation of anti-idiotypes thus represents a pathway for applying the empirical selection processes of the adaptive immune system toward creating a useful molecular structure, itself a mimic of another defined structure.† While this is a potentially powerful approach, a caveat must be noted. An

---

*Since T cell help in germinal centers promotes somatic hypermutation, affinity maturation, and immunoglobulin class switching, thymus-independent antigens produce deficient responses in these respects (although not absolutely lacking). Somatic hypermutation in T cell-independent germinal centers is reduced relative to corresponding T cell-dependent centers, but still present.

†This is not to say that such peptide mimetic compounds cannot be obtained in other ways. For example, the power of immune-based empirical selection can be applied *in vitro* by screening for phage-displayed peptides that bind to anticarbohydrate antibodies, thus mimicking the original carbohydrate antigens. (See the next chapter for phage display details.) "Rational" design of peptide mimics has also advanced (Chapter 9).

antibody directed against an anti-idiotypic internal image will often recognize the original antigen (as Ab_3 of Fig. 3.6), but structural studies have shown that such Ab_3 antibodies will tend to bind their anti-idiotypic targets more effectively than the desired antigen itself.[295] As such, the Ab_3 (or anti–anti-idiotype) antibodies are not optimized for the target antigen binding, and may therefore be imperfect in implementing the desired goals of a vaccine strategy. This has been borne out by certain comparative experimental studies,[296] and these potential limitations of anti-idiotypic strategies must be balanced against their advantages (as outlined above) and the advantages achieved with molecular mimetics in general.

What we have considered so far with respect to the adaptive immune system serves as an introduction to that aspect of it which is of primary interest for evolutionary molecular discovery. At this point we should look at some general issues with respect to the evolutionary generation of immune recognition proteins and their artificial counterparts.

Antibody Libraries and Iterative Selection: Natural and Artificial

Much of this chapter has been dedicated to the theme that cumulative natural selection, in response to environmental selective pressures, has evolved somatic systems that themselves are striking demonstrations of the power of evolutionary selection for molecular design solutions. For natural selection, the raw material upon which selection can act is germline genetic diversity resulting principally from mutation and varying forms of recombination (as discussed in Chapter 2). For an empirical somatic engine of evolutionary selection, the diversification mechanisms must be programmed into the system by natural selection itself. But beyond these fundamental messages, there are certain other lessons to be gleaned from the adaptive immune system, especially if we compare selection of antibody combining specificities from the "library" of naturally generated somatic antibody diversity with artificial molecular libraries. The latter could be antibodies themselves expressed as large libraries of variants in artificial systems (as we will see in Chapter 7), but the same principles apply to any type of library that can be systematically diversified.

The first issue to consider in this light is the two-tiered approach to antibody diversification as used in certain mammalian adaptive immune systems. As we have seen, with this kind of process primary immunoglobulin diversity arises from somatic rearrangements and junctional variation, such that B cell clones expressing immunoglobulin receptors of suitable specificities are initially selected and amplified. Subsequently, the variable regions of such immunoglobulins undergo affinity maturation, where the diversification process of somatic hypermutation allows antigen and T cell-driven selection for B cell clones expressing immunoglobulins with high-affinity binding properties (Fig. 3.3). We can thus distinguish two rounds of diversification and selection in the mammalian adaptive immune system, which can be analyzed in very general terms.

A recurring theme in the study of evolution by natural selection is its cumulative nature, where compounding incremental changes lead through time to major alterations in the morphologies and functioning of organisms. A corollary of this is that large single "jumps" in the latter qualities, especially those involving complex coordinated supramolecular systems, are unlikely. For the arm of an adaptive immune system concerned with producing soluble recognition proteins, one could envisage an arrangement where a single round of selection events (from a large library of diverse variants) is the only selective process used for obtaining target-binding proteins of the desired characteristics. In effect, such a strategy relies on single "jumps" toward achieving the desired molecular ends, but a two-tiered system (as outlined earlier) has numerous theoretical and practical benefits and corresponds with a somatic model for evolutionary cumulative change.

At this point, the other lesson of natural adaptive immune systems is introduced: the practical limitation of the screenable molecular "library size." All vertebrates are necessarily limited in the numbers of lymphocytes expressing variable immune receptors that can be deployed at any one time, which is usually much less than the total potential diversity. *In vitro* artificial molecular libraries too are necessarily constrained by numbers (as detailed in the next chapter). So, while in principle a sufficiently large and diverse library might directly yield a suitable binding molecule, in practice library size limitations will usually make this ideal an impracticality. One potential way of side-stepping this is to successively screen separate libraries of equal maximal sizes but independent diversities. Thus, an organism will necessarily have an upper limit to its numbers of circulating immune cells bearing variant receptors, but with lymphocytic turnover and continuous arbitrary generation of new variants, the available repertoire will be in constant flux. (Such that at any given time even two genetically identical animals will not share identical somatic immune repertoires.) By this process, after a time period in excess of the average lymphocytic lifetime has elapsed, the available repertoire should be significantly divergent from the starting-point repertoire.[*] In practice, such "replaced available repertoires" are unlikely to be completely unrelated to earlier available repertoires in the same organism due to inherent biases (such as nonrandom use of germline V regions in immunoglobulin gene somatic rearrangements) and the dictates against self-recognition. Such constraints would not in principle apply to artificial *in vitro* molecular libraries.

Yet this "alternate library" scenario carries a significant cost for both natural and human-engineered molecular libraries. For the latter, the cost can be literal in economic terms, if exhaustive library screening is undertaken. (The limiting factor in this case is the maximal library size that can be scanned per round of screening, whether these are "independent" libraries or subfractions of a larger single library.[†]) To provide protective function against infection, natural immune systems do not have the luxury of waiting until a suitable receptor specificity turns up through arbitrary generation from the total possible theoretical pool. Even though innate immunity represents the front-line defense, the adaptive immune system must respond as quickly as possible to often burgeoning growth of invading parasites. So, either the immediately available receptor pool must suffice or an alternate selective approach must be found. As for the iterative process of natural selection, a solution for optimizing antibody responses is iterative antigen-based selection, in the form of the two-tiered adaptive immune response (Fig. 3.3). (And as we have seen, this is combined with a germline antibody repertoire with increased combining site flexibility over mature antibodies.) To illustrate the efficacy of this iterative selection, consider the following metaphorical scheme.

Iteratively Speaking

Once again, let us use Lucy as our metaphorical researcher, where her work involves selecting a recognition element with the desired properties from a large combinatorial library. In order to find the desired species, she has been provided with a "probe" that defines the desired target specificity for the recognition element in question. In this particular circumstance, she is sifting through sets of elements composed of strings of 26 possible building blocks, in any sequence. These 26 constituent linkable groups just happen to be denoted by the same symbols as those used by the English alphabet, and Lucy can generate random libraries of short

[*]This is not including long-lived memory B and T cells that are selectively maintained after infection to rapidly control future challenges with the same agent.
[†]This kind of limitation is actually contracting with steady advances in high-throughput technologies and robotics (Chapter 8).

concatenations of these "letters," as she refers to them. In this universe she is presently in, she has prior knowledge that a suitable recognition sequence can be found as a string of eight building-block letters.

She is accordingly searching random strings of eight letters in length, but for practical reasons the maximum number of these that she can evaluate at any one time is one billion, and this constitutes the upper limit of her library size. This could be unfortunate, since this figure is less than 1% of the total number of possible 8-"letter" elements (8-mers) in any sequence combination (26^8; about 2.10^{11}). At the outset, of course, she does not know what the desired letter sequence of the recognition element is, but we will use the benefit of hindsight to divulge it: "ANTIBODY." The chances of pulling this specific sequence out from a library of only 10^9 members do not then seem very favorable, but Lucy proposes that certain suboptimal recognition element sequences of less than eight letters may be sufficient for selection with her target/probe molecule (itself a letter string that happens to form the sequence "ANTIGEN[*]"). She can also adjust the "stringency" conditions during library probing to favor either weaker or stronger interactions, as desired. Again with our preview knowledge, we know that certain combinations of five letters will permit detectable binding of the target. (In other words, a minimally detectable recognition element must contain five of the eight letters in the optimal configuration.) We can therefore see that while finding complete ANTIBODY in her library is a forlorn hope, she might well be able to isolate "ANTxBOxx," or some other 8 mer with five specified letters from the optimal target sequence (where $x =$ any other letter). Any one of such sequences will occur approximately once every 10^7 random letters (26^5), compatible with the size of Lucy's library. So Lucy succeeds in finding ANTxBOxx, which interacts with her probe with low but sufficient affinity to enable its isolation.

But this is not the end of the story. Lucy realizes that she can take this suboptimal recognition element and "tune" its affinity. Remember, we are seeing this element as "almost-ANTIBODY," but while Lucy has determined the sequence of this isolated candidate, she has no way (at present) of knowing that ANTIBODY is the *optimal* sequence she is striving for. She can, however, take her primary candidate element (ANTxBOxx) and subject it to a second round of controlled and limited mutagenesis. Some such secondarily mutated elements will be rendered worse than before, but there is a good chance that some will be improved binders. Lucy's efficient selection conditions with her ANTIGEN probe-target ensure that only improved binders are isolated, and all others can be ignored. Under her limited mutagenesis conditions, the probability of finding the three specific single-letter changes required to convert ANTxBOxx to ANTIBODY is ($[1/8.26]^3$) or about 1:9 million, easily achievable if her secondary library is the same billion-member scale as her first library. Even if only one such iterative mutagenesis/screening round is implemented, the prognosis for molecular improvement by incrementally selecting for superior binding is excellent. And given the constraints under which Lucy was operating, finding the best recognition element would be most unlikely with a single round of library probing. Someone suggests to Lucy that her iterative tuning process for homing on her desired optimal molecule could be called "affinity maturation," and she can but agree.

This metaphorical diversion serves to illustrate the power of iterative selection, but has a deficiency for generalizable comparison to the real world at the selection level. The library element that Lucy seeks acts as a "perfect recognition agent," which is not possible with actual molecules. Versatile though antibodies are, for some recognition problems an optimized

---

[*]Note again the terminology for "target/probe" as opposed to the library members. Here ANTIBODY is the specific library member (out of a large number of variants) that is eventually fished out, and ANTIGEN corresponds to the invariant target used to probe the library. See Chapter 1 (Fig. 1.3c) for a recap of this point.

antibody will only return a local binding optimum, which may be surpassed by alternative recognition frameworks. (In this context we might recall the earlier consideration of antibodies versus dedicated proteins for the recognition of specific DNA sequences.) If we extended the above metaphor to include this kind of limitation, it would be as though even the best possible recognition element within her 8-mers library had an inherent limitation in recognizing the probe. For example, "ANTIGEN" might not be distinguishable from "ANTIGEx" or "xNTIGEN."

It might be noted in this metaphorical sketch that the methods of mutagenesis, or primary and secondary diversification mechanisms, were not detailed. The reason for this is that the essential message is the iterated process itself, rather than the particular pathways to diversity that are used. Although it is obviously important to understand the specific types of diversification mechanism used in real adaptive immune systems, both in principle and practice multiple different approaches can be taken toward such ends. Thus, humans and mice use combinatorial and junctional mechanisms for primary diversity and somatic hypermutation for secondary diversification, but this is not a universal feature even of other mammalian species. As noted earlier, rabbits have limited germline segment combinational diversity, but rely on gene conversion and somatic hypermutation for achieving both primary and secondary GOD. By the same token, an artificial mutagenesis procedure with iterated rounds of selection need not be limited in its methods of choice for creating molecular diversity. It is important to also point out that this empirical approach is not incompatible with the coapplication of rational design methods. Pulling out a candidate molecule as a solution to a specific goal by initially empirical means can go a long way toward pointing one in the right direction for future design. Studying the functional properties of a primary candidate molecule may allow rational steps to be taken toward its subsequent improvement, as we will consider in more depth in Chapter 9.

Such Selfish Systems Send Selfless Sermons: The Largesse of Immune Systems   We have concentrated thus far on the utility of antibodies and the lessons and inspiration for molecular design that may be drawn from gaining an understanding of their origins, but we should not forget equivalent lessons from the T cell and MHC recognition and the innate immune system underpinning the whole enterprise. But even before we were able to appreciate the lessons of the immune system for evolutionary molecular design, the antibody arm of immunity was serving us in a variety of ways, which we will pick up again and extend in Chapter 7.

This brief reflection serves also as a reminder of the amazing complexity and practical functioning of biological immune systems in general. In their capabilities and astonishing organization, immune systems are things of wonder and indeed of special beauty. With reference to evolution and the panoply of its results through time, Charles Darwin famously stated, "There is grandeur in this view of life. . ." Surely the truth of this becomes more and more clear as we gain, through painstaking and meticulous research, an understanding of the power and elegance of the molecular and cellular systems that have arisen through natural selection. In the case of adaptive immunity, the blind watchmaker creates another watchmaker, no less blind, but in its own way capable of delivering sophisticated molecular design solutions on a vastly accelerated timescale. There is only one more complex biological system, and that too has some molecular design messages for us.

GOD IN THE BRAIN AND ITS EXTENSIONS

In the previous chapter, the apparent paradox was drawn between the complexity of the human genome in gene number and the complexity of the organism that arises as a result of the unfolded expression of this genome. Even an isolated aspect of the whole organism, the immune system, is a good exemplar of this, and the brain/nervous system is in another league again. We also noted that the totality of the genome's information content, including its multilayered regulatory control systems, is highly parsimoniously encoded. Many levels of complexity unfold as the genome is expressed during the development of an organism. Expressed proteins themselves can be modified in numerous ways, either by direct self-modification or through the agency of other proteins again, in the form of many different enzymes. But the immune system demonstrates the programmed generation of somatic diversity. This type of GOD is subject to genetic, environmental, and stochastic influences, such that the specific immune repertoire of each individual organism is unique. As such, the total informational content of a mature immune system cannot be directly encoded in the genome.

Even without specific neurological information, the same conclusion applies to the brain with equal conviction.[297] This stems from the elementary observation that learning and memory are specific to the individual and not innate, and therefore must be somatically acquired and stored in some manner. Such a conclusion applies not merely to human beings but also to much simpler organisms,[*] if to a lesser degree. We will not stray long into the diversification mechanisms (electrochemical, cellular, and molecular) involved in the recording of memories, not least because there is a great deal still to be learned despite much recent progress. For long-term memory storage, the involvement of specific signaling pathways and transcription factors in synaptic remodeling[†] has been demonstrated, and epigenetic mechanisms have been proposed to have a role in at least some types of memory formation. While neural diversification processes are a logical necessity, given the information-processing role of the brain, there is to date no definitive evidence of neural diversity arising from somatic DNA rearrangements as seen in the immune system (detailed in ftp site).

We will not pursue these interesting neural processes, but they can serve to remind us of the field of sensory recognition. The neural mechanisms of sensory input can be considered as extensions of the brain, and it is in the brain where sensory information is finally decoded as perceptions. In this area, it is the ability of the chemical senses (taste and smell) to respond to a very wide diversity of molecules, which is of the greatest interest for further understanding molecular recognition in general, and ultimately molecular design. Immune systems, which have been the major preoccupation of this chapter, themselves can be regarded in a sense as additional physiological sensors.[298] All immune systems are concerned with the recognition of molecules, and this feature is shared by the mechanisms of chemosensory perception: taste and smell (or if you prefer, gustation and olfaction, respectively). Chemosensory processes do not appear to use somatic diversification in the same manner as for the immune system, but olfaction at least involves stochastic somatic gene expression processes. Yet chemical sensing is of particular interest for molecular design at the system level, where different natural solutions have arisen in response to different signaling problems. In addition, the chemosensory systems can show when somatic diversification is and is not required, and also is informative with respect to empirical versus rational molecular design approaches. For these reasons, we should

---

[*]A workhorse for memory studies has been a mollusc: the Californian sea hare (*Aplysia california*). Eric Kandel was awarded a Nobel Prize in 2000 for his seminal work in synaptic memory mechanisms in this organism. Recent genetic studies of memory mechanisms often use mice due to the availability of sophisticated methods for manipulating their genomes.

[†]A *synapse* in a general biological sense is a gap junction between two adjoining cells across which secretory signaling occurs, and which thus serves as a functional connection between the cells. It is most commonly applied in neurology, to refer to adjoining neurons or adjoining neurons and muscle or glandular cells, but also has usage in immunology in reference to interactions between T cells and antigen-presenting cells.

consider chemosensing in a little more detail, as follows. Much of this knowledge is of very recent origins.

Molecular Sensing

In thinking about molecular chemosensory mechanisms, let us continue to use the immune system as a benchmark for comparison. All adaptive immune responses are concerned with molecular recognition and are also "trained" to avoid reacting to self-molecules. If we think about molecular recognition in a general sense, we inevitably recall that recognition and interactions between specific proteins and other proteins or nucleic acids are the very stuff of cellular life. A pantheon of "self" small molecules (including a variety of hormones) is also involved in such transactions and initiates signaling pathways through binding to specific cognate receptors. Of course, as integral parts of the "operating systems" of organisms, these interactions simply proceed through molecular complementarities, and self and nonself distinctions are irrelevant to their functions. (An invariant nonimmune cell surface receptor will be invariably "fooled" by a foreign viral structure* or artificial drug that mimics its normal ligand, and no somatic evolution of such an ordinary receptor (in contrast to an immune receptor) is possible in the lifetime of an individual organism.)

Chemosensing also involves small molecules, but from the point of view of molecular recognition it would seem to overlap the domains of immunity (since chemosensing also involves recognition of environmental nonself molecules) and nonimmune "internal" systems (since self-perception of tastes and odors is clearly possible†). Obviously the evolutionary justification for this goes beyond recognition *per se*. Immune systems necessarily require recognition and interaction with foreign antigens, but many powerful effector mechanisms then follow for removing or inactivating the unwanted threat. Auto-reaction through the immune system is therefore highly damaging. But if the recognition involves only a neural sensory stimulus, self-reactivity has no special consequences, and no selective pressure would be expected to lead to the evolution of control mechanisms for self-chemoperception.

Interestingly, though, an overlap of sorts between one aspect of the immune system and chemoperception does appear to exist, since odorant perception of MHC peptides is at least one mechanism for recognition of non-kin members of some mammalian species (a form of allorecognition). This facility for sensing genetic relatedness between individuals has been proposed to have preceded the advent of adaptive immunity itself, rather than the other way around. In other words, by this proposal the MHC–peptide binding initially evolved as a carrier system for chemosensory molecular recognition, which was afterward co-opted by the adaptive immune system.[216, 299] The relatively recent demonstration of a link between MHC peptides and olfaction brings to mind an old proposal to train dogs to sniff out histocompatible donors for human transplants,[300] which does not seem to have been much pursued.

At least some aspects of biological chemosensing can thus span the boundaries of self and nonself recognition. As always, the "design" by cumulative natural selection of such biological systems and their recognition boundaries will reflect needs that maximize an organism's fitness. This principle also results in wide differentials in chemosensory abilities between different species, with the vastly less sensitive olfactory abilities of humans in comparison with many

---

*This applies to invariant receptors even on immune cells themselves, as with the targeting of helper T cells by HIV through the CD4 surface molecule and associated coreceptors.

†The nature of the perception might vary between an individual and other members of the species (some body odor challenged individuals would seem to be also challenged in their abilities to perceive their self-odorants, but this is most likely related to the well-known effect of odorant sensory paralysis with continued signaling). But a self-odorant is clearly not ignored by the chemosensory system in the same way self is ignored by a normal immune system.

other mammals (such as dogs) standing as a case in point. Another interesting example along these lines is the indifference of obligate carnivores of the cat family toward sweet substances, recently shown to be due to mutational loss of sweet taste receptors in these animals.[301]

Based on the immune system, one could easily imagine a chemorecognition biology that used a combinatorial or mutagenic adaptive diversity process for extending its range toward recognition and sensing of virtually any molecular species, but this is certainly not the reality. Why is this, when we know what the immune system is capable of? Several answers could be suggested for this question, but an obvious and ever-present issue is the requirement for speed of perception of environmental chemical stimuli. Perception of tastant and odorant molecules does not have the luxury of time for an adaptive response to develop. A rapid response to such chemical environmental cues will often be a life-or-death matter of survival, and in consequence the mechanisms for such perception must be innate. Chemorecognition is thus much more analogous to the innate immune system than for adaptive immunity, although only up to a point. Both gustation and olfaction have certain unique features of special interest from a design point of view, so it is worthwhile looking at these a little further.

First let us savor the mechanisms of taste, so to speak. Tasting of salt and sour (acid) is mediated separately by proteins forming specific ion channels, and is relatively poorly characterized, although very rapid progress has been made in this field in general.[*] The three remaining defined taste categories (sweet, bitter, and umami [amino acid or savory taste]) are mediated by specific receptors for appropriate target molecules, and for the present illustrative purposes these will serve as exemplars of taste mechanisms.

A Bittersweet Worldview   The recognition of carbohydrates in general is biologically important and widespread. Within this chapter we have already encountered lectins, the generic class of proteins equipped for specialized binding of sugars. A "sugar code" is believed to be an important pathway for many intracellular and cell–cell recognition processes,[303] and sugar-based recognition may also be an important factor in guiding the specificity of neuronal wiring.[304] Many simple sugars (such as glucose, fructose, sucrose) are of course important as foods and serve as identifiers of natural beneficial nutritional sources such as fruits. Recognition and signaling of the presence of these carbohydrates is thus of survival benefit for many animals, including primates (but at least excluding cats, as noted above). The fact that taste perception of sugars provides a (usually) rewarding "sweet" sensation thus is explicable from an evolutionary point of view. By the same logic, savory or "meaty" umami tastes (indicative of food) are also worth a perceptual reward in many species (especially cats, presumably). At the opposite end of the perception scale are the most aversive tastants, which are substances perceived as having a highly bitter taste. As the latter are very often highly toxic compounds such as plant alkaloids, this too makes evolutionary sense. Earlier in this chapter we briefly looked at the "xenobiotic immune system" involving the SXR/PXR proteins and other nuclear receptors, leading to the activation of detoxification mechanisms. Since prevention is better than cure, the bitter taste system and its linkage to aversive behavior toward "bad" chemicals can be considered a front line screening process to reduce the load on innate detoxification processes.

While the perceptual assignment by natural selection of certain types of environmental tastants as "good" or "bad" is therefore logical, it has quite distinct requirements in each case. Simple sugars and amino acids indicating useful foods are a reasonably restricted set of molecular target types, where from first principles one might envisage that a limited number of receptors will suffice. On the other hand, noxious compounds cover a huge range of chemical

---

[*]A likely candidate for the sour (H^+ acid) ion channel has been obtained.[302]

space, creating what would appear to be a much more challenging recognition problem. So what are the actual natures of these taste receptors? It is now known that with the exception of the above-mentioned salt and sour taste ion channels, taste-related receptors are members of a very large family termed G-Protein-Coupled Receptors (GPCRs). As a whole, these are the most abundant type of receptors,[*] representing >1% of the expressed mammalian genome and covering diverse forms of signaling responses. This large GPCR receptor family is also evolutionarily ancient.

A subset of the GPCR universe, the taste receptors are then constituted by two major types, TAS1 and TAS2 (or T1, T2), each of which is subdivided into specific member receptors (R1, R2, etc.). Sweet and umami taste receptors are GPCR heterodimers encoded by a small family of genes, with one protein chain (T1R3) shared for both gustatory functions. Sweet receptors are expressed as T1R2/T1R3 heterodimers, and umami receptors as T1R1/T1R3.[305] In contrast to this limited number, bitter taste signaling uses considerably more receptors of the T2R type (~30). This increase in the number of receptors signaling aversion (bitterness) is consistent with the above-noted need for a wider scope for recognition of potentially dangerous compounds, but it is still nowhere near the level of natural diversity of such problematic molecules. How is this conundrum resolved?

First, we must again consider the special signaling needs in chemosensory perception. In all such cases, it is not necessary to distinguish each detected chemical species (out of a huge potential multiplicity) as a unique perceptual signal, since grouping them into aversive/ beneficial categories will achieve the required ends from an evolutionary point of view. This could be attained by arranging for the coded signaling of specific tastes (sweetness, bitterness, and so on) to issue from cells separately dedicated to each signal category. In other words, an individual "bitter" cell (for example) is precommitted to bitter signaling and no other. With this arrangement, activation of a specific taste cell type by receptor engagement produces the same coded signal even if such a cell expresses multiple different receptors (Fig. 3.7). This scheme is essentially the "labeled-line" model of taste signal encoding, and there is much experimental evidence in favor of it.[305]

Each bitter-signaling taste receptor cell thus expresses all (or most of) the available repertoire of about 30 T2R proteins, such that the cells are "broadly tuned." But only to 30 separate molecules? This is not the case, owing to the capacity of the bitter taste receptors to respond to multiple compounds, which is assumed to account for the ability of a limited set of bitter receptors to respond to thousands of different chemical species. Depending on the required end-point readout, high specificity is thus not always the best molecular solution. For bitter taste perception, fine specificity toward a wide variety of potentially toxic compounds is not required and might even be counterproductive. This consideration also provides another answer to the previously posed question regarding whether a need exists for generation of somatic diversity toward chemosensory receptor systems.

One implication of the "labeled line" model is that the specificity of molecular recognition of taste receptors does not determine the nature of the perceived taste; it is the specific taste cell neural linkage that provides the appropriately coded downstream signal in response to receptor engagement and G-protein activation. In turn, it would be predicted that installing a taste receptor into a heterologous taste cell would couple the foreign taste receptor's specificity with the new taste cell signaling code. That is to say, putting a sweet taste receptor into a bitter-taste cell should enable the latter cell to provide a "bitter signal" in response to a sweet tastant. And vice versa for a bitter taste receptor in sweet-taste cells. (For Fig. 3.7, this would be as if the cells

---

[*]GPCRs also represent the largest single category of drug targets, and continue to be intensively studied in pharmacology (Chapter 8).

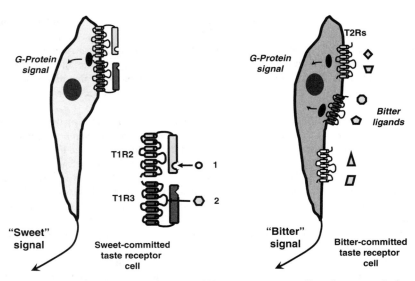

FIGURE 3.7   Schematic depictions of sweet and bitter taste receptor cells and receptors. Left panel: Sweet taste cell and receptors (T1R2/T1R3 heterodimers of 7-transmembrane segment G-protein-coupled receptors). Ligand 1: sweet tastant N-terminal site; ligand 2: sweet tastant extracellular transmembrane site. Right panel: bitter taste cell and receptors. T2Rs may also function as dimers; ~30 different T2Rs are believed to be expressed per cell.

in the left and right panels were kept constant but their receptors and ligands swapped over. The output signal would be predicted to remain the same.) Remarkably, this indeed appears to be the case. Gene-targeted mice[*] expressing specific bitter taste receptor genes (T2Rs) in sweet taste cells show strong attraction to the normally repulsive tastants that the specific T2R receptors recognize.[306] Even a taste receptor engineered to recognize an unnatural (normally tasteless) chemical ligand confers sweet or bitter taste signaling when expressed in corresponding cells in gene-targeted mice.[306]

Since the taste receptor system does not use somatic diversification, it might be predicted that "holes" in the repertoire exist. We have seen that the bitter taste system appears to rely on about 30 multispecific receptors to signal a wide variety of compounds, but the fraction of chemical space this could cover is undefined. Certainly, there are well-known genetic determinants of taste perception, such as allelic variation of humans as tasters or nontasters for the compound phenylthiocarbamide. Also, as poisoners throughout history could attest, many highly toxic elements and compounds fail to register as tastes for human perception. An example of a tasteless organic poison of economic importance is the complex ciguatera toxin (ciguatoxin), synthesized in marine dinoflagellate single-cell protist organisms, and passed up through the food chain into tropical fish commonly consumed by humans.[307] It must be presumed that holes in aversive-chemosensing most likely reflect lack of evolutionary selective pressures until relatively recent times (exposure of humans to ciguatoxin, for example, is probably a latecomer in evolutionary terms). With recent advances as described in this section, mice in principle could be engineered to taste many such otherwise-tasteless substances, but

---

[*]Gene targeting by homologous recombination has been referred to earlier in this chapter and Chapter 2.

this service is unlikely to be offered for human application any time soon. On that note, let us now complete our short survey of chemosensing by moving on to olfaction.

Smelling a Rat and Many Other Things   As for the sense of taste, there are clear fitness benefits to an organism proficient in odorant detection. The "bad" and "good" classification likewise applies to the latter as well as taste, where animal behavior can be modified from appropriate smell signals signifying either beneficial or potentially harmful environmental cues. Although the sense of smell can deliver much information beyond this simple dichotomy, odorant chemosensing is not a strong point for humans, in comparison with many other mammals. We do not even have more than a vestigial representation of what is in effect a whole accessory chemosensory apparatus, the vomeronasal organ. This is located at the bottom of the nasal cavity and well represented in mice and many other mammals, but almost absent in humans and other primates. Vomeronasal chemosensory receptors are concerned with detection of pheromones and constitute a distinct class from the odorant receptors of the main nasal olfactory epithelium (although both are subtypes of G-protein-coupled receptors[308]).

For the purposes of this brief overview, though, we can focus on the main olfactory receptors. Expressed in olfactory neurons, these are one of the largest single gene families in the mammalian genome. Mice have >900 separate functional olfactory receptors and humans have >300 functional equivalents,[309] although almost 300 nonfunctional receptor pseudogenes have been identified in each case. A very important feature of the regulation of olfactory signaling is tightly controlled allelic exclusion, such that only one olfactory receptor is expressed per single olfactory neuron.[309] It has also been convincingly demonstrated that (quite unlike immunoglobulins and T cell receptors), this allelic exclusion effect is not accompanied by somatic DNA rearrangements.[*] Olfactory receptors appear to recognize target odorant molecules by means of the receptor transmembrane regions, with broad but selective binding for sets of odorants, and a single odorant in turn can usually be recognized by more than one such receptor.[312] Binding of an odorant to defined subsets of these receptors then establishes a combinatorial code for the overall olfactory signal.[312] Interaction of an odorant with a olfactory receptor can also result in blocking of the receptor signal. A single odorant molecule can therefore potentially act as either a trigger or an antagonist toward different receptors,[313] which introduces another layer of complexity into the combined encoding of signaling. By these arrangements, the diversity of odorants that can be distinguished by means of differing patterns of receptor signaling or blocking is vastly higher than the net number of receptors alone.

In common with the immune system, chemosensing is confronted with the problem of recognizing environmental molecular diversity. As we have seen, the respective solutions taken have been quite different, and reflect the divergent biological requirements and constraints imposed in each case. In its simplest terms, chemosensing is expressable as molecular recognition → signaling → perception, but for the immune system it is molecular recognition → signaling → activation, with potentially dangerous consequences if not tightly controlled and limited. The adaptive immune system must carefully distinguish self from nonself, and must neutralize or destroy dangerous foreign invaders, as noted earlier. Fully adaptive immunity necessarily involves somatic changes to produce the required receptor novelty. Chemosensory recognition, on the other hand, can attain high levels of diversity at the level of combinatorial

---

[*]This has been elegantly done by cloning mice from olfactory sensory neurons. If an irreversible somatic change is accounted for olfactory allelic exclusion, then such mice would have a monospecific olfactory repertoire, but this is not the case.[310] In contrast, mice cloned from B or T cells show immunoglobulin or T cell receptor somatic rearrangements in all tissues.[311]

signal encoding for higher level perception. This requires allelic exclusion of receptor expression, achievable through genetic control rather than via somatic rearrangements. Moreover, apart from kin/non-kin recognition associated primarily with the vomeronasal system, chemoperception need not have the same preoccupation with selfish matters, which is so central to the immune system's functioning. With the signal-encoding solution for perceiving molecular diversity and no requirement for a highly complex system for distinguishing self from nonself, biologically adept chemosensory systems are achievable with innately encoded sets of genes.

Multirecognition

An interesting aspect of chemosensing is the linkage between a ligand–receptor molecular interaction and conscious perception. Obviously, a vast number of other molecular interactions are likely to influence consciousness, but with the exception of the signaling of pain sensations by certain chemicals, these are not directly reported to higher level awareness. Other senses involve signaling induced by physical inputs (photons, pressure, etc.) rather than specific molecules. If compound A is sweet, it necessarily follows that sweet taste receptors are engaged with Compound A and signaling has occurred. (The tastelessness of compound B might in principle be more complicated than simply through a failure to bind any taste receptors, but it would be a primary working hypothesis.) The subjective nature of sweetness has meant that the discovery of artificial sweeteners has historically been a serendipitous process, and this applies to other applications of chemosensing. Certainly empirical screens for chemosensory properties of molecules can be conducted. Yet unlike an *in vitro* empirical screen of a chemical library for a objectively assayable effect, comparable screening of compounds for taste or odorant properties has had to rely on the direct sensory effects of receptor engagement. With a wide variety of chemical structures inducing the range of chemosensory stimuli, deciphering a chemical/sensory "code" has been difficult, although progress has been made for odorant classification.[314]

But this aspect of chemosensing might prompt another comparison with the immune system. Just as one can legitimately ask why chemosensing does not use somatic receptor diversification, one can turn this comparative mirror back on the immune system and ponder why immune receptor engagement does not produce a sensory signal. (Clearly we do not receive a direct sensory input from B cell or T cell receptor engagement, even after profound cellular amplification.) This question could also arise from known links and analogies between the immune and nervous systems,[315] and the above-mentioned portrayal of the former as another form of "physiological environmental sensing." Would it be an advantage to "taste" (or directly experience in some unfathomable way) the effects of immune-receptor binding of antigen? Second-guessing the selective forces molding natural evolution can be difficult, but we can assume this kind of direct communication between the immune and nervous systems would have no survival benefit.

But there is another reason for highlighting these distinctions, beyond mere speculation. Molecular recognition has been an undercurrent throughout this chapter, and indeed continues throughout this whole volume. But the profound differences between the signaling events that occur in the chemosensory and immune systems remind us forcefully that in a complex biosystem, a primary recognition event is only the first step. A single molecule can in fact interact with multiple natural recognition systems, through different receptors and with very different consequences.

Let us consider a specific example to illustrate this. The simple compound coumarin (Fig. 3.8), found in many plant sources, has long been noted for its chemosensory properties. It has a specific vanilla-like taste, and a "cut-grass" or almond-like aroma. Its binding and

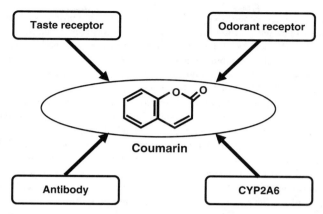

FIGURE 3.8   Recognition of the coumarin molecule by four different natural molecular frameworks. CYP2A6 denotes cytochrome P450 2A6 isoform, which catalytically hydroxylates coumarin.

triggering of both gustatory and odorant G-protein-coupled receptors can be inferred from these effects. Coumarin is not suitable for direct use as a flavoring agent due to its toxicity, and this raises another recognition process. The "xenobiotic immune system" processes many foreign chemicals, and coumarin is exclusively recognized as a substrate by the cytochrome P450 isoform 2A6. This enzyme converts coumarin into its 7-hydroxy derivative.[316] Finally, antibodies can be raised, which bind the coumarin molecule when it is presented to the immune system as a hapten conjugate.[317] The latter is dependent on an artificial manipulation, but the process that enables the recognition molecule to emerge still derives from the natural immune system.

So, even in natural circumstances, a variety of protein scaffolds can recognize the same small molecule target. Human ingenuity can potentially extend this indefinitely, assuming that within our imaginary OMspace a vast number of alternative binding solutions exist for any small molecule, including coumarin. This need not necessarily involve proteins at all, as we will see with respect to nucleic acid aptamers in Chapter 6. A "hole" in one repertoire (whether through an absolute structural dictate or not) may be filled in another. For example, earlier we noted the tasteless poison ciguatoxin. Even if it proved an intractable problem to engineer taste receptors to accept this molecule, specific antibodies have already been generated against it.[*]

These thoughts lead us to a general consideration of alternative natural solutions for different molecular recognition problems, with which we will conclude this chapter.

MESSAGES IN BIOLOGICAL MOLECULAR RECOGNITION AND EMPIRICAL DESIGN

In thinking about immunoglobulin, MHC, and T cell receptor compartments of the immune system, as well as neural diversity and chemorecognition, we have covered a lot of ground in the

---

[*]More specifically, antibodies have been derived against a part of the relatively large ciguatoxin molecule, which enables the whole molecule to be recognized. Even so, to accommodate this ciguatoxin moiety, an unusually deep binding pocket was selected in the antibody, as revealed by structural studies.[318]

field of natural molecular recognition. Although each of these are products of natural selection operating over vast amounts of time, we have at several points considered them as solutions to molecular design problems. As we have seen in our comparison of the adaptive immune and chemosensing systems, different functional requirements will dictate often divergent solutions to such challenges. Table 3.1 accordingly provides a list of the major types of molecular recognition phenomena described in this chapter, as though they presented initially as molecular design problems in search of a solution. These outcomes have been hewn and honed by uncompromising evolutionary forces, and their long-term persistence is potent evidence that they "work." Yet artificial design of multicomponent molecular systems in general will become more and more important, and from this point of view alone, natural exemplars of molecular design are well worth understanding, even if human ingenuity can eventually improve upon their design features wrought through evolutionary pressures. (Of course, the evolutionary "design features" of natural systems will not necessarily match human needs in the first place, but can serve as useful starting points.)

TABLE 3.1   Natural Solutions to Molecular Design Problems Involving Recognition of Wide Molecular Diversities

"Molecular Problem"	Natural Solution Type Example	Cell Expression	Specificity of Single Receptor
Bind a vast array of ligands unspecified in advance	Somatic variable diversity *Immunoglobulins*	One functional immunoglobulin molecule per B cell	High toward defined target (cross-reactions known)
Bind a vast array of ligands unspecified in advance but always in a specific presentation context	Constrained somatic variable diversity *T Cell Receptors*	One functional T cell receptor molecule/T cell[a]	High toward definedtarget (cross-reactions known)
Bind large numbers of diverse peptides with high affinity with a limited number of receptors	Constrained ligand binding; somatic constant genes; high polymorphism *MHC*	Six alleles (usually different through polymorphisms) expressed per cell	Binds a large family of peptides defined by anchor residues
Bind a variety of molecules, channel signals into a limited set for perception	Somatic constant genes *Taste Receptors*	Single or Multiple receptors per cell committed for preset signal type	Multispecific/multiple binding mode/cell-specific readout
Bind a vast array of chemical species in complex mixtures for diverse perception	Somatic constant multigene families *Olfactory Receptors*	One functional receptor per olfactory sensory neuron	Cross-reactive specificities, complex signaling

[a] Exceptions to the one receptor/cell rule are known (as noted earlier).

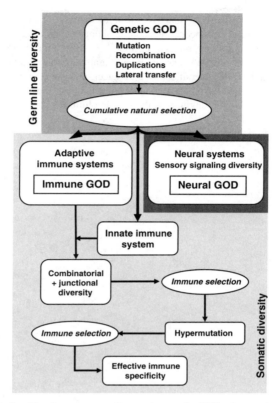

FIGURE 3.9   Relationships between germline and somatic GODs (generation of diversities) in vertebrate phylogeny and ontogeny.

Apart from this, it is worthwhile reiterating a "software" message principally from the study of the adaptive immune system and also from the study of evolution itself, as we noted in the previous chapter. This is the power of reiterated empirical selection from a large set of variants, or "library," toward achieving a desired molecular end. And in order to have a library of variants, some form of diversification must exist. In Fig. 3.9, the different levels of selection are shown leading from evolution of adaptive immune and neural systems, and their application of specific somatic diversification mechanisms. As we have previously noted, a major difference between the two levels is the time scale. Evolution can act over gigayears and can thus rely on slow natural diversification as the raw material for selection, while somatic diversification within an individual organism must by its nature operate within the lifespan of the organism and in response to environmental dictates.

We can leave these considerations about diversification and selection with a question, which we will address in succeeding chapters: How can we best implement *in vitro* processes for molecular discovery that have parallels with natural processes? Biological GOD has been one of the major themes of this chapter at a number of levels, but a major aim of this book as a whole is how molecular diversity can be artificially generated and exploited for obtaining useful products. By definition, "usefulness" is only meaningful for a human observer, and the concept of molecular utility is likewise a product of the human mind. Natural products conferred through biological evolution have been a rich source of empirically obtained useful molecules for diverse applications, but for truly novel molecular discovery, human innovation is called for. Such creativity certainly includes the harnessing and manipulation of analogous processes to natural evolution and somatic diversification and selection, whose power we have paid our respects to. This thought links us to the remainder of this book, and the appropriate place to begin is the vast field of human-engineered directed evolution.

4 Evolution While You Wait

OVERVIEW

A person contemplating and appreciating the dazzling array of molecular design solutions that have arisen by natural selection might dream of applying the same principles, even if he or she had no prior knowledge of current basic and biotechnological research. In fact, this kind of "applied evolution" has been intensively studied at a variety of levels, and this chapter is devoted to this general theme. The optimal outcome of an artificial evolutionary experiment is by definition not predicted at the outset, or it would fall into the ambit of "rational" design. Evolution in the laboratory is very often given the label "directed," but (as noted previously) it is an empirical process nonetheless. Still, while the algorithm for directed laboratory evolution has a clear-cut parallel with natural selection, it is not identical. As already pointed out in Chapter 2, human direction can dictate the precise nature of the starting target molecule(s) or systems and the selection pressure consistent with the desired type of modification or improvement. The ways in which these experimental parameters can differ from their natural equivalents will be spelt out in more detail below.

Both the underlying principles of directed evolution and the practical technologies that enable its implementation should be considered, to gain an appreciation of its elegance and power. Though much of the subject matter within this chapter accordingly has a technological focus, it is not possible to cover all variants of these experimental strategies. (Indeed, whole volumes have been devoted to this task, and without doubt this need will grow in the future.) Yet not all methods are created equal. Some relatively new strategies that are applicable to directed evolution are sufficiently innovative that they usher in new directions of pure and applied research. As such, these approaches may be termed "metatechniques," and they share the keynote feature of spawning a large range of additional conceptually related methods. Technologies for *in vitro* evolution (and molecular discovery in general) themselves evolve at a fast pace, and it is often the case that if the origins of multiple experimental procedures are traced back, they converge on a single seminal concept. We will accordingly give more attention to the principles underlying such metatechniques in this field than the full range of experimental approaches that they have inspired. Two examples of metatechniques included within this chapter are DNA shuffling and display technology. A third such technique, SELEX, is placed within Chapter 6 owing to its direct applicability to RNA, although its principle is generalizable.

PUTTING EVOLUTION TO WORK

Humans have used artificial selection to accentuate desirable characteristics in domestic animals and plants for thousands of years, often gradually introducing startling changes

Searching for Molecular Solutions: Empirical Discovery and Its Future. By Ian S. Dunn
Copyright © 2010 John Wiley & Sons, Inc.

relative to the wild-type parental species. Classic examples are the variety of canine forms arising from wolf ancestors, or the development of high-yielding wheat from wild cereals. Although seemingly performed with rational intent when viewed from a long perspective, the implementation of this artificial evolution was in reality a rule of thumb, haphazard affair. Deliberate attempts to recreate comparable changes in wild mammals using modern genetic rationales have indicated that significant behavioral and morphological changes can be developed surprisingly quickly.[319] Even so, the entire history of the "selective breeding" of domestic organisms by humans has been but a blink of the eye by the standards of natural evolution. If we could act as evolutionary guides (using only judicious selective breeding) for far longer periods of time, it is inevitable that radical changes in animal and plant structures and forms could be engendered. The historical track record nonetheless indicates that whatever shortcomings existed in its implementation, domestic animal and plant selective breeding has been very useful for many human societies. Yet as you would probably expect, this is not the level of directed evolution with which this chapter is preoccupied. Although artificial selection applied to a whole organism can certainly work, on timescales of interest to humans it has significant limitations stemming from the complexity of the "target" and how the selective pressure can be applied.

Consider the selective breeding of cows. Screening based on milk yields would seem straightforward enough, and this in fact has resulted in greatly increased milk production over the natural norm. This "low-tech" approach could not be directly applied if one wished to increase the protein content of milk, since a protein assay would be necessary to enforce appropriate selective breeding based on screening for milk protein content. Where such tests are available, though, a "protein-enrichment" breeding program is feasible if sufficient time is devoted to the task. (Protein enrichment might occur as a beneficial side effect of selection for some other phenotype, but this is distinguishable from a direct attempt to select on this basis.) In principle, one could propose to move to a different level and improve an enzyme in milk by assaying for a particular feature (such as its substrate range or catalytic rate) and selectively breeding cows based on this feature. This is conceivable theoretically, but would not have a high probability of success in practice. Severe limitations are likely to exist in the available natural diversity of such enzymes, which might have low levels of genetic polymorphism and low mutation rates.[*] It is obvious also that big organisms with slow generation times present logistic and cost problems for the evaluation and processing of large numbers of individuals for novel phenotypes. For much of history, selective breeding, whatever its level of sophistication, was restricted by these fundamental constraints.

Levels of Selectable Replicators

So on a human timescale, direction of the evolution of large organisms comes up against practical limits, even if it is (through stringent and consistent selection criteria) operating much faster than under natural circumstances. Whatever the validity of arguments concerning the levels of selection for natural evolution (Chapter 2), for its artificial counterpart, selection is necessarily at the level of individual replicators. From bovines to bacteria, artificial selective pressures can pick out desirable traits and allow the reproductive amplification of individuals possessing the appropriate phenotypes. But most crucially, single-celled organisms offer a very different picture in terms of generation times to cows and even mice, and accordingly are a vastly more facile system for evolutionary studies.

---

[*]This can be contrasted with the somatic generation of diversity that leads to the selection and production of antibodies (Chapter 3). Most proteins, of course, are not subject to such somatic alterations.

A mammalian organism consists of many different cells of the same genotype, but with distinct differentiation expression patterns (cellular phenotypes). It is often possible to isolate specific differentiated cells from an organism expressing a protein of interest and grow them in culture *in vitro*.* By doing so, the multicellular mammalian replicator is in effect reduced to a single cell replication system that contains the proteins of interest, and this can in principle be used as the vehicle for directed evolution. The advent of recombinant DNA furthermore meant that a wide variety of proteins from mammalian organisms could be expressed in xenogeneic (foreign) hosts, enabling in effect a transfer of the protein of interest from a slow and cumbersome replication system to a fast and efficient one.

Biological replication systems used for directed evolution share the fundamental feature of allowing the amplification of the expressable (and screenable or selectable) phenotype, and bear the information enabling the identification of the relevant evolutionary changes associated with useful phenotypic alterations. At the molecular level, the earliest *in vitro* evolutionary study in fact used RNA from a bacteriophage as an *in vitro* replication system (in the 1960s) and we will return to this when we focus on nucleic acid molecular evolution in Chapter 6. We will see later in this chapter that the genotype (information)—phenotype linkage can be usefully exploited in various display technologies. On the other hand, there is no reason why a library of mutant enzymes (for example) cannot be evaluated by systematically screening organisms expressing individual molecular clones of such variants (problematic in cows, feasible in bacteria or yeast). The nature of what is necessarily such *high-throughput screening* was introduced in Chapter 1 and will be extended in Chapter 8.

Directed evolutionary experiments with fungi and bacteria have been conducted for decades, exploiting natural diversity arising within very large populations of cells. Also, by their natures it is not difficult to treat microorganisms with chemical mutagens to globally increase their mutation rates, often an advantage. A long-standing industrial aim has been the betterment of fungal strains producing useful compounds (especially antibiotics). Directed evolutionary approaches in such organisms prior to the advent of recombinant DNA technology constitute one area in the general field of "classical strain improvement."[320] But it is really the ability to manipulate and clone DNA segments that has empowered the full development of artificial evolution, whose range and significance we should now consider.

The Scope and Implementation of Evolution as a Workhorse

Since we want to look at the theme of artificial directed evolution in a broad sense, it is important to define the scope of what is to be covered. A purist definition of an "evolutionary" study in the laboratory will insist that the only processes that qualify for this label must apply *repeated* rounds of artificial diversification followed by selection or screening, resulting in *cumulative* desirable molecular changes through successive generations. After each round, a candidate is chosen for subsequent additional diversification, and this process is then compatible with the evolutionary algorithm (as shown in Fig. 2.1). By such a strict interpretation, screening a *single* random mutant molecular library is not directed evolution even if a beneficial mutation is identified, and the cyclical aspect of applied evolution is often emphasized.[321] While acknowledging this important distinction, for the present purposes it will be simplest to include, within the same chapter devoted to "directed evolution," any artificial molecular discovery process that uses a directed means for diversification combined with a selection or screening

---

*Normal cells have limitations on how many division cycles they can undergo in culture, but if they are "transformed" (as for a tumor cell), growth can continue indefinitely.

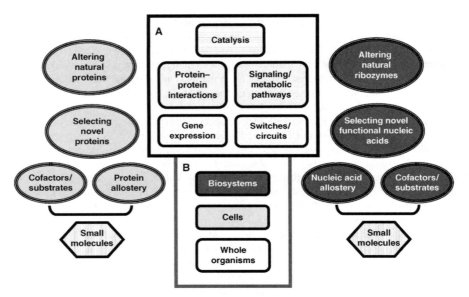

FIGURE 4.1 Scope of directed evolution in a wide-ranging sense to include artificial evolutionary processes employing protein, nucleic acid, and small-molecule interactions. In the central box (A), the items indicate some of the most important general areas where evolutionary approaches can be applied, and box (B) shows the levels in which the evolved molecules or systems can be screened or selected. Analogous operations on proteins and nucleic acids and their interactions with small molecules are shown on the left and right, respectively The various categories noted in this figure are referred to at the appropriate points throughout this chapter and Chapters 6–8.

process. Selection or screening from a single library of molecular diversity can be considered as a "snapshot" of an ongoing evolutionary process, and any such study has the potential for evolutionary reiteration, even if it has not yet been applied. Some processes involving diverse collections of biological molecules (such as selection from expressed peptide libraries) are often exploited and duly reported without being considered as "directed evolution" as such. Nevertheless, they use technologies that can indeed be applied toward directed evolution, and for that reason are considered within the confines of this chapter. In practice, almost all cases of significant molecular improvement that use artificial diversification and screening or selection require reiteration to fine-tune the beneficial mutations. As shown in Fig. 4.1, directed evolution itself can take place at a variety of different levels with different biopolymers, and these can interact with each other in complex systems.

Many studies of the nature and mechanisms of evolution have also been performed in the laboratory with model organisms (usually viruses or bacteria), as briefly noted in Chapter 2 ("Complexity and Molecular Discovery" section). Whatever creationists might want to believe, the study of evolution is a thriving experimental science, and has a very respectable history. Yet (despite the pioneering work noted earlier) evolutionary experimentation specifically aimed at producing improved proteins or nucleic acids has really only taken off since the 1990s. One way to look at this is to check the publication rate. A search for the terms "directed evolution" or "*in vitro* evolution" in PubMed since the 1980s reveals a plot as shown in Fig. 4.2. Now, of course it is true that restricting a search to these specific word usages is not comprehensive, but it reveals a meaningful trend. What then is the major cause of the marked growth in directed

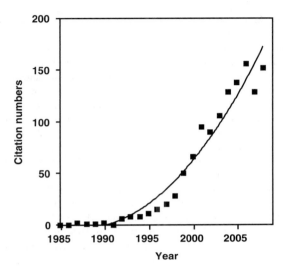

FIGURE 4.2   Publications in PubMed found with "directed evolution" or "*in vitro* evolution" search 1985–2008. Each point represents the citation numbers for these combined searches for each year. Curve fitted as a second-order polynomial.

evolution work over the course of a decade? Like any form of human endeavor, when new ideas emerge, scientific research can sometimes be directed by waves of enthusiasm that might cynically be called fads, but this we can confidently reject in this context. A far more convincing case can be made for the application of increasingly sophisticated tools for both generating molecular diversity and screening or selection of desired products, which in combination have both increased the power of directed evolutionary strategies and significantly improved their prospects of success. These new technologies also put the last decade of directed evolution into a different league to the "classical" era.

We can repeatedly hark back to earlier chapters for comparisons of directed artificial evolution with natural evolution and the somatic Darwinian basis of the adaptive immune system. Any evolutionary process requires a pool of diversity upon which selection can act, and as for the discussion of natural selection in Chapter 2, we will begin analyzing artificial evolution by looking at strategies for the generation of diversity.

GOD in the Laboratory

In our earlier overview of natural evolution, the drivers of diversification could be broken down into the two major categories of mutation and recombination. And so it remains for evolution directed by humans. Both can be further subdivided into processes that target the entire coding sequence of interest, or those which focus on discrete regions. The central issues are then practical and technological: how are the mutational and/or recombinational changes elicited, what is an optimal mutational level (or "load"), and how "random" are the introduced mutations in actuality? The latter question is very significant with respect to the sizes of libraries of diversified molecules that are required to have a reasonable prospect of finding useful variants, and we will return to it later in this chapter. But first, we should examine the immediate practicalities of making mutations.

Mutations at Will   As we have seen, early experiments in directed evolution could simply make use of natural mutations arising from the fact that all DNA polymerases have a definable error rate. Previously, we considered the proofreading functions of many (although not all) polymerases (Chapter 2), which reduce nucleotide misincorporations to a low but measurable level. Although useful results can be obtained in certain cases if sufficiently large populations can be screened, the very fidelity of normal DNA replication acts as a limiting barrier to directed evolution, at least on a timescale that is convenient. Replicative fidelity and the overall level of accuracy in genomic replication in fact depends on three parameters: the inherent accuracy of nucleotide incorporation by DNA polymerases, the action of proofreading (if present), and postreplication mismatch repair.[322] Measured as mutations per base pair per generation, such fidelity in *Escherichia coli* is approximately 10^{-7}, and $\leq 10^{-10}$ in most eukaryotic cells.[322]

So it is highly desirable in most cases to adjust the natural mutation rate upward for artificial evolutionary purposes. We will say more about what constitutes an "optimal" mutation level in a short while, but first let us consider how the mutation rate can be modulated in target DNAs encoding proteins to be subjected to directed evolutionary change. The oldest method makes use of chemical or physical mutagens, which originated in Muller's 1927 discovery of mutagenesis by ionizing radiation (X-rays).[*] Practically by definition, such mutagenic influences damage or modify DNA, although not necessarily completely indiscriminately (depending on the type of mutagen, there may be biases toward specific base substitutions). The efficiency of mutagens in changing a sequence (whether as point mutations or insertion/deletions ["indels"]) depends in turn on DNA repair pathways, and mutations will be stably fixed in place after replication of the mutagenized DNA target. Since processing of replication errors and DNA repair is a crucial factor in determining the net replicative fidelity, it stands to reason that organisms with lesions in such repair genes should have accelerated mutation rates. This has long been known in *E. coli*, and a variety of "mutator" genes are known encoding replicative or repair functions.[323] When the products of these genes are absent, increased mutation rates result, usually with a characteristic specificity. Judicious exploitation of mutator bacterial strains is thus a viable strategy toward acquiring libraries of mutant genes with directed evolutionary intentions in mind.[324]

While simple to implement, mutagenesis approaches with chemical or physical treatments or the use of mutator strains both suffer from the "global" nature of the induced mutations. That is to say, sequences outside of the region(s) of interest will also be targeted, with potentially deleterious effects if the growth of the host organism (as a replicative vehicle) is itself compromised. Some steps can be taken to reduce this problem, such as using mutator gene combinations that preferentially favor mutations in plasmids (bearing the genes of interest) rather than the entire host genome.[325] Nevertheless, other mutational techniques are available which by their natures are restricted to the sequences of interest.

It would seem reasonably evident that targeting mutations with finesse upon a specific sequence segment is superior to a global battering-ram philosophy of mutagenesis. But this in turn is a subset of the larger question of how mutations should be optimally distributed throughout the region of interest. As we will see, a decision regarding the general targeting of a region versus more selective mutagenic focusing will greatly influence which technologies should be chosen from the available tool kit. A popular method for "sprinkling" a whole defined region with mutations is error-prone PCR,[†] where a polymerase is required that does *not* have a

---

[*]H. J. Muller (1890–1967) has also been noted in Chapter 2 in the context of "Muller's ratchet." He was awarded Nobel Prize in 1946 for his mutagenesis discoveries.

[†]The polymerase chain reaction (PCR) is based on repeated cycles of primer annealing to a target template, extension by a thermostable DNA polymerase to form duplexes, and denaturation. Denaturation allows the primers access to single-stranded complementary sequences in their target strands and thus a repeat of the annealing–extension process (further details in ftp site). The most commonly used polymerase is Taq, from the thermophilic organism *Thermus aquaticus*.

high level of fidelity. Since Taq polymerase is widely used for PCR, lacks a $3'$ to $5'$ exonucleolytic proofreading function and has a significant error rate, it might be thought that Taq in itself would fit the bill here. In fact, under normal conditions Taq still has insufficient infidelity for most evolutionary mutagenic requirements, but this can be overcome by tinkering with the reaction conditions. Alternatively, nucleotide analogs can be included to tune mutagenic misincorporation to desirable levels.[326] Engineering of other thermostable polymerases is another route toward obtaining an ideal mutagenizing PCR system.[327] Most of the sequence changes elicited by error-prone PCR are point mutations, which limits the range of diversification that can be achieved from a parental coding sequence. This follows inherently from the nature of the triplet and degenerate genetic code, where some codons will have a higher probability of formation from single-base changes than others.[326] With this in mind, some innovative amplification methods have been devised that result in random insertions and deletions,[328] or frameshifts.[329]

When conducting any program of mutagenesis, it is important to evaluate the presence of systematic biases (the deviation of the mutational distribution from true randomness, especially where there are measurable patterns in the observed nonrandom mutational changes). This must be obviously checked at the level of the types of mutations themselves, but for a technique nominally placing mutations randomly across a desired sequence, the positioning of the mutations is also a valid question. The base distribution of mutations resulting from conventional error-prone PCR is itself nonrandom (for example, A $\rightarrow$ C mutations tend to be under-represented[327]), and although some engineered polymerases reduce this problem, it is not entirely eliminated.[327] This in turn will result in certain codon biases when the mutagenized sequence is expressed. An approach termed sequence saturation mutagenesis has used the incorporation of the base deoxyinosine, which has more relaxed hydrogen-bonding requirements in DNA compared to the normal bases, although still with a preference for cytosine.[330] Although a way around polymerase biases, this method still falls short of an ideal mutagenesis scheme. It has long been known that "hot spots" of mutation can arise from problems encountered by polymerases in reading through sequences with special characteristics, including palindromes* or near-palindromes. Tandem repeat sequences can result in polymerase "slippage" (or "stuttering"), which is one contributing factor in the polymorphism of ubiquitous "microsatellite" repeat sequences in eukaryotic genomes.[331] Different polymerases *in vitro* will probably show differing susceptibilities to such effects, but sequence context must be taken into consideration in the evaluation of the randomness of mutational patterns in artificial libraries.

Often it will be desirable not to mutagenize an entire coding sequence, but to restrict the artificial diversity to discrete regions of the sequence of interest. For example, one might wish to randomize specific residues at (or in the vicinity of) an enzyme's active site, when the enzyme's mechanism is defined. To accommodate such needs, a number of techniques are available, all of which have one thing in common: the application of synthetic oligonucleotides ("oligos"). (The latter are of course also required as primers in any PCR application, error prone or not, but in this context we are referring to more direct roles in the mutagenesis itself.) With oligonucleotide mutagenesis, sequence changes can be introduced as triplet blocks corresponding to codons, and classically these can be inserted into the desired parental sequence on single-stranded templates[332] or via "cassette" mutagenesis into duplexes[333] (Fig. 4.3). Numerous variations on

---

*A palindrome in conventional terms is of course a reversible word or sentence, as with "AND NO GAG ON DNA" (which might be a protest against either DNA-based humor or patent-related restrictions on the use of specific DNA sequences). In molecular biological terms, however, a palindrome refers to a sequence whose complement on the opposite strand of a nucleic acid duplex is the same (that is, when read in the same $5'$ to $3'$ direction). Thus, the classic recognition site of the restriction enzyme *Eco* RI ($5'$ GAATTC $3'$) is a palindrome, but a sequence such as $5'$ GTCCTG $3'$ is not.

(a)

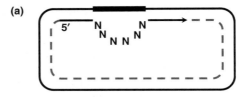

(b)

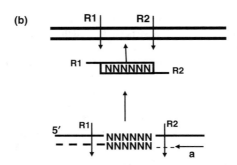

FIGURE 4.3 Depiction of insertion of randomized codons (2 × NNN in this example) into circular single-stranded template (A) or duplexes (B). (a) A single-stranded oligonucleotide with a randomized tract is annealed to single-stranded template containing the gene of interest (usually derived from a single-stranded filamentous phage) and extended with an accurate DNA polymerase. (b) A single-stranded oligo is first extended with an appropriate primer (a) and then inserted into a defined region in the target DNA by means of restriction enzymes recognizing sites R1 and R2.

the theme of random mutagenesis using oligonucleotides have been subsequently reported,[326] and these procedures in general can be applied iteratively. The more information regarding the initial function of the target protein, the more "rational" the choice of sites for such comprehensive mutagenesis can be.[334] The most commonly used chemistry for the synthesis of oligonucleotides makes use of a solid-phase support on which derivatized nucleoside phosphoramidites are coupled successively.[*] There is no inherent requirement to use a monomeric nucleoside phosphoramidite during each addition step, as blocks of trinucleotide phosphoramidites (usefully corresponding to amino acid codons) can also be used.[335]

There is a perceived need to explore alternative means for random mutagenesis, owing to demonstrable deficiencies in "traditional" methods such as error-prone PCR,[336] which largely stem from codon biases as noted previously. But this in turn raises the earlier question of what constitutes an optimal level of mutagenesis throughout a protein coding sequence. Initial studies (with error-prone PCR) found a trend for exponential decay of proteins retaining their original functions as average numbers of mutations per expressed coding sequence increased

---

[*]This kind of chemical synthesis proceeds in a 3′ to 5′ direction, which means that successive nucleoside phosphoramidites are coupled to the 5′ end of the growing oligo chain (hence, the chain extension begins from the 3′ end (the 3′ nucleoside is coupled to the support) and proceeds *toward* the 5′ end). This is the opposite direction of chain extension to all known nucleic acid polymerases, which add bases to the 3′ end and thus extend in a 5′ to 3′ direction. Thus in this case, humans do what is never found in nature (compare with the ruminations at the end of Chapter 2), but it does not follow from this alone that an enzyme capable of 3′ to 5′ polymerization is impossible, only that there is (evidently) no selective reason for its evolution.

from low (<2 mutations/coding sequence) to moderate levels (<10 mutations/coding sequence).[337,338] Selecting or screening for functional changes in a low to moderate mutational background has the theoretical benefit of reducing the probability of irrelevant second-site mutations. Nonsynonymous second-site mutations that are "neutral" for the screened activity might nonetheless have unwanted and unexpected effects in protein stability or immunogenicity.

These factors would seem to indicate that high levels of mutation are inappropriate, and it would seem logical that increasing loss of function should go hand in hand with increasing mutation burden. Yet when high mutational rates have been examined, libraries have yielded unexpectedly elevated levels of both functional and improved protein variants. (That is to say, such levels upwardly deviated from the predicted exponential-loss trend.[337,339]) What would explain this apparent paradox? Initially, it was suggested that high mutation rates allowed a more profitable exploration of sequence space, or that epistatic effects (mutational interactions) became significant beyond a threshold mutation background. A detailed study of this issue[338] noted that it was important to keep in mind the nature of the (low fidelity) PCR amplification and the resulting distribution of mutations that result at high error levels. While many template copies are indeed heavily mutated through successive rounds of error-prone replication, a fraction of the total final population also results with lower levels of mutation compared to the average, and these have a much higher probability of functional retention.[338] This study also stressed the important factor of *effective* library size, which is simply the number of unique mutants screened, rather than the total number *per se*. For libraries with low mutation rates (especially where coding sequences are relatively short), the recurrence of mutants can limit the effective library size. The much higher rate of unique sequences in high-mutation libraries in turn increases the chances of finding improved variants.[338] Finally, this same study considered the issue of optimal random mutation rates, and concluded that no absolute recommendations for directed evolution could be made, since optima (highest fraction unique and functional) were dependent on both library sizes and PCR cycling conditions.[338]

Diversification through mutation is a primary means for enabling *in vitro* directed evolution, and mutational processes are also a fundamental source of the raw material for natural selection (Chapter 2). It is interesting to compare directed evolution using high mutational rates with natural circumstances of transient hypermutation. The significance of natural low-fidelity polymerases in overcoming DNA damage has become increasingly recognized as applicable to all kingdoms of life.[340,341] Also, transient induction of some of these polymerases has the potential to act as a somatic mutagen, of probable significance in the hypermutation of immunoglobulin genes.[340] When control of error-prone polymerases is faulty in multicellular eukaryotes, an increased risk of somatic tumorigenesis may be a consequence.[342] Transient elevated rates of mutagenesis engendered by polymerases with low fidelity also have evolutionary implications, at least in prokaryotic organisms.[341] So the parallels between hypermutation as a tool for directed laboratory evolution and its natural counterparts are clear enough, but what of genetic recombination, the other major driver of diversity as a raw material for natural selection?

Sex for Biotechnologists

The fundamental biological importance of recombination has been highlighted in Chapter 2, along with its significance in sexual processes. Recombinational processes are also key contributors to the diversification mechanisms of the adaptive immune system (V|D|J re-arrangements and gene conversion; Chapter 3). A biotechnological surrogate for recombination emerged in the 1990s, termed *DNA shuffling*. This process is the recombinational accompaniment to mutation in the context of directed evolution, and is important enough to qualify as a "metatechnique" as noted at the beginning of this chapter. The key word in this context is "shuffling" (or "breeding" if you prefer), which in its general sense can be taken to refer to *in vitro* genetic recombination. This technology is sometimes known as "sexual PCR,"[343] which certainly catches the eye, and is quite accurate if we recall that sex at its most fundamental is simply the exchange of genetic information. And finally, we come to the relevance of sex in molecular discovery, as promised in Chapter 2. When used without qualification the term "DNA shuffling" is often intended to refer to the original innovation,[344,345] and thus a quite specific procedure. Yet "shuffling" has also served as a springboard for numerous other techniques for *in vitro* recombination that are inspired by the same underlying principle.

The DNA shuffling (or molecular breeding) technique itself uses a panel of coding sequences derived either from PCR amplification of a specific target (with an associated significant error rate) or from a panel of homologous genes from different sources. Fragmentation of such panels with deoxyribonuclease I (DNase I) is then performed, after which a thermal cycling procedure analogous to PCR but initially without primers is used to allow fragments to prime onto each other by means of their homologous sequences (Fig. 4.4). By such means, mutations can be rapidly "shuffled" together in a variety of combinations, and arrangements that confer superior phenotypes upon expression are then identified by appropriate screening or selection. The shuffling process itself introduces additional diversity at a significant rate in the form of point mutations, owing to the extensions mediated by error-prone Taq polymerase. Following repeated rounds of shuffling, one can go a step further and remove most irrelevant mutations by "backcrossing" a candidate recombinant against an excess of wild-type coding sequence, and repeating the original selection/screening process.[344] In the original report, a dramatic increase in the efficacy of a β-lactamase enzyme resulted from shuffling and selection,[344] and since then numerous other cases testifying to the power of this technology have emerged. A notable feature of shuffling is that the most favorable molecular chimeras are not necessarily the closest in sequence to the best of the contributing parental sequences.[346] In such circumstances, it can be seen that it would be quite difficult to approach the sequence of an optimal shuffled chimera by conventional mutagenesis from the starting point of any one of the parental contributors.[347]

DNA shuffling can be compared with natural occurrences of genetic lateral transfer where a gene received by a recipient organism is homologous to sequences within its own genome. In such circumstances, the novel exogenous sequence can participate in homologous recombination with its host cognates, and act as an evolutionary diversifier. Artificial shuffling can greatly accelerate this kind of process. While rare natural lateral transfer events will usually involve only one set of incoming homologous genes, the experimentalist can introduce as much sequence diversity from multiple sets of related gene family members as desired (Fig. 4.4). Simultaneous shuffling of an extended set of homologous genes in a family allows efficient sampling of genetic combinations already favored by natural selection.[348] Although recombination is a fundamental process in all living organisms (Chapter 2), and DNA shuffling has been billed as the "natural approach" to protein design,[349] "lateral transfer" by DNA shuffling can truly constitute an acceleration of events likely to be encountered in nature under normal circumstances. At the same time, it should be noted that this does not necessarily imply "the more the better" in terms of the number of contributing parental gene sequences that are used at the starting point. This involves trade-offs between aiming for maximal recombinational diversity and the ability to effectively screen the recombination library, which we will consider in a little more detail later on.

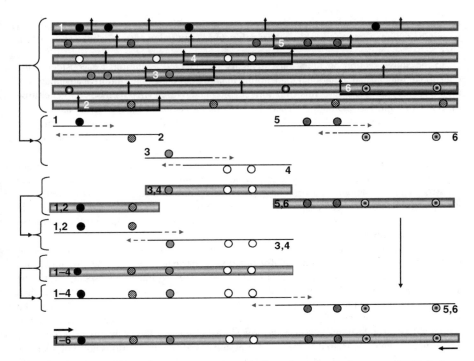

FIGURE 4.4   Basic DNA shuffling procedure. A panel of homologous coding sequences (DNA duplexes shown as solid bars, with sequence differences shown by dots of varying shading) is fragmented by nuclease digestion (short vertical arrows). Denaturation and strand extension by mutual homologies is then instituted. Horizontal lines correspond to single-stranded sequences from indicated fragments. (For example, single strands from fragments 1 and 2 prime on each other and extend to form a fused duplex shown as 1,2.) An example (of many possible outcomes) is shown for recombination between a set of fragments (boxed) from the original coding sequences. (Polymerase extensions from fragment 3′ ends depicted with dashed arrows.) After a suitable number of cycles of this process, a full-length sequence is reconstituted with effective recombination between the diverse sequence point mutations within the initial segments. The polymerase-mediated recombination process itself contributes a background mutation rate giving further diversification. After initial (primer free) cycling, the product can be amplified with primers defining the full-length sequence (solid arrows), and fed back into the shuffling process for further rounds of diversification.[11,12]

Since the inception of *in vitro* recombination by the shuffling strategy, a number of alternative strategies have been reported, some of which have been specifically designed to overcome limitations in the original shuffling protocol. In one of the original DNA shuffling reports, it was noted that suitable oligonucleotides could be included in the *in vitro* shuffling reaction and incorporated into chimeric products by virtue of short segments of homology at their flanking sequences.[345] More recently, a number of oligo-based shuffling techniques have appeared that offer significant advantages. Variations on the theme of "synthetic shuffling," these approaches can in effect direct recombination at will to any desired region, and all shuffled variants should theoretically have an equal probability of formation irrespective of the linkage between sites of interest.[350] The more closely that sites are physically linked, the greater the probability that they will cosegregate during a shuffling experiment using random fragmentation (analogously to natural genetic recombination). Oligonucleotide-mediated shuffling offers an escape route from this problem, simply by designing oligos whose overlapping ends fall within the region of interest, enabling the combinatorial separation of tightly linked mutational sites.

In the above section on mutational diversity for directed evolution, we considered the question of the optimal mutational load, noting in particular the counterintuitive advantages of high levels of mutagenesis. Is this situation recapitulated with *in vitro* methods for generating diversity through recombination? There are two major factors with respect to the optimal diversity requirements for DNA shuffling: the source of the starting coding sequence panel, and the nature of the available screening or selection process. With a powerful selection in hand enabling the rapid evaluation of a large number of functional variants, high levels of diversification are generally useful. In such circumstances the additional random mutations occurring through the original shuffling procedure (as noted above) are a welcome source of extra diversity (as with the original report where antibiotic selection was exerted on a β-lactamase gene[344]). Accordingly, if one's experimental design is geared for functional searching in a panel of randomly mutated coding sequences, additional shuffling-mediated diversity is a probable bonus.

But one of the secrets of the success of DNA shuffling is its ability to take families of sequences that have been "prevalidated" for function (either naturally or artificially) and move them around into novel combinations. In such cases, *de novo* mutations arising during the shuffling process are in effect "wild cards" that in a majority of cases may impede rather than enhance function. Again, such wild cards may be useful if a direct selection is available or if a very large number of variants are screenable, but failing these circumstances, the *de novo* mutations will be an undesirable add-on to the procedure. Fortunately, this problem can be relatively easily circumvented as desired by the use of alternative proofreading thermostable polymerases during the primer-extension steps of shuffling protocols.[351] A shuffling technique entirely using PCR has been developed which likewise can adjust the background mutation rate by means of appropriate polymerases.[352] (This "multiplex" procedure uses internal pairs of primers to generate overlapping and cross-priming fragments for formation of molecular chimeras.)

Thus far, the targets for shuffling that we have been referring to are sets of coding sequences with significant homology. Yet if we go back and look again at the precedents set by natural evolutionary change, nonhomologous recombination processes have the ability to generate higher degrees of true novelty (as considered in Chapter 2). Indeed, nonhomologous recombinational hybridization of proteins may be the rate-limiting step in the evolution of novel protein folds.[353] Several procedures have been developed for making chimeric proteins that are completely independent of sequence homologies. A key step these techniques have in common is blunt-end ligation, where any DNA duplex segments can be joined together. Nonhomologous recombination techniques by definition can use unrelated sequences in producing chimeric genes, but they cannot in themselves shuffle sequence blocks in all possible permutations. In our consideration of evolutionary mechanisms in Chapter 2, we briefly touched upon the role of nonhomologous recombination in the context of exon shuffling,[44] and a role for such "block" assortment in evolution has long been proposed.[354] Protein functional domains and exon boundaries are often (although not always) coincident, and recent evidence is consistent with an ancient role for exon shuffling in the formation of the first functional protein modules.[355] It has also been noted that natural recombinational events are inherently more likely to occur outside of exons than within them, since intronic sequences are on average longer, and tend to contain tracts of repetitive DNAs that can promote recombination between (otherwise) nonhomologous partners.[356] The potential utility of artificial exon shuffling for directed protein evolution is inspired by such observations and has been extensively discussed.[356] Oligonucleotides can serve as tools for directing desired recombinational events at exon or domain boundaries.[356]

The shuffling of exons is an application of the principle of design modularity, which is a recurring theme in both natural and artificial evolution of proteins.

But despite these observations and developments, exon–intron boundaries are not a robust predictor of protein functional modules in general. It is perfectly possible to ignore such concerns and shuffle random coding sequences arbitrarily as short duplex segments (reassembled through efficient blunt-end ligation), and screen for function. The problem in such circumstances is that the vast majority of resulting recombinants are nonfunctional or retain only very low residual function.[357] Arbitrary juxtaposition of unrelated protein sequences thus has a high probability of fatally disrupting functional folding. If one could accurately predict protein sites where chimera formation would be tolerated, the efficacy of low or nonhomologous shuffling would be concomitantly boosted. This has indeed been attempted, either by applying structural knowledge of folding motifs in proteins targeted for shuffling ("structure-based combinatorial protein engineering"[358]), or through the application of algorithms for quantitating interactions disrupted (or "clashing") near the site of a hybrid protein junction.[359–361]

A focus on human molecular design solutions is necessarily technological at the core, but the above scientific shuffling applications should make us pause and take the opportunity to examine the interplay between science and technology in this context. Shuffling technology arose as an application of recombinant DNA and PCR, with roots in evolutionary and sexual processes.[343,344] With its maturation as a routine tool for laboratory molecular evolution, shuffling has fed back into other technological applications and more basic scientific studies. Let us then consider an example along these lines. We have previously noted the profound influence of synthetic oligonucleotide chemistries in the development of many facets of molecular biology and biotechnology, with a particular reference to the enabling effects of continual improvements in synthetic cost-effectiveness. An important factor in the uptake of such a technology is the quality of synthesis, which in this case equates to sequence fidelity. For short oligonucleotides, this is rarely if ever a problem, and accuracy is often taken for granted. On the other hand, for projects involving gene assembly with large sets of oligonucleotides, a significant error rate of at least 10^{-3}/base is evident.[362] This effect is made dramatically clear in projects synthesizing whole viral genomes, where a growth selection is necessary to identify functional full-length synthetic products from large inactive backgrounds.[363] With an approach analogous to DNA shuffling and employing *E. coli* DNA repair proteins that bind to mismatches, significant improvements in synthetic sequence accuracy has been reported.[362] Before moving on, we should emphasize that shuffling and any mutagenesis procedure are not at all mutually exclusive. Again in parallel with natural evolution (and its product, the adaptive immune system) mutation and recombination can work cooperatively to achieve high diversification prior to selection or screening. A common strategy has thus been the use of error-prone PCR followed in tandem by standard shuffling to maximize diversification.

The practice and optimization of artificial evolution has commonly involved computation at many levels, such as the above-mentioned programs for calculating the best sites for generating sequence blocks for nonhomologous recombination. But in parallel with this, in the past few decades, programmers have used evolutionary strategies themselves for solving many computational problems. The progenitors of these are genetic algorithms, which were cited as inspirational for the original DNA shuffling process itself.[344,345]

Artificial Evolution Becomes HOT

In Chapter 1, we considered that rational design can make use of nondeterministic computational methods toward its ends, and this strategy has become very important in many areas. It may seem somewhat paradoxical that computer algorithms based on evolutionary methods should underpin a significant portion of current rational design, but this is consistent with the knowledge-based definition of rational molecular discovery as Chapter 1 introduced. Yet the evolutionary inspiration for such algorithms suggests that they should be briefly noted within this chapter where evolution is a keyword. In fact, one can cast a wider net to include computational methods that are not strictly speaking evolutionary, but which remain non-deterministic and useful in design contexts. These collectively can be referred to as "heuristic optimization techniques" (HOT), where the meaning of "heuristic" is closely allied to "empirical," referring to a practical set of rules for reaching the solution to a problem. These algorithms can be more specifically referred to as "adaptive," since they approach an optimum by iteration of cycles of evaluations of best "fitness" in populations of encoded solutions to problems.

Evolutionary computing essentially began with the introduction of genetic algorithms, and these can be considered as a computational application of Darwinian evolution for the optimization of algorithmic efficiency.[13] In common with natural evolution, genetic algorithms are accordingly iterative nondeterministic search processes. The operation of these algorithms requires representation of problem solutions as digital strings ("chromosomes"), with selection for the best in each round determined by a fitness function. Both mutations and recombination among the "chromosome" population can be programmed, such that iterative rounds of selection move toward an optimum. Genetic algorithms are in fact a subset of a larger field of evolutionary algorithms, which includes genetic programming, where the architectures of program populations themselves are subject to variation and selection.[14] In comparing evolutionary algorithms with the real thing, it is notable that natural evolution almost always operates through variation and recombination at the nucleic acid (informational) level, and selection at the protein-based higher level functional phenotype. In contrast to this, with conventional evolutionary algorithms the "genotype" and "phenotype" coincide.[*] But both natural evolution and evolutionary algorithms share the feature that while they produce solutions, such answers can never be considered as global optima, at least in the absence of any further information. Nevertheless, the "good enough" blind evolutionary philosophy of the competitive natural world delivers exquisite design solutions, as we have seen previously. And for many problems, computational evolutionary approaches are likewise creative.

Evolutionary algorithms have found useful applications in diverse areas,[†] and (as noted) can be encompassed within a rational design approach for molecular discovery. But they can also augment the prospects for empirically based methods as well. This can be at the level of chemical or protein library design[15] or by optimization of library synthetic procedures.[365] Again, the distinction between rational and empirical approaches is made at the level of the available information (Chapter 1). This ultimately enables rational design by complete virtual screening on an accurate simulation of the target molecule, which may include the implementation of evolutionary algorithms. In contrast, even though the synthesis and deployment of a specific physical library might rely on optimization by evolutionary algorithms, a target molecule in principle can be empirically screened by the same library in the absence of structural knowledge.

Additional heuristic adaptive techniques have been developed more recently, and a notable case in point is termed particle swarm optimization.[††] In this computational method, problem solutions are represented as a population of particles that are programmed to move within a

---

[*]There is a notable exception to the genotype–phenotype distinction in the real molecular world, and this comes in the form of functionally selectable nucleic acids, as we will see in more detail in Chapter 6.
[†]As examples, this range of utility includes "evolutionary electronics"[14] and the stock market.[364]
[††]"Swarm intelligence" approaches collectively have been inspired by natural examples of mass social behavior in animals such as bees and ants, or flocking birds.

virtual search space. (It is the position of a particle in the search space that defines that particle's solution after each iteration of the program, and the solution value can be identified and ranked by the fitness function.) During the running of the program, each particle is linked with a memory of its best existing solution, and after each step, particles are attracted toward a different search space region based on a weighted combination of their "personal" memories and the "social" memory of the particle population as a whole.[366] Many iterations of this process result in a convergence of particles toward the optimal solution in the search space. The chief advantage of this type of approach is that it has less chance of convergence toward a suboptimal fitness peak than evolutionary algorithms, and can thus theoretically offer a broader sampling of solutions when used for adaptive molecular design purposes.[367]

In some respects, these optimization approaches overlap with the rapidly expanding field of "artificial life" and "digital genetics," where digital "organisms" can be used to study a range of different phenomena. An early application of artificial life (A-life) was in fact the study of emergent swarm behavior, as in models for the "spontaneous" emergence of patterns in bird flight formations.[368] A-life has also been applied toward addressing many other issues relating to self-organizational phenomena of complex systems.[124] Advanced digital organisms can be used for the investigation of a variety of evolutionary questions, owing to the ability to precisely control variation and selection conditions, and the obvious high speed with which experiments can be conducted and analyzed.[213] But for the present, we will return to the "wet" world of real molecules and molecular systems, bearing in mind that the power of computational models and simulations is steadily increasing, and the mutual feedback process between the results of virtual and real experimentation will continue to be highly productive.

EVOLUTION ON THE LABORATORY BENCH

Continuing the parallel between natural and artificial evolutionary processes, we arrive at the critical level of selection, or screening as is often the case for evolution *in vitro* (remember the important distinctions between these terms first pointed out in Chapter 1). Again, to attempt a comprehensive and fully detailed coverage of the range of alternatives available would risk drowning in an avalanche of information. But just as a limited number of examples of laboratory GOD have served to illustrate the main principles involved, we can focus on some representative screening and selection issues to provide an picture of the field as a whole. Many of these will emerge in the next chapter specifically dealing with the directed evolution of proteins.

Screen (Properly) and Ye Shall Find

Our previous consideration of selection versus screening may have given the impression that a selection process will always be the preferred option, should it be available. When a powerful selection can quickly and directly "pick out" desired candidates from a huge background of nonfunctional species, this point would seem self-evident. Yet things are not always as simple as this, as the stringency of screening (the range of conditions under which it is instituted) can be more easily varied than for selection procedures.[369] From first principles, *in vivo* screening is usually easier to institute than selection, since for the former it is only necessary to link the desired functional activity with a read-out signal that can be assayed and quantitated.

Nevertheless, the nature of the screen (or selection) determines the nature of the read-out, and consequently it is very important to design a screen or selection effectively. So important, in fact, that the "first law of directed evolution" states "you get what you screen for."[370] This sounds a little like a variant of the "wizzy-wig" principle (WYSIWYG: What You See (Screen)

Is What You Get), which is applied in many other contexts. Alternatively, we have "DESIGN" principles (Directed Evolution: Screening Intelligently Gains Novelty/Screening Incompetently Gives Nothing). At the heart of the "what you screen is what you get" principle is the meaning of the informational read-out obtained from a screening process. For example, if an artificial fluorogenic enzyme substrate is used for high-throughput screening of a large number of mutant enzyme candidates, then the immediate and direct read-out is activity on that specific, convenient substrate. In principle, activity on other chemically related but ultimately more important substrates could be suboptimal with the candidate enzyme variant obtained by the primary fluorescent screening. Hence, while it may not be feasible to screen directly for activity with the most important substrate of interest, there could be potential trade-offs with the use of more tractable and screenable (but structurally divergent) substrates.

Sometimes, there are practical problems for instituting a screen based directly on the desired improved protein property. For example, a search for an enzyme with enhanced thermophilic properties (improved stability at elevated temperatures with retention of activity) can be difficult to perform directly on a high-throughput basis owing to thermal limitations on most ordinary types of screening equipment.[347] It is important in this context to distinguish an increased temperature optimum for enzyme activity from thermostability itself. Thus, an enzyme might be capable of resisting high temperatures (even boiling), but only function effectively when temperatures are lowered. (A natural example in this regard is the RNA-degrading enzyme ribonuclease A,[*] which resists temperatures of 100°C, but only has significant activity at <70°C). Still, it might seem a reasonable proposition that if an enzyme's thermal stability was enhanced, so too its optimum reaction temperature might increase. In some systems, this is exactly what has been found. Directed evolution of the bacterial proteolytic enzyme subtilisin E and screening for both thermostability and retention of activity at 37°C resulted in concomitant elevation in the ability of the variant enzymes to function at high temperatures.[371] In a likewise fashion, engineering of another bacterial protease toward thermal stability[372] was shown to positively correlate with enzymatic performance at increasing temperatures.[373] But these kinds of stability/activity correlations are certainly not universal, and clear-cut cases where thermostability and catalytic optima are "uncoupled" have been documented.[374,375] Another way of circumventing temperature-related screening difficulties in searching for heat-stable enzymes has been to exploit the relationship between resistance to proteolysis and thermal stability. In other words, this assumes that as the resistance of a protein to proteolysis increases and so too will its resistance to unfolding and thermal denaturation. Nevertheless, exceptions to the linkage between thermostability and protease resistance also exist.[376,377]

As exemplified by indirect approaches to enzyme thermoactivity, screening can thus be instituted by means of a surrogate measurement that approximates the quality that is actually sought. This might seem to be tempting fate, in accordance with the first law of directed evolution, but the real message is to enter into such an arrangement with one's eyes open. (Hence, minimizing the "blind" aspect of evolution in the laboratory.) As in so many other areas of human activity, trade-offs are often an essential and useful aspect of "doing business," as long as it is clearly understood what the terms of the trade-offs are. A surrogate screen for thermoactivity can save much time and trouble, and can often (but not always) bear fruit. The existing successful research track record here is an important element in making a decision to pursue this kind of approach.

---

[*]Owing to the great stability of this enzyme and its near-universal presence in ordinary circumstances, it is important to take extreme caution in removing it from experimental situations involving RNA, and to include potent ribonuclease inhibitors.

The principle of "you get what you screen for" was phrased with enzyme directed evolution specifically in mind, but it is valid to extend it to artificial directed evolution in general. If you are clever enough, and have sufficient molecular and cellular tools already at your command, you may be able to design increasingly intricate screening or selection systems to satisfy increasingly demanding molecular design requirements. The natural example of the vertebrate adaptive immune system provides wonderful examples of powerful and highly sophisticated processes resulting in the Darwinian selection of functional molecules and systems, as described at some length in Chapter 3. Antibodies are selected for affinity and specificity, but the T cell receptor shows the most exquisite balance of "constrained recognition" between MHC molecules and foreign antigen peptide fragments. A mature T cell repertoire is itself generated by coordination of both positive and negative selection pressures that mold it to potentially recognize a highly specific subset of the peptide universe; the nonself component that is bound and presented by self-MHC molecules.

A consideration of immune systems might serve to remind us of another critical issue with respect to directed evolution, and inevitably screening as well: library size. At several points within Chapter 3, it was noted that repertoires were inherently limited at any one time by the number of circulating lymphocytes within a vertebrate organism, which is in effect an immune "library number." With a randomly generated adaptive immune system undergoing constant turnover, the specific receptor representation will change with time but the upper size limit imposed by lymphocyte levels will not. (As we have also seen previously, the full potential GOD of the adaptive immune system therefore cannot all be functionally deployed at once.) One solution to this is at the heart of evolutionary design, in the form of iterated cycles of selection and the accumulation of beneficial incremental improvements. The analogy with the logistics of directed evolution screening or selection is clear enough. While we have seen the unquestionable case for carefully choosing the correct and most efficient screening process for the evolutionary goal in mind, the size and quality of the relevant library are also important.

Assuming that a protein library with random mutational changes has been constructed in a technically competent manner, one can easily enough calculate the required library size to "cover" the known number of randomized specific residues, or the known average random mutation rate per coding sequence. The trouble is, for generalized mutagenesis, as the mutation rate increases, the numbers get out of hand too rapidly for "coverage" to be anything like a practical proposition in most cases.* As we have seen, mutagenesis can be localized to distinct regions of interest based on pre-existing structure–function information, which may take the numbers into the manageable range. Even so, for a small cassette-mutagenesis experiment with five residues, 3.2×10^6 (20^5) variants would need to be screened, which may be straightforward or arduous depending on the technical dictates of the screening protocol. In practice, related sequences are likely to share functional qualities, potentially greatly reducing the numbers of variants to evaluate before finding a promising candidate. Finding a primary "hit" within a library then allows its optimization by examining in more detail localized structure–function relationships, through reiteration of random mutagenesis at specific sites. As we have seen earlier, combinations of such candidates can be subsequently combined into favorable configurations by means of DNA shuffling.

---

*In Chapter 2, we saw the vast numbers of sequences possible even for relatively small proteins. For a 100-amino acid residue protein with one residue altered to any of the remaining amino acids, 1900 variants are possible (19 remaining amino acid alternatives at each of 100 positions), but numbers rise rapidly for increasing numbers of mutations per coding sequence (for 2 mutations per sequence, 1.8×10^6 variants; 3 mutations per sequence; 1.1×10^9 variants; 4 mutations per sequence, 1.2×10^{13} variants. This is calculable from a general formula: variant number $= a^k(n!/k![n-k]!)$, where a = number of possible mutant variants at one residue position (19 for normal amino acid sequences), k = average number of mutations per sequence, and n = length of sequence (in amino acid residues for proteins).

Then what about optimal size requirements for libraries obtained from DNA shuffling experiments themselves? Ideally, a shuffled library would cover all possible permutations of the sequence divergences within the original parental genes, which rapidly escalates as the numbers of both increase. Again, however, the iteration principle comes to the rescue. Repeated rounds of shuffling and screening have yielded novel and useful enzymatic variants with increasing performance after each diversification cycle, when the numbers of recombinants screened for each round is only on the order of 10^4.[378,379]

We have seen that the specific means for sequence diversification in artificial directed evolution is significant, to avoid mutational biases and to maximize the coverage of sequence space. Nevertheless, if the numerical diversity that can readily be generated is greatly in excess of the numbers that can be effectively selected or screened, then library processing becomes the limiting factor in the overall *in vitro* evolutionary progress. It is accordingly important to carefully analyze the optimal means for performing selection or screening, and this in turn raises the question of how to ensure that all aspects of an evolutionary experiment are likewise conducted in an optimal manner. A rational person should do no less, and this topic we should now explore in more detail.

A Highly Rational Empirical Process

We have already made note of attempts to render directed evolution into an efficient process capable of delivering the desired outcomes. This is clearly a rational aim, even though the evolutionary process does not qualify as "rational design" *per se*. At its heart is the implicit understanding, backed by empirical evidence, that completely unguided artificial evolutionary molecular design can be inefficient. Although various sources of inefficiencies can be identified in this regard, either directly or indirectly they relate to the time required for laboratory success. An obvious hallmark of natural evolutionary change is that it has operated over geological time periods, but it is also obvious that such time constraints are rather inconvenient from the transient viewpoint of we humans. Microbial evolution can operate on a vastly accelerated timescale to this, but to some extent this is dependent on the molecular system in question. Evolution in the laboratory by definition must be at least as rapid as natural microbial evolution, and again the specific evolutionary goal can have a major impact on this. In some cases, an empirical approach relying on random variation combined with the application of appropriate selection pressures can yield rapid results, but in others the time required may be prohibitive by such means alone. Therefore, "knowing where to look" can make a difference, which translates in protein evolution to having knowledge of a protein's mechanism, and thereby understanding the functional sites that may benefit most from random evolutionary change. On the other hand, the potential fitness effects of "global" changes must be kept in mind also.

In a sense, resorting to an evolutionary approach could be seen as a pragmatic answer to design problems when full information is lacking. But "full information" in this context can be a tall order indeed, depending on the nature of the molecular design which is required. Usually what is needed is both detailed structural information for the target molecule(s) (certainly achievable with modern technologies) and the means to predict the functional consequences of sequence divergences from the parental (a much more challenging prospect, though also feasible in many cases). This of course brings us to the boundary of rational design, about which we will have much more to say in Chapter 9. The "bottom line" for the continued use of evolutionary approaches is that there is no short-term prospect of their replacement by rational design, if indeed replacement in an absolute sense is possible (as we will also discuss later on). In lieu of a "rational" alternative, the most rational course is surely to pursue the most effective alternative. Ever a pragmatic exercise, intelligent application of directed evolution will surely make use of whatever informational tools are available to increase the probability of success.

There is a widespread and entirely accurate perception that the rational practice of directed evolution involves far more than simply aiming to increase the numbers of screenable clones, and this has been presented as a "quality not quantity" philosophy, or even as an "ideological" shift.[380] In any case, a brute-force approach rapidly encounters the "big number" problem of screening where variant populations escalate into impractical levels, as we have seen. A comparison could be made between the screening of mutant protein libraries and development of computer capabilities at games of skill such as chess. Even with an advanced "number crunching" facility, exhaustive calculation of all possible chess moves in a recursive decision tree becomes unfeasible, so "pruning" of nonuseful classes of moves has to be instituted.[381] Significantly, while "Deep Blue" required customized logic chips and an analytic capability of 2.10^8 moves/second to defeat Grandmaster Gary Kasparov in 1997, the later-generation "Deep Fritz" performs at very high levels (defeating champion Vladimir Kramnik in 2006[382]) with far less raw processing power, but superior algorithms for cutting down the numbers of dead-end searches. The game of Go presents even more of a challenge in this regard with such a hyperastronomical number of possible moves that advanced pattern recognition for focusing on productive strategies is essential. Analogously with this, "smart" approaches to directed evolution rather than brute-force "library crunching" are in order, to find a more efficient "search tree" toward desirable improved molecules.

For some time, then, theoretical biologists have striven to model various aspects of directed evolution with mathematical and computational tools, to better understand its operation, and naturally to understand how to apply it in the most effective possible manner. The fitness landscape (which we encountered in Chapter 2) is a frequently used device in these kinds of endeavors, which lends itself to a strikingly evocative "geological" parlance of ranges, peaks, valleys, and ridges, where small peaks represent local fitness optima, and with all over-shadowed by the "Mt Everest" of a global optimum.[23] But the terrain of a specific fitness landscape depends on how it is defined. With the *NK* model of Kauffman[18,124] where (for protein sequence space) N = number of amino acid residues of a protein and K = number of fitness-related interconnections between each residue, a completely "smooth" fitness landscape results when $K = 0$ (Fig. 4.5). In such hypothetical circumstances, a single global optimum rises smoothly from a low-fitness plain, leading to its christening as a "Fujiyama" landscape[18,23] (geology once more raising its pretty head). An "adaptive walk" in this kind of landscape can be made by progressively selecting the optimal residue at each point in a polypeptide chain, and many possible different independent paths would converge toward the top of the metaphorical Mt. Fuji. If real proteins behaved in this manner, the task of both natural and human-directed evolution would simply require identifying the optimal amino acid residues at each position, since each site would have a "best-fitness" residue independently of all others, and their total combination would sum up to the global fitness peak for that particular protein.

In the real world, residues of a protein interact in very complex ways to determine the net fitness of the macromolecule. Changing the K value for residue interconnectivity results in very different fitness landscapes, culminating in total "ruggedness" when each residue is connected in a fitness context with every other residue (i.e., when $K = N - 1$; Fig. 4.5). In a landscape with many local fitness peaks separated by steep valleys, a protein can in principle become "marooned" on such a minor peak and be prevented from attaining the global optimum, since in such circumstances any single amino acid residue change will decrease fitness and cannot be directly selected for.

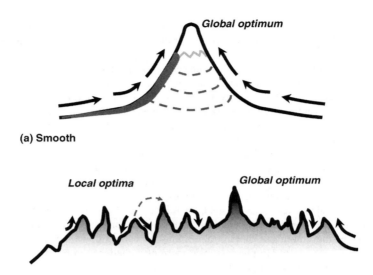

(a) Smooth

(b) Rugged

FIGURE 4.5 Fitness landscapes for protein sequence space. In protein sequence space, each point is a unique amino acid sequence that in turn determines protein fitness. A "walk" in the fitness landscape then proceeds by mutations or recombinational processes. In the *NK* model, if each amino acid residue determines overall fitness in isolation from all others ($K = 0$), a "smooth" landscape (a) results, where progressive optimization for each residue leads inexorably to a global fitness peak. If $K > 0$ (increasing numbers of residue interconnections determining fitness), a "rugged" landscape (b) results with the end state reached where each residue interacts with every other residue. Proteins can be "trapped" at local optima when any single-residue change leads to a loss of fitness (solid arrows). Multiple mutations or recombination are needed to "jump" to another fitness peak (dotted arrow).

The dilemma of the rugged fitness landscape can be then used to help explain the great success of recombinational approaches to directed evolution. We have seen that DNA shuffling has been notably (and sometimes amazingly) successful in deriving chimeric proteins with improved functions. A protein at a local fitness optimum has a low probability of further improvement if only mutational change is permitted, since >1 amino acid residue must be favorably mutagenized to "jump" across a low-fitness valley to a new peak. If we bring in recombination, the picture is quite different, as combinations of local optima can be rapidly assessed,[349] raising the prospect of a jump to another fitness peak (Fig. 4.5), and perhaps allowing the global optimum to be accessed. Modeling has accordingly demonstrated the ability of recombination to "tunnel" across low-fitness barriers.[383] One can also view mutagenesis as a local sampling of protein sequence space radiating out from a single parental form, whereas recombination allows sampling of the volume of sequence space encompassed by the parental donors in a shuffling experiment, with an accompanying increase in the exploration of alternative functional forms (Fig. 4.6). Recombination between structurally related proteins also has a higher probability of preserving folds and function than random mutation alone.[384] If mutation and recombination are combined, the two processes can thus sample both local and more long-range aspects of the fitness landscape.[18] Yet theoretical considerations show that recombination will not work on just any fitness landscape, and indeed in certain cases would be counterproductive. In a landscape where fitness is randomly assigned, recombinant products

Protein sequence space

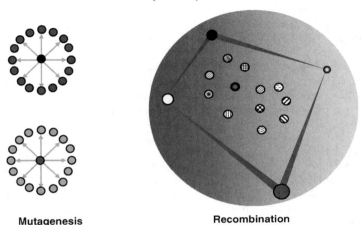

Mutagenesis   **Recombination**

FIGURE 4.6 Mutagenesis versus recombination (DNA shuffling) as sampling either local sequence variation or a much greater volume of protein sequence space, respectively. For mutagenesis, the parental protein is represented at the center of a cluster of mutants close in sequence; for the shuffling, four parental proteins allow a range of recombinants more scattered in sequence space.

will have reduced fitness relative to parental donors, but with a "correlated" landscape[*] itself shaped by natural selection, recombination will be effective.[18,124] The *K* value (as in Fig. 4.5) set by natural evolution itself is predicted to be low, or else recombination (and evolvability) would not be effective.[37]

Apart from providing a cogent rationale for its effectiveness, what can rational principles do for the methodology of DNA shuffling itself? We have divided shuffling into "homologous" and "nonhomologous" categories, but obviously homologies between parental coding sequences in a shuffling experiment can vary continuously from near-perfect matching toward complete nonhomology. Several groups have sought to put the experimental process of DNA shuffling onto a more rational footing, and one of the questions initially was the nature of the favored sites for crossovers. Not surprisingly, fragment cross-annealing leading to template switching and recombinational crossovers correlate with regions of strong sequence identity,[385] and this in turn is observed to be temperature dependent as expected.[345] A cautionary note has come from computational analyzes that find that sequence identity as such is not always the most accurate guide for crossover prediction. Crossovers are favored by contiguous blocks of identity rather than more dispersed sequence matches, even though both sequence arrangements could be defined as regions of significant homology. The thermodynamic measure of free energy of annealing between fragments was found to be a better measure of crossover likelihood.[386] Modeling studies have resulted in algorithms for predicting crossover numbers,[387,388] which reflect the extent of diversity of the recombined library. Yet trade-offs exist here, since the higher the crossover number, the lower the general efficiency of reassembly (subject to the stringency of the reaction conditions and fragment concentration). At low stringency and high concentrations, undesirable "junk" recombinants may become significant.[388] Such types of

---

[*]In a "correlated" fitness landscape, peaks nearby to each other tend to have similar heights of fitness, rising to a global maximum. These types of landscapes arise with specific *K* values in the Kauffman *NK* model.[124] In this view, evolution itself has "tuned" the *K* value toward fitness landscape evolvability.

reassemblies resulting from low homology fragment extensions ("out of sequence" events) have a high rate of failure to form functional proteins, as with generalized nonhomologous shuffling (as noted earlier). Efforts to identify the factors that contribute to such undesirable reassembly effects have also been made.[389]

Mutations and Libraries   Despite the successes of recombination, it is still important to understand, improve, and rationalize mutation-based directed evolution. Mutation, as we have seen, is useful for "fine-tuning" of function from sampling of local sequence space, and can be used in conjunction with recombinational shuffling techniques. Also, controlled mutagenesis may be more useful for detecting completely novel single-residue improvements,[390] as beneficial mutations may sometimes be masked in a recombinational background of other pre-existing mutations from parental sequences. In studies of mutational effects in proteins, several key concepts emerge. Amino acid residues that are *coupled* are (as the term implies) subject to interactions whose disruption is detrimental to protein stability or function; "pairwise" interactions between residues that confer enhancement of protein stability are a classic case in point.[23] (The coupling state of a residue is equivalent to the K value explained in Fig. 4.5.) Mutation of a single coupled residue thus has a high probability of resulting in a protein with reduced function, and a low probability of conferring an adaptive phenotype. (Finding an improvement for coupled residues thus involves their simultaneous variation and optimization.) Residues with low coupling, on the other hand, may have much higher *tolerance* for mutational substitution. We saw earlier that with a hypothetical completely smooth fitness landscape, the fitness of the whole protein is the sum of the individual fitness contributions from each residue; in such cases there is maximal *additivity* of mutational effects[391,392] (Fig. 4.7). It hence follows that highly coupled residues cannot show simple additivity in their contributions to global protein stabilization or function. A starting strategy of focusing mutational changes on residues with the best mutational tolerance and the least coupling could accordingly be a decided advantage in principle. One might reflect that this is logical enough, but how well does it operate in practice? The qualities of coupling, tolerance, or additivity do not immediately leap

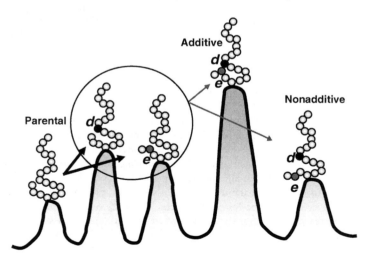

FIGURE 4.7   Depiction of additivity effects of mutations shown as relative fitness peaks. Mutations d and e in a parental protein improve fitness, but in combination either further enhancement (additivity) or deleterious effects (nonadditivity) can be observed.

out from a knowledge of a protein sequence alone. Of course, if protein three-dimensional structures are known, a wealth of information emerges concerning the likely status of residues as mutational sites, especially if accompanied by structure–function results from specific site-directed mutagenesis experiments.

But it is possible to use empirical means themselves to gain information for effectively targeting mutations for directed evolution. Recursive ensemble mutagenesis[393] uses initial rounds of mutagenesis to gather information that feeds back into subsequent rounds and progressively increases the efficiency of the process. The primary random mutagenesis steps may show what types of residue changes are most compatible with positive functional results, allowing rational focusing of later rounds of mutagenesis to specific sites or classes of amino acids, thereby increasing the probability of achieving improved function. With comparable general aims, a statistical algorithm based on primary analysis of combinatorial libraries has been developed for correlating protein sequence–activity relationships and refinement of fitness predictions for subsequent rounds of directed evolution.[394,395] When put into practice, these types of approaches require considerable sequencing to obtain the necessary initial data, a drawback when sequencing was relatively expensive, but less of a consideration as high-throughput sequencing technology steadily progresses.

But progress in computational analysis of proteins and combinatorial protein-encoding libraries may in the end obviate any need for multitiered levels of empirical mutagenesis (as with recursive mutational approaches), and allow application of directed evolution in the most efficient manner. (This is a high-level example of rational deployment of the capacity to mutagenize or recombine proteins at will, or of rationally steering an empirical strategy toward the best possible outcome.) We could divide these types of analyses into two broad categories: strategies for improving on library design, and characterization of libraries once they have been prepared.

A number of computational methods have been derived for predicting sites within proteins of high functional or structural substitution tolerance. Ideally, for such purposes, all possible interactions between protein residues would be surveyed, but this is not feasible even computationally (let alone experimentally), as a complete analysis would include all alternative residue conformational states (rotamers), incurring hyperastronomical numbers even with small proteins.[23] In lieu of this, statistical methods have been developed for assessing residue coupling by means of the structural tolerance of residues for substitutions.[23] Here the thermodynamic value of entropy can be used as a measure of potential information content,* which is correlated with the degree of coupling of an amino acid residue. The higher the residue coupling, the higher is its informational content and the lower the associated entropy. (As entropy rises, informational content falls and disorder increases.) Residue sites with calculated high entropy values are consequently tolerant of substitution and become favored targets for directed evolution,[23,396] and as mutational optimization proceeds, the overall sequence entropy will fall. A related aim is the identification of the range of protein sequence features that remain compatible with core structural folding in a given protein.[397] With such information in hand, the targeting of the most promising residues for directed evolution is rationally achieved.†

When you have constructed your rationally devised library of protein variants, it is not time to cease operating in a rational manner. In addition to the stern screening dictate of the first law of directed evolution, there is the quality and size of the library itself to consider, and all these aspects are interlinked. As we have seen, rational screening is highly correlated with the assay

---

*Flowing from the work of Claude Shannon, the pioneer of Information Theory.[148]

†Such knowledge is also clearly important if predictions of function resulting from sequence changes are made, as required for rational protein design (Chapter 9).

system and the nature of the sought-after improvement, but the appropriate level of screening is also rationally determined by the status of the library itself. Knowledge of not only the actual mutational profile of the library but also the mutational redundancy is important for assigning realistic screening numbers. Once mutational statistics have been experimentally derived through sequencing (average mutation rates for global approaches such as error-prone PCR, or confirmation of random codon insertions for oligonucleotide-mediated mutagenesis), then it is possible to run programs that enable one to derive useful information concerning the diversity makeup of the library,[398,399] which is the "bottom line" for *in vitro* directed evolution. But the best questions to ask in this area, and suitable library sizes, are heavily dependent on the type of mutagenesis that has been instituted.

For example, with random mutagenesis where each of 20 triplet codons is a single "block" (using trinucleotide phosphoramidites as noted earlier), the number of possible sequence variants V is a simple combinatorial product of 20^n mutagenized positions, assuming no other experimentally introduced bias (as tested by sequencing). For two such codons, 400 variants are thus possible, but obviously higher numbers than this must be screened to know that all variants have actually been sampled with a high probability. It can be shown from a mathematical expression derived from the Poisson distribution that the size of a library with a defined probability of being 100% complete is dependent upon the value of V.[*,398] But if alternative random codon strategies are used where the number of possible codon variants increases because of codon degeneracy, the required library size accordingly increases in turn. With libraries derived from distributed mutagenesis throughout a coding sequence (as with error-prone PCR), the number of possible variants increases rapidly with the measured mutation rate (as we noted earlier) to the point where it becomes impractical to consider a library representing all such possibilities. It is nonetheless important to be aware of the expected number of variants as a fraction of the total possible, which (along with the associated level of redundancy) can also be ascertained.[398] These calculations make certain assumptions for simplification purposes, such as postulating that all types of mutations have the same probability. More complex computer analyses have been described aimed at accurate simulation of libraries generated under a range of conditions, allowing simultaneous characterization of their diversity profiles.[399]

Protein libraries for *in vitro* directed evolution currently have an upper size boundary of $\sim 10^{13}$ (for ribosome or mRNA display techniques, to be considered in more detail later in this chapter.)[400] If the number of variants theoretically obtainable from any mutagenic and/or recombinogenic procedures approaches or exceeds the effective screenable limit, does this imply that one should simply aim for as large a physical library as possible? Though this might seem eminently logical, some theoretical considerations and experimental results have suggested that an effective "adaptive walk" strategy should use reiterated mutagenesis and screening of relatively small library samples.[23,401,402] But in theoretical adaptive walks on a fitness landscape, a law of diminishing returns appears to operate. In such cases, as a global optimum is approached, the numbers of mutants that must be screened to identify improvements in fitness increase in a linear fashion, since the numbers of fitter mutants steadily decrease toward final optimization.[23] Large libraries may thus be beneficial if the protein of interest for evolutionary studies is nearing a fitness peak, but again the staggering theoretical numbers

---

[*]This is given by $L = -V \ln(-[\ln Pc/V])$ when $V \gg -\ln Pc$ (as applies here), where L = required library size; V = total number of possible variants; Pc = desired probability of 100% coverage of variants.[398] For a desired probability of 0.95 (95% chance of a complete library) and 400 variants in this example, the L value is ~ 3600, or an approximate ninefold redundancy. But if three codons are likewise randomized (8000 variants), the L value becomes $\sim 96,000$ for the same probability level (12-fold redundancy).

required in many specific circumstances render attainment of global optima by mutation alone unfeasible.[*][,403] Maintenance of library diversity throughout screening rounds is important to avoid "premature convergence" or movement toward a local fitness peak that is an overall suboptimal solution.[23]

An implicit advantage of rational strategies for evolutionary mutational targeting is that library sizes can be drastically reduced over random-unstructured approaches, with concomitant elevation of screening efficiency. Comparison of separate types of computationally targeted small libraries of fluorescent protein variants (with random mutations at similar frequencies) awarded significant superiority to most of the "rational" approaches, particularly with a recent computational design innovation.[404] Thus, a "take-home message" from this survey of library design and characterization is that many rational mathematical and computational tools have been derived and continue to be produced, and have demonstrable value in the real world at the laboratory bench. In short, the best library designs involve complex patterns that are unlikely to be discerned without computational assistance.[405]

In certain areas of protein design, such as the thermal stabilization of enzymes, rational approaches will become increasingly competitive with evolutionary strategies as our general knowledge base advances. Yet the key appeal of directed evolution remains its potential for finding functional molecules that elude rational identification. Here optimism must be tempered by the vast size of the sequence space for proteins alone, which brings home the limits of both natural and artificial evolution. Jumping distances in a "design space" of any description (or local peaks in a fitness landscape) is a problem if incremental advances do not increase fitness. Apart from recombinational solutions, relatively improbable multiple sequence changes may allow a transition to a new pathway in design space if sufficient time is available. Time, of course, is not on the side of the *in vitro* experimenter in the same sense as it has been with natural selection. Let us now consider how directed evolution can attempt to overcome such limitations and seek true novelty in molecular design.

Crossing the Valley of the Shadow of Low Fitness: Journeys across Molecular Spaces

Fundamental to the understanding of an evolutionary process is the concept of incremental change. Through repetitive cycles of diversification and selection, an evolutionary "prototype" molecule can be "tuned" to increasingly sophisticated levels of performance. But a molecule or molecular system can only be considered a "prototype" in retrospect, with the benefit of hindsight and by comparing specific functional features of the parental molecule(s) with those of the resulting evolved progeny. At any given time, each molecule in an evolutionary series must have a fitness advantage in order for its selection and for the propagation of the gene(s) that encode it. Through the agency of our mythical Lucy, in the previous chapter we looked at the benefits of iterative processes in evolutionary molecular development, especially in the context of antibody affinity maturation. Patient iteration of the diversification/selection procedure, the "Rome wasn't built in a day" philosophy of molecular design, certainly has its rewards, especially if one's "day" is an open-ended segment of time. And the two-tiered adaptive immune system has taught us that high-affinity binding proteins with a huge range of specificities can be derived by Darwinian processes in a time span much shorter than the lifespan of animals that host such systems, provided one starts with a suitable molecular framework.

But at the same time, therein lies the rub with iterative molecular solutions. While antibodies are supremely versatile recognition proteins (and have been evolutionarily selected for such

---

[*]And again, recombination is suggested as an escape route.[403]

properties), they are not universally effective at binding all conceivable targets relative to alternative protein designs.* (We compared antibodies with specialized DNA-binding proteins in Chapter 2 to make some of these points.) Selection and maturation of an antibody proceeds as an adaptive walk toward an affinity optimum, but one that is constrained by the general framework structure of immunoglobulins. In most cases, this is not an issue, and a better alternative binding molecule than a natural antibody may be hard to find. But while antibodies against certain specific DNA sequences may be derived and selected for as high an affinity as possible, in most cases they will fall short of the performances that DNA-binding proteins with different protein folds can manifest. Or in the language of fitness landscapes, in such cases, the peak scaled by antibody optimization is local and is overshadowed by the global optimum of a dedicated DNA-recognizing protein.

We have already come across the idea of an evolutionary trend "stagnating" on a local fitness peak. Once at such a peak, any single-residue change for a protein is deleterious (reducing fitness), and thereby in theory the only future evolutionary progressions with such a molecule are highly improbable multiresidue "jumps." These considerations are even more pertinent for artificial evolution than its natural counterpart, since certain unlikely events may become probable if the time factor is long enough,† but this option is not available in the laboratory, as already noted. Earlier, we compared screening of protein mutant libraries with computer strategies for game playing, and this analogy can be further extended in this context. In chess, the knight's L-shaped move and ability to jump over other pieces is often used as a metaphor for lateral thinking, and by analogy we could use it as a metaphor here for jumping to a new fitness peak over the "linear" moves of the other players on the board. Evolutionary molecular maneuvers akin to the knight's move are then required to leap over valleys of low fitness into new realms of molecular function.

Given the importance of the issue, let us spend a short time comparing the potential "escape routes" encountered before, and some additional implications. A simple pathway toward optimal improved fitness based on additivity of single cumulative mutations (Fig. 4.7) can falter at a local fitness peak if all possible subsequent single-residue changes decrease fitness and are accordingly nonselectable (Fig. 4.8a). In such circumstances, a relatively improbable multi-mutational "jump" is required to cross the valley of decreased fitness. Experimental models of *in vitro* β-lactamase evolution indeed find that many potential multimutational pathways from wild type to enhanced function are inaccessible due to reduced fitness.[406]

But once again this assessment does not take into account the agency of recombination, which (especially in the case of prokaryotes) is further promoted by lateral gene transfer. Also, we have already seen the power of *in vitro* recombination in the form of DNA shuffling and related techniques for leaping across low-fitness valleys and extending the exploration of protein sequence space. Yet recombination can only be as good as the initial net input diversity. If one or more mutational changes required to attain full optimization are absent from the entire parental input population (natural or artificial), then these must be introduced as concomitant or secondary events. Nevertheless, by crossing at least some fitness barriers, recombination can

---

*And natural proteins themselves may be suboptimal for certain tasks. In Chapter 5, we will consider efforts toward extending the genetic code by the addition of nonnatural amino acids. In Chapter 7, we will further consider alternatives to proteins altogether.

†But caution is needed before simply declaring that a low-probability event will inevitably happen in the fullness of time; it is entirely dependent on what kind of "jump" one has in mind. Thus, a simultaneous beneficial mutational event involving two residues is feasible on a long timescale, but the chances of greater numbers of mutations occurring which are both beneficial and simultaneous in either one protein or an interactive system rapidly decreases to effectively zero even in the temporal history of the Universe as a whole. (See Chapter 5 and *de novo* empirical protein design). Iterative selection, rather than sudden jumps, has the power to "tame probability."[17]

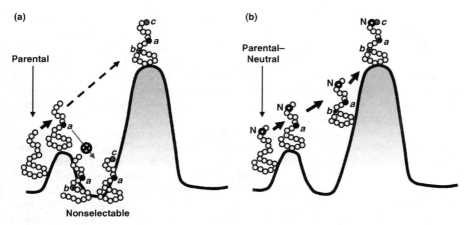

FIGURE 4.8 Evolutionary progressions of a protein toward improved fitness (or in principle any molecule constituted from combinations of a limited set of monomeric building blocks that can be selected and used as the basis for further incremental improvement). (a) An improved variant *a* cannot undergo subsequent single-mutation changes without loss of fitness, requiring two simultaneous mutations (at much lower probability, dotted arrow) to reach the next peak. (b) The same protein with a neutral mutation (N) allows a single-mutation pathway up a fitness peak.

greatly accelerate molecular evolution. Indeed, analysis of an evolutionary transition of β-glucuronidase into a β-galactosidase elicited by DNA shuffling has shown that some enzyme intermediates have reduced mutual substrate specificities[407] that would presumably not be directly selectable.

A more subtle scenario arises if one of the transitional mutations are not directly selectable, but not deleterious either, and thus by definition neutral. As we considered in Chapter 2, the evolutionary consequences of initially neutral mutations in the longer term may be significant, since a secondary mutation in the context of a neutral mutation may behave differently than when placed in the equivalent parental context.[*,408] Different populations with different backgrounds of neutral mutations can by such means explore different regions of sequence space. For example, mutations have been identified that are neutral in a wild-type protein background but beneficial when the protein topology is altered in a specific manner.[83] A neutral or nearly neutral mutation can thus offer an alternative pathway for crossing a fitness valley onto a new peak (as depicted in Fig. 4.8b), a viewpoint that is supported by theoretical modeling depending on the volume of protein sequence space covered by a "neutral network" extending from a parental sequence.[23] Let us take a moment to look at a metaphor for evolutionary jumps based on neutral mutations.

Colorful Neutral Springboards   Lucy returns as a director of evolution in a word universe. In a similar manner to a common game, in this universe words can change by single-letter substitutions, additions, or deletions, provided the new word is "functional." (Any nonfunctional letter combinations cannot be "reproduced" and are rapidly removed from the dynamic universe of words.) Changes of more than one letter at a time are not technically forbidden, but are only allowed so rarely that they are not usually a productive pathway toward

---

[*]In Chapter 2, we also saw that mutations that are silent at the coding level can nonetheless be significant as springboards for secondary mutations which shift codons to a specific amino acid change.[32]

new words. But how are new letter combinations ever preserved? A novel group of letters can be rapidly seized upon if it meets a previously unfulfilled need, and thereby is "functional" in the word universe. In such circumstances, the novel word is "selected and reproduced." Some combinations of letters can be rejected out of hand for failing to conform to certain basic rules (and thereby fail to "fold" properly); other combinations may technically be acceptable, but still fail to find a useful function. Lucy finds "PRTEIN" as an example of the former, and throws it in the trash heap. In contrast, "BROTEIN" seems to "fold" adequately, but remains unemployable, and eventually meets with a similar fate.

The word universe that Lucy is charged with developing is in an early stage of its evolution, and Lucy is confident that many novel functions are still "out there" that have not yet been described by a word. Sometimes, quite different words evolving independently converge on the same function, and both are preserved, although often not at the same frequencies. She notes that such words with similar or identical function ("synonyms") may compete, with the "fitter" word predominating and its vanquished competitor eventually dying out. An example passes through her hands: the word "PROTEIN" has triumphed over its near-extinct synonym "PROTIDE." She also pays heed to a more subtle effect, where very similar versions of the same word exist, with both being equally acceptable. The letter differences between such words appear to be completely "neutral" with respect to their functions. For reasons outside of Lucy's knowledge base, one large group of such "neutral" changes are labeled "American" or "English," and she inspects a sample pair: COLOR and COLOUR. She suspects that the "neutrality" of such single-letter differences is not absolute, and each version may have differential fitness levels within the mysterious "American" and "English" environments (whatever they might be), themselves encompassed within the word universe.

Be that as it may, in Lucy's hands they appear to be more or less interchangeable, and she proceeds to use them as the starting material for some directed evolution experiments aimed at generating novel useful words. With the "English COLOUR," though, she makes little headway. Even using a large library of "mutants," only one single-letter derivative can be found that is "foldable" or functional.* Could the "neutral American" alternative COLOR be any better? The usual array of useless trash is ignored (CCLOR, CODOR, etc.), but she finds something extra of note: "COLON." This interests her, since this word already has a function describing a part of the human intestine, but it eventuates that "COLON" can also make a "punctuated" jump into a completely novel function (:). Lucy then performs a second round of directed evolution on "COLON," and emerges with the highly functional derivative "CODON." She concludes that even if the ascribed neutrality of her two initial alternatives is correct at the outset, very different sets of functional evolutionary progeny can result from different "neutral" starting points. Also, she notes that the American word string COLORFUL (but not the COLOURFUL English equivalent) contains the triplet "ORF," which she believes stands for "open reading frame." She wonders whether this is related to the "CODON" mutation, but there she is making an unwarranted "jump" between an imaginary universe and the real one.

Punctuating Sequence Space, Neutrally or Otherwise   The theory of neutral mutations allows for the "punctuated" appearance of proteins† with novel structures and functions,[23] through the abilities of neutral variants to access different areas of sequence space (as simply depicted in Fig. 4.8b). In Chapter 2, we also noted the ramifications of "neutrality" in terms of

---

*And sadly enough, this is "DOLOUR"; also obtainable from the American version as "DOLOR."

†When we invoke "punctuation" in evolutionary processes in general, we are reminded of the theory of punctuated equilibrium referred to in Chapter 2, but a jump involving multiple sequence changes in a protein is a different prospect to rapid whole-organism phenotypic change.

protein robustness to sequence changes. Can neutral mutations be exploited by real directors of evolution *in vitro*? By its nature, this might seem a difficult proposition, since a specific neutral mutation provides no rational basis for choosing it, and a vast range of alternative possible neutral mutations may be found in an average-sized protein. In other words, in the absence of additional information, one cannot know in advance which neutral mutations may lead to improved outcomes in the context of novel secondary mutations. Increasing the mutation rate may be theoretically beneficial,[23,338] but this is accompanied by the usual accelerating multiplicity of variants and the associated screening difficulties. Yet it is possible to prepare artificial libraries of neutral variants for a functional protein as a resource for directed evolution. Since many random mutations will interfere with folding and stability, simply screening for retention of folding in a large collection of mutants[409] can allow the assembly of a "neutral-drift" library for subsequent screening for novel variants of interest. When all starting-point library members deviate from parental sequence but retain folding and functionality, no assumptions are made as to which neutral mutations will prove to be important. But if a lead into an area of novel function is close in sequence space to any one of the neutral-drift library members (as in Fig. 4.8b), then there is real prospect for finding it, and then exploring further down that new avenue.

In Chapter 2, it was noted that natural evolution is highly likely to be "algorithmically incompressible," where the evolutionary process cannot be anticipated by any shorter set of instructions other than running through the complete algorithm itself.[18] If applied without any other input, the artificial counterpart of natural evolution operates under the same premise. But as also noted in Chapter 2, human agency can modify at least the starting point of the algorithm in an unnatural way, potentially advantageous for molecular improvement. The "instructions" for natural evolution proceed along the lines of "take the most beneficial (fittest) variant at the end of a selection cycle and repeat the cycle" (as depicted in Fig. 2.1). A natural protein is already a product of a vast number of such cycles, and would thus seem the normal starting point for further artificial progressive evolutionary improvement (as is usually indeed the case). It is possible, though, for human artifice to choose a defective protein to initiate directed evolution, even though this would not occur naturally. It has long been known that "down mutations" in proteins can be compensated by certain second-site mutations, which thereby act as suppressors of the initial deleterious mutational effect.[410,411] Moreover, some mutations appear to act as "global" stabilizers or suppressors of deleterious protein sequence mutational changes.[412] Deliberately starting with a defective protein can accordingly act as a sensitive means for isolating such beneficial mutations. If transferred to the original wild-type protein, mutations which boost activity in the defective background may show additive enhancement of the parental activity, but only if the compensating mutation acts independently of the deleterious mutation. This has been demonstrated with β-lactamases.[413]

But nonadditive effects (Fig. 4.7) in proteins have also long been observed.[410] Coupling between two mutations that are deleterious in isolation can in principle create a positive effect when the mutations are combined. (Two wrongs making a right?) Such an arrangement can arise, for example, when two mutant residues in a protein are spatially coupled in a stabilizing charge-based interaction that is reversed if either of the partner mutant residues is absent. Reaching such an improved double-mutant configuration would then require either a simultaneous specific double mutation or the deliberate choice of a single (defective) mutant as the starting point for progressive evolution. Unfortunately, since by definition the nature of the improved variant is not known at the outset, a huge number of variants proceeding from any given defective origin would be complete dead ends in terms of overall fitness improvements. The second-site suppression strategy for artificial molecular evolution is thus highly contingent on the specific background in which the initial deleterious and secondary suppressor mutations occur, and has not been widely applied.

The issue of the "starting point" for directed molecular evolution might remind us of the notion of evolutionary "lock-ins," where precommitment to a specific design renders a shift to a completely novel (and possibly superior) design unfeasible. (We considered the well-worn "QWERTY" keyboard metaphor for this effect in Chapter 2.) In such situations, the optimal solution is defined relative to starting-point molecule or system, and adaptation moves to a fitness peak. But this peak is not necessarily a global peak. An improved variant obtained at the end of the walk through sequence space is superior to the parental form, yet this improvement in itself does not preclude the possibility that a radically different molecular design might show further functional enhancement. Such a radical change would require the crossing of a wide fitness valley and is thus of low feasibility for natural evolution. Human agency, nevertheless, can in effect "move the goal posts" and rationally choose a different starting framework that natural selection could not jump to. Such alternative molecular origins can then be used for progressive directed evolution toward potentially better solutions than the optima obtained from evolution of a natural starting-point molecule.

So far, we have thought about the artificial generation of diversity for directed evolutionary purposes, and certain aspects of the implementation of these approaches for library design. The technological focus of this chapter should now be brought to bear on another "metatechnique" mentioned at the outset, which has wide-ranging ramifications for directed molecular evolution and molecular discovery in general, as well as additional applications.

THE POWER OF DISPLAY

In many different fields of both basic research and biotechnology, the term *phage display* is often encountered, and although phage-mediated molecular display is both historically and practically important, it is only a subset of display technology in general. For the more generalizable keyword here is indeed "display" rather than "phage." Display of a molecule in the broadest sense can occur in a variety of circumstances, but all molecular display systems relevant to directed evolution have a singular feature in common. This attribute is conceptually simple yet very powerful: a physical linkage of some kind between genotype and phenotype. If we think again about natural evolution, we can recall that selection is exerted at the phenotypic level (Fig. 2.1), controlled by an organism's genotype. For this to be a viable arrangement, an implicit assumption is that there must be a linkage between the controlling genotype and the phenotype upon which selection is exerted. This may initially seem a trivial point, for why would the genes and expressed proteins of an organism ever be separated? But in fact it is not trivial, from a very fundamental point of view that is directly relevant to laboratory molecular evolution. (It is also a basic issue in the origin of life, but let us leave that aside for the time being.) Genes and their expressed protein and RNA phenotypes are typically physically embedded together in a common cellular environment bounded by membranes, and in turn at higher levels of cellular organization leading up to whole organisms as replicators. The linkage between genotype and phenotype is thus an effect of common compartmentalization, where positive selection for a specific phenotype automatically ensures the survival of the *corresponding* informational macromolecules.

So far, this principle may be accepted but still seem of underwhelming importance for any practical applications. Biotechnology routinely makes use of bacterial (or other) host cells for expression of libraries of mutant proteins. In such circumstances, clearly the bacterial cells are functioning as the necessary compartments that link genotype and phenotype, which enables a screening process to be carried out in principle. Improved mutants are identified through appropriate screens, which are typically done at the level of individual bacterial colonies, even if automated to a very high rate of throughput. If we want to search for a mutant protein-binding ligand A, we would thus screen the bacterial library for clones that expressed an "A-binder," as already outlined in Chapter 1 (Fig. 1.2). When promising candidate A-binders are identified, the corresponding bacterial clones provide the genetic information which specifies the A-binder candidate proteins, which is usable for further rounds of directed evolution. To get this information, an amplification step of some kind is required: either directly on the DNA from the identified bacterial clone(s) (via the polymerase chain reaction) or through growth of the relevant bacteria themselves to provide sufficient DNA for sequence determination.

Yet the key feature of interest in this hypothetical process is the interaction between the candidate A-binder and ligand A; why can we not obtain the important sequence information from the A-binder proteins directly? In principle, the A ligand could be used as an affinity reagent for helping the purification of sufficient A-binder proteins for their characterization by protein sequencing, but (despite great advances in the sensitivity of protein sequencing methods) this remains cumbersome, time consuming, and unsuitable for high-throughput screening. Imagine, though, if we could amplify the protein of interest to circumvent the purification process and greatly assist its sequence identification. Alas, we cannot do this directly. But the corresponding genetic information that encodes the A-binder proteins can indeed be amplified, either naturally (bacterial growth) or artificially (the polymerase chain reaction again). So if, by a suitable screening assay that signals the presence of an A-binder, we find a bacterial clone that is A-binder positive, we can identify the sequence of the protein by analyzing the mutant gene that specifies it. The bacterial "compartments" thus fulfill the genotype–phenotype linkage enabling the feasibility of the screening process. So we have now dissected certain key screening issues, but the genotype–phenotype linkage still seems simply a preset necessity that has little direct experimental bearing. But wait...

In Chapter 1, an alternative possibility termed (for illustrative purposes) "direct library selection" was also outlined (Fig. 1.2). In this strategy, binding molecules from a library are "pulled out" by means of their affinity for a probe (target) molecule and then amplified prior to their identification. One detail that was not "amplified" in Chapter 1 was exactly how such an (essential) amplification could be brought about. If we attempted to directly screen a library with a lysate (solubilized fraction) of the large number of bacterial cells within which the mutant proteins have been synthesized, we would have no way of getting an informational "handle" on any A-binder proteins within this complex mixture (Fig. 4.9a), even though somewhere within it the desired A-binding protein might exist. But what if we engineered the library of mutant proteins such that they were expressed on the surface of each bacterial cell of the library? Then we could visualize the entire bacterial cell as simply a "tag" that contains the vital genetic information, which specifies the surface protein of interest. Also, and very importantly, we could select for binding to the A-ligand without requiring cell lysis (Fig. 4.9b). This arrangement then constitutes a *surface display* system, and we can see that both the "surface" and "display" terms in this context are very significant in meaning.

When we speak of a surface, we generally refer to a complex structure possessing an "inside" and an "outside," which certainly is still applicable at the molecular level. Display technology, though, can be taken to the most basic level of engineering a direct linkage between the expressed proteins of interest and their encoding nucleic acids, as depicted in Fig. 4.9c. In this example, the linkage is affected by incorporation into a phage virion of a fusion protein bearing the foreign polypeptide of interest. Selecting for binding by the displayed surface structures in Fig. 4.9b or c therefore also provides their coding sequences and a means for amplifying such sequences through phage or bacterial growth.

Finally, and returning to the above question of amplification from "direct library selection," it is the informational tag that is amplifiable, not the protein itself. In this respect, we also return

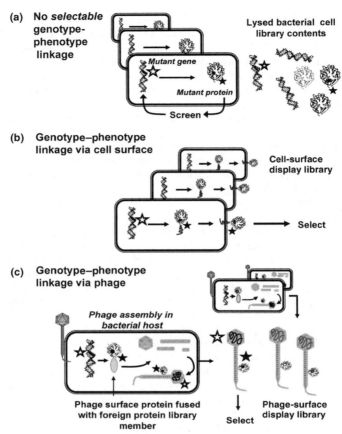

FIGURE 4.9 Depiction of certain types of genotype–phenotype linkages. Open stars indicate locations of DNA sequences encoding specific mutant library members; black stars indicate corresponding expressed proteins or peptides. (a) Expression of a library of mutant proteins within bacterial cell compartments, with a single such compartment shown. Screening clones of individual cells, each with a separate mutant protein, yields the corresponding genetic information as long as clonal library members are evaluated separately. If library cells are lysed *en masse*, no linkage between mutant proteins and the gene(s) that encode them is maintained. (b) If mutant proteins are engineered such that they are stably expressed on the surface of the bacterial cell, physical selection for a binding interaction with the surface protein also delivers the accompanying "attached" whole cell with the appropriate genetic information. (c) Linkage between a mutant protein and the genetic information that specifies it, via phage proteins. Phage DNA entering a bacterial host cell is engineered to encode a library of fusion proteins between the protein of interest (mutagenized as desired) and a constant phage structural gene. (All genes required for the normal phage lifecycle are left intact.) Upon assembly of phage virions, the fusion protein is incorporated into the phage structure (in this example, a phage tail tube); phage DNA that includes the specific mutant protein sequence is incorporated into the phage head structure. In both (b) and (c), selection can be based on the properties of the specific expressed surface phenotypes.

to the analogy with natural selection, where selection is exerted on phenotypes, which are linked to and encoded by genetic replicators. We can see too that the concepts of display and genotype–phenotype linkage are also not unrelated to the deeper meanings of screening versus selection, as described in Chapter 1. After this rumination on the fundamental aspects if display technology in general, we are now in a better position to look at some aspects of this important tool for molecular discovery, specifically in the world of phage biology. (Some other types of display approaches are provided in ftp site.)

Show Me Your Coat: Phage on Display

In 1985, a paper appeared in *Science*[414] describing the ability of a type of bacteriophage called fd to accept foreign peptide sequences into one of the proteins making up its outer coat. This involved making a corresponding change to the phage gene ("gene III") encoding this coat protein. The *E. coli* phage fd and its close relative M13 are in fact prototypical examples of *filamentous* phage, which have single-stranded DNA genomes wrapped in a coat composed of several different proteins. Even apart from their description as "filament like" (an accurate reference to their shapes), these phage differ from most other bacterial viruses in some important ways. They selectively infect *E. coli* only when the latter host cells express a special structure called the F-pilus, which allows bacterial "conjugation," or sexual exchange of genetic material between bacterial cells. (Humans are painfully familiar with parasites that exploit sex for their transmission; this principle can be generalized across kingdoms of life, or so it seems.) Filamentous phages are notable in another manner as well. Unlike so many other phages, they do not actively kill their hosts, but convert them into continuous factories for producing phage particles, which are constantly extruded through the bacterial cell membrane once infection has occurred.[415]

In any case, the demonstration that foreign peptide sequences could be displayed on filamentous phage was of considerable significance. Obviously, the fusion of a phage protein with an additional exogenous sequence would be of no value if phage infectivity was compromised. The gene III product (gp3) of filamentous phage binds to the bacterial F-pilus and thereby mediates entry of phage DNA into the host cell. This protein has three domains (designated from the N-terminus as N1, N2, and CT (or D1, D2, and D3, respectively)[415]) separated by glycine-rich unstructured linkers,[416] and most peptide fusions to the N1 domain experimentally do not affect phage growth.* Analysis of the structure of the N1 and N2 domains of the gp3 protein have provided a rationale for this, as the protein site that is believed to interact with the F-pilus is in the N2 domain,[418] which is distant from the N-terminus (Fig. 4.10a and b).

Although the first to be tinkered with, the product of gene III is not the only filamentous phage protein that can be used for display purposes. In fact, all five of the separate types of proteins that comprise the mature phage particle (Fig. 4.10a) have been successfully exploited for this purpose in one application or another.[417,419–421] The gene VIII product (gp8) is a small protein that forms the tube of the filament itself, and consequently has thousands of copies per phage, with the associated potential for high-multiplicity display (Fig. 4.10a). This property is useful for detecting weak binding interactions.[422] Note, though, that a high copy number of displayed structures per phage (that is to say, a high *valency*) is not necessarily a good thing in all possible circumstances. Here we need to recall the distinction between avidity and affinity introduced in Chapter 3 in the context of antibody binding. While a phage particle with ligands

---

*Certain larger polypeptides can nonetheless reduce infectivity.[417]

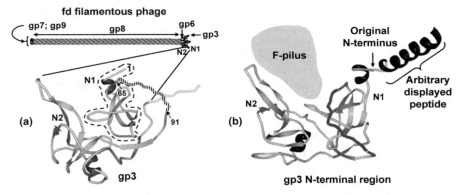

FIGURE 4.10   Structure of the filamentous phage fd (and relatives M13, f1). (a) Phage virion showing the filament (in which the single-stranded phage genome is contained) composed of the major coat protein gp8 (product of gene VIII), with the four minor proteins as shown at each end. The end mediating contact with the host bacterial F-pilus and infection contains gp3 (product of Gene III), whose N-terminal domain structure is also shown; five copies are believed to be present in the phage filament[13] (N1 domain encircled with dotted line). Numbers show positions in amino acid sequence. Thick dashed line depicts unstructured linker sequence between domains N1 and N2 (residues 65–91 as shown) also referred to as D1 and D2[14]; light gray depicts the linker region joining to the C-terminal remainder of gp3. (b) The N-terminal region of gp3 depicting an arbitrary helical peptide sequence fused and displayed at the N-terminus. The position of the N-terminal fusion is such that it does not interfere with binding of the N2 domain to the (schematically represented) F-pilus[15] on the bacterial host. Source of gp3 structures: Protein Data Bank[2] (http://www.pdb.org) 1G3P. Images generated with Protein Workshop.[5]

displayed at high valency on gp8 may show strong overall avidity for a receptor target, the affinity of an individual ligand may be relatively low. Consequently, *monovalent* display is a useful feature in certain situations, especially where high-affinity interactions are sought. Some additionally useful features of gp3 display can be brought to bear on this problem. Although five copies of gp3 are believed to be present in the wild-type filamentous phage particle, 3–4 copies are sufficient for good phage infectivity.[423] Also, instead of fusion of a foreign segment to the very N-terminus of gp3 ("long fusions"), fusion to only the "CT" (D3) domain is sufficient for incorporation into the phage. Infectivity of such phage bearing a limited number of copies of the CT domain gp3 fusions will still result as long as a source of full-length gp3 is also present. (This is usually provided by a "helper phage" vector.) By controlling the levels of expression of the foreign fusion with gp3, monovalent display can be achieved.[423]

A major application of display in filamentous phage has been the generation of displayed peptide libraries, from which functional peptides can be selected and analyzed.[424–426] Another is the display of antibodies, which is considered in more detail in Chapter 7. Since the maturation of filamentous phage requires extrusion through the bacterial cell membrane rather than lysis, displayed peptides must be compatible with this process, and certain peptide sequences are consequently under-represented in filamentous display libraries.* This prompted the development of a number of alternative lytic phage display systems, including the phage

---

*At least some polypeptides that are normally poorly displayed on filamentous phage can have their display levels greatly improved by altering signal sequences for protein translocation,[427, 428, 429] or through mutations in a component of the *Sec* secretion pathway.[430] Also, certain mutations in gp8 themselves improve display of proteins.[422] Overexpression of certain periplasmic chaperones may also be beneficial in some circumstances.[431] Therefore, the exclusion of polypeptides from display by filamentous phage is conditional for at least a subset of the total.

lambda. Broadly speaking, the morphology of lambda is shared with many types of lytic phage, where the phage genome is contained within an icosahedral protein capsid "head" attached to a tubular tail through which the phage nucleic acid is ejected during passage into the host cell (Fig. 4.9c). Both head and tail proteins of lambda can be used for display purposes separately,[432–434] or together.[435] Other lytic phage display systems have also been developed, and in the phage T7 system, less sequence bias has been noted in peptide display libraries than for corresponding libraries in filamentous phage.[436]

But what of the applications of phage display toward directed evolution in the laboratory? At the beginning of this chapter, we noted that selection for binding from a single peptide library[*] could be considered a "snapshot" of an ongoing evolutionary process, and many phage-displayed peptide libraries are used (at least initially) to isolate primary candidate binders for analysis in other ways. It is certainly possible, though, to subject the coding sequence for a primary peptide selected by display to directed mutations that subsequently allow selection of optimal binders by an *in vitro* evolutionary process. This can be compared with an "affinity maturation" process,[437] and inherently enables sampling of diversity that may not be present in the initial starting-point library. But while peptides can certainly be evolved, larger protein structures, including enzymes, are of obvious interest for analogous applications. A variety of proteins, including enzymes and transcription factors, have been successfully expressed on phage surfaces for such applications, with more details provided in ftp site. All the diversity of phage display vectors share the requirement for a bacterial host for replication, but the display methods with the greatest power in terms of sheer library numbers dispense with host cells altogether.

Display Without Cells: Doing It *In Vitro*

Consider the preparation of bacterial libraries for laboratory molecular evolution. In the majority of cases, directed diversification is performed *in vitro*[†] (whether by mutagenic techniques, recombinational DNA shuffling, or a combination of both) and the library of variant sequences (in a suitable vector) is transferred to the host cell for clonal replication, expression, and screening. It is this transfer process ("transformation" or "transfection") that is a major limiting step in the total efficiency of directed evolution in general.[††] *In vitro* diversification of coding sequences by mutagenesis, shuffling or synthetic approaches can generate huge numbers of variants. (A collection of sequences containing only 10 fully randomized codons will have up to 20^{10} ($\sim 10^{13}$) different members.) Yet diversity on this impressive scale has to pass through a relatively very narrow bottleneck to enable screening or selection in bacterial (or other) host cells. A range of techniques exist for getting the variant library of interest (in a suitable replicative vector) into cells, but a typical single transformation

---

[*]Note that a "single peptide library" refers to a single source of molecular diversity from which candidates are extracted. This will almost always involve multiple rounds of selection with a binding target molecule of interest to remove nonspecific background (where bound candidate display phage after each round are propagated for their enrichment and then rescreened). Progressive *in vitro* evolution involves secondary mutations based on primary candidate(s) and reselecting for improved performance. Yet the amplification of the primary candidates for each rescreening round of a single library can also introduce new mutational diversity, depending on the replication fidelity. This is especially relevant to PCR-based amplification, as noted for the SELEX procedure in Chapter 6.

[†]As opposed to global *in vivo* mutagenic strategies noted earlier in this chapter, an example of directed *in vivo* mutagenesis with an evolutionary aim is the exploitation of immune system somatic hypermutational mechanisms,[438] but this is not widely used.

[††]Note that directed evolution itself is very often referred to as "*in vitro*" evolution, but to be precise the latter term often overlooks the expression of libraries which is usually *in vivo* within a bacterial host cell.

might yield 10^7 independent clones, or only a randomly selected one-millionth of the theoretical total molecular diversity in this example.

So devising a means for bypassing *in vivo* delivery of a molecular library prepared *in vitro* would of considerable benefit. To see how this has been achieved, we must first be aware that it has long been possible to perform transcription and translation of nucleic acid sequences *in vitro* using extracts from prokaryotic or eukaryotic sources. Translation in bacteria requires special sequences within mRNAs for binding to ribosomes, which are close to the codons from which protein synthesis is initiated. An mRNA that is being translated on a ribosome is in effect "read-out" for its encoded protein sequence, as the ribosome begins at a ribosome-binding site/ initiation codon and moves along the mRNA in a 5' to 3' direction, with a polypeptide chain progressively growing at the same time. In both prokaryotes and eukaryotes, more than one ribosome can simultaneously be engaged in translation from a single mRNA, and the resulting complex is termed a *polysome*. As a polypeptide elongates during synthesis, it becomes accessible to interaction with external factors. Owing to this, nascent translated products can often be bound by specific antibodies (especially if the recognized epitope is not close to the C-terminal end of a polypeptide) and *immunoprecipitated* as polysome complexes. In this manner, a newly formed polypeptide can be "pulled down" along with its encoding mRNA, and this process was long used as a means for enriching specific mRNAs and to assist specific cDNA cloning. It was recognized that if random DNA libraries encoding peptides were transcribed and translated *in vitro*, polysome immunoprecipitation could be used as a selective process (or "partitioning agent"[439]) for isolating peptides with desired properties and identifying them via their accompanying encoding RNAs. (Indeed, any receptor protein could be used for this selective purpose in lieu of antibodies.) Since RNA molecules can be reverse-transcribed into complementary DNAs, the process can be reiterated in an evolutionary manner[440] (Fig. 4.11).

This therefore constitutes a fully *in vitro* display technology, overcoming the loss of diversity through transformation procedures. But before the full potential of protein display on ribosomes could be realized, several important additional innovations needed to be introduced. By now we have become familiar with the "genotype–phenotype linkage" principle for generalized display, and it can readily be understood that for display based on ribosomes to work, both the translated polypeptide *and* its mRNA must remain stably attached as a ribosomal complex during affinity-mediated selection (Fig. 4.11). Yet for an entire protein to be displayed in a folded state, translation must normally be completed at a stop codon, and this results in disengagement of the nascent protein chain from the ribosome itself. Solutions to this can be devised by the addition of a polypeptide spacer to the C-terminus of the desired protein(s) (to allow tethering to the ribosome and room for folding to occur), combined with removal of the stop codon.[441] The latter trick itself comes at a price, because *E. coli* has a dedicated system for labeling and degrading proteins that fail to properly terminate during translation, by means of a special RNA molecule (*ssrA* or 10Sa RNA) that acts as both a transfer RNA and a messenger RNA encoding a specific peptide tag.[442] One way of overcoming this problem uses an antisense RNA molecule to interfere with the 10Sa RNA.[441]

With the inclusion of these technical modifications, "ribosome display"[*] of proteins was enabled.[441] As well as sidestepping the transformational inefficiency problem, ribosome

---

[*]Note that although both the earlier polysome display and the later ribosome display are of course based on ribosomes, the nomenclature difference is meaningful. A peptide segment encoded near the 5' end of a coding sequence can be displayed by multiple ribosomes progressing along the corresponding mRNA, and hence can constitute a polysomal display system. A full-length protein, on the other hand, has to be translated over the entire coding sequence and thus only one ribosome can display it per mRNA.

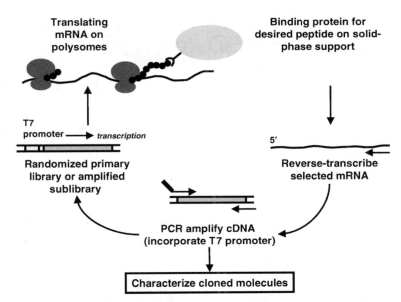

FIGURE 4.11   Depiction of general principles of polysome or ribosome display. A synthetic randomized DNA coding sequence library is transcribed *in vitro* (from a suitable promoter; the specific promoter for bacteriophage T7 RNA polymerase is convenient as shown), and coupled with (or separately followed by) *in vitro* translation with an *E. coli* extract. Cycles of selection for a specific translated binding activity are followed by cloning and sequence characterization of the relevant amplified cDNAs. The replicative fidelity of the PCR amplification can be adjusted if desired to promote additional mutations during cycles of ribosome display for directed evolution.[16]

display has certain other advantages that come inherently with its cell-free milieu. Some proteins, including receptor domains, are prone to aggregation when expressed in bacterial systems, and thus are refractory to conventional phage display. Since aggregation occurs only after release of nascent proteins from ribosomes, ribosome display has been shown to circumvent this problem.[443] Many applications of this display technology have concerned the maturation of high-affinity antibodies[444] (Chapter 7), but other binding proteins have been subjected to evolutionary selection in an analogous manner.[400,445] Still, not every aspect of ribosome display is ideal. It has been pointed out that the very large size (>2 MDa) and complexity of ribosomes increase the chances of spurious and interfering interactions with displayed peptide or protein library members.[446] Display techniques performed *in vitro* that result in a direct linkage between either encoding RNAs or DNAs and their expressed polypeptide products have been developed, which go a long way toward reducing or eliminating this problem. Let us look at what can be done for genotype–phenotype linkage at the RNA level.

An antibiotic inhibitor of protein synthesis has been deftly exploited to allow a novel means for covalent attachment between an mRNA and its translated polypeptide product. The structure of *puromycin* is such that it has the ability to mimic a tRNA molecule and "trick" the ribosome into joining the puromycin molecule to the C-terminal end of a nascent polypeptide, thereby terminating translation. Moreover, the resulting conjugate is stable. It was reasoned that covalently linking puromycin to the 3′ end of an mRNA would result in covalent conjugation of the mRNA to its translated product.[447,448] The labeling of mRNAs with puromycin is effected by means of a DNA linker with a puromycin molecule chemically attached to its 3′ end

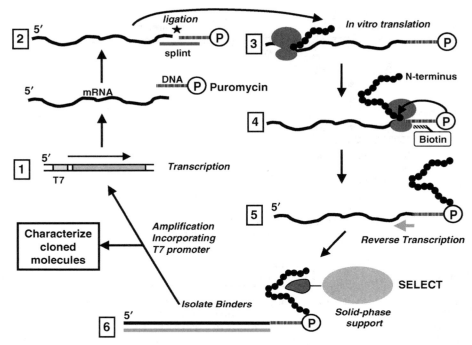

FIGURE 4.12   mRNA display for directed evolution.[17,18] (1) Transcription of a DNA coding sequence (usually with the T7 RNA polymerase promoter as for Fig. 4.11), either a natural cDNA or an artificial construct (provided both are engineered to remove stop codons). A DNA oligonucleotide is chemically labeled at the 3′ end with puromycin. (2) The oligo-puromycin adduct is ligated to the 3′ end of the mRNA, either by means of a complementary oligonucleotide acting as a "splint,"[17,18] or via a photochemical conjugation process.[19] (3) *In vitro* translation is then initiated on the mRNA ligation product library. (4) When the ribosome reaches the RNA–DNA boundary, translational pausing occurs and the puromycin can access the "A site" of the ribosome and bond with the nascent peptide. The population of adducts can be isolated with a complementary biotinylated oligonucleotide (as shown) or in other ways. (5) To avoid potential interference with selection from the single-stranded RNA, it is rendered as double-stranded mRNA: cDNA hybrid by reverse-transcription. (6) Affinity-mediated selection for a polypeptide of interest can then be performed, and the associated mRNA amplified as double-stranded cDNA and the cycle resumed, or the products cloned and sequenced.

(Fig. 4.12). Since the puromycin reaction is slow compared with normal translation elongation,[*] the puromycin attachment is favored at the boundary of the mRNA with the ligated DNA segment, where translation pauses.[447] The result of this is the desired result of linking the puromycin moiety to the end of the translated polypeptide (Fig. 4.12).

This novel linkage between mRNA and polypeptide translated from the same mRNA affords a powerful approach for directed evolution.[450] As with other approaches above, mRNA display has been applied toward antibody affinity maturation (Chapter 7), and has also been used for

---

[*]Owing to the inefficient competition between puromycin and natural aminoacyl-tRNAs, at a low concentration, free puromycin does not inhibit translational elongation but can still bond to the C-terminal residue (translational termination being a slower step than progressive elongation, allowing puromycin to compete).[449]

searching expressed genomes for encoded receptors toward a desired protein or ligand.[451,452] Since the "identity tag" for translated proteins in mRNA display is relatively small (an mRNA molecule itself), this display technology is well suited for general searches for protein interaction partners,[453] and can be viewed as an alternative to other methods such as the two-hybrid system (ftp site). Among such applications are the identification of partner proteins for DNA-binding proteins that function as heteromers.[453,454] As a consequence of its great selective power, one of the most exciting applications of mRNA display is the selection of *de novo* function from vast polypeptide libraries, which is examined further in Chapter 5.

Other *in vitro* display methods have also been developed, including display of polypeptides on their encoding DNAs (detailed in ftp site). It should be noted here that the term "DNA display" has also been applied to another process involving the use of DNA sequences as "tags" for chemical syntheses, which is discussed in more detail in Chapter 8. Although not to be confused with the above biological processes, the chemical-tag display has clear parallels with the gamut of biological display systems we have covered in this chapter to date, and is therefore not a misnomer from this point of view. There is another important area to cover, which is included in this section even though it is technically not a display technology, insofar as that is defined by physical genotype–phenotype linkage. Nevertheless, the maintenance of genotype–phenotype correspondence is crucial for this process also, and it highlights the limitations of physical display. Let us then continue down this pathway. . . .

Making Compartments   Living cells act as compartments that maintain the local correspondence between genotype and expressed phenotype. An artificial way of generating cell-like compartments could in principle be used for the purposes of directed evolution, and this has indeed been achieved. But what is the potential advantage of this kind of exercise, especially when we consider the multiplicity of display systems already available? The best way of explaining this is to understand what all the other systems cannot do.

Up to this point, we have considered display technologies that are diverse yet share a common principle of selection through molecular interaction affinities, or "finding by binding." In these circumstances, an expressed variant polypeptide selectively binds to a protein receptor or ligand of interest, along with its physically linked encoding nucleic acid. This is clear enough when (for example) proteins displayed on phage are sought for binding to a specific ligand. To be sure, some ingenious schemes have been devised to broaden the applicability of display by physical linkage to include certain enzymatic activities (ftp site), but the end read-out is inevitably dictated by binding effects. In filamentous phage display, rather than a positive selection for ligand binding, sometimes selection is negatively based on retention of infectivity after an appropriate treatment. This applies to methods selecting for protease resistance (and indirectly thermostability) of displayed proteins on gp3,[455] where only protease-resistant sequences survive digestion and preserve both intact gp3 and infectivity. Yet while physical display is a very powerful strategy that can be cleverly adapted to a variety of applications, it has limitations in certain areas. Some proteins remain poorly expressable in known display formats, and complex eukaryotic multisubunit proteins are especially problematic in this regard. Most fundamentally, though, directed evolution of a complete interactive multiprotein system is an intractable stumbling block for conventional display.

But if genes and their expressed phenotype are encompassed within a discrete and stable compartment, linkage between genotype and phenotype need not involve a direct physical association (as we have already seen from the initial discussion of display principles [Fig. 4.9a]). It has been demonstrated that tiny aqueous droplets (or globules if you prefer) within an oil phase can serve as *microcompartments* for biological reactions, and thus in effect mimic whole cells.[456] A necessary (and fairly obvious) precondition for the application of this process to directed evolution *in vitro* is to ensure that no more than one copy of an expressible gene of interest is encompassed within each microcompartment.* The desired compartments are achievable via *emulsions*, which in themselves should be familiar at least superficially to consumers of mayonnaise. An emulsion is defined by two nonmixable ("immiscible") liquids, one of which forms small discrete droplets (the discontinuous phase) within the other liquid, which then constitutes the continuous phase.† *In vitro* compartmentalization of expression libraries within water-in-oil emulsions in this manner then becomes analogous to the use of conventional libraries in bacterial or other host cells.

But does this mean we have simply come full circle, back to nothing more than an equivalent of the traditional approach of using bacterial cells as the compartments for directed evolution? Not at all, because we must remember that a major impetus for the development of *in vitro* evolutionary methods was the limitation on gene transfer into host cells, whether bacterial or otherwise. *In vitro* compartmentalization could be viewed as "transformation in reverse," since in effect the "cell" is formed around the target gene, rather than trying to transfer such genes into a bacterium. The efficiency of the partitioning of gene libraries and transcription/translational factors into microcompartments within emulsions is high.[456,457] But of course, this is not the only factor in the overall picture. A protein can be synthesized within an emulsion droplet in its normal full-length form, free of any fusion peptide sequences or covalent conjugation as found with other display technologies. For enzymatic evaluations, it is not necessary to physically couple product formation with the relevant enzyme (as in phage-based selections), since each variant enzyme, its substrate, and resulting product can all be encompassed in discrete microdroplets. Evolutionary studies on large multisubunit proteins or interactive systems difficult or impossible for physical display are compatible with compartmentalized expression systems.

Within an artificial microcompartment, one has the opportunity to control the participating proteins and nucleic acids to a precise degree that is not available for *in vivo* systems. Exclusion of nonessential factors can also effectively eliminate potentially interfering influences, and (in common with other *in vitro* systems), proteins that are highly toxic for intact cells can be processed and studied. Artificial microcompartments are also particularly amenable to very high-throughput screening approaches (Fig. 4.13 and Chapter 8). And certain selection systems that are applicable to *in vitro* compartmentalization would be difficult or even impossible to implement *in vivo*.

Now we can refocus on the issue of selection by binding or by other means. Certainly, *in vitro* compartmentalization evolutionary systems have often used molecular binding interactions in diverse ways, but in principle, this is not a necessary condition for screening of such compartments. Consider a library of enzymes in microcompartments supplied with all necessary cofactors and a substrate that a desired enzyme variant can act upon to generate a fluorescent product. The individual compartments in which such enzyme variants have been expressed will accordingly harbor an accumulation of this product, which will become detectable by virtue of its fluorescence. Screening and sorting of such compartments can then be performed by flow cytometry.[458] In this technological arrangement, the desired variant enzymes can therefore be isolated by virtue of their product formation rather than through a binding interaction. Even a single enzyme molecule in principle can turn over many substrate molecules within a single compartment, with concomitant increased sensitivity. (This can be contrasted with physical display methods for enzyme libraries using substrate-based strategies for maintaining

---

*In practice, to enforce this state of "no more than one per compartment" an excess of microdroplets will be present, where many have no internal library DNA, but virtually none have more than one copy of a library member.
†Both water-in-oil or oil-in-water emulsions can be prepared, and the former type has been widely used for the presentation of antigens to the immune system as immunological adjuvants.

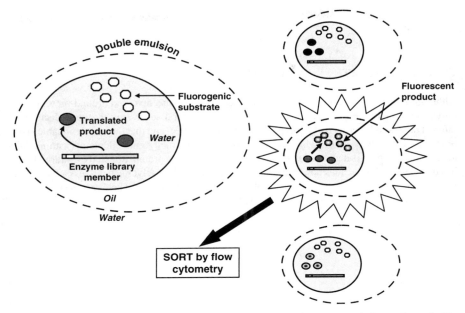

FIGURE 4.13   *In vitro* compartmentalization system for directed enzyme evolution, using a double emulsion for fluorescence-based sorting of microdroplets based on the formation of fluorescent product by the desired enzyme. Though on an approximate scale of a bacterial cell, *in vitro* double-emulsion microdroplets are large compared to their molecular contents, and are amenable to analysis by the fluorescence-activated cell sorter.[20]

genotype-phenotype linkage. In such cases, only a single round of catalysis is possible, and selection is usually based on specific binding of product.[459])

But have we forgotten something in this rosy picture? We have not only compared *in vitro* microcompartments to bacterial cells in some aspects but also noted crucial differences. One such distinction not yet highlighted is the rather obvious point that bacteria can replicate themselves, but aqueous emulsion droplets cannot. At least, an entire microcompartment within an emulsion has no direct means for replication, but this is unnecessary when we consider that all that must be amplified is the information specifying the novel library member of interest. Hence, in the above example, fluorescence-sorted compartments of course cannot "grow" like bacteria, but their internal genetic information is indeed replicable, by the polymerase chain reaction (PCR).

As well as providing a means for linking genotype with corresponding phenotype, micro-droplets serve as vehicles for the maintenance of suitable concentrations of participating molecules to foster efficient biological reactions.[456] This issue raises in turn the problem of delivery of reagents into microdroplets that have been formed previously, sometimes a highly desirable feature. Just as living bacteria need to have specialized systems for regulating the uptake of molecules from the environment, so too is it important to have a means for "communicating" with the microenvironment within *in vitro* emulsion droplets.[457] Of course, simple microcompartments alone cannot have the sophistication of bacterial transport systems, but both water-soluble and hydrophobic compounds can be delivered into the artificial internal microenvironments of emulsion droplets. Soluble hydrophilic molecules can be transported by use of even smaller *in vitro* compartments ("nanodroplets") and hydrophobic molecules through the oil phase itself.[457] A variety of other physicochemical methods for "communication" with microdroplets also exist.[457,460]

TEST IT, REPLICATE IT, EVOLVE IT

Although much of our thinking about molecular evolution has revolved around proteins, the only requirements for directed evolution in the most general sense are that the molecules of interest possess a functional phenotype which can be assayed, and that they are replicable. This obviously includes proteins, which replicate by means of their "entanglement" with nucleic acids, and Chapter 5 is dedicated to the wide-ranging theme of protein directed evolution. But it has been known for decades now that certain nucleic acids themselves can possess a functional phenotype, and thus show a convergence of genotype and phenotype. The directed evolution of nucleic acids and related themes is presented in Chapter 6.

5 The Blind Protein Maker

OVERVIEW

The evolutionary processes that have unfolded on this planet for the last three billion years have been incredibly creative, but the resulting intricate molecular designs did not emerge from any predetermined plans. Repeated iteration of the blind algorithm of natural evolution proceeds in response to variable environmental and biological selective pressures, and complex features can be lost as well as gained if a selective advantage accrues. But while biotechnologists can draw inspiration from the messages of natural evolution and its products for artificial evolutionary aims, we have seen that the parallels between the natural and artificial processes are not perfect. From the human point of view, the obvious aim of the exercise of directed evolution is to achieve a positive and clearly defined functional molecular gain, and any additional information that can assist the implementation of this will be fed into the process as much as possible. While natural mutation may deviate from randomness in many cases (depending on a number of factors), it does not have a teleological focus on specific genomic regions. As we have seen in the previous chapter, humans on the other hand can use "semirational" *in vitro* evolution to target mutations or recombinational boundaries at will. Human rational decision making can also dictate a starting-point choice for evolutionary direction that would not occur in nature. The screening or selection processes imposed by human agency on an *in vitro* evolutionary pathway can be as convoluted and essentially unnatural as is considered appropriate. Directed evolution can also be useful in "bootstrapping" approaches toward protein improvement, by initial rounds of mutagenesis and selection to define crucial sites for specific protein functions. Following identification, these sites can subsequently be subjected to saturation mutagenesis to test the functional effects of all possible substitutions.

Having thus briefly visited the mechanisms and process of *in vitro* evolution in the last chapter, we can now address some specific subsections of this large area of human activity, beginning with proteins. An exception needs to be made for proteins of the immune system, owing to their functional importance and the large amount of effort that has been devoted to them and conceptually related recognition molecules, and this material is found separately in Chapter 7. Although it is necessary for clarity to define subfields within the broad topic of artificial molecular evolution, there are many examples of overlap. For example, enzymes and antibodies are here treated as distinct themes, but in reality they are not completely mutually exclusive, as antibodies themselves can exhibit catalytic activities in some circumstances, which we will also explore further in Chapter 7.

From these considerations, the tag "directed" as applied to artificial evolution certainly seems appropriate, as the general process is clearly being guided as much as possible by human intervention. Earlier we used computational approaches to games such as chess as a metaphor for the "smart" application of directed evolution, by comparing brute-force number crunching with increasingly sophisticated decision-making algorithms. To mix metaphors, let us now think in more detail how best to play chess with the blind protein maker.

APPLYING NATURAL PROTEIN EVOLUTION PRINCIPLES

Some features of proteins believed to be evolutionarily significant are themselves relevant to applications in protein engineering and *in vitro* evolution. A useful place to start in this respect is the tolerance of proteins for amino acid residue changes, which has obvious implications for both natural and artificial evolutionary change. The concept of designability, introduced in Chapter 2, is defined as the numbers of sequences that will fold to the same (lowest energy) structural state. A "designable" sequence thus has many closely related sequences that are structurally equivalent, which reflects the mutational "robustness" of a protein (or the ability of a protein to resist sequence changes without loss of structure and function). A related measure is the "substitutability" of a protein or its tolerance for random codon changes, which is inversely associated with the degree of coupling between residues. Theoretical models suggest that in an evolving population, tolerance for sequence substitutions will itself be selected for, as manifested by progression toward more highly tolerant regions of protein sequence space.[23] (Protein sequences with low tolerance will tend to become under-represented through a higher fraction of mutations that ablate function.) Mutational tolerance of a protein is an index of evolvability (again first mentioned in Chapter 2), or the capacity of a protein to undergo evolutionary modification. Other studies[155] predict that evolvability itself is selectable in a rapidly changing environment, despite the fact that selection for the ability to evolve superficially smacks of "anticipatory" change. Under some selective pressures (including selection for thermostability), proteins are predicted to evolve folds with increasing designability,[461] and the trend toward robustness has been experimentally associated with evolving polymorphic protein populations of large size.[462] The latter effect was considered a plausible evolutionary factor in large bacterial or viral populations.[462]

Increasing thermal stability has been predicted to correlate with enhanced protein robustness,[463] and this has been demonstrated with natural thermostable versus mesostable enzyme homologs by measuring their respective tolerances for random substitutions.[464] Similar effects have also been observed with engineered versions of cytochrome P450 with differing starting-point thermostabilities.[174] (Modeling and practice have also indicated that it is easier to enhance stability while maintaining high function than vice versa.[465]) Enhancement of mutational robustness in turn promotes evolvability, by allowing sampling of a range of residue changes that would be fatal in a nonrobust version of the protein.[174] The ability to explore an increased volume of sequence space with a more robust protein therefore greatly improves the chance of finding a beneficial mutation. We also noted in Chapter 4 that acceptance of neutral mutations as a consequence of robustness allows the preparation of "neutral-drift" protein libraries for functional screening.[409]

What do these findings indicate for the best practice of directed protein evolution? Performing multistage modifications on a target protein has theoretical advantages, to first render the target more thermophilic with the concomitant prediction of enhancement of general evolvability toward other desired functions.[174] Yet this may be an onerous rate-limiting step, as is preliminary garnering of extensive experimental data with respect to the mutational tolerances of candidate proteins. It has been accordingly suggested that it is easier in practice to identify the best natural targets for *in vitro* evolution by virtue of their likely evolvability characteristics.[466] This in turn is predicated on the view, strongly supported by experimental evidence (as above and Chapter 2) that protein evolvability is far from uniform and that specific evolutionary circumstances will determine the degree of evolvability of proteins. "Generalist" proteins in organisms facing rapid environmental change are one such broad category of proteins with good prospects for evolvability and artificial evolutionary modification.[466]

Many natural proteins thus appear to be "primed" for further evolution, which can be exploited by humans. On the other hand, some proteins are likely to exhibit poor evolvability, including those that are highly expressed (and resistant to misfolding through translational errors) and those with very high "connectivity" in metabolic pathways or genetic circuits. While all these considerations are undoubtedly useful and foolish to ignore, the ability to choose a natural protein with optimal evolvability may often be limited by the specific experimental goals at the outset. In such cases, preliminary work such as initial rounds of evolutionary selection for thermostability, or mutational analysis of alternative candidate proteins, might be worthwhile despite the required time investment.

PRODUCTIVELY PLAYING WITH ENZYMES

Certainly, the artificial evolution of enzymes can be driven by practical motives. Given the importance of enzymes in a wide variety of research and industrial applications, improved enzymatic products of directed evolutionary manipulations may often gain economic importance. But it should also be noted that although the laboratory evolution of enzymes often has such pragmatic underlying goals, its scientific relevance by no means stops at the level of enzymatic improvement alone. Directed protein evolution in general can be a powerful tool for advancing basic knowledge, by testing theoretical evolutionary models. Importantly, though, directed evolution can be also used to differentiate between structural or sequence features of natural proteins that are physically necessary for maintenance of a specific property or function, and those which are replaceable by alternatives.[467] When an enzyme is artificially evolved toward a new catalytic specificity, certainly the patterns of mutant residue substitutions can be significantly divergent from natural homologous enzymes with the corresponding specificity.[468]

Room for Improvement

To seek an improvement in anything, it is clearly a fundamental requirement to know the initial properties and performance of what one is starting out with, whether we are referring to a tool at the macroscopic or nanoscale levels. Searching for improvements in a tool may mean going beyond the attitude that proposes "if it ain't broke, don't fix it," and building a better figurative mousetrap. But in some cases, a tool may not function at all under the desired circumstances, rendering the former aphorism inapplicable. With enzymes in mind as the molecular tools, we can cut through their great diversity to define three major areas where meaningful performance improvements might be sought. These are: (1) Improving the ability of the enzyme to withstand unusual environments without loss of high-level function. This is synonymous with "stability," since "environment" is used here in a broad sense to include both physical (for example, temperature) and chemical factors (for example, salt concentration); (2) improving natural catalytic performance; (3) changing the natural substrate specificity.

Clearly, these goals are not necessarily mutually exclusive, and all three could be combined, as where an enzyme is evolved toward a novel substrate, with improved catalysis over existing alternatives, and at elevated temperatures. Other enzyme attributes will usually fall to some degree within one of these improvement goal categories, although some overlap is possible.

For example, evolutionary engineering of an enzyme cofactor requirement would naturally fall within "catalytic performance," but such an alteration might in principle also affect the original substrate specificity. An additional fourth aim for *in vitro* enzyme evolutionary design can be also be included, the most ambitious of all. This is the generation of biocatalysts with completely novel activities.

Tougher Enzymes

Let us examine the first of these aims, enzyme stability, in a broader context. A typical globular protein sculpted by natural evolution is not any more stable than it needs to be in its normal environment. Most globular proteins have such "marginal stability,"[469,470] and computer modeling of competing evolving proteins suggests that proteins with marginal stability have an intrinsic evolutionary advantage.[470] In Chapter 4, we noted that a limited correlation exists between thermostability and resistance to proteolysis[376] based on the observation that partial thermal denaturation of a protein can increase its exposure as a proteolytic substrate. Since efficient proteolytic turnover is essential for normal cellular metabolism, this may be one factor selecting for marginal protein stability within an organism's normal operating temperature range.[471] The possible trade-off between stability and functionally significant structural flexibility has also long been noted.[472] Many natural precedents exist for enzyme homologs optimally active at low (psychrophiles; <20°C), midrange (mesophiles; ~20–50°C), high (thermophiles; ~50–80°C) and very high temperatures (hyperthermophiles; >80°C*). Although the upper boundary of thermal survival for whole organisms is ~110°C,[471] some individual proteins from hyperthermophiles can survive at considerably higher temperatures. The denaturation temperature of the protein rubredoxin from the hyperthermophilic organism *Pyrococcus furiosus* (whose name alone suggests a fondness for hellish conditions) has been estimated at near 200°C.[473] Given the marginal stability of most natural globular proteins, improving the thermostability of familiar mesophilic enzymes by evolutionary approaches is a realistic prospect in principle, and very often in practice as well. Studying the physical basis of the stability of enzymes in hyperthermophilic organisms would seem a good place to start, and information obtained in this manner should be relevant for both rational and semirational design approaches.

Improved polypeptide side chain packing can contribute to thermal stabilization, but the packed cores of both mesophilic and thermophilic proteins do not significantly differ in their average packing efficiency,[471,474] suggesting that further improvements in this area may rapidly reach a point of diminishing returns. But if proteins are considered in total, it has also been known for some time that strong local interactions and minor changes in sequence can significantly affect thermostability.[472] Nevertheless, there do not seem to be any universal set of general rules that can be applied toward the thermostabilization of mesophilic proteins.[347,475] Thermophilic proteins are apparently stabilized by a variety of mechanisms, and defining these interactions is complicated by sequence context-dependent effects.[475] To be sure, some general guidelines have been defined, such as the roles of electrostatic interactions, hydrogen bond networks, and optimization of hydrophobic packing (to the extent possible, as noted above),[476–478] but their implementation through the application of any general formula has remained elusive. Thermostabilizing mutations have somewhat surprisingly often been located on the surfaces of proteins, although such sites would on average have a lower tendency

---

*These are only approximate temperature ranges for each category. The enzyme temperature optima obviously reflect the natural environments of their host organisms, ranging from subzero (where a liquid state is maintained by dissolved salt or pressure) to ~110°C (superheated thermal sites for "extremophile" organisms.

to disrupt secondary structures.[347] In the special case where a close thermophilic homolog exists for a mesophilic protein of interest, a detailed sequence comparison may reveal key residues that can be subsequently altered by "rational" design,[372] but such handy pairwise comparisons are not always possible. Obviously, if reliable and generalizable rules for thermostabilization existed, protein engineers would not hesitate to use them, regardless of their inherent complexities. While the lack of such step-by-step instructions is frustrating from the rational design viewpoint, these observations highlight the ability of directed evolutionary processes to leap over complex structural conundrums in the search for design solutions.

Despite the difficulties in routinely applying thermostabilization as a rational design process, in most cases the possibility of improvement *per se* is a good prospect at the outset. The marginal stabilities of normal mesophilic proteins, combined with the basic knowledge that thermostable proteins are widespread in nature, has suggested that desirable thermostable variants of mesophilic proteins should be attainable. The problem of pinpointing such variants out of the huge number of possibilities in sequence space has been successfully addressed by directed evolution, using mutagenesis and/or recombination, with a variety of enzymes.[371,479–481] A useful feature of this pursuit has been the frequent additivity of individual mutations conferring degrees of thermostability, such that a combinatorial approach may be profitably adopted. Combining thermostable mutations can have impressive results, raising moderately stable enzymes into the hyperthermophilic range.[482]

The difficulties of making precise generalizations regarding thermostabilization do not mean that pre-existing structural information is without value. Any knowledge can be fed into a plan for directed evolution of a specific protein, such that the approach enters the "semirational" domain.[483] Depending on circumstances, this can take the form of phylogenetic sequence comparisons between the protein of interest and thermostable homologs, or targeting of potential "hot spots" by initial rounds of random mutagenesis (as noted earlier). It is to be also expected that computational methods for basing predictions on stabilizing residue interaction networks will continue to advance.[347]

In a case of directed evolution of a bacterial lipase for thermostability, pre-existing structural information was indeed useful for choosing combinations of mutations derived initially from mutagenesis and selection for thermal denaturation resistance.[484] Analysis of the 3D structures of mutant versus wild-type enzyme allowed the physical basis of thermostabilization to be rationalized, and indicated subtle structural differences that enabled indirect stabilizing hydrogen binds to form (Fig. 5.1). After the initial directed evolution steps, additional specific point mutations were chosen for further projected improvements, but some of their stabilizing effects in the lipase were not predicted. Such subtle effects illustrate the challenge of complete rational design for thermostability.[484]

It is important to realize that thermostable enzymes are isolated from organisms that flourish at high temperatures as integrated systems, which means in practice that the thermostability of some natural enzymes results (at least in part) from *in situ* interactions with other factors present in their normal environments. Ionic stabilization of enzymes (and proteins in general) is well known,[485,486] and by directed evolution it has been possible to reduce the dependence on sodium chloride for the thermal stability of an enzyme.[487] Although the importance of "environmental" stabilization mechanisms is clear, a specialized class of proteins themselves are vital regulators of protein folding and stability. These *chaperones* have received this label by virtue of their essential roles in assisting the correct folding of many proteins.[488] Many chaperones are themselves induced by stresses such as heat and are thus included among the numerous proteins found in the "heat-shock" response, which is virtually universal among cellular organisms.[489,490] One might then wonder whether chaperones from hyperthermophilic organisms have undergone adaptations that improve correct protein folding and suppress

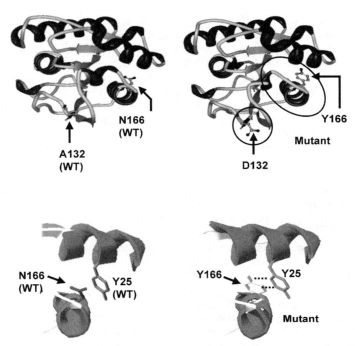

FIGURE 5.1   Structures of wild-type (WT; on left top and bottom) and mutant lipases from the bacterium *Bacillus subtilis*. Top, whole enzymes comparing wild-type[21] and a double mutant with 100-fold improvement in its thermal denaturation time,[22] showing the side chains for the A132D and N166Y mutation in the mutant. Dark gray ribbons: α-helices; light gray ribbons: β-strands, turns, or coils. *Sources*: Protein Data Bank[2] (http://www.pdb.org) 1I6W (wild type) and 1T4M (double mutant). Images generated with Protein Workshop.[5] Bottom, regions of wild type (left[23]) and mutant[22] lipases compared to show an example of a stabilizing interaction (aromatic ring stacking between mutant tyrosine 166 and wild-type tyrosine 25) that enhances interhelical packing. *Sources*: Protein Data Bank[2] (http://www.pdb.org) 1ISP (wild type) and 1TM4 (mutant). Images generated with Swiss-PdbViewer[24] (http://au.expasy.org/spdbv/).

denaturation at high temperatures. Consistent with this supposition, the presence of chaperones from hyperthermophiles can indeed enhance the thermal survival of both mesophilic and thermophilic proteins.[491,492] This can in effect be an indirect way of eliciting at least partial heat resistance at temperatures above a protein's normal upper limit, which may have practical ramifications. One such application may be the extension of the normal half-life of Taq polymerase in the polymerase chain reaction.[492]

Thermophilic organisms have been used as hosts for the selection of proteins with improved heat resistance, and mutational stabilization has been successfully observed with such a strategy.[493] In such cases possible effects of thermophilic host chaperones on the exogenous target protein need to be excluded, although this will normally fall into place during *in vitro* analyses of purified candidate proteins (where additional host factors are absent). The ability of chaperones to buffer foreign proteins against thermal stress also raises the interesting issue of their specificity. At least some chaperone pathways appear to be based only on the recognition of hydrophobic tracts of unfolded proteins, and it might seem quite improbable that for every protein requiring such folding assistance, a distinct chaperone should exist. This view would in turn predict at least a level of chaperone promiscuity[494] and is consistent with the action of chaperones on completely exogenous proteins, such as jellyfish green fluorescent protein (GFP) in bacteria.[495] Yet directed evolution itself has had a role in showing that the activity of molecular chaperones is not completely nonspecific. It has been possible to use *in vitro* evolutionary approaches to alter the GroEL chaperonin[*] of *Escherichia coli* such that it greatly enhanced the folding of GFP, but this gain came at the cost of decreased activity toward its normal cellular targets.[497] Such results highlighted an inherent conflict between the natural evolution of chaperones toward increasing target specificity, balanced against a need to remain active for a wide range of proteins.[497]

Many industrial incentives exist for producing enzymes with enhanced thermal stability.[498] Some work in enzyme thermostability that might appear purely research-based also has a high probability of feeding into later practical applications, such as the directed evolution of antibiotic selectable markers for hyperthermophilic organisms.[499] But so far we have been looking at "stability" as though resistance to thermal denaturation was the only factor worth considering, but of course enzymes have been naturally selected to resist a variety of environmental challenges. The interactions controlling these different types of stabilizing influences are of considerable interest to human enzyme engineers and are also amenable to *in vitro* evolutionary developmental strategies. In the same manner as for thermal stability, one can look toward organisms in extreme environments as potential sources of enzymes with enhanced tolerances toward conditions such as (but not limited to) high acidity, alkalinity, salt concentrations, pressure, desiccation, and ionizing radiation.[500] Recalling the above consideration of chaperone-mediated protein stabilization, the tolerance of extremophile organisms for "extreme" conditions does not automatically indicate corresponding intrinsic resistance of their enzymatic repertoire, since their intracellular environments may be maintained as privileged sites by specific mechanisms. For example, organisms tolerating extremes of acidity use high-efficiency proton pumps to preserve a pH differential between the external and intracellular environments.[501] But even with this caveat acknowledged, "mining" of extremophiles for enzymes has been a profitable exercise.[502] Ideally, these organisms at the outer limits of life on earth would also provide information enabling rational design of comparably resistant enzymes at will. Although understanding of the mechanisms of their tolerance of extreme environments has greatly advanced,[500] this knowledge is not easily translatable into general principles, again analogously with thermostability. And once more, directed evolution can step into this gap between theory and practice. In any case, there is no reason why artificial evolutionary processes cannot be applied toward a variety of "extreme" conditions, including those not likely to be encountered in nature. Various enzymes have been artificially evolved toward improved pH activity ranges[503] and also resistance to surfactants,[504] organic solvents,[401] and oxidizing agents.[505]

Some extremophiles exhibit resistance to more than one stressor, or resistance to a specific environmental stress is also accompanied in some cases by thermal resistance.[500] In the earlier section on the importance of well-designed screening (Chapter 4), a correlation was noted between thermostability and resistance to unfolding, which might be manifested in additional modes of stabilization. In the context of the above-noted screening examples, a convenient assay design scheme accordingly used protease resistance as a surrogate measure of thermostability. The general concept of "stability correlation" can be simplistically visualized for a globular protein as the tightness of packing of a coiled ball, where the tighter the packing, the greater the resulting stability. According to such a model, it might be expected that resistance to thermal denaturation would augment resistance to other physical or chemical denaturants as well.

---

[*]Chaperonins are a specific subset of chaperones that do not directly interact with newly formed polypeptides from ribosomes, and that form large ring structures where protein folding guiding takes place.[496]

Evidence in favor of this notion has accumulated with enzymatic thermal stability shown to correlate in specific cases with alkaline pH tolerance,[506] oxidative stability,[505] and resistance to organic solvents.[507] But in spite of such observations, the mechanisms of stabilization to such varied denaturing influences result from diverse local structural changes, and there is no firm suggestion that correlation with thermostability is a universal rule. As noted earlier, thermostability and at least protease resistance do not always go hand in hand.[376]

We have seen that the marginal stability of most mesophilic natural proteins renders their stabilization a feasible prospect beyond the "settings" of natural selection. But what of the enzymatic performance *per se*? How does artificial evolution shape up as a means for further improving enzymes as biocatalysts?

Better Enzymes

The second item on the list of enzyme improvements above cites catalytic performance, or how well an enzyme does its job. This raises some questions that do not automatically arise when stability is the target for improvement. It is easy enough to accept that an enzyme naturally accustomed to mesophilic temperatures will not necessarily be stable to temperatures well outside its normal operating range, since there have been no natural selective pressures exerted in this direction. But if one aims to make an enzyme derivative catalytically superior to its natural precursor at the *same* temperature range, this surely poses the question of why natural enzymes should not already be perfected in this regard through biological evolution.

Before continuing, there are some basic issues to note. In order to define the relative efficiency of enzymes, it is absolutely essential to accurately measure their catalytic performances, and for the theory and practice of this we must look to quantitative enzyme kinetics. In this context it is useful to keep in mind the parameters k_{cat} (catalytic turnover number), K_m (the Michaelis constant, equal to the substrate concentration at $V_{max}/2$, where V_{max} is the maximal reaction velocity obtainable in the limit as substrate concentration increases indefinitely), and k_{cat}/K_m (the catalytic efficiency*).[508] Natural enzymes differ in this measure of catalytic performance. When an enzyme attains "catalytic perfection," the rate of catalytic turnover is high enough that it becomes limited only by the rate that the enzyme can bind substrate, and this in turn is determined by substrate diffusion in solution. Enzymes that have evolved to this level are hence "diffusion-limited." (as noted in Chapter 2), and some (but by no means all) natural enzymes have attained this status.[189,509]

If an enzyme naturally evolves to become diffusion-limited, it is inferred that no further selective pressure could be exerted for additional improvement, and the enzyme has reached a final evolutionary peak. Yet passive diffusion of substrate in solution is not the end of the story. Where charged residues create strong electrostatic fields, substrate may be "guided" toward the active site in a more efficient manner than via diffusion alone.[510] Some enzymes have been taken to a state of "superefficiency" by mutations at a distance from the catalytic site, attributed at least in part to enhancement of substrate attraction.[509] So at least some enzymes that are theoretically perfect in their chemical catalytic mechanisms may be further improvable through mutations that augment the rate of substrate uptake beyond passive diffusion.[511]

Catalytic efficiency, though, is not the final word on the capability of an enzyme. The catalytic tasks that enzymes perform vary enormously in their "severity," as defined by the rate

---

*K_m gives an indication of the tightness of substrate binding by an enzyme, and the lower the K_m value, the tighter the binding. However, the K_m value depends on the catalytic rate constant and is thus not a true enzyme–substrate dissociation constant.[508] Reaction velocities with different substrates can be compared with their respective k_{cat}/K_m values, and thus k_{cat}/K_m also provides an index of reaction specificity.

at which the chemical change engendered by the enzyme will occur spontaneously. Catalysis of an event that occurs spontaneously at an extremely slow rate is therefore a more "severe" undertaking than one where the spontaneous change is relatively rapid.[512] The spontaneous (uncatalyzed) rate in many cases may be so slow as to approach zero, but "practically zero" is not totally zero. Extremely gradual uncatalyzed rates of chemical change at room temperature can be assessed by extrapolating measurable rates at elevated temperatures.[513] With this information available, the "catalytic proficiency" for an enzyme can be derived by comparing the rate constants for the catalytic and spontaneous reactions.* Quantitated enzyme proficiencies vary over a huge range spanning more than fourteen orders of magnitude.[513] The enzyme orotidine 5′-phosphate decarboxylase can accelerate uncatalyzed change in its substrate (decarboxylation) by a factor of 10^{17} and wins a prize for catalytic proficiency.[512,513] Not only is orotidine 5′-phosphate decarboxylase an exceptionally proficient enzyme, but it performs its catalysis without the assistance of cofactors.[512] We should recall from Chapter 2 that it is possible in at least some cases to artificially evolve enzymes toward natural cofactor independence,[200] but this aspect of enzyme "improvement" to date has received less attention than attempts to improve catalytic efficiencies *per se*.†

Many directed evolutionary experiments have indeed resulted in enzymes with enhanced catalytic performance, and such information itself has been informative. As might be expected, changes at or near the active catalytic site are often found in enzymes with enhanced activity generated by *in vitro* evolution,[395,514] but mutations remote from the active site can also positively modulate enzymatic performance.[515,516] Although the active sites of enzymes are obviously of central importance to catalysis, it is increasingly apparent that the whole (often very large) enzyme protein molecule has a dynamic role in the optimization of the catalytic mechanism.[517,518] Certainly, mutational changes not directly involved with the catalytic center can dramatically affect the activity of an enzyme.[515] Nevertheless, in the absence of additional information, a useful primary strategy would direct random mutations to the vicinity of the active site of an enzyme. The prospects of success in such endeavors will continually improve through better definition of specific local residue positions that accommodate diversity and mediate active site "plasticity."[519] Of course, as this kind of knowledge and its attendant predictive power advances, "semirational" enzyme engineering is progressively transferred into the unequivocally rational domain. But the global effects of protein structures on catalysis at a number of different levels (as noted above) serves to remind us that tinkering with the catalytic site alone is insufficient for the realization of the full potential of enzymes. In turn, the greater difficulty of anticipating the effects of distant mutations on catalysis (especially in complex combinations) renders directed evolutionary strategies a strongly viable alternative for the foreseeable future.

These considerations tell us about the range of natural enzyme capabilities, and strategies through which enzymes may be catalytically improved. Yet we have still not addressed the question of why some enzymes are *not* naturally diffusion-limited. Why are some enzymes apparently "less than perfect" to start with? If changing a limited number of residues (whether through directed evolution or "rational" design), can improve enzyme catalytic rates under normal conditions, surely these would have already been tested and selected naturally? A more convincing case can be made that a novel beneficial combination of residues that have hitherto escaped "natural evaluation" can be generated by recombinational DNA shuffling. Yet with the

---

*Catalytic proficiency has been defined as $(k_{cat}/K_m)/k_{non}$, where k_{non} = rate constant for the noncatalyzed reaction.[513]
†This also brings up the issue raised in Chapter 2 of how essential cofactors may be for optimal enzyme catalysis in the first place, and the catalytic limitations of the 20 natural amino acids. We will return to this topic later in this chapter when considering efforts to incorporate unnatural amino acid residues into proteins.

ancient origin of most enzymes, the existence of natural horizontal gene transfer, and vast amounts of time, it may seem surprising that so many novel catalytically beneficial enzyme sequence combinations have been found by artificial evolution. We can recall that an evolving protein may reach a local fitness optimum from which further progress is constrained, but when an apparently simple evolutionary pathway for further improvement is revealed artificially, one is prompted to ask why the heights of such a fitness peak seem not to have been already naturally scaled.

Here, we must step back and consider that natural enzymes work as interlocking components of highly complex systems, but the human protein engineer usually is asking for enzymes to perform better in isolation. We have noted previously that selection for enzyme stability may be counter-balanced by the essential need for proteolytic turnover (which may correlate with increased conformational flexibility and marginal stability). As also noted above, it is also known that protein dynamics is a very important contributor to the catalytic efficiency of enzymes, and regions not necessarily close to the active site are often critical for such motions.[517,518,520] Complex networks of intracellular protein–protein interactions may then provide counter-balancing selective forces that limit the natural evolutionary movement of enzymes in sequence space, such that compromises must be reached that may fall short of catalytic perfection. Different enzymes will also have differing levels of selective pressure placed upon their efficiencies in the first place. Enzyme-catalyzed biological activities where speed is at a special premium might be expected to undergo strong pressure toward maximal rates of catalysis. One such example is acetylcholinesterase, a highly efficient diffusion-limited enzyme[521] that is essential for rapid turnover of the neurotransmitter acetylcholine at synaptic junctions.[*] Natural enzymes will thus tend to evolve toward an optimum that balances a variety of factors *in vivo*, one of which is obviously catalysis, but not the only one. In contrast, such additional constraints are unlikely to hinder the laboratory director of artificial evolution. When screened in isolation for improved catalysis, some enzymes may thus be capable of traveling an evolutionary pathway that is circumscribed under natural conditions.

If global protein flexibility and movement is an integral component of optimal catalysis, and flexibility itself appears to be inversely correlated with enzyme thermostability (as we have seen above), how are catalysis and the temperature optima of enzymes linked in general? Since chemical reaction rates increase with temperature, it might be expected that thermophilic enzymes would show generally higher rates of catalysis than their natural mesophilic counterparts, but their activities tend to be roughly similar at their respective temperature optima.[347] This has been explained as flowing from the observed inverse relationship between stability and flexibility (or increasing "rigidity" with thermostabilization), since reduced flexibility might have a cost in catalytic rate. With this view, increasing stabilization comes with a trade-off in catalytic rate that counter-balances the benefits of increased temperature.[347] Conversely, to operate effectively at low temperatures, a degree of additional enzymatic flexibility should logically be required. Studies on cold-adapted psychrophilic enzymes are generally in support of this notion, frequently showing loss of weak residue interactions that serve to increase flexibility (or "plasticity") at low temperatures.[523] Indeed, when thermophilic enzymes have been adapted to activity at low temperatures by experimental directed evolution, this has been associated with a gain in protein flexibilities.[524] In the opposite direction, artificial evolution of enzymes toward thermostability has been accompanied by reduced flexibility.[525]

For natural psychrophilic enzymes, the downside of increased flexibility is a decrease in temperature stability.[526] It is well established that natural cold or heat-adapted enzymes

---

[*]The physiological importance of acetylcholinesterase is chillingly demonstrated by the rapid and deadly effects of its inhibition, as with the highly lethal organophosphate "nerve gases."[522]

(psychrophiles and thermophiles, respectively) work poorly if they are at unaccustomed temperatures,[527] but evidence suggests this is not an intrinsic restriction of proteins, but a consequence of natural evolutionary selective pressures. Thus, a psychrophilic subtilisin protease can be taken by artificial evolution to mesophilic stability with acquisition of a broad temperature range of activity[527] and conversely for a mesophilic subtilisin in the reverse direction.[528] Experimentation has clearly shown that conferring thermostability on an enzyme is certainly not necessarily detrimental to catalytic function,[372] and thermostabilization by directed evolution has in many cases been compatible with improved activity[529,530] or at least retention of approximate original activity levels.[531,532]

Yet as we have previously seen, thermostability and catalytic optima are not necessarily coupled,[374,375] and there is no necessary linkage between stability and activity.[371] Indeed, thermostabilizing mutations may come at a high cost in catalytic function unless both stability and activity are carefully selected for.[533] Since mutations that augment thermostability but simultaneously maintain activity at reduced temperatures are rare, evolution will tend to favor activity at a narrow temperature range unless appropriate selective pressures are exerted.[534] Again, this highlights the observation that natural evolution will not necessarily explore the full potential of a protein's capabilities. If no selective pressure exists for an enzyme to operate across a wide range of temperatures, then its natural evolutionary progression will tend to be thermally restricted. Nevertheless, human directors of evolution in the laboratory decide in advance the desirable temperature/activity profile of the enzyme of interest, and screen or select accordingly, and often with considerable success.

Altered and Novel Enzymes

The theme of the preceding section was simple in concept (if not necessarily in its execution): take an enzyme that performs a specific task, and change it such that this task is done more efficiently, without sacrificing any other useful attribute. In practice, it is the efficiency not just of a catalytic process that is of both scientific and potential commercial interests, but also of its specificity and how to tinker with it. This itself can be further divided into two broad interrelated categories: altering the specificity of substrate recognition and thereby eliciting the formation of new products, or altering the specificity of product formation without necessarily changing the initial substrate preferences. Where practical applications are sought, the desirable target substrate specificity of an enzyme is necessarily linked with the ultimate envisaged task. For industrial degradative enzymes (such as proteases or amylases), the precise specificity may be of limited importance, and indeed broad specificity may be a desirable trait. In such cases, the factors of enzyme stability and catalytic preferences are of much greater significance.[498,535] On the other hand, for biosynthetic and organic chemical applications, the specificity of both substrate binding and product formation are paramount.[536,537] If an enzyme's initial substrate specificity and/or catalytic mechanism is altered radically enough, the new enzyme might be considered truly "novel."

The artificial manipulation of the substrate targeting of enzymes should prompt us to think about the natural evolution of enzyme catalysis to cover the huge range of tasks observable in the biosphere of the present day. Certainly, one can identify natural precedents for using specific enzyme framework scaffolds for evolving divergent catalytic properties.[190,192,538] It is both expected from evolutionary first principles and borne out in practice that enzymes are dividable into families based on shared properties, and furthermore that an understanding of these relationships should be helpful for studies aiming to change enzyme specificities. Using sequence and structural information, enzyme superfamilies with fundamental underlying common features can in fact be identified, and their conserved key residues defined. When

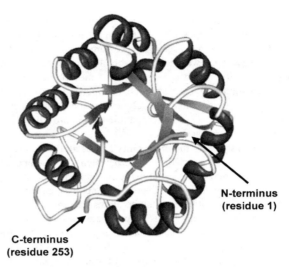

**N-terminus
(residue 1)**

**C-terminus
(residue 253)**

FIGURE 5.2   Structure of a representative TIM $(\beta/\alpha)_8$ barrel scaffold protein, the cyclase subunit of imidazoleglycerol phosphate synthase (product of the *HisF* gene, a component in the histidine biosynthetic pathway) from the hyperthermophilic prokaryote *Thermotoga maritima*.[25] View of the β/α barrel: α-helices are shown in dark gray and β-strands are shown in light gray, with N- and C-termini of protein (253 residues) indicated. (In this enzyme structural motif, the active site is always configured at the C-terminal end of the barrel's β-strands.[26]) *Source*: Protein Data Bank[2] (http://www.pdb.org) 1THF.[25] Image generated with Protein Workshop.[5]

mechanistic information is available for a number of members of the extended superfamily, it is feasible to further define certain common features of the superfamily enzyme chemistry that are shared at a fundamental level.[*,538] It is also possible to identify nonenzymatic proteins that share certain structural and primary sequence features with specific enzymes, suggesting an evolutionary relationship.[539]

We have already encountered (in Chapters 2 and 3) the important observation that the number of natural protein folds is vastly less than the total diversity of natural protein sequences (and the latter in turn constitute a tiny subset of all possible protein sequences). This evolutionary parsimony is certainly reflected in the enzyme universe, where a limited number of folds form the basic scaffolding for a very large number of enzymes with diverse catalytic activities.[538] The most common enzyme fold in protein databases is the β/α_8 barrel or TIM barrel[†,540,541] (so named for the enzyme triosephosphate isomerase, where this structure was first detected). The barrel-like structural fold of these enzymes is defined by an inner group of eight parallel β-strands encompassed by an outer set of eight α-helices[540] (Fig. 5.2).

The active sites of all TIM-barrel enzymes are at the C-terminal end of the β-strands, comprising the inner "barrel," consistent with a common evolutionary origin for this structural motif that has diverged extensively in function.[542] It is thus evident that protein-based biocatalysis has used a relatively limited number of folds to explore "reaction space,"[192] the

---

[*]Basic chemical steps can be shared in common even though the detailed catalytic mechanisms and overall substrate and product natures have diverged as the superfamily members evolve. One example is the amidohydrolase superfamily, where the common chemical process is metal-assisted hydolysis.[190]

[†]Approximately 10% of all enzymes to date conform to the TIM-barrel fold.[540] Another example of a TIM-barrel enzyme is the champion rate enhancer orotidine 5'-phosphate decarboxylase,[513] which we have noted earlier.

most common of which is the TIM-barrel. During evolution, an ancestral TIM-barrel pocket may have served as a low-affinity site for binding and concentrating a wide range of potential substrates, before protein divergence and the acquisition of catalytic specificity.[192] Experimental evidence indicates that inactive "half-barrel" $(\beta/\alpha)_4$ proteins can assemble into a catalytically active complex,[*] supporting the proposal that the TIM-barrel remote ancestor evolved by duplication and fusion of two such "half-barrel" domains.[544]

The evolution of diverse catalytic activities via a "generic binding site" raises the issue of natural enzymatic substrate specificity, or its relaxation into promiscuity, which we also have encountered in previous chapters. Promiscuity has in fact emerged as an important consideration for both natural and artificial evolution of enzyme specificity. A mutation that has no significant effect on the function of an enzyme would seem to warrant categorization as a "neutral" change, and thus not susceptible to selective pressure. Yet, there is evidence that mutations in some enzymes that are "neutral" in terms of their activity toward one substrate can confer "promiscuous" activity toward another substrate(s).[545,546] Should the additional substrate recognition later become selectable, rapid evolution of a new primary enzyme activity could ensue, especially if accompanied by gene duplication.[545] We saw in Chapter 3 a precedent for antibody cross-reactivity toward alternate antigens being mediated via conformational diversity,[277] and enzyme catalytic promiscuity also has been associated with conceptually analogous conformational variability,[547] often in flexible loop regions.[548] Catalytic specificity can show a considerable degree of "plasticity" in which a limited number of mutations can have a dramatic effect on the induction of promiscuous function.[542,547] Practically speaking, the prediction of sites for mutation to facilitate promiscuity is not necessarily straightforward, as equivalent site changes in different proteins within the same superfamily can have very divergent effects.[190,542]

From this point of view, studying natural examples of enzymes showing catalytic promiscuity is very worthwhile[549] and overlaps with the study of enzyme superfamilies with common structural motifs (as for the TIM-barrel enzymes considered above). Analyses of specific cases of enzyme evolution may permit directed alteration of catalytic activity in related enzymes,[†,549] and naturally promiscuous enzymes are promising starting points for rational engineering or directed evolutionary approaches for further changes in the existing patterns of enzyme catalysis.[549,550] Alternatively, directed evolution itself can be used to enhance the promiscuity of an enzyme,[551,552] which can be a useful feature of certain biosynthetic applications, as noted earlier.

If we look at the track record in general of directed evolution in altering enzymatic catalytic specificity (including semirational strategies for targeting of mutations), we find a multiplicity of successful experiments, many of which involve minimal changes of 1–2 residues[553] (invoking the "plasticity" concept noted above). Some studies have used random mutagenesis as the source of diversity for the artificial evolutionary process, while many others have used DNA shuffling or a combination of shuffling and mutagenesis. Each case study is a story in its own right, requiring a detailed level of inspection to appreciate the changes that have been invoked through directed evolutionary pathways. Let us return to the TIM-barrel fold to consider some cases in more detail. Single-residue mutations at the C-terminal end of certain bacterial TIM-barrel enzymes can enhance substrate promiscuity. (The positioning of such mutations can derive either from rational choice through cross-enzyme structural comparisons, or via directed evolution experiments.[542,554]) A conclusion has been that a mutation that

---

[*]This work has been done with the same (*HisF*) TIM-barrel enzyme as in Fig. 5.2.[543]
[†]This strategy stands as an example of gaining knowledge from the empirical process of natural evolution rather than through directed artificial evolution, such that "rational" design can be instituted.

promotes the binding of a novel substrate in a "productive geometry" with respect to key catalytic residues will enable promiscuous reactivity with a distinguishable mechanism.[542] Rational mutation of a TIM-barrel enzyme (*E. coli* L-Ala-D/L-Glu epimerase) for enhancement of cross-reactivity toward a substrate of a different TIM-barrel enzyme (*o*-succinylbenzoate synthase) led to low-level promiscuous activity toward the latter substrate.[542] This catalytic change in turn could be further enhanced by single residue mutations obtained via directed evolution,[555] and it constituted an effective change of an isomerase into a lyase; a "new" enzyme specificity.

A TIM-barrel enzyme of special interest is phosphotriesterase, which hydrolyses organophosphates, including certain insecticides and the highly toxic "nerve agents." There is a good case for the view that this bacterial enzyme has naturally evolved within the last four decades or so, since its "native" substrate is the synthetic compound and insecticide paraoxon, and a natural target has not been identified.[556] Despite this evident recent origin, phosphotriesterase is a remarkably efficient catalyst toward paraoxon and related organophosphorus compounds.[557] While phosphotriesterase can thus act on a range of organophosphate esters, its activity toward specific substrates is variable, leading to interest in "tuning" its activity by directed evolution.[558] It has been shown that the promiscuity of phosphotriesterase can be enhanced by a specific residue change (wild-type histidine-254[559]), and additional mutations near the active site can confer substrate specificity changes, including toward the deadly nerve agents of chemical warfare.[560]

An important issue for directed enzyme alteration is that of substrate *chirality* or stereochemistry (ftp site). We noted at the start of this section that for synthetic applications, the specificity of enzyme activity is crucial, and included within this dictate are the precise stereochemical structures of the substrates and products of enzymatic reactions. Usually, the latter are of key interest, and from a stereochemical point of view "enantioconvergence" is a desirable feature.[536] This occurs when an enzyme can utilize either of the pair of enantiomers (mirror-image stereoisomers that cannot be superimposed on each other and that are optically active) as substrates, but where the product corresponds (or "converges") to one specific enantiomeric form. Although it is known that changing enzymatic reaction conditions, solvent conditions, and the precise structures of substrates in some cases can affect the outcome of product chirality,[536] it is also possible to adjust stereochemical selectivity at the enzyme level itself, with potentially greatly improved efficiencies. Directed evolution has been applied toward this end, usually beginning with an enzyme with limited or no "enantioselectivity" (stereochemical discrimination); most of which have been hydrolases.[537] Various enzymes have been derived by artificial evolutionary processes and screens to achieve enantioselectivity[537,561] or reverse the pre-existing enantioselective specificity.[562,563]

For additional examples of enzymatic chiral engineering, we can once again return to TIM-barrel enzymes, and more specifically the above phosphotriesterase example. Substrate recognition by bacterial phosphotriesterase has been shown to be strongly influenced by stereochemistry, through the specific arrangement of bound substrate with respect to active site residues.[564] By mutagenesis of rationally selected residues at the active site of this enzyme, tightening, relaxation, or reversal of enantioselectivity can be elicited.[564] Although these phosphotriesterase examples show the value of changes near active site for manipulating chiral recognition, in general mutations remote from the center of catalysis can still engender large effects on this enzymatic property.[565] Once again, this underscores the usefulness of global directed evolution of proteins as a "no-assumptions" approach that can reveal unanticipated long-distance interactions. It should also be noted as an aside that at least in some cases, enzyme engineering for precise chirally selective organic synthesis will face competition from increasingly sophisticated approaches using small molecule or inorganic catalysts.[566]

Some impressive changes in enzyme activities have thus been brought about by evolutionary approaches, combined in certain cases with the application of pre-existing information. Many of these involve a shift in substrate recognition specificity or an alteration in the nature of the resulting product. But as we have seen, at least in some circumstances a surprisingly limited number of residue changes can have profound effects on enzyme activity, to the extent that we might consider "new" catalysis to have been wrought. This kind of novelty builds on pre-existing enzymatic scaffolds much as nature does, and novel biocatalysts of this type will continue to pour out of research laboratories. Forever pushing the boundaries of the possible, many researchers still aim higher, toward an ultimate end point where a biocatalyst can be designed at will "from the ground up" to efficiently match any catalytic task, whether previously observed in nature or not. Although this proposal is the most ambitious enzymatic design goal of all, and one that is still relatively in its infancy, it is not at all an idle dream or another example of hubris. We therefore need to consider the current status of "first principles," or *de novo*, enzyme generation.

Enzymes from Scratch

A "truly novel" enzyme might be a completely new protein sequence with a conventional enzymatic activity, or an enzyme with completely novel catalytic properties. In principle, this could involve radical modification of an existing protein scaffold or a totally new protein design for biocatalysis. This area falls within the general sphere of *de novo* protein design, which we will further consider later in this chapter, and also overlaps with the topic of rational enzyme design (the ambit of Chapter 9). It is obviously helpful for a would-be *de novo* designer to understand as much as possible about the general structural requirements to make specific enzymes work, so let us first consider these issues in a little more detail.

We have seen that a wide range of catalytic tasks have arisen from natural application of a relatively limited number of protein folds. A relevant question for enzyme design that flows from this becomes, When *different* enzyme scaffolds are employed for certain distinct activities, are these necessary arrangements for optimal efficiency or are they additional examples of "frozen accidents"? In other words, if structurally distinct enzyme scaffolds A and B are used to naturally catalyze reactions *a* and *b*, respectively, can scaffold A be artificially adapted to perform reaction *b* (and vice versa)? Conversely, one could ask if the same enzymatic reaction can be catalyzed by very different protein framework structures. The latter question at least can be answered by looking to natural precedents. Many cases of analogous enzymes[*] are known where equivalent catalytic tasks are performed by entirely distinct enzymes with presumed widely divergent evolutionary histories.[568] A pair of such analogous enzymes is shown in Fig. 5.3, where one (from barley) is yet another example of our friendly neighborhood TIM-barrel fold, but the other (from a bacterial source) has an obviously different fold based on β-strands. This leads to another general question that probes the relationship between structural folds and catalytic tasks: When the *same* scaffold is used for different natural catalytic tasks, is this divergent evolution (multiple functions arising from a common ancestor) or convergent evolution (independent evolutionary design converging on the same or very similar scaffold)?[541] The occurrence of the latter would support the proposal that the "best" design is in fact associated with a specific scaffold. Evolutionary relationships between many TIM-barrel

---

[*]Note that this usage of "analogous" refers to the functional analogies with diverse structures. "Analogy" is also used to refer to structural similarities between proteins (arising from limited ways of accommodating secondary structural motifs into folds), which may have quite distinct functions.[567]

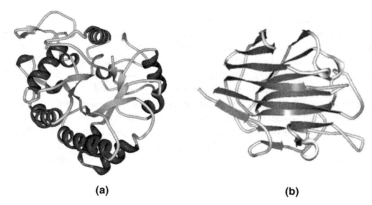

(a)   **(b)**

FIGURE 5.3   Examples of analogous enzymes catalyzing the same glycosyl hydrolytic reaction (both 1,3-1,4-β-ᴅ-glucan and 4-glucanohydrolases; from[27] and online supplement http://www.ncbi.nlm.nih.gov/ Complete_Genomes/AnalEnzymes.html). (a) Enzyme from barley with a TIM-barrel fold[28]; (b) Enzyme from *Bacillus macerans* (showing one subunit of a homodimer), with a very different fold ("sandwich jellyroll"; Concanavlin A-like lectin fold), almost all β-strand.[29] In both panels, α-helices are shown in dark gray and β-strands are shown in light gray. *Source*: Protein Data Bank[2] (http://www.pdb.org). (A) 1GHR; (B) 1MAC. Images generated with Protein Workshop.[5]

enzyme superfamilies are evident, but it is not certain whether all have diverged from a single primordial precursor.[569,570]

Looking at this general issue from another perspective, we have seen that the TIM-barrel fold (for example) is widely applicable to many types of enzymatic reactions, and folds of this type accordingly possess a high degree of "biochemical plasticity."[192] All of the broad enzyme catalytic reaction classes in the Enzyme Commission nomenclature (Chapter 2) include TIM-barrel enzymes, with the notable exception of ligases.[570] We can take the issue of ligases and TIM-barrel enzymes further to consider the interplay between structure and catalysis and in turn the structural dictates of specific catalytic tasks. (This of course is another way of rephrasing the question regarding the roles of chance and necessity in natural enzyme design.) Though the substrates of ligases in general are chemically diverse, the common feature of these enzymes is molecular joining, or ligation. This in itself, depending on the specific substrates involved, may place constraints on how ligases are structurally arranged. Some DNA ligases have deep clefts within their protein structures for substrate binding (Fig. 5.4a) or wrap around DNA in a "protein clamp" (Fig. 5.4b). Additional modes of ligase interaction with DNA are known,[571] but the observed natural exclusion of TIM barrels from utilization in a ligase capacity suggests that at best this structural fold is inefficient as a scaffolding for this type of biocatalysis. The word "suggests" in the previous sentence should be stressed, since once again, absence of evidence is not any type of formal proof. Yet a more general point can be made to the effect that a "universal enzyme protein scaffold" would be unlikely to deliver equal efficiency for all catalytic tasks, since the scaffold geometry itself will not be irrelevant for some mechanisms of catalysis.

In Chapter 3, we considered the generality of antibodies as universal binding reagents, and the possible circumstances when alternative designs could prove superior. The conclusion reached was that despite the exquisite versatility of antibodies, some specialized binding tasks (such as the sequence-specific binding of long DNA sequences) could be better performed by alternate protein structural designs. An analogous conclusion can then be proposed with respect to enzyme design. It is additionally interesting to note that the fields of antibody binding and

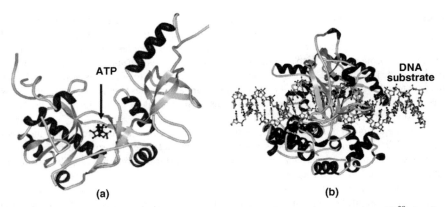

FIGURE 5.4 Structures of DNA ligases. (a) DNA ligase from *E. coli* bacteriophage T7[30] showing substrate binding cleft with cofactor ATP within it. (b) *E. coli* DNA ligase with a (nicked) DNA substrate enveloped within a "protein clamp." In both the panels, ribbons depict α-helices in dark gray and β-strands in light gray. *Source*: Protein Data Bank[2] (http://www.pdb.org). (A) 1A0I; (B) 2OWO. Images generated with Protein Workshop.[5]

enzymatic catalysis actually overlap, since antibodies themselves can act as enzymes. We will address the subject of *catalytic antibodies* (or "abzymes") in more detail in Chapter 7, but for the present purposes we can simply note that antibodies can be engineered to act as catalysts by binding and stabilizing transition-state intermediates of chemical reactions (Chapter 2).

If a specific antibody can be obtained toward any low-molecular weight organic structure when accompanied by appropriate immune presentation, selection, and affinity maturation, then in theory could not an antibody exist for any transition-state intermediate? In turn, could not antibodies be considered as hypothetical "universal biocatalysts"? This would seem to contradict the above proposal that different fundamental scaffoldings are required for certain enzymatic tasks. In reality, though, most catalytic antibodies are inefficient in comparison to natural enzymes, and binding of a transition-state intermediate *per se* does not guarantee optimal catalysis, as additional specific placement of residues may be required for catalytic acceleration.[572] It was noted earlier that whole-enzyme dynamics are crucial for efficient catalysis,[517,518] and antibodies selected for transition state analog binding alone will clearly be suboptimal in this regard.[572] When independent antibodies selected with the same transition state analog are structurally examined, convergence toward similar binding pockets and catalytically relevant residues is found,[573] suggesting that the range of possible solutions for enzymatic tasks with the immunoglobulin framework is limited.[*] It is also interesting to note that while it is indisputable that the immunoglobulin fold can foster catalysis, natural enzymes have not made use of this specific structural motif.[572] The immunoglobulin β-sandwich motif may also be incompatible with certain catalytic tasks involving large substrates, by analogy with the ligase/TIM-barrel considerations above. So, while both antibodies and TIM barrels can be used as structural scaffolds for diverse catalytic processes, it is unlikely that they will offer the best possible solutions in all circumstances.

It would clearly be more productive to use the best available scaffold to meet the desired task, but if the task itself is novel this may be a formidable challenge. If an enzyme can be successfully derived *de novo* purely by *in vitro* evolution, then the structural scaffolding will not

---

[*]We will consider certain strategies for increasing the efficiency and repertoire of antibody-mediated catalysis in Chapter 7.

be rationally chosen in itself, but will simply emerge as a consequence of the evolutionary selection/screening process. We will return to this issue under the broader umbrella of general *de novo* protein design a little later in this chapter. The somewhat less ambitious aim of using pre-existing protein scaffolds to generate novel enzymes then would appear a practical option to explore in the first instance. Here too are numerous opportunities to use pre-existing knowledge, to the extent that most work using directed evolutionary approaches to generate completely new enzymes can more accurately be classified as semirational at some level, or as broad applications of "rational evolution."

A good example of this is the process of enzymatic change in a desired direction by means of "grafting" designed active site loops (using structural and other information derived from enzymes with a particular activity) onto a related structural framework provided by a different enzyme. Directed evolution can then be applied to enhance the desired novel catalytic activity. An example of this procedure has been demonstrated by the replacement of the original activity of the metallohydrolase enzyme glyoxalase II with β-lactamase catalysis.[574] This "loop grafting" approach has been likened to the changing of antibody combining specificities by the transfer of appropriate specific complementarity-determining regions (a topic considered in more detail in Chapter 7), although altering enzyme catalysis in an analogous manner is a much more challenging undertaking.[575] Even with the application of adjunct directed evolutionary random mutagenesis and DNA shuffling, the novel enzymes have been much less efficient than corresponding natural equivalents.[574] Nevertheless, the combination of rational design for setting up the framework for novel catalysis and directed evolution to tune-up the desired activity is generally considered a productive partnership at this time.[576] The rational "set-up" stage can involve extensive framework manipulations and restructuring, as exemplified by the generation of new TIM $(\beta/\alpha)_8$ barrels from deduced $(\beta/\alpha)_4$ "half barrels."[577]

A discussion of the future of novel enzyme design would not be complete without reference to the very wide possibilities opened up by the incorporation of unnatural amino acids into enzymatic active sites. This work, based on artificial extensions to the natural genetic code, will be considered in a whole separate section later in this chapter.

Finally, we should not forget that natural enzymes are "team players" and operate in complex intracellular environments. The modulation of enzyme activity by activating signals and/or protein–protein (or protein–ligand) interactions is the rule rather than the exception. Also, many enzymes perform specific tasks as components of extended biosynthetic pathways and are of limited usefulness in the absence of a series of their enzymatic partners. While these issues are relevant to any enzyme engineering project, they are especially pertinent to *de novo* enzyme design where "additional" features mediating signaling or control mechanisms may be absent. Obviously, the intended application of an engineered enzyme is important. For industrial applications, often an improved enzyme will indeed be required to function in isolation, but protein–protein interactions may be significant factors if enzymes within a complex system (up to the whole organism level) are manipulated. This is not to say, however, that novel enzymes generated *in vitro* cannot function *in vivo*, and many of the above examples have had detectable effects when expressed in bacterial hosts lacking the appropriate enzymatic activity.[*] The "bottom line" here is efficiency, both for novel enzyme catalytic reactions themselves (for which, as we have seen, there is much room for improvement) and their optimization in a complex milieu. *In vivo* expression of designed and evolved enzymes in hosts lacking the corresponding activity can be a powerful selection method (at least up to a point).

---

[*]These include the designed/evolved β-lactamase from glyoxalase II.[574] Some catalytic antibodies (when appropriately expressed) can also allow the growth of bacteria lacking an enzymatic activity that the "abzymes" catalyze.[578]

In turn, this topic leads us to the next section that deals with a powerful concept directly applicable to selection.

DE NOVO PROTEIN DESIGN: THE EMPIRICAL WAY

There is an approach to protein design in general that can be performed as an *in vitro* experiment along the following lines: With no preconceptions, generate a long random string of amino acids and search for a predetermined function among a huge library of these variants. Since a functional polypeptide isolated from the random-string population has no evolutionary precursor, this strategy can be termed a *de novo* ("new") functional search. Some opinion would hold this idea as at best naïve and at worst idiotic, bearing in mind the tiny subset of functional proteins out of all possible amino acid sequences (refer back to Chapter 2 for more on this theme). Is *de novo* empirical selection for protein function consequently a dreamer's folly? Not necessarily...

First let us remember earlier thoughts about the size of protein sequence space (Chapter 2) and that "hyperastronomical" is not an exaggeration in reference to the number of possible sequence combinations for even average-sized polypeptides. Just as the number of possible moves in the board game Go is so large ($\sim 10^{170}$) that a computational approach to playing the game requires pruning strategies, so too it would seem that some kind of "pruning" of nonuseful sequences from the enormous range of possible protein sequences is necessary for any prospect of success with empirical selection. "Nonuseful" in this context refers to sequences that cannot undergo orderly folding (the vast majority), but even the set of all "foldable" sequences is vast. Given these elementary facts, the thermodynamic problem of protein folding confronts us with full impact. The Holy Grail of protein chemistry is achieving the ability to predict protein folding and function from primary sequence, and its full realization would indeed render empirical approaches redundant. Protein-folding mechanisms have been intensively studied and many advances clocked up, but despite significant recent advances (Chapter 9), the quest for this particular prize yet remains open. But some aspects of protein folding are indeed worth considering a little further before returning to the issue of *de novo* protein selection processes.

Return to the Fold

Decades ago, it was established that protein tertiary structure is determined by its primary amino acid sequence,[*,579] and self-assembly is accepted as a fundamental property of many biological systems.[580] This knowledge is implicit in searches for *de novo* protein structure and function by empirical means, in that it is assumed that there will be a one-to-one correspondence between a primary sequence "out there" in sequence space and a specific protein structural property. In fact, the full story for protein folding is somewhat more complex than the self-assembly principle would indicate. Earlier in this chapter, we noted that specialized proteins termed chaperones are important for assisting the folding of many proteins. Although crucially important for cellular life, chaperones promote folding and minimize competing reactions of undesirable protein aggregation (especially in the crowded intracellular environment), rather than transmit "folding information" to their target proteins. In other words, a protein can fold without its normal chaperone, but inefficiently.[581]

---

[*]Christian Anfinsen was awarded a component of the 1972 Nobel Prize for Chemistry for this and related discoveries in protein folding.

To an extent, the need for chaperones correlates with a protein's size and domain structure. Small, single-domain proteins can often efficiently fold without chaperone assistance.[582] Multidomain proteins may require folding facilitation, although separate domains of such proteins can often sequentially fold as translation proceeds[583] without the apparent need for chaperones (especially in eukaryotes). As previously noted (Chapter 2), eukaryotic proteomes have a higher percentage of multidomain proteins than proteomes of prokaryotes. Bacteria also appear to be less efficient in folding random multidomain proteins than eukaryotes, attributed to their lower capacity for cotranslational folding.[584] Yet in both prokaryotes and eukaryotes, other chaperones are actually ribosome-associated and protect "newborn" proteins from aggregation events.[585] Despite these different levels of chaperone assistance, the above principle of the determination of final protein structural conformation by its primary sequence is still not altered.

Having accounted for this type of chaperone function, we still find that protein folding reveals additional complexities. Enzymes exist that catalyze either the formation of disulfide bonds (protein-disulfide isomerases) or the isomerization of proline residues (peptidyl–prolyl isomerases).[*] In turn, these enzymes catalyze the folding of some proteins and have been termed "foldases" as a result.[587] Nevertheless, slow spontaneous reactions or isomerizations in their target polypeptides will eventually allow enzyme-free protein folding (in the absence of interference from unrelated proteins),[588] analogously to proteins folding inefficiently in the absence of ordinary chaperones. Certain proteins, though, receive folding help that cannot be so readily classified as merely rate-enhancing or blocking of interference. In general, the proteins that provide such folding help are termed "steric chaperones"[589] or "steric foldases,"[588] where "steric" refers to structure. These kinds of foldases in essence transmit a structural imprint onto their protein targets to divert them into a folding pathway that would be otherwise difficult or virtually impossible to reach. (Indeed, for target proteins of steric foldases, spontaneous folding into their correct structures may be so slow in biological terms as to be nonexistent for any practical purpose.[588]) Once having attained their correct folds, these proteins are generally very resistant to both unfolding and proteolysis,[†] since a large energy barrier must be crossed to reverse the enzymatically directed folding.[588]

One class of such proteins has been termed "intramolecular foldases," since following translation they contain a "prodomain" that serves to direct the correct folding pathway of the rest of the protein. After folding, the prodomain is proteolytically removed, a consequence of which is that a mature protein of this type cannot spontaneously refold if denatured.[589] A very interesting phenomenon observed in this context is that mutations in the prodomain of the subtilisin protease can "imprint" a stable alternative conformation into the remainder of the protease, which has been considered as a kind of "protein memory."[590] Since this kind of foldase activity is intramolecular, it has been interpreted as still consistent with the Anfinsen principle of protein self-assembly.[581] Unfortunately nonetheless for "Anfinsen's rule," steric foldases exist which are wholly distinct entities from their protein targets and which consequently appear to violate this self-assembly dictum.[588] Within the universe of protein sequences,

---

[*]Proline residues have a unique structure where their side chains cyclize with the α-carbon chain itself, which means that a polypeptide with a proline residue can exist in two different isomeric forms (*cis* and *trans*). The proline *cis–trans* isomerization status obviously affects the structure of a polypeptide chain and thus has a strong impact on protein folding. Nonenzymatic interconversion between these isomeric forms is possible, but slow relative to enzyme-catalyzed events. Interestingly, proline isomerization in proteins is also used by nature for the control of molecular switches and signaling processes.[586]

[†]Here, we might recall the inverse correlation between denaturation resistance and susceptibility to proteolysis, noted earlier in this chapter. The steric foldase systems may in fact have evolved as a way of generating proteins that are very durable in protease-rich environments.[588]

some then are only usefully foldable in biological terms with informational input from other proteins (Fig. 5.5a).

The fact that few natural proteins exist that require obligatory foldases does not rule out the potential existence of many others (in the entirety of nonbiological protein sequence space), which could attain a stable fold if provided with appropriate catalytic assistance.* Yet widespread evolution of proteins that require specific folding direction would be more cumbersome and energetically costly than for self-folders, since many more specific foldases would be required. "Guided folders" would then be selectively disfavored with the exception of certain special circumstances such as high-protease environments.[588] *De novo* protein design would also become enormously more complicated if for every protein sequence another unspecified sequence was required to enable successful folding. Self-folding is thus an implicit (and inevitable) requirement for proteins selected from random sequence libraries. (Of course, any polypeptides requiring foldases in such a library could not functionally fold and by escaping selection would be "invisible," despite their hidden functional potential.) Another factor to consider is that *de novo* protein design endeavors are currently limited in their feasible size ranges to relatively small single-domain proteins, which are usually foldable without chaperone assistance[582] if they are foldable at all.

An additional implicit assumption of selection for functional protein sequences *de novo* is that each primary amino acid sequence will fold only in one functional way. The "Paracelsus Challenge" (Fig. 5.5b) asked if protein engineers could produce a pair of proteins with distinctly different folds yet sharing at least 50% sequence identity, and these conditions were indeed met.[591,592] Taken to its logical conclusion, this challenge would ask if stable divergent structures could be obtained at 100% sequence identity, by which point we are obviously referring to the same polypeptide chain (as depicted in the middle panel of Fig. 5.5b). Certain small peptides can clearly exist in quite distinct stable conformations, with an interchange between these states possible under specific conditions.[593] It has also long been known that short peptide segments within proteins can be components of conformationally quite distinct secondary structures.[594,595] But if one considers alternative conformations for proteins as a whole, we must remember the domain structure of proteins and domain modularity. There are clear cases of stable and functionally significant variation in local protein conformations† that retain the global folding pattern for the protein involved. Apart from the above "protein memory,"[590] we can note the instance of alternative conformational states in certain immunoglobulin molecules determining antigen-binding specificity.[277] (While of obvious significance for antibody function, this conformational isomerism does not constitute a radical structural reworking of the entire antibody β-sandwich framework.) Another example of a functionally significant but nonglobal change to a protein conformation has been seen with the membrane efflux transporter P-glycoprotein. Synonymous but rare codon changes in the P-glycoprotein coding sequence result in translational pausing, which alters the protein folding pathway and substrate specificity.[31]

Some proteins have been characterized with "chameleon domains" that can assume very distinct alternative conformations. One such example is Mad2 (involved with cell division), whose C-terminal region can undergo structural re-organization with functional implications.[78,596] Another class of proteins appears to lack a definable folded structures in the first

---

*As with natural substrates of steric foldases, such hypothetical proteins would be trapped by "guided folding" into a state above their global free energy minima. Further discussion of folding pathways is provided in Chapter 9.

†Note also in this context that stable alternative protein conformational isomers are quite distinguishable from conformational changes induced after ligand binding. The latter effect is a common feature of many protein transactions, including enzyme *allostery*, which is featured further in an RNA context in the next chapter.

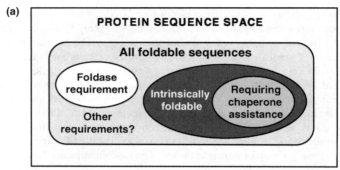

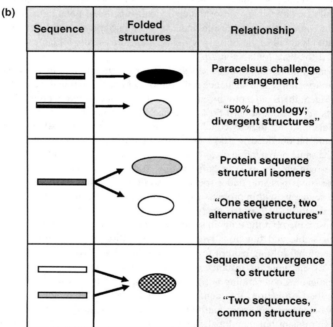

FIGURE 5.5 Protein sequence space and folding and sequence-structure relationships. (a) Schematic representation of protein sequence space and categories of foldable proteins. (No relative scaling is intended; the size of the total foldable subsection of sequence space is small.) "Intrinsically foldable" sequences self-fold by the "Anfinsen rule" through the informational content of their primary sequences, but some require assistance from chaperones to be usefully produced under cellular conditions. An (apparently) minor set of proteins can only fold into thermodynamically unfavored states via input from foldases. "Other requirements" leave open the formal possibility that some sequences could fold with assistance from some other (as yet undefined) molecular mechanism(s). (b) Some broad types of sequence-structure relationships of at least theoretical possibility. The top panel depicts the "Paracelsus Challenge" where $\geq$50% sequence identity is specified in designing proteins with divergent folds[31,32]; the middle panel depicts true protein isomers (identical sequences but different stable conformations), and the bottom panel represents the converse situation of differing protein sequences converging toward the same structural folds.

place, even under normal cellular conditions. These *intrinsically unstructured proteins* lack distinct tertiary or even secondary structural character, but gain structural order upon interaction with target proteins.[597] The same intrinsically unstructured protein can interact with different target partners, with divergent structural and functional consequences. This phenomenon is believed to be biologically important for many "moonlighting" proteins that perform multiple functional roles.[597] Such plasticity approaches the "one polypeptide, two structures" scenario of Fig. 5.5b, but requires the target proteins to mediate the alternative conformational shifts. As such, specific targets of these initially unfolded proteins can in a sense be viewed as specialized chaperones themselves, as well as fulfilling other cellular functions.

But perhaps the most significant alternative conformational states known in proteins are *prions*, initially characterized as normal mammalian cellular proteins that can convert from a state rich in α-helices to another pathogenic fibrillar form rich in β-sheets.[598,599] The most interesting aspect of the latter is their ability to transmit conformational information onto the normal form, and thereby replicate as "infectious proteins." Biologically useful roles of prions have been defined in fungi and yeasts,[600] and other normal prion activities may exist in mammals.[601]

While interesting, the above cases of structural dimorphism* of a small subset of proteins is unlikely to have much impact on *de novo* protein design for the near future. What then of the converse situation in this regard? This refers to the concept of multiple sequences converging upon a common fold (Fig. 5.5b bottom panel). If a sizable number of different sequences could fold into comparable shapes, it might be proposed that the chances of finding at least a prototype from random sequence are accordingly increased. It will be recalled from Chapter 2 that natural proteins can be distinguished by their robustness or the capacity of their folded structures to tolerate sequence change. This can be viewed as a consequence of designability (the numbers of variant sequences that converge onto the same fold type), which is the issue of interest for *de novo* protein selection. In this context, it might also be recalled that there is an inherent asymmetry between protein sequence space and natural protein structure space owing to the convergence of different sequences toward similar folds,[8] a consequence of evolutionary designability in action. (This is also evident from observations of protein "structural analogy," arising from restrictions on the ways that secondary structural motifs can be packed into a protein fold.[567]) Fold diversity is then much less than sequence diversity. We have previously seen examples of certain folds being used repeatedly in many different natural contexts (TIM barrels could pop up again here, for instance). The commonest folds appear to be also the most designable,[602] and folds with relatively low designability have been associated with diseases caused by protein aggregation.[602] Selection may then operate at multiple levels in favor of designability.

Does designability of natural proteins then have a message for artificial *de novo* protein selection? Evolution tends to be parsimonious with successful protein designs, and a common designable fold would have a large number of functional variants in protein sequence space (that is to say, many sequences would converge toward such a fold). Commonness of a fold type in nature is therefore a reflection of its functional utility as favored by evolutionary selection, especially with regard to deployment of such a structural feature in numerous different functional capacities with relatively small sequence changes. But successful natural folds may themselves be rare (atypical) even within the universe of possible protein structures,[603] indicating that commonness in nature has no direct bearing on the likelihood of "pulling out" a

---

*We will consider this further in the comparison with RNA enzymes with the same sequence but distinct conformations and functions (Chapter 6).

prototypical structural motif from random sequence. Again, natural protein sequences and their self-folding structures are a very special, and very small, subset of all possible sequences.

Has all this thinking about protein sequences and folding then merely reinforced the view that selection for function from random sequence is a waste of time? Here there is another point to consider. We must remember that of course one does not select directly from a random-sequence expression library at either the sequence or structural level, but by functional criteria. We need not assume that there is only one unique solution to a given functional problem, or in turn that there is only one protein sequence (or family of related sequences) in sequence space that satisfies the functional quest. The existence of analogous enzymes with common catalytic functions but divergent structures (as described earlier in this chapter) shows that at least in some cases, there is more than one structural way to "skin a cat" for the achievement of catalysis. Divergent DNA-binding protein structures directed at the same DNA sequence targets* serve as exemplars of alternative folding structures recognizing and binding a common ligand. While this observation of multiple structural pathways to a prechosen protein function gives hope for *de novo* protein isolation, it is not particularly useful in practice. At the outset of screening or selecting from a random library, it is not going to be clear how many different structural motifs for a single function are "on offer" in sequence space. These considerations should help to make the point that the determination of the structures of *de novo* selected proteins is of very great interest, as we see further below.

In the end, the pragmatic issue is not so much whether *de novo* selection is "naïve" or not, but whether there are any realistic prospects for something functional emerging from the other end of a random-sequence screen. Certainly, pulling a fully functional multidomain protein in one "hit" from random sequences would indeed be reminiscent of the somewhat overused metaphor of a tornado in a junkyard producing a jumbo jet. But just as this notion is a fundamental misunderstanding of how iterative selection processes operate, so too we could expect that iterative selection can take us from a crude protein functional prototype into a highly efficient derivative down the evolutionary line. The "trick" then is to find such a prototype in the first place.

Ultimately, we should let experimentation speak. Even if a very specific theory predicting the frequencies of functional proteins within random sequence libraries were to be developed, it would still need to be confirmed in the real world. And this applies even if the theory predicted failure, though as we will see, progress to date has been far more interesting than that dismal conclusion. But many important factors intrude at the practical experimental level, such as library size and selection methods. Here size would indeed seem to matter rather a lot, so the combination of very large libraries and the most powerful selection methods on hand would appear to be logical choices in this kind of endeavor. But we have seen earlier in this chapter that "smart" libraries and screening methods are very good moves for increasing the effectiveness of directed evolution.[380] Given the daunting prospects for selection from purely random sequence, it is accordingly not surprising that workers in the field have devised ways of "stacking the odds" by semirandomization strategies, which we should now consider.

Alphabets and Patterning

One way to narrow the field for *de novo* protein design is think of the number of building blocks involved, and ask whether this can be simplified. Can the normal "alphabet" of 20 amino acids

---

*We have seen (Chapter 3) that antibodies can bind to some sequences recognized by natural DNA-binding proteins, albeit relatively poorly. Many natural precedents exist for different proteins binding the same DNA sequence, such as numerous endonuclease "isoschizomers," some of which have very divergent structures.[604]

be reduced while still allowing functional proteins to be selected from a random library? Written passages deliberately missing one or more of the letters of the alphabet are termed *lipograms*, so a "molecular lipogram"[*] would correspond to a biopolymer deliberately restricted in its alphabetic repertoire.

For the usual set of natural amino acids, their corresponding "alphabet" is conventionally rendered in a code based on single letters of the roman alphabet. Since only 20 letters are needed in this regard, six letters (B, J, O, U, X, and Z) are excluded for description of any natural polypeptide sequence. Since some human authors (perhaps with too much spare time on their hands) have written lipogrammatic novels excluding even the common letter "e," it should be possible to write an extended piece of work using only the 20 letters of the "protein alphabet." Regardless of the ultimate verbal semantic meaning or literary value of such a "Protein Book," the total letters contained within it could represent a gigantic protein sequence when strung together (disregarding punctuation and spaces). If we then arbitrarily cut up this continuous letter string into pieces corresponding to a small-to-moderate protein size (say, on average 200 "letters"), what would be the chances of obtaining at least one piece with "meaning" at the utterly different level of protein folding, if the sequence was genetically expressed or chemically synthesized? (Note that we are only asking that a string of letters from the book (corresponding to an amino acid sequence) should fold up as a natural protein, not possess any immediate function.)

The answer is that the prospects for this are very poor indeed. Obviously this would be linked to the size of Protein Book in question, so let us assume a moderate length of 200,000 words, and on average 5 letters per word. We then have a giant protein sequence of 1 million letters, which we cut up into 5000 pieces of 200 letters each; a trivial number for a random protein library. (If the "Protein Book" has linguistic meaning, the total letter string would obviously be nonrandom, but we can overlook that for the present purposes.) Even if we allow permuted cutting to yield any possible set of contiguous overlapping 200-letter strings, we still have less than 10^6 "library" members, a completely inadequate number by many orders of magnitude (as we will see further below with *in vitro* selection from random protein libraries). If the imaginary Library of Babel (Chapter 1) holds a Protein Book that contains any foldable protein sequence (let alone all such sequences), it will need to have an extraordinary amount of room.

Consider then if we go back to our metaphorical but hard-working Master of Lipograms and ask him or her to write another such book, only this time to be limited to less than 10 letters from the protein alphabet. No matter what level of wizardry our lipogrammaticist possessed, we would likely be told that the proposed undertaking was impossible, that no conceivable linguistic meaning could survive such severe restrictions. Yet when such reduced "letters" are used for real proteins, generation of "meaning" may not be so formidable...

In Chapter 2 it was briefly noted that certain amino acids (and their associated encoding at the nucleic acid level) were widely believed to be of more ancient origins than others. In other words, this proposal holds that the most ancient prebiotic proteins were constituted with a simpler alphabet than the one we are familiar with. And if we continue with the lipogram analogy, this in turn suggests that protein and linguistic lipograms have a point of divergence in this regard. While increasing alphabetic restrictions causes literary lipograms to soon grind to a

---

[*]This use of "lipogram" should not be confused with a corresponding (but infrequently used) medical term referring to a lipid profile for a patient. A lipogrammatic sentence could also be "molecular" through its subject matter, such as this terse commentary upon "selfish" parasitic genetic elements: "The rebel genes seem ever greedy, then self-centered," where all vowels are excluded except "e." (This "e-centric" lipogram is thus technically also a monovocalic.) Hopefully, there is little chance of confusing this with a real molecular lipogram.

halt, much simpler protein alphabets may be compatible with folding and certain functions.[*]
For this to hold, though, it is important to note that the right types of amino acids must be
involved, or in other words the type of reduced alphabet is crucial to the likelihood of secondary
structures and folding arising. This principle at least also applies to linguistic lipograms. The
fact that the "protein alphabet" contains a majority (20/26) of the letters of the roman alphabet is
not the main reason why it could be potentially used for a lipogrammatic novel. It is rather the
fact that the 3/5 of the essential vowels (A, E, and I), including the most common (E), are
included. If all vowels were excluded, nothing at all could be done, even though one would still
have a majority of the total letters. Analogously for protein sequences, some (but not just any)
limited combinations of amino acids can be compatible with folding.

What is the evidence for protein folding with reduced alphabets? An early study in this area
constructed random polypeptide libraries comprised of just three amino acid residues:
glutamine (Q, representative hydrophilic), leucine (L, representative hydrophobic), and
arginine (R, representative charged residue). A surprisingly high proportion of such QLR
libraries (80–100 residues) appeared to be foldable into helical structures showing resistance to
proteases, although requiring denaturants to maintain solubility.[606] This was attributed to
excessive hydrophobicity, and when the leucine content of the libraries was reduced, folds with
more native-like properties and improved solubility were identifiable.[607] Nevertheless, these
artificial QLR proteins still deviated from natural proteins in significant ways, and computational folding models have failed to find truly native-like sequences from QLR alphabets.[608] On
the other hand, certain limited alphabets of four amino acids were modeled as foldable,[608] and
these fall within the "early" or "primitive" subset of the extant protein alphabet.[609] Experimental investigations with one such set of five "primitive" amino acids (V, A, D, E, G; See
Glossary for single-letter amino acid code) found that random polypeptides composed only of
these "letters" showed higher solubility levels than for corresponding full-alphabet random
proteins.[610] These limited alphabet proteins did not, however, exhibit significant secondary
structures.

An alternative approach exists for "alphabetic reduction" of proteins. Instead of proceeding
with a "bottom-up" strategy (attempting to make entirely novel proteins with reduced
alphabets), one can take an existing protein and attempt to reduce its amino acid alphabet
while retaining function. The latter might then be termed a "top-down" strategy in comparison
with the former. If we again invoke the above lipogram analogy, it would be akin to taking an
existing book and rewriting it where appropriate to conform to the chosen lipogrammatic
constraints, while retaining its meaning and intent.[†] As an example, one could take "To be or not
to be, that is the question" and render it into "I CAN LIVE. ALTERNATIVELY, I CAN DIE.
WHICH IS RIGHT?" in the protein alphabet. This is quite different to its original letter
combination, but the meaning (if not the poetry) is more or less retained. For a protein, to retain
the "meaning" simply means retaining structural folds and function without unacceptable
losses of efficiency. Proteins with simple secondary structural motifs can certainly be

---

[*]Of course, just as a linguistic lipogrammatic novel would necessarily be stilted and suboptimal in certain ways, it might be
expected that proteins with reduced alphabets would likewise be diminished in their ranges of possible functions. A long-
standing and interesting observation has been that protein secondary structural classes tend to have similar amino acid
compositions, even if their sequences are unrelated. This composition-similarity has been extended to include even
proteins sharing a fold type (but divergent in sequence), interpreted as evidence for the role of specific residues in
intramolecular interactions stabilizing folds.[605] If certain residue combinations are required for specific folds, then a
limited-alphabet protein (lacking one or more of such residues) might be unable to efficiently accomplish certain folding
arrangements. Also, in a functional context, we have already noted (Chapter 2) the importance of specific residues for
catalysis, particularly histidine.

[†]This too has been done in reality in various instances by dedicated (if arguably eccentric) people.

"alphabetically" simplified. A small β-sheet protein could be reduced to using only five different amino acids.[611] In a powerful demonstration that this principle can be applied toward an enzyme, chorismate mutase was progressively mutagenically simplified down to an alphabet of only nine amino acids without loss of catalysis (albeit with some loss of stability).[612,613] These and similar studies also supported the notion that relatively simple protein alphabets occurring during early molecular evolution could still be catalytically effective.[613]

Between completely empirical selection of proteins from random sequence libraries and the ideal of completely rational protein design lies an intermediate zone where predetermined patterns are superimposed upon other random sequence tracts. An implicit assumption in this approach is that folding and packing of proteins has a considerable amount of leeway; that many potential solutions in protein sequence space will converge onto a specific fold. The view that the packing of a protein's hydrophobic core must occur with precise shape-fitting complementarity, or a "jigsaw" model, is at an opposite extreme to this. There is evidence, though, for the compatibility of a variety of alternative packing solutions with folding,[614–616] albeit at the expense of stability if residues with large and small hydrophobic side chains are extensively interchanged.[615,617] The other rationale for design by protein patterning is the observation that amphiphilic* secondary structures call for a regular alternation of hydrophilic and hydrophobic residues, and this "binary patterning" principle can be applied by means of quite simple guidelines. Thus, an amphiphilic β-sheet design can be specified by hydrophilic (H) and hydrophobic (ψ) alternating every other residue (HψHψHψ ...), whereas for α-helices the required hydrophilic/hydrophobic periodicity (HψHHψψHHψHHψψH) must take into account the α-helical property of 3.6 residues/helix turn.[618,619]

Libraries of binary-patterned novel proteins are thus constrained by their hydrophilic/hydrophobic periodicities, but otherwise unspecified. Screening of library members for appropriate secondary structural properties can be carried out by biophysical methods, and with α-helical patternings, initial findings indicated a high content of α-helical structures.[618] With the above alternating β-sheet pattern, proteins have been produced with a high propensity for intermolecular aggregation,[620] as seen in nature with amyloid fibrillar production in pathological conditions such Alzheimer's and prion diseases. This highlighted the importance of "negative design," or paying heed to the importance of attempting to design against unwanted interactions, as well as those that are indeed favorable.[621] Design of β-sheets is inherently more difficult than α-helices, owing to the need for sheet stabilization by interactions between different strands in the same molecules.[622] Intermolecular β-sheet interactions can be minimized by the additional design feature of providing a charged residue at a hydrophobic face in the sheet structure, which disfavors oligomerization and promotes stable monomers.[623]

Patterning can clearly be successful in deliberately biasing an otherwise random sequence library toward specific secondary structures, but there is no reason why this principle cannot be taken a step upward toward a specific protein fold. Patterned libraries have been based on the TIM-barrel fold (see Fig. 5.2), where loop regions (between the α-helical and β-strand components defining the fold topology) are randomly varied.[624] All members of such a "fold-pattern" library are thus theoretically variant TIM barrels, and one can thereby probe the functional limits of this very versatile folding family. The trade-off, of course, is that by so focusing on a single fold, one cannot simultaneously screen other regions of protein sequence space.

We have already seen the power of shuffling in general for protein diversification. Since protein folds are constituted of specific secondary structural regions, what if one made a set of

---

*"Amphiphilic" structures have sides (or "faces") that are distinguishable by their net hydrophilicities and hydrophobicities. The hydophilic face is typically solvent-exposed, and the hydrophobic face buried within the protein core structure.

coding sequence "modules" corresponding to specific secondary structural motifs (α-helices, β-sheets, turns, coils, etc.), shuffled them arbitrarily as discrete blocks, and screened the expressed library? This combinatorial strategy for *de novo* protein generation has in fact been instituted,[625,626] and ribosome display has been exploited for selection of folded proteins by protease resistance.[627] This approach has the potential to yield novel folds for subsequent functional "tuning" by targeted mutagenesis and directed evolution.

A conceptually similar, yet distinguishable, concept to secondary structural patterning is the proposal that protein function can emerge from selection from large numbers of repeat structures. At its heart, this notion is one of the two fundamental alternatives for tracing the ultimate origin of functional proteins: Did proteins derive from molecular selections from pools of random sequences,[628] or have they primarily originated from repeated polypeptide blocks?[629,630] In the latter view, primordial gene replication inherently tended to produce repetitive sequences that in turn manifested extensive periodicities in the corresponding encoded primitive polypeptides.[630] Repeat units lacking stop codons in all reading frames are a potential source of diversity, by skipping between reading frames during imprecise polymerization.[630] Expression of repetitive periodic DNA blocks of this open-frame type has been noted to result in a high frequency of structured proteins, even when all of the possible coding reading frames were arbitrarily chosen.[630] Natural proteins with repeated structural motifs are not uncommon and have certain biotechnological applications that include reconfiguring as binding proteins,[631] as we will consider further in Chapter 7. But let us now look at the fruits of *de novo* protein empirical searches, in their various forms.

Finding and Evolving Folds

Finding foldable patterns in a huge background of useless sequences is a challenge and would not be a feasible proposition were it not for experimental advances that essentially act as enabling technologies for this kind of endeavor. These are the display systems we have surveyed in Chapter 4, combined with certain clever screening techniques for folding, generally based on differential proteolytic susceptibility[455,632] or retention of binding by a host protein after insertion of a novel sequence into a loop region.[633,634] Solubility and absence of nonspecific "stickiness" (exposure of unfolded hydrophobic cores) has also been used as a broad index of the folded state.[627] If direct selection for protein function is instituted, then folding can be inferred indirectly (function is unlikely to arise in a disordered structure).

Regardless of the elegance and ingenuity of selection procedures that can be applied to phage display, to achieve the largest library sizes, *in vitro* display is needed to circumvent transfection losses (as discussed in the previous chapter). Impressive results have been achieved with mRNA display, where independent adenosine triphosphate (ATP)-binding proteins of novel sequence could be selected from a random library.[628] Based on this result, the frequency of functional ligand-binding proteins in random libraries was estimated at 10^{-11}.[628] It was subsequently demonstrated that an alternative approach can be taken by evolving an existing scaffold (the zinc finger motif) toward binding the same ATP ligand, also by mRNA display.[635] This finding was of additional interest given that one of the ATP-binding proteins selected from random sequence had a zinc ion-stabilized fold.[628] The random-library ATP-binding proteins initially possessed poor solubility in the absence of high ligand concentrations, but could be substantially improved in this regard by subsequent rounds of directed evolution.[636] Folding stability of the zinc-binding random-library ATP-binding protein was also further progressively evolvable, and structurally interpretable.[637,638]

As noted above, the structures of functional proteins from random libraries are of intense interest, since they can have no possible natural evolutionary antecedents. Determination of the structure of the functional core region of the zinc-stabilized ATP-binding protein* revealed an overall α/β fold that appeared to have no biological precedent.[639] It was subsequently pointed out that the zinc-binding motif in this protein significantly resembled the natural "treble-clef" zinc-binding structure.[640] The obvious lack of any evolutionary relationships between the abiotic selected protein and natural treble-clef proteins was then hailed as a clear-cut example of truly "structurally analogous" proteins.[640] Mutations that optimized folding, stability and ligand binding of this protein were extensive, but did not alter the basic fold plan.[638] Interestingly, surface residue changes were important for stabilization of the abiotic protein,[638] a finding mirrored in directed evolution for stability of natural proteins (as discussed above).

The structures of *de novo* proteins generated through hydrophilic/hydrophobic binary patterning are theoretically constrained to a large degree by the nature of the patterning itself. Structural studies with proteins patterned as four-helix bundles confirmed their α-helical content.[641] Furthermore, a significant proportion of these showed stabilizing tertiary interaction and native-like structures,[619,641] where stability was associated with longer helices than used in early patterning studies.[618] "Blind" prediction of detailed patterned helical protein structures before biophysical structural analysis has also been successful.[642]

An accumulating database of *de novo* protein structures will be a very valuable resource for protein evolutionary studies and structural studies in general. When selecting from an abiotic library, it might be asked whether the selection process itself might impose any biases on the types of folded proteins that are "pulled out." If selection was based on binding to a biological protein (or even certain biological cofactors and metabolites), the selected abiotic fraction of protein space might conceivably be biased in favor of "biologically compatible" folds, but this too would be of considerable interest to investigate further. For binding of inorganic metal ions, no such hypothetical bias should be present in any case. Current evidence (though sketchy) is consistent with a limited number of structural solutions for certain tasks, as with zinc binding and the treble-clef motif.[640]

Thus far, we have considered *de novo* selection of protein folding and binding function. Is it possible to go beyond this into selecting directly for enzyme function from a random library?

De Novo Enzymes

Experience with selection for *de novo* enzyme activity has curiously provided evidence superficially supporting both of the diametrically opposed views that catalytic function in protein sequence space is relatively commonplace[643] or exceedingly rare.[612] The latter stance is based on calculations for the frequency of obtaining both an appropriate scaffold and positioning of active-site residues, with estimates from (10^{-24})[612] to as low as (10^{-77}),[644] vastly lower than the claimed frequency for ligand-binding alone (10^{-11})[628]. But on the other hand, catalysis (specifically esterase activity) has been found in relatively very small samplings of random protein sequence[643] and in unselected α-helical binary patterning libraries.[645] A likely resolution to this seeming paradox[644] lies in the ability of certain peptides without fixed tertiary structures to exhibit functional activities. Some peptides can show ligand-binding[646] or catalytic properties,† which may have implications for the ultimate origins of biological

---

*The functional core was determined through assessing function of N- and C-terminal truncation mutants.[639] The original selected protein bound adenosine diphosphate (ADP) with slightly lower affinity than ATP,[628] and ADP co-crystallized with the core protein used in the initial structural study.[639]

†The history of catalytic peptides has been contentious, with early claims of high-level catalysis in relatively small peptides (compared with corresponding enzymes) discounted.[647] Many cases of low- to mid-range peptide catalytic efficiencies have been nonetheless reported.[648,649,650]

catalysis.[651] Certain chemical reactions can likewise be catalyzed by "nonspecific" sites on proteins such as serum albumin[652] at rates that are significant but much less than a "proper" enzyme. Relatively short catalytic peptides that do not take up an ordered tertiary structure are very likely to be more frequent in sequence space, and more easily isolatable, than a well-structured fold with appropriately positioned catalytic residues. At the same time, a sequence isolated from a random library with "peptide-type" catalysis may only allow a direct optimization pathway up to a local activity peak, much less than the potential global optimum achievable with an enzymatic fold.[644] Many (although not necessarily all) candidates with low-grade catalytic activity from random sequence libraries could therefore represent evolutionary "dead ends." The nature of the catalytic task is also very relevant to the possibilities for peptide catalysis, and only certain reactions may be feasible in this regard. Esterolytic activity with p-nitrophenyl ester substrates, for example, has been noted as an "easy" task to achieve.[*,645,653] Selection for demanding catalytic tasks has itself been noted as an indirect, though stringent, means for assessing correct protein folding.[654]

If a structurally defined enzyme could be obtained de novo purely by in vitro evolution, a scaffold would "automatically" emerge along with the de novo-selected activity. With sufficient library size and selective power, interesting questions pertaining to analogous enzymes, general requirements for catalysis, and enzyme evolution could be addressed. Unfortunately, the numbers game in random libraries that hits hard upon "real" enzymes with specific folds (as noted above[612]) is a big limitation here. So where does this leave us? If fully de novo enzymes are still in the "too-hard basket," the next best thing is randomization based upon a prechosen scaffold as the most appropriate route toward novel enzyme design. This, of course, then overlaps with the earlier section on "Enzymes from Scratch." But since the scaffold of choice can be initially devoid of any catalytic function, an enzyme resulting from this process is certainly novel, if not completely de novo by the strictest definition. One such test of this kind of approach used sequence randomization upon a noncatalytic zinc finger scaffold and mRNA display selection to generate novel RNA ligases.[655]

Those who would create complete novelty in proteins, whether with an eye toward catalysis or any other protein function, sometimes consider the 20 natural amino acid building blocks and look beyond them. We spent some time in this section considering reduced protein alphabets; the opposite concept of expanded alphabets is attractive on many grounds. While reduced-alphabet proteins may provide a foothold for selecting a novel enzymatic function, by their natures they have inherent limitations. An expanded alphabet conversely may offer new functional potential beyond nature's repertoire and is also a considerable scientific challenge. It is logical then that we too should look in more detail at this burgeoning field in basic science and applied biotechnology.

NATURAL AND UNNATURAL ACTS

Until this point we have often referred to the "20 natural amino acids" (or words to that effect) as the basic building blocks of proteins. We also noted (Chapter 2) that these amino acids and functional side chains were not sufficient for all biological tasks, especially in the field of catalysis. Cofactors, whether inorganic metal ions or themselves synthesized by proteins, can come to the rescue in a great variety of contexts. While acknowledging this often-used

---

*This might be compared with the notion of "severe" enzymatic tasks (noted earlier, as with orotidine 5′-phosphate decarboxylase), with an extremely low spontaneous rate. Some substrates for esterases (especially p-nitrophenyl esters) have an easily measurable spontaneous hydrolysis rate, which can be accelerated significantly even by free derivatives of the histidine imidazole side chain.[645]

biological solution, there may be numerous situations where reaction efficiencies are improved, or even made possible in the first place, if the "cofactor" was an integral part of a protein's polypeptide chain. From first principles, this could be achieved in nature in either of two ways: direct enzymatic modification of a specific amino acid residue at defined site(s) in a target protein, or by extending the 20-amino acid genetic code. Both of these effects have been observed, and the former category can be generalized to cover diverse post-translational modifications to proteins or natural peptides. Such chemical modification is commonplace across all divisions of life, with the involvement of a variety of functional groups including sugars, lipids, and other peptides. Phosphates are of supreme importance as informational chemical tags mediating cellular signaling events. Some modifications can have either regulatory or structural roles, as exemplified by the hydroxylation of proline residues (by prolyl-4-hydroxylase) in control of specific gene expression[656] or collagen biosynthesis.[657]

Post-translational protein modification as a general concept has been understood for a long time, but the recognition of natural extensions to the "regular" genetic code is of much more recent vintage. The first such addition (or the 21st naturally encoded amino acid) was selenocysteine, bearing the same side chain as cysteine except for the replacement of sulfur by selenium.[*] The natural genetic code has been expanded in this case by the utilization of the codon UGA (TGA at the DNA level), which is normally a stop codon. In order to appreciate this and subsequent material in this section, we should have a quick look at some aspects of protein synthesis.

We saw in Chapter 2 that the "entanglement" of informational nucleic acids and their encoded functional proteins is a ubiquitous biological feature, and one that a satisfactory theory of the origins of life must account for. It was hypothesized in the 1950s by Francis Crick, purely on logical grounds, that synthesis of proteins encoded by nucleic acids must use an adaptor molecule, itself very likely to be a nucleic acid. A vast amount of information since then has vindicated this elegant speculation. These "adaptors" are in fact transfer-RNA (tRNA) molecules, which provide the interface between the genetic code and the amino acid building blocks of proteins. Each triplet codon within an mRNA molecule in recognized through the *anticodon* of a corresponding tRNA molecule (Fig. 5.6). If (at the RNA level) the codon is GCA (specifying alanine, in this example), then the anticodon is the complementary UGC (read from the 5′ end, by convention). In order to function as the adaptor molecules for protein synthesis, tRNA molecules have to be "charged" with their specific amino acids, and the enzymes that perform this essential function are *aminoacyl-tRNA synthetases*. Sets of tRNA molecules that are substrates for the same aminoacyl-tRNA synthetases are termed isoacceptors,[†] and therefore become charged with the same amino acid. Protein synthesis then proceeds as programmed by the codons of mRNA molecules, which in the ribosomal "factory" successively direct tRNAs with the corresponding anticodons into place and transfer the amino acid onto the growing polypeptide chain.

Normally, any of the triplets UGA, UAA, or UAG are recognized as stop codons, and terminate translation at that point.[††] But under specific circumstances, the UGA codon is used for the alternative role of acting as a codon for insertion of selenocysteine.[663] Logically, some way of distinguishing UGA codons that are "real" stops and those designated for insertion of seleno-cysteine must be in place, or else the latter seleno-amino acid would be indiscriminately inserted

---

[*]Selenium is an essential trace element for both prokaryotes and eukaryotes. Computational analysis of the human proteome has revealed 25 selenoproteins.[658]

[†]These include tRNAs encoded by replicate genes within the same organism and tRNAs with different anticodons corresponding to degeneracies within the genetic code. For example, in *E. coli*, there are five distinct genes encoding tRNAs for the amino acid alanine: two with the anticodon GGC (for the codon GCC) and three with the anticodon TGC (for the codon GCA).[659,660]

[††]This does not involve tRNAs but rather specific protein release factors.[661,662]

MODIFIED BASES

G ★: 2-Methylguanosine; U ⊙: 5,6-Dihydrouridine; G ▼: N²-Dimethylguanosine; C ⬧: O²′-Methylcytidine;

G ◆: O²′-Methylguanosine; W: wybutosine; U ✦: pseudouridine; C ○: 5-Methylcytidine; G ●: 7-Methyl
guanosine; U ✚: 5-Methyluridine; A ◗: 1-Methyladenosine

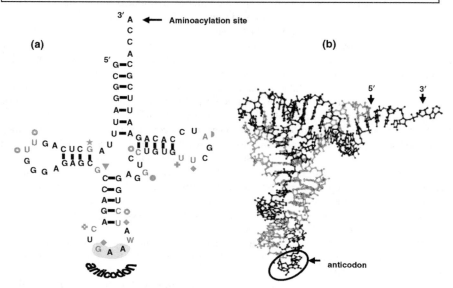

FIGURE 5.6 tRNA structural arrangements with example of 76-base yeast phenylalanine-tRNA. (a) Classic "cloverleaf" diagram of this tRNA secondary structure, showing conventional hydrogen bonding (lines), enzymatically modified unusual bases, and GAA anticodon. Note that the third "wobble" position G in the anticodon (with respect to the UUC codon) is itself a modified guanosine residue. (b) Tertiary (crystal) structure of this same tRNA. Bases 1–33 are shown in gray; Bases 34–36 (anticodon, circled) and 37–76 are shown in black. *Source*: Protein Data Bank[2] (http://www.pdb.org) 1EHZ.[33] Image generated with Protein Workshop.[5]

wherever a UGA codon occurred. This operates through special sequences forming defined secondary structures in mRNAs where selenocysteine UGA triplets are found, termed seleno-cysteine insertion sequences.[664,665] Although the tRNA mediating selenocysteine (sec) insertion is unique, it is not charged by an aminoacyl-tRNA synthetase that is dedicated to this task. The selenocysteine tRNA is instead charged by serine aminoacyl-tRNA synthetase, and the resulting serine-tRNA[sec] is then enzymatically modified to form the correct sec-tRNA[sec]. Nevertheless, selenocysteine qualified as the twenty-first amino acid to be represented by a genetic code.[663]

This was not the last word in the area of natural code extensions. More recently, certain archaea and bacteria have been found to have a genetic encoding system for the modified amino acid pyrrolysine[666] using UAG (normal stop) codons. This differed from selenocysteine in certain important ways. Pyrrolysine insertion during translation uses a dedicated tRNA and aminoacyl-tRNA synthetase,[667] and recognition of pyrrolysine UAG codons is also divergent from recognition of selenocysteine UGA codons.[668] While selenoproteins are ubiquitous, proteins with encoded pyrrolysine appear to be restricted to a subset of the archaea and a very limited set of eubacteria.[666] It is clear that pyrrolysine itself must be present within an intracellular environment before it could be utilized by the appropriate pyrrolysine-specific aminoacyl-tRNA synthetase, and organisms with the latter accordingly also possess the enzymatic machinery for pyrrolysine synthesis. If these synthetic enzymes, pyrrolysine tRNA, and the cognate aminoacyl-tRNA are transferred to *E. coli* in unison as a polyfunctional cassette, the host *E. coli* cells acquire the ability to produce and translate UAG codons by pyrrolysine insertion.[669] This successful "lateral transfer" also has interesting evolutionary implications for the early development of the genetic code.[669]

Rare or not, pyrrolysine takes its place as the twenty-second naturally encoded amino acid. Are there more of these out there still awaiting discovery? More surprises like this may lie in wait, but it has been argued from searches of prokaryotic tRNA genes bearing stop signals as anticodons that additional fully genetically-encoded amino acids are unlikely to be widely distributed.[670] Natural genetic code extensions have great inherent interest and are likely to ultimately benefit biotechnology. What then are the best options for human workers for the insertion of unnatural amino acids into proteins? Back to basics again, we can note three broad alternatives. Two of these (direct *in situ* site-directed modification of specific amino acids, and genetic code extension) also apply as natural options, but for human operators a third possibility of complete (or partial) chemical synthesis exists, at least in principle. Synthesis of normal and chemically modified peptides has a long history,[671] and synthesis (or partial synthesis) of moderate-sized proteins has more recently become possible. But the engineering of unnatural expansions to the genetic code has relied heavily on evolutionary methods in combination with rational decisions regarding the targeting of mutations. As such, it is an interesting example of the interplay between empirical and rational approaches in bootstrapping an entirely new technological option, which will have extensive future ramifications. Not the least of these is the future of directed evolution itself, and for these reasons it is appropriate to include within this chapter a separate section covering this topic.

Unnatural Insertions

In order to manipulate the genetic code for the controlled insertion of unnatural amino acids, it is clearly necessary to designate specific codons for this application. In a completely defined *in vitro* system, in principle any codons that were not prechosen for the normal portion of the expressed coding sequence(s) of interest could be redirected for the purpose of encoding unnatural amino acids. *In vivo*, matters are not so simple. Consider the case of an organism with a unique tRNA gene whose transcribed tRNA molecule recognizes a specific amino acid codon in protein synthesis (as with tRNA[tryptophan] in *E. coli*[659]). If such a tRNA gene is switched for another tRNA with the same anticodon but which is only charged with an unnatural amino acid, then the organism's proteome will be globally affected by the replacement of the original natural amino acid by the artificial substituent. Even if the novel amino acid is a structural analog of the original, cellular viability is likely to be significantly perturbed. Yet transient global incorporations of unnatural amino acids have been performed by using bacterial mutants that cannot synthesize specific amino acids. In a minimal medium containing the "bare necessities" for growth but lacking the amino acid that the bacterium cannot make of its own accord, no growth can occur. If structural analogs of the missing amino acid are provided and globally tolerated at least weakly, then some cell growth can proceed.[672] Another "global" strategy has been the overexpression of specific aminoacyl-tRNA synthetases, which can drive "mischarging" of their target tRNAs with structural analogs of their cognate amino acids.[672] While useful for some purposes, these approaches cannot give site-specificity of insertion *in vivo*.

Since the genetic code is degenerate, some amino acids are specified by multiple codons. As a result of this, in principle one of the alternate codons for a natural amino acid could be commandeered for the purposes of unnatural amino acid insertion. Again, all proteins in the host genome whose coding sequence contains the "earmarked" alternate codon would be affected by such a global change. But some alternate codons are much rarer than others, and it has been suggested that certain rare codons could be rechanneled for such coding purposes.[673] This kind of engineering would involve deleting the cognate tRNA molecules from the *E. coli* genome and also the chosen codons from any essential coding sequences.* For the time being, a much easier option is to use one of the natural stop codons, which (as we have seen above) is just what nature has done for genetic code expansion. The principle (if not the practice) is simple enough: place a stop codon (by standard site-directed mutagenesis) in a specific coding sequence at a desired site, and express this gene in an environment containing a tRNA whose anticodon is complementary to the stop codon, and which is charged with an unnatural amino acid. If the latter tRNA charging is performed *in vitro*, it can be achieved chemically,[677–679] but *in vivo* an aminoacyl-tRNA synthetase that will transfer the unnatural amino acid only to the desired tRNA molecule is required. (In current parlance, an "orthogonal" tRNA/aminoacyl-tRNA synthetase pair.)

A part of the above scenario has already been enacted in natural systems. In the early days of molecular biology, a particular class of mutations in the bacteriophage T4 were found to have interesting properties. Some *E. coli* strains could support the growth of phage with such mutations, while other strains would not. The former "permissive" strains could over-ride the original mutation and allow continued production of the affected gene, by means of a second-site mutation. The original T4 variants were later shown to bear "nonsense" mutations that caused loss of gene activity and premature termination of protein synthesis. These mutations were the result of single-base changes that converted a normal codon into one of the three stop codons, thus providing a signal for mRNA translation on ribosomes to cease at that point, before completion of the full-length protein. For example, a C → A mutation could convert a TCG serine codon at the DNA level into a TAG stop codon. Detailed analysis of the second-site *suppressor* mutations in permissive strains showed that they were mutant tRNA molecules. Once a tRNA molecule has become correctly covalently coupled to its cognate amino acid by the appropriate aminoacyl-tRNA synthetase, the specificity of its "decoding" into protein sequence is dictated by its anticodon only. Consequently, a simple mutation in the anticodon of a tRNA molecule can change its coding specificity, and in specific cases allow the "reading" of stop codons as sense codons (depicted in Fig. 5.7† with suppression of "amber"†† UAG codons). Any other "missense" mutation in a protein coding sequence can also be suppressed by

---

*This is technically feasible but still a challenge. The rarest *E. coli* codons (AGG/AGA[674,675]) encode arginine (two of the six arginine codons in total), but still constitute 1692 and 2896 codons out of the total *E. coli* expressed proteome (1,360,164 codons[659]), respectively. Only a subset of these rare codons will be in essential genes, but a complicating factor is the role of rare codons as a translational regulatory mechanism,[674,676] such that global codon replacements might affect cell growth (at least under specific circumstances). A way to circumvent this is the engineering of specialized ribosomes, as noted in ftp site.

†It is worth noting that for cell viability, a normal copy of the mutated tRNA (as with the tyrosine-tRNA of Fig. 5.7) must also be present, or else no protein could have the original amino acid (tyrosine in this example) correctly inserted. A protein that has a nonsense mutation suppressed through insertion of a different amino acid to its normal sequence (as with mRNA2 of Fig. 5.7) may potentially lose some or even all activity, but is very often functional. This is equivalent to "mutational robustness" as discussed in Chapter 2. Note also that the recognition of specific tRNAs by aminoacyl-tRNA synthetases very often involves the anticodon, so an anticodon mutation may greatly affect the efficiency of tRNA charging.[680] The efficiency of a suppressor tRNA depends in part on the effects of the anticodon mutation and whether any additional mutations are present.

††UAG codons are called "amber" codons for whimsical reasons dating back to early studies with phage T4 mutants. The UAA and UGA stop codons have accordingly received in turn the color-related names "ochre" and "opal", respectively. One might call this kind of nomenclature "proceeding logically from a whimsical origin," but its converse ("proceeding whimsically from a logical origin") also has precedents in molecular biology. Thus, the important technique of Southern blotting (for DNA) was named for its originator (E. M. Southern[681]), but subsequent blotting technologies were whimsically given the "directional" names of Northern blotting (RNA) and Western blotting (protein).

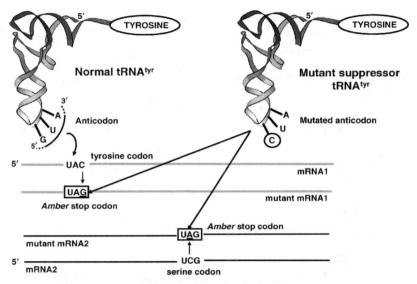

FIGURE 5.7 Mechanism of suppression of nonsense (translational-terminating) mutations. A normal tyrosine tRNA (tRNA^{tyr}) charged with its cognate amino acid recognizes its normal UAC codon. If the latter mutates to an "amber" UAG (at the mRNA level) translation will stop prematurely. A point mutation in the anticodon of the tRNA^{tyr} can alter its codon recognition such that UAG codons are translated with insertion of tyrosine. This applies to any UAG, such as the second example of a mutated serine codon for mRNA2. In both cases, the expressed protein will have a tyrosine residue corresponding to the position of the amber codon. This restores a wild-type protein in the case of mRNA1 or creates a protein with a serine → tyrosine mutation in the case of mRNA2.

analogous tRNA mutations that enable a "wrong" codon to be coupled with a "right" amino acid, with the tRNA as the intervening adaptor.

Now, it has been useful to look at the phenomenon of suppressor tRNAs in the context of the engineering of genetic code extensions. An artificial system where a novel tRNA is charged with an unnatural amino acid could "suppress" a stop codon deliberately inserted into a target protein of interest, where the resulting expressed protein then receives the desired unnatural residue at the predetermined site. Yet nonsense (or missense) suppression is not the only known way coding sequence mutations can be overcome through changes in tRNAs. Another category of tRNA mutations can act as *frameshift suppressors*. Consider an mRNA sequence (AUG|UCC|GUG...), beginning translation in the normal triplet reading frame as shown (giving methionine-serine-valine...). If a single-base (+1) mutational (underlined) insertion is made in the coding DNA to give (AUG|UUC|CGU|G...), the reading frame changes to give (methionine-phenylalanine-arginine...), obviously completely destroying the correct expression of the original protein. But a mutant tRNA with a four-base quadruplet anticodon can suppress this defect and restore the normal reading frame, as depicted with a hypothetical example in Fig. 5.8. Real tRNA frameshift suppressors have long been known, and these too can be co-opted as the adaptors for the insertion of foreign amino acids.[682] Since four-base codons can unquestionably work, what are the limits of codon/anticodon sizes? This has been systematically explored, with the conclusion that three–five bases spans the functional range.[683] Five-base codon suppressors can act as (−1) suppressors, and thus an appropriate suppressor tRNA mutation in principle can overcome any insertion or deletional mutation.

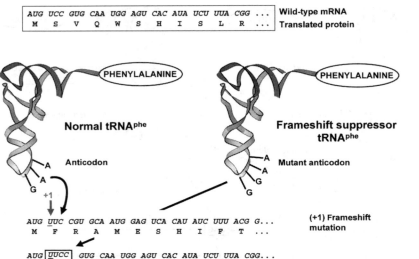

| AUG | UCC | GUG | CAA | UGG | AGU | CAC | AUA | UCU | UUA | CGG | ... | Wild-type mRNA |
| M | S | V | Q | W | S | H | I | S | L | R | ... | Translated protein |

FIGURE 5.8 Frameshift suppression by tRNAs with four-base anticodons. A wild-type mRNA (at top) suffers an insertional single-base (+1) mutation (vertical gray arrow) that destroys the normal reading frame, resulting in expression of aberrant protein sequence "...FRAMESHIFT ..." In this example, a phenylalanine frameshift suppressor tRNA recognizes a four-base codon, and thereby restores the normal reading frame, with a serine → phenylalanine single mutation.

We have already noted that charging of tRNA molecules can be performed *in vitro* by chemical means, which then allows testing of *in vitro* translation systems with the altered adaptor. Let us now look at this area of work in more detail.

In Vitro *Opportunities* When one is interested in charging tRNA molecules *in vitro*, usually the aim is inherently "unnatural," since the exquisite specificity of normal aminoacyl-tRNA synthetases (which aminoacylate only their cognate tRNAs) is deliberately by-passed. Owing to this elementary fact, the process has been termed "mischarging," or more frequently "misacylation." Thus the "wrongful" prefix refers to the natural charging process, but it is a little misleading in the sense that the chemical operation involved is exactly what the experimenter desires under such circumstances. Accordingly, one can prepare *in vitro* large quantities of a defined nonsense or frameshift suppressor tRNA, and then chemically link it with the desired nonnatural amino acid. Although the tRNAs in such cases have mutated anticodons (Figs 5.7 and 5.8) that enable their recognition of four-base or stop codons, *in vivo* they are faithfully and specifically aminoacylated by their cognate aminoacyl-tRNA synthetase. The *in vitro* chemical charging of such tRNAs is therefore still a "misacylation" event. Such unnatural but highly useful charging of tRNAs has actually been performed for many years, by enzymatic ligation (with a ligase not specific for any particular tRNA[677]) or chemical conjugation.[679,684]

There are many experimental pathways offered by *in vitro* misacylations, but before walking down any of these, we should first consider an *in vitro* option that *does* use aminoacyl-tRNA synthetases. The latter natural enzymes discriminate well among the 20 usual amino acids (as they are "designed" to), but often will use chemically related analogs to their natural amino acid substrates. Systematic screening efforts have thus been undertaken with large panels of amino acid analogs and specific aminoacyl-tRNA synthetases, in order to define empirically which compounds can be used as substrates for these enzymes.[685] Although there are limitations on the specific range of unnatural amino acids that can be utilized by the natural panel of aminoacyl-tRNA synthetases, there are also considerable advantages to performing *in vitro* aminoacylations enzymatically where possible. One such benefit is convenience, since an *in vitro* translation experiment using natural synthetase enzymes to charge tRNAs with unnatural amino acids is of little greater difficulty than conventional translations, once the potential substrate range has been defined.[685] Also, the enzymatic aminoacylation can continue during a translation experiment, with improvement in yields, although an off-set to this gain is the potential for a background level of tRNA charging with the incorrect chemical specificity.[685] An approach that also depends on catalysis but entirely waives the use of aminoacyl-tRNA synthetases exploits RNA enzymes (ribozymes, which we will look at in more detail in the next chapter). Initially it was shown that ribozymes catalyzing the aminoacylation of tRNAs could be selected by directed evolutionary techniques.[686] Subsequently, ribozymes capable of directing tRNA aminoacylation with a range of unnatural amino acids have been generated as useful tools for *in vitro* generation of modified peptides and proteins.[687,688]

While yield is obviously an area for improvement, for applications of *in vitro* translation of unnatural amino acids involving certain display technologies, a relatively low level of efficiency is acceptable since translational turn-over is not required.[685] For example, during mRNA display, an mRNA molecule is "read" by the ribosome in a single translational round and then coupled to its own polypeptide product via puromycin (as described in Chapter 4). *In vitro* mRNA display and selection of polypeptides bearing unnatural amino acids has great potential for generating novel molecules for a variety of purposes.[689,690] The variant of ribosome display termed "pure translation display" uses a purified translational system and desired modified tRNAs with unnatural aminoacylation, and is likely to be applicable to the screening of large libraries completely composed of unnatural peptidomimetic compounds.[691] Repeated rounds of such screening with modifications based upon primary candidates could lead to "peptidomimetic evolution,"[691] taking directed evolution into a new phase of complexity and productivity.

We should at this point pause and consider in more detail just what is being referred to when we speak of "unnatural amino acids," since there are degrees of straying from what is "natural" as far as biological polypeptides are concerned. To have any chance of incorporation in the first place, a compound must be capable of forming a peptide bond, and the minimal requirements in principle are therefore the amino (NH_2) and carboxylic acid (COOH) groups. For α-amino acids, a huge range of alternative side-chain "R" groups beyond the natural set can be contemplated, or di-substitutions at the α-positions (Fig. 5.9). Not all chemical α-R groups may be compatible with protein synthesis, but it is well established that many are, albeit at varying levels of efficiency.[673,685,692] Other more radical possibilities affecting the nature of the α-carbon protein backbone itself are stereoisomers of natural amino acids (D-amino acids), amino acids with *N*-substituted amine groups, or β-amino acids (Fig. 5.9). Although a simple rearrangement converts an α-amino acid to a β-isomer (as with alanine; Fig. 5.9), β-amino acids are a very different prospect for biological peptide synthesis than the normal α-amino acid set. Indeed, attempts to use β-amino acids for incorporation into biological polypeptides have met with little if any success,[685,693] as have most other "altered-backbone" insertions.[679,693] This is not the end of the story for such radical protein engineering, since there are prospects for changing matters at the level of the ribosome itself, as we will see shortly.

Thus far, we have been considering *in vitro* and *in vivo* systems for unnatural amino acid insertion in proteins as if they were mutually exclusive, but in reality this is not at all the case.

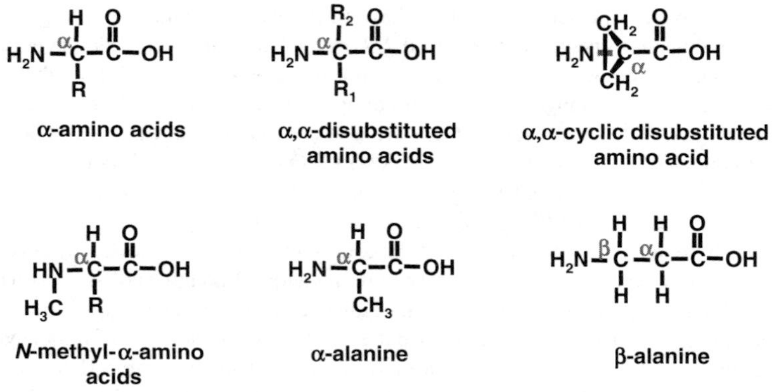

FIGURE 5.9   Amino acid structures, showing α-amino acids where R = any chemical substituent; α/α di-substituted amino acids (where both hydrogens on the α-carbon atom are substituted, and which can form a cyclic group as shown), amine substituents (as with N-methyl-α-amino acids), and β-amino acids, shown by comparing the normal protein constituent amino acid α-alanine with its isomer β-alanine.

A conceptually simple process is to chemically charge a tRNA molecule *in vitro*, and then ensure its transport into the desired host cell for *in vivo* participation in protein synthesis. Of course, it is still necessary to decide the appropriate codon/anticodon system for use, and for *in vivo* application this in practice means tRNA suppressors of the three stop codons or frameshift mutations. Bacterial suppressor tRNAs for amber (UAG) codons charged with amino acid analogs function in mammalian cells (suppressing amber mutations) and are not substrates for eukaryotic aminoacyl-tRNA synthetases.[694] It should be obvious that when tRNAs are chemically charged *in vitro*, no possibility of continuous recharging of such tRNA *in vivo* with the desired unnatural molecules can occur.

In both fully *in vitro* and hybrid *in vitro in vivo* systems, protein yields tend to be low,[673] but may suffice for many analytic purposes. For some biotechnological and scientific aspirations, though, only a fully *in vivo* system of genetic code reprogramming could provide the necessary yields, which was one factor prompting the development of *in vivo* genetic code extensions. So, let us now go through it in more detail. . .

New Codes **In Vivo**   The ancient statement, "There is nothing new under the sun" can easily be refuted in the modern age, since many thousands of organic and inorganic molecules have been created through human artifice that have never previously existed under this or any other sun, in all probability.[*] Until very recently, though, the dream of changing living things in any truly fundamental way remained as a dream. But not any more, as what was once the domain of the supernatural has become merely unnatural, and rigorously logical. And this venture is just beginning.

There are a series of general requirements that must be satisfied before a true expanded genetic code can be realized *in vivo*. As before, the codons for reassignment to an unnatural code

---

[*]An obvious qualification to making a universal statement regarding the true uniqueness of any molecule or molecular system produced by humans is whether other intelligent life in the universe exists or not, which cannot be answered definitively.

and also the cognate tRNA molecule must be chosen. The latter must fail to be charged by any host cell aminoacyl-tRNA synthetase, but perform as a substrate for a novel aminoacyl-tRNA synthetase that must charge it only with the desired unnatural amino acid. The unnatural amino acid in turn must fail to be coupled with natural (unaltered) tRNAs by host enzymes, and this unnatural building block must be capable of importation into the host cell from the external medium, in order to maintain its intracellular levels. (If enzymes capable of synthesizing the desired amino acid exist (or are engineered) and are expressed in the host, it would be unnecessary to supply the new amino acid in the growth medium. A situation analogous to this has been achieved with the full transfer of biosynthetic and genetic-coding capability for pyrrolysine in *E. coli*, as referred to earlier.[669]) When these conditions are met (as depicted in Fig. 5.10), an "orthogonal" tRNA and aminoacyl-tRNA sythetase have been created and the code extended.

How can these complex conditions be realized? A simple answer is through judicious choice of starting-point materials, knowledge of critical regions in molecules of interest, targeted mutagenesis, and sophisticated selection procedures. At the beginning, what is the best codon

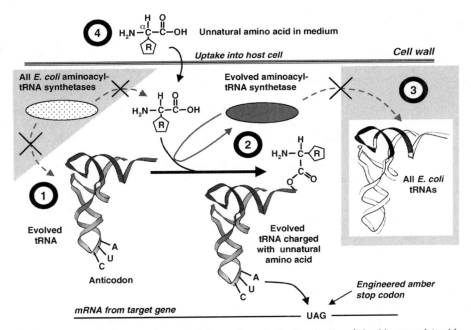

FIGURE 5.10 Requirements for expanded genetic code *in vivo* (in *E. coli* in this example) with "orthogonal" tRNA and aminoacyl-tRNA synthetases.[34,35] (1) A tRNA with an appropriate anticodon for suppression of a target stop codon (amber UAG in this example) must be altered (by directed evolution) such that it is functional in *E. coli* but not charged by any host cell aminoacyl-tRNA synthetases. Also, the latter host enzymes must not charge any *E. coli* tRNAs with the desired unnatural amino acid. (2) An aminoacyl-tRNA synthetase must be obtained (again by targeted mutagenesis and directed evolution) that charges its orthogonal tRNA with a desired unnatural amino acid (and no other natural amino acid), but at the same time (3) fails to charge any host cell tRNA. (4) The unnatural amino acid must be transportable into the host cell. (An alternative available in some cases is to transfer exogenous biosynthetic machinery such that the unnatural amino acid can be synthesized in the host cell from common precursor molecules.)

for reassignment? Initial work in this area used *E. coli* as the host, and amber UAG codons were chosen since these are the least frequent stop codons in this organism (only 7.6% of all *E. coli* translational stops, from annotated *E. coli* sequence[659]). Pilot experiments found that reassignment of endogenous *E. coli* tRNAs and aminoacyl-tRNA synthetase specificities toward "orthogonal" status was difficult, so an alternative starting pair was derived from an archaean organism. (Archaean tRNAs are usually poorly charged by *E. coli* aminoacyl-tRNA synthetases, but show improved translational function in *E. coli* compared to eukaryotic tRNAs.[673]) An archaean tRNA for a specific amino acid can then be altered into an amber suppressor by changing its anticodon. (This has been done for an archaean tyrosine-tRNA, analogously to the suppression depicted in Fig. 5.7.[673])

But this anticodon change might also alter the recognition of the mutant suppressor tRNA by its cognate synthetase,[680] although for archaean tRNAs the anticodon region contribution to such recognition is believed to be small.[695] In any case, directed evolution can be applied toward deriving orthogonal suppressor tRNAs and aminoacyl-tRNA synthetases by means of appropriate selections, both negative and positive. The first step is to ensure that the chosen tRNA is truly orthogonal in the desired host, such that it is both translationally functional and charged specifically by its cognate aminoacyl-tRNA synthetase. Since it is necessary to preserve tRNA recognition by its cognate synthetase, logically random mutagenesis should be targeted to regions in the target tRNA molecule *excluding* those known to directly interact with its cognate synthetase enzyme.[673] From such a mutant library, specific tRNAs that are translationally functional as amber suppressors and charged only by the cognate aminoacyl-tRNA synthetase can be "pulled out" by application of the right selective pressures. The most powerful way of achieving this is the combination of both positive and negative selections, sometimes referred to as a "double sieve" strategy.[369]

A negative selection is designed to actively delete library members that possess an undesirable property, whereas an appropriate positive selection allows members possessing a useful property to differentially grow and amplify. An amber codon in a lethal gene thus allows the removal of *E. coli* cells (from within a library), which bear foreign (archaean) amber-suppressor tRNAs that can act as substrates for host aminoacyl tRNA synthetases, an undesirable state of affairs for orthogonal coding extension. In this setup, any archaean amber-suppressor tRNAs which become charged with amino acids will suppress the lethal gene defect and allow its protein expression, killing the host cell. This then acts as a negative selection in favor of library tRNA members that *cannot* be charged by the host enzymes. Positive selections for desirable tRNA/aminoacyl tRNA synthetase combinations use amber mutations in antibiotic resistance genes. (A summary of positive and negative selection methods for code expansion is provided in ftp site.)

But the ability of a host cell to make use of such an extended code is limited by the availability of the unnatural amino acid itself, which can be provided in the external medium as long as it is transportable into the cell (as in Fig. 5.10). We noted earlier that in principle, the host cell can also be equipped with the enzymatic machinery to synthesize the novel amino acid itself, as has been shown with the natural genes for pyrrolysine and its genetic encoding in *E. coli*.[669] In an analogous but "unnatural" manner, *E. coli* host cells have been engineered to both synthesize the nonstandard amino acid *p*-aminophenylalanine and encode it genetically, such that it is inserted at amber codons.[696] By this feat, the host cells achieved autonomous usage of the novel amino acid, and this work enhances the future prospects for sophisticated evolutionary studies on the advantages of an expanded genetic repertoire.[696]

Success in bacterial systems prompted work aimed at establishing analogous systems for unnatural amino acid incorporation in eukaryotic cells. Some features of eukaryotic tRNA transcription and processing must be heeded initially before proceeding further. Unlike prokaryotes, eukaryotes use a separate RNA polymerase[*] for the transcription of tRNA genes.[697] This enzyme, RNA polymerase III (Pol III) has a complex multisubunit structure and notably requires sequence elements downstream from its transcriptional start site (within the transcribed region itself), which are essential components of the Pol III promoter.[697] Also, the universal 3' CCA of tRNAs (Fig. 5.6) is genomically encoded in most prokaryotes but never in eukaryotes.[†] Additional eukaryotic tRNA sequence elements are required for recognition by specific export pathways from the nucleus to the cytoplasm.[699] It is therefore important to accommodate these combined features in the design and selection of modified tRNAs for eukaryotic expression irrespective of their original source.

Nevertheless, prokaryotic tRNAs and cognate aminoacyl-tRNA synthetases can be source material for design and selection of genetic systems for unnatural amino acid incorporation in eukaryotes. Toward this end, eukaryotic *in vitro* translation systems have been used to selectively adapt an *E. coli* aminoacyl-tRNA synthetase toward recognition of an unnatural tyrosine derivative,[700] and in mammalian cells this enzyme was capable of selectively charging a heterologous prokaryotic amber-suppressor tRNA with this unnatural amino acid.[701] Conceptually analogous positive/negative selection schemes to that described above (for *E. coli*) have been devised for eukaryotic (yeast) systems with *E. coli* tRNA/synthetase pairs.[702]

Having briefly inspected the generation of novel genetic codes, let us now take a glance at how this dramatic new technology can be applied and where its shortcomings might lie. . .

Unnatural Rewards and Limits

Many different unnatural amino acids have been successfully genetically encoded ($>30^{673,703,704}$), a trend that will inevitably accelerate. The incorporation of unnatural amino acids into proteins has so many potential applications that the cliché "limited only by the imagination" would seem fairly accurate (a range of applications are provided in ftp site). From the point of view of this chapter as a whole, the application of encoded unnatural amino acids with the greatest potential is their incorporation in directed evolution experiments toward the alteration and improvement of protein function. To this end, in principle any of the directed evolutionary technologies we have considered in this chapter can be taken to another level by the inclusion of one or more unnatural amino acids. Phage display with such unnatural incorporation has been reported.[287,705] Enzyme activity modulation through encoded unnatural amino acids has tremendous potential, and improvement in the enzyme nitroreductase from *E. coli* has in fact been achieved in this manner.[706] This *in vivo* study thereby confirmed that enzymatic catalysis can be taken to new levels if the natural amino acid tool box is extended. An effective application of directed evolution has used the incorporation of unnatural sulfotyrosine residues into a phage-displayed antibody variable region library, allowing the *in vitro* selection of more effective binders against a target antigen.[287] The related and overarching issue of whether an extension of the 20 conventional amino acids provides an evolutionary advantage to an organism can now be systematically addressed by codon reassignment technology.[673,707] The enzymatic efficiencies and repertoires of catalytic antibodies (Chapter 7) may also be targeted with unnatural amino acids.

As we have seen previously, fully *in vitro* evolutionary methods (such as mRNA display) have the greatest "number-crunching" selective power from extremely large libraries. Since the

---

[*]RNA polymerase I transcribes the large ribosomal RNA while protein-encoding mRNAs are transcribed by RNA polymerase II.[697] Pol III also transcribes the small ribosomal RNA subunit and various other RNAs including regulatory microRNAs.[698]

[†]Where the 3'CCA is not genomically encoded, it is transferred onto the 3' end of the tRNA precursor by means of specific tRNA nucleotidyltransferase enzymes.

purpose of *in vitro* selection is to identify the best performers (usually by binding affinity) from a vast background, the issue of yield is less important than obtaining the desired information. Powerful display methods can thus be combined *in vitro* with tRNAs charged with unnatural amino acids and assigned to specific codons to generate novel modified polypeptides.[685] *De novo* protein selection including unnatural amino acids can thus be envisaged in the same manner, as an extension of work with the natural amino acid repertoire. Taken to its logical conclusion, a point will be reached where polypeptides with fully unnatural amino acid structures are generated and selected, although their encoding and translation would remain ribosomally directed.

But "the sky is the limit" enthusiasm must be tempered by a consideration of potential pitfalls and limitations. We have already noted problems that have been encountered with certain unnatural amino acids (such as β-isomers), and problems of this nature derive from the requirements of the protein synthesis factory, the ribosome. Just as it has proved feasible to generate orthogonal pairs of tRNAs and aminoacyl-tRNA synthetases, so too it is possible to generate orthogonal pairs of altered mRNA molecules and specialized ribosomes (detailed in ftp site). This strategy has great promise for further modifying ribosomes to accommodate a wider range of unnatural amino acids, without interfering with normal ribosomes and risking compromising global cell function. A more fundamental issue may be the tolerance of protein folding for certain unnatural residue substitutions,[708] especially within hydrophobic core regions, although unfoldable variants within large libraries will automatically fail to be functionally selectable. The machinery of genetic encoding of proteins has been amazingly tolerant of diversification, but this is not to say that certain limits may yet be defined. From this point of view, it is interesting to note that when directed evolution of an aminoacyl-tRNA synthetase has failed to yield a variant specific for an unnatural amino acid of interest, it may be possible to overcome this limitation through evolution of an intermediate synthetase that recognizes a structural analog of the final desired side chain.*

Another issue to consider is the specificity of evolved aminoacyl-tRNA synthetases, as a single synthetase may be capable of charging its orthogonal tRNA with more than one distinct unnatural amino acid.[709] This is not a problem, and may even be convenient, if only single unnatural residues are desired in a target protein at any one time. In such cases, the specific incorporated amino acid can be altered by simply changing the environmental supply from one to another out of the set that an orthogonal tRNA/synthetase pair can process. On the other hand, if multiple unnatural insertions using this same amino acid set are desired at precise sites, clearly more specific synthetase recognition is required, with a corresponding number of independently assigned codons and orthogonal tRNA/synthetase pairs.

The genetic code by its nature interfaces between information-carrying RNA and proteins. As we have seen, RNA/amino acid interactions have been proposed as pivotal in the evolution of the code itself. Functional RNA molecules have been an important experimental approach toward probing these questions that echo back to the "RNA world" hypothesis, when a proto-biology was dominated by a single type of molecule for both carrying information and performing effector functions. The RNA world is indeed a world of its own, which we will explore in the next chapter.

---

*A specific case in point in this regard is the evolution of an aminoacyl-tRNA synthetase recognizing a bipyridyl amino acid, via secondary evolution of a synthetase initially evolved to recognize structurally related biphenyl.[673]

6 The Blind Shaper of Nucleic Acids

OVERVIEW

Directed evolution by the agency of humans can be targeted in principle to any molecule or system that can be diversified and replicated. For the purposes of illustrating some of the relevant general principles of "applied evolution," we have concentrated on protein-based themes thus far. At the same time, it was noted briefly in Chapter 4 that in fact an RNA-based system stands as the earliest example of laboratory investigation and direction of molecular evolution. Since that pioneering work, a revolution in biology has arisen with the realization that nucleic acids (especially RNA) can be functional effector molecules as well as repositories of encoded genetic information specifying proteins. The combination of selectable effector phenotypes and direct *in vitro* molecular amplification (replication) have made functional nucleic acids powerful subjects for directed evolution, with both theoretical and practical ramifications. At this point, it is therefore appropriate that we should take a closer look at the wide world of artificial nucleic acid evolution and its consequences. This also leads us to think further about more general possibilities for directed evolution, and how other molecules can be usefully inducted into this process.

FUNCTIONAL NUCLEIC ACIDS AND THEIR DIRECTED EVOLUTION

Early Days

Bacterial viruses (bacteriophages) have been primer movers of the rise of molecular biology, as well as rich sources of biotechnological reagents. *In vitro* molecular evolution constitutes yet another example of the benefits of their study. It was in fact a simple phage of *Escherichia coli*, called Qβ, which provided the first and one of the most celebrated instances of a molecular evolutionary model system. Qβ is a single-stranded positive-sense RNA virus (or positive strand; meaning that the encapsidated viral RNA corresponds to the coding strand for expression rather than its antisense sequence) of 4215 bases. Replication of the viral RNA is dependent on a replicase tetrameric complex composed of one virally encoded and three host proteins. If this replicase is isolated and mixed *in vitro* with viral RNA and ribonucleoside triphosphates (in an appropriate buffer solution), replication of the RNA ensues. The Qβ replicase acts on both positive and negative strands, resulting in growth of product copies in an exponential fashion.[710] This system was developed by Spiegelman and colleagues to study changes in the RNA substrate through iterative rounds of replication. *In vitro* replication of the Qβ genome with its own replicase resulted in selection for truncated variants with improved growth rates, culminating with a defined replicable form of only 218 bases.[711] This molecule (or

Searching for Molecular Solutions: Empirical Discovery and Its Future. By Ian S. Dunn
Copyright © 2010 John Wiley & Sons, Inc.

any related truncated RNA selected through the *in vitro* replication system) is sometimes called "Spiegelman's monster" in popular accounts of this work. In this simple *in vitro* system a Darwinian selection process in favor of efficient replication thus resulted in a stripped-down version of the original genome. The 218-base RNA derived in this manner was a superior replicator in its *in vitro* setting, but completely defective as a bacteriophage. Natural selection likewise will relentlessly inactivate genes or whole genetic systems if they no longer serve functional purposes and come at a fitness cost, as we have seen in Chapter 2.

These pioneering experiments are now rightly regarded as classics in the field of *in vitro* evolution,[712] and set the stage for consideration of the other major domain of directed evolutionary studies, with replicable nucleic acids.

Protein-Free Enzymes

Once there was a biochemical aphorism "All enzymes are proteins, but not all proteins are enzymes." Some aphorisms stand the test of time, but this one patently has not. Certainly, it is true that not all proteins have catalytic activity, but proteins are not the only possible biocatalysts. This dates back to the discovery in 1982 of self-splicing RNA in the protozoan organism *Tetrahymena*.[713] Within this organism, an intervening (intronic) sequence in a ribosomal RNA precursor was found to be capable of splicing itself out, with the appropriate rejoining of the ends of the exonic RNA sequences to form the final processed product. To mainstream molecular biology at the time, catalytic RNA was largely an "off the wall" notion. Accordingly, detailed evidence was necessary to demonstrate convincingly that the self-splicing phenomenon could indeed operate in the absence of protein, requiring only guanosine and metal ion cofactors.[713] Soon after this another group working on RNase P (required for processing tRNA precursor transcripts) showed that this ribonucleoprotein enzyme had an essential RNA component that itself constituted the actual catalytic center.[*,714] While these combined results showed that the chemistry of RNA had hitherto unrecognized subtleties, a true RNA enzyme should show the usual properties that describe an enzyme as a catalyst, remaining chemically unmodified after a catalytic cycle and capable of turning over substrate into product at a measurable rate. *In vitro* experiments with the essential RNA moiety of RNase P[715] and the isolated self-splicing *Tetrahymena* intervening sequence[716] indeed demonstrated true enzymatic properties with such RNA species. As a result, "RNA enzyme" perhaps inevitably became condensed into the term "ribozyme," which (if not quite a household word) has had a steadily rising profile in the fields of molecular biology and biotechnology.

A major reason for this has been the recognition that ribozymes are not merely peripheral entities in cellular functioning, or renegade offshoots of molecular evolution. One of the most interesting features of RNA catalysis was the realization that a replicable molecule capable of carrying genetic information could also possess effector functions. This in turn led to the proposal that RNA actually preceded the appearance of proteins in the sequence of chemical developments leading to the origin of cellular life. In this hypothetical "RNA world" (as discussed in Chapter 2), RNA molecules replicated themselves and performed all catalytic functions.[133,717] This model then postulates that the entanglement of informational nucleic acids and proteins allowed the more structurally and functionally diverse protein universe to take over all effector activities. We have already noted, though, (also in Chapter 2) that ribozyme function is alive and well and at the very heart of a fundamental activity of all cellular life, the protein synthetic factories called ribosomes. Other instances of essential and

---

[*]The 1989 Nobel Prize for Chemistry was awarded to Tom Cech and Sidney Altman for their work in revealing the existence of RNA catalysis.

widespread ribozymes are also very likely to be confirmed. Although splicing of introns can be mediated by self-processing ribozymes (as in the case of *Tetrahymena* ribosomal RNA and a different category of intervening sequences termed Group II introns), the splicing of nuclear introns from higher eukaryotes requires a complex multisubunit structure termed the spliceosome, composed of >200 protein and several small RNA subunits. Here also, there is accumulating evidence that the catalytic splicing activity is directly mediated by the small nuclear RNA molecules within the spliceosome.[718] A precedent for an endogenous ribozyme regulating expression of a human gene (β-globin) has been found,[719] and ingenious screening processes have revealed the existence of other ribozymes encoded within the human genome.[720] While the RNA world may indeed have preceded the "protein revolution," RNA and protein function have evidently developed in parallel as an "RNA–protein world."[721]

The interplay between functional RNA and proteins is directly observable in the activities of some natural ribozymes.[*] In particular, protein "facilitators" can have a role in assembling RNA secondary structural elements into configurations appropriate for efficient ribozyme catalysis, which has been observed with Group I introns[723,724] and the ribosome itself.[725] Attempts to find ribosomal RNA peptidyl transferase activity in the absence of accompanying ribosomal proteins have not yet been successful,[726] but ribozymes with comparable activities have been independently selected[727,728]). While proteins are certainly the major and most versatile biocatalysts, and (in the RNA world view) have superseded their RNA counterparts in almost all roles, it is interesting that basic cellular functions enabling protein synthesis, tRNA processing, and nuclear splicing have retained ribozyme function. It is not clear at present whether this is due to an inherent advantage of RNA catalysis for certain specific functions, or whether some fundamental cell functions are difficult to alter by evolution once they have become established. If the latter is true, such ribozyme activities would represent another "locked in" (or "frozen") arrangement for which any evolutionary tinkering has an unacceptable fitness cost. Some of these questions can in principle be addressed experimentally if proteins capable of performing the same activities as for specific natural ribozymes can be artificially derived.[†]

Of the natural RNAs catalyzing self-cleavage/ligation reactions, "hammerhead" ribozymes are the smallest, reducible to <40 bases with retention of activity.[730] By the nature of the hammerhead catalytic and cleavage target sites, it was possible to separate self-cleaving hammerhead ribozymes into distinct catalytic and substrate molecules (Fig. 6.1).[731] The only substrate requirement for efficient *trans*-cleavage in these circumstances is a specific two or three-base sequence[††] immediately 5' to the cleavage site, and H-bonding complementarity with arbitrary ribozyme "arms."[731] Additional auxiliary ribozyme sequences as well as the catalytic core are required to optimize folding, and for the ability of hammerhead ribozymes to efficiently operate *in vivo*.[**,733] In fact, sequences outside the hammerhead catalytic core have

---

[*]While a great many natural ribozymes have been described, they can be slotted into a handful of classes based on sequence and structural similarities: hammerhead, hairpin, Group I and II introns, hepatitis delta virus ribozyme, RNase P, varkud satellite ribozyme, and the ribosomal peptidyl transferase ribosome.[194,722]

[†]In the previous chapter, we briefly noted that ribosomal engineering may improve incorporation of unnatural amino acids. A high challenge for future protein and biosystems engineers would be the generation of ribosomes whose peptidyl transferase activity is provided by an engineered protein rather than the natural RNA. If this was possible with comparable efficiency to native ribosomes, the "locked-in" hypothesis would receive support. Certainly, nonribosomal peptide bond formation by protein peptide synthases[729] is well characterized.

[††]Initially, this target sequence was considered to be 5' GUC↓[731] (arrow showing cleavage point), but most sequences with 5' VUH↓ (where V = A, U, or G and H = C, A, or U) are also cleavable.[732]

[**]Such auxiliary sequences can reduce the ribozyme dependence on metal ions such that *in vivo* ion levels (as for Mg^{2+}) are compatible with efficient activity.

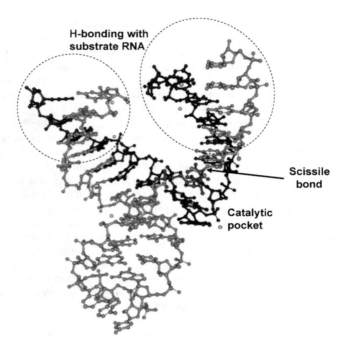

FIGURE 6.1 Structure of a hammerhead ribozyme (black) in complex with RNA substrate (gray).[36] The C residue 5′ to the scissile bond site is indicated with arrow. Regions of conventional base-pair hydrogen bonding that direct the ribozyme to the target sequence specificity are indicated, as is the ribozyme catalytic pocket that contains invariant base residues. (The catalytic pocket is essential for activity, but the H-bonding regions can be arbitrarily altered for designed complementarity with a substrate bearing a GUC triplet.) *Source*: Protein Data Bank[2] (http://www.pdb.org) 488D. Images generated with Protein Workshop.[5]

been crucial in resolving questions concerning structure–function relationships in these types of RNA catalysts.[734]

Many studies have aimed to adapt ribozymes for biotechnological and clinical applications, especially the silencing of genes through specific mRNA cleavage.[735,736] In recent times, the exploitation of small interfering RNAs (RNAi; as discussed in Chapter 3) has emerged as a favored technological competitor in this important area. Although RNAi has generally put ribozymes in the back seat for the purposes of knockdown of specific gene expression, ribozyme applications in biotechnology remain very prominent, and we will discuss this briefly in Chapter 9. But since *in vitro* selection and evolution have gone beyond natural precedents and revealed numerous new functional types of ribozymes (and other functional nucleic acids), we should initially consider how these processes are carried out. Some of these evolved ribozymes challenge the efficiency of the hammerhead ribozyme for RNA cleavage, but many other types of ribozyme activities have been selected and characterized.

Finding Functional RNAs

In Chapters 4 and 5, we looked at artificial evolutionary strategies chiefly from a protein point of view, but the general principle holds for any replicable molecule. As we have seen at the beginning of this chapter, *in vitro* evolution was pioneered with RNA, and indeed this holds true

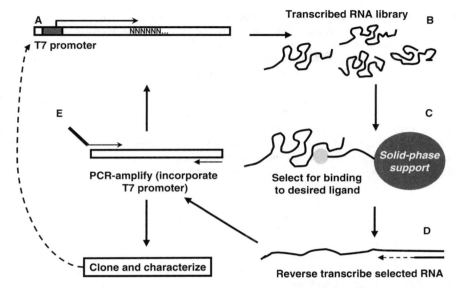

FIGURE 6.2   Basic SELEX principle for RNA. (A) A library of DNA duplexes encoding RNAs with a designated variable tract (NNNN...) is transcribed by a suitable RNA polymerase (T7 RNA polymerase in this diagram, with its cognate promoter provided in *cis*). (B) The transcribed RNA library is used to select for RNA variants binding to a ligand of interest, such that nonbinders can be removed (often by immobilizing the ligand on a solid-phase support) (C). The selected RNAs are then reverse transcribed (D) and amplified such that the products have incorporated T7 promoters (E) to allow the cycle to be repeated. If appropriate, a specific binding RNA can be selectively rediversified and the process renewed (dotted arrow).

for more advanced techniques where the molecular diversity is artificially engineered. In August of 1990, two groups reported a technique for selecting functional RNA molecules *in vitro*.[439,737] One of these groups referred to this general procedure as "SELEX" (systematic evolution of ligands by exponential enrichment[439]), which was noted at the beginning of Chapter 4 as an example of a "metatechnique," by virtue of its extremely broad applicability. The basic principle of SELEX (Fig. 6.2) is familiar in the context of directed evolution in general: diversification, selection, and amplification. But the advent of SELEX and comparable procedures preceded most laboratory-directed evolution experiments at the protein level. As we will see shortly, SELEX is not limited to RNA, but is applicable to any replicable nucleic acids. Indeed, in the original SELEX publication, it was suggested that an analogous procedure could apply to a polysome-based technique,[439] anticipating the subsequent development of polysome display.[440]

A typical SELEX experiment can begin with a suitably diversified population of RNA molecules, and then apply multiple rounds of selections to enrich for functional RNA molecules of interest. Although each round involves an amplification step, the process can be viewed as a purification strategy for a population of active binding molecules, with each selection round reducing irrelevant background "noise." Indeed, with a powerful purification technology, the requirement for amplification during each selection round can be waived.[738] A purist might raise the question of whether application of SELEX with only a single diversification step truly constitutes *in vitro* evolution (recall this kind of issue discussed in Chapter 4). Certainly, a molecule derived from a SELEX operation can in principle be used for a subsequent round of rediversification and selection cycles, in a true evolutionary mode (Fig. 6.2). A caveat here, though, is that choosing a promising candidate from an early selection round as a new starting point may limit one to a local fitness optimum only, and exclude the possibility of finding better "global" molecular solutions to the defined problem.[739]

In practice, it is indeed valid to refer to SELEX (and related procedures with nucleic acids) as "evolutionary," as additional diversity is factored in (and can be adjusted at will) with each "generation" of selective cycling, through the PCR amplification step.[*,739] Another issue in this context is the size of the diversified nucleic acid molecules. Given the limited nucleic acid alphabet in comparison to proteins, a "complete" library of 25-mers (4^{25} or $\sim 10^{15}$ molecules) is feasible with modern oligonucleotide synthetic capabilities.[739] If an ideal functional molecule of interest is already contained within such a library of 10^{15} different members, then the task comes down to an exercise in *in vitro* selection rather than evolution. In practice, many nucleic acid effector functions will require significantly longer sequences than this, and an *in vitro* library will then fall short of full representation of all possible sequence combinations. Usually, a functional nucleic acid will tolerate a certain degree of sequence substitution while retaining measurable activity, increasing the probability that a first-generation prototype can be isolated from a diverse library even if it is theoretically suboptimal.

If an arbitrary DNA sequence is successively amplified through repeated rounds of PCR, an effect tends to occur that is analogous to the "Spiegelman's monster" of early Qβ studies (as above). Molecular species that have "an edge" in their polymerase-mediated replication rates will progressively accumulate, and artifacts have a high probability of rearing their ugly (or at least highly undesirable) heads. In the directed evolution of nucleic acids involving repeated rounds of PCR-mediated amplification, the same problem potentially exists, but is mitigated against by the continuous application of the chosen selection procedure. This has the action of channeling the nucleic acid population into a desirable functional subset, provided, of course, that the selection itself is effective. ("You get what you screen/select for" is still applicable here.)

As we have seen, RNA can exhibit catalytic activities, but specific interaction with desirable target molecules *per se* is also a very useful property. Experimental observations have shown that relatively short single-stranded RNA molecules (often <30 bases) can bind to a wide variety of different molecules, and are selectable on this basis through SELEX-based procedures. RNA molecules selected for such specific binding to a target molecule of interest have been termed "*aptamers*," from the Latin "to fit."[†,737] It might be noted that an aptamer can be regarded as a ligand itself (if a protein is the target[439]) or a binder of ligands if small molecules are the interactive partners.[737] Though this distinction is generally made on the basis of respective molecular sizes (the ligand tending to be the smaller interactive molecule), more specifically a ligand-binding molecule usually provides a "docking site" or "pocket" for its cognate ligand, and is definable on this basis. Aptamers can then function in either roles, by donating a binding site, or fitting into one.

But before saying more about aptamers, let us return to the topic of ribozymes, and how to specifically apply the powerful techniques of *in vitro* evolution toward their selection. The derived products of ribozyme evolution in the laboratory are both theoretically fascinating and of real practical import.

---

[*]As we noted in Chapter 4, error-prone PCR is a useful mutagenic approach for a variety of directed evolutionary applications. Other means for introducing diversity, including recombination (Chapter 4), are applicable to nucleic acid directed evolution as for proteins,[739] a theme continued further in this chapter.

[†]One could praise this choice as an apt descriptor, but that would be somewhat circular, as the word "apt" itself comes from the same Latin origins. Fittingly nonetheless, perhaps. A strategy that might be helpful in reminding us of the core meaning of "APTAMER" is to reinterpret the word as a "backronym" to reflect the SELEX process itself: Adaptive Process That Applies Molecular Evolution Recursively.

NOVEL RIBOZYMES

Selecting for a desired binding activity is conceptually quite straightforward. Treat a randomized molecular library with the partner molecule of interest, remove the unbound species, amplify the bound ("partitioned") fraction, and repeat the cycle as required (Fig. 6.2). But if enzymatic activity is the desired property, the sophistication of the selection process itself must be duly increased. Many different types of ribozymes have been obtained "from scratch" by *in vitro* evolutionary methods.[739] For present purposes, it is not necessary to exhaustively describe the selection processes for each, but a few examples are worth examining to illustrate some general points.

A great many searches for novel ribozymes have been inspired by the RNA world hypothesis that we have noted previously. If a self-sustaining metabolic and replicative biosystem existed based entirely on RNA, it should be possible to demonstrate that specific ribozymes can fulfill all of the necessary roles in this regard. Ultimately, this would lead to comprehensive models for how the hypothetical RNA world could have logically operated, reproduced, and persisted.[740,741] Of course, the potential complete range and limits of ribozyme catalysis are important questions in their own right, with possible practical ramifications as well.[742] A fundamental requirement for a self-sustaining RNA world is the existence of ribozyme-based replicators. With this in mind, let us take a look at progress toward recapitulating these entities, extinct in the natural world for billions of years.

Ribozyme Replication and Other Things

An ultimate aim of those who study the RNA world is to create a fully functioning biosystem based entirely on catalytic RNAs.[741,743,744] One of the most fundamental properties required for such an entity is that of *replication*, and this would necessarily pre-date the advent of RNA/ protein entanglement. Many workers have therefore sought to find (through the agency of directed evolution) ribozymes that can act as RNA polymerases. Where do you begin such a quest? In principle, by directed evolution from random sequences, but (as noted above) the sampling of RNA sequence space becomes "sketchy" above 25-mers (with existing capabilities), and a complex polymerization function is likely to require a longer sequence to assume a suitable catalytic conformation. So it is a logical premise to build as much as possible upon pre-existing ribozyme activities, and natural self-splicing by ribozymes (involving cleavage–ligation events) has been used as a platform for developing low-level ribozyme RNA polymerization.[745–748] Another starting point has been the use of certain evolved ribozyme RNA ligases, based on some similarities in their activity with known protein polymerase enzymes. Some artificially evolved ribozyme RNA ligases can catalyze the reaction of a 3′ OH group with a 5′ triphosphate on another RNA, resulting in release of pyrophosphate and formation of a joined 3′–5′ phosphodiester bond.[*,751]

De novo ribozyme ligase activity in fact has been selectable by *in vitro* evolutionary methods, perhaps surprisingly successfully. Despite such ribozymes having relatively large catalytic domains (>90 bases in some cases), multiple classes of ribozyme ligases (based on sequence similarities) were isolatable from random libraries large in size (typically with >10^{15} members) but vastly smaller than those required for full sequence coverage.[751,752] This

---

*This 3′OH–5′ triphosphate reaction is shared with protein RNA polymerase enzymes. In contrast, protein RNA (and DNA) ligases join 3′OH groups to 5′ monophosphates, with the assistance of cofactors (often adenosine triphosphate). Ribozyme ligase activity is in fact known naturally, as with self-splicing RNAs that require a ligation step to rejoin exonic RNA sequences following excision of introns,[713,749,750] but this also does not involve 3′OH–5′ triphosphate reactions.

finding was interpreted as evidence that active configurations for catalysis were frequent in RNA sequence space, and that the appearance of complex ribozymes in the RNA world was perhaps easier to account for than hitherto suspected.[751] The selection process for ribozyme ligation uses the transfer of an affinity- tagged nucleic acid substrate to the ribozyme itself, where the "tag" is usually the small biotin molecule, selectable by its high-affinity binding to avidin proteins. By the design of the process, theoretically only useful products can be both affinity selected and amplified. (Such products in this context are molecules that have both the biotin tag (accompanying the substrate) and ribozyme sequences with the desired activity.) But if one needs to find a *trans*-acting ribozyme directly, a problem arises, for a ligated substrate (or the product of any other *trans*-catalytic reaction, for that matter) will not provide you with the required ribozyme sequence information itself. A solution to this problem is to devise a means for linking genotype (ribozyme sequences encoded as DNA segments) and phenotype (the results of catalysis), an issue treated in some detail in Chapter 5. *In vitro* compartmentalization has been directed toward this problem, to search for *trans*-acting ligase ribozymes.[753]

While some ribozyme ligases share functional similarities with polymerases, they cannot fill the catalytic role as agents of replication. To function as a generalizable replicase, a ribozyme RNA polymerase should be able to act on any other RNA template-primer system. Self-replication of ribozymes itself has in fact been accomplished[754] based on ribozyme-mediated ligation of two RNA subcomponents, which upon joining form the ribozyme ligase itself. Yet this kind of system (Fig. 6.3a) is inherently dependent on the supply of relatively complex subunits (in this case, composed of many nucleotides). It has been further possible to split ribozyme self-replication into two coupled cross-catalytic reactions,[755] as depicted in Fig. 6.3b. But what of the evolution of true RNA ribozyme replicases? Recent progress has seen the isolation of additional novel ligase-derived ribozyme polymerases,[756] and improvement in the maximal ribozyme RNA polymerase extension to ~20 bases.[757] In the latter case,[757] *in vitro* compartmentalization was used as a superior system for preserving the necessary genotype–phenotype link to identify *trans*-acting ribozymes. Still, this is clearly nowhere near enough for the replication of a complex ribozyme of 200 or more bases, and other issues such as fidelity and substrate discrimination also remain before a viable contender for a reborn RNA world replicase

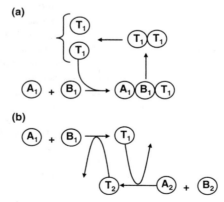

FIGURE 6.3   Self-replicating molecular processes. (a) T_1 catalyzes the joining of A_1 and B_1, by first forming a trimolecular complex. The union of A_1 and B_1 results in the formation of T_1 itself. The resulting T_1:T_1 dimer then dissociates into two T_1 molecules, which can re-engage with the same cycle. (b) Here, T_1 and T_2 cross-catalyze the reactions of $A_1 + B_1$, and $A_2 + B_2$, respectively, to form a coupled loop.

can be proffered.[758] Those at the cutting edge of the field remain confident that the remaining hurdles can nonetheless be overcome.[756–758]

Directed evolution in principle also allows questions to be asked about the number of potential solutions to a given catalytic task (or any function in general). This can be said of any artificial evolutionary approach (including those targeting protein enzymes, as in Chapter 5) but it tends to be a more practical proposition with ribozymes, owing once again to their relatively constrained alphabet. Two extreme positions can be defined: (a) an *in vitro* evolution experiment is performed, and all "molecular solutions" (out of a statistically significant number) share a common sequence or structural feature; and (b) the same experimental design yields at least two classes of active molecules, which appear to be unique within themselves and unrelated to each other. The ramifications of such hypothetical clear-cut findings are also straightforward. The second scenario suggests that if multiple random and diverse sequences satisfy the original design problem, then multiple design solutions at the molecular level are possible, and finding any one solution from a random library will be commensurately easy. On the other hand, the first position suggests that the possible molecular solutions (at least within the dictates of the experimental *in vitro* evolution specifications) are very limited, if not manifested by only one realistic active configuration (although this cannot be proven by this means alone).

Some investigations have been performed with the selection of self-cleaving ribozymes that are interpretable along these general lines. From purely random sequence, the hammerhead ribozyme motif (Fig. 6.1) emerges as the most common solution to the selected target function of self-cleavage, and some short tracts of these selected RNA catalysts showed significant sequence conservation.[759] A "take home" message from this study is that natural evolution in different settings may converge upon similar molecular solutions to "design problems" based on both their relative simplicity and chemical constraints themselves.[*,759] In other words, while there may indeed be more than one way to skin a cat, there are not an infinite number of ways, and in some cases natural molecular evolution may be channeled into specific pathways by such real-world factors.[†] Somewhat ironically, when *in vitro* selection does indicate multiple solutions to a design problem, it can complicate attempts to trace the nature of putative equivalent catalysts in the primordial RNA world.[740]

Two broad types of catalytic activities are known to be mediated by natural ribozymes, at least those that are still extant in the modern biological universe. These reactions are certain types of phosphate group transfers involved in cleavage/ligations in splicing and RNA processing, and peptide bond formation in ribosomes.[194,760] Of such natural cases, less than 10 different structural and mechanistic classes have been defined (as noted earlier). In contrast, artificial evolution and selection have yielded a rich harvest of 30 or more novel ribozyme classes,[761] some of which also catalyze phosphate transfer reactions (including the polymerase and ligase ribozymes mentioned previously). Transfer of a phosphate group is also an inherent feature of the polynucleotide kinase reaction, well-defined in specific protein enzymes, and also observed in artificially selected ribozymes.[762] Some of the more interesting cases of artificial RNA catalysis have been those that catalyze carbon–carbon bond formation, since such bonds (and their making and breaking) are fundamentally important in biochemical reactions. To date, this very broad class of reactions has been represented by ribozyme catalyzes of two separate organic processes (the Diels–Alder reaction and the aldol condensation),[763–765] both of which

---

[*]This is not to ignore the contributions of regions outside the catalytic core of natural ribozymes for folding and stabilization enhancement.[733]

[†]This is consistent with "replaying the tape" theoretical evolutionary considerations noted in Chapter 2, where some modeling studies find that the freedom of evolutionary "movement" may be constrained by various factors. At the same time, within such dictates vast possibilities for alternative forms exist.

are of great synthetic importance in organic chemistry. In one of these approaches, *in vitro* compartmentalization was used to isolate a Diels–Alder ribozyme capable of multiple enzymatic turnover.[764] These results too have been cited as consistent with the proposed existence of the RNA world.

We should think a little more about ribozyme mechanisms in general, but before doing so, let us take a look at some general comparisons between ribozymes and proteins. In particular, we can compare them with respect to some of the general areas of artificial evolution discussed in the last chapter.

Comparisons with Proteins

An important aspect of the directed evolution of proteins (and protein engineering by any means at all) is the derivation of enzymes with improved robustness toward adverse environments. (A significant portion of Chapter 5 was devoted to this general topic.) Can RNA enzymes be likewise adapted in an analogous manner, to acquire improved thermal stability, or resistance to other environmental extremes? As might be expected, given the relatively recent discovery of RNA catalysis in general, there has been less work in this area. For the majority of desired catalytic activities, a mesophilic protein already exists, which can then be evolved or engineered toward increased robustness. In contrast, the range of all functional ribozymes (at any temperature) is far more limited, so the challenge has initially been to obtain and "tune" a desired ribozyme activity in the first place. Of course, this is not to suggest that the stability of ribozymes has been ignored, since this is part and parcel of a comprehensive understanding of the mechanisms of RNA catalysis.

Aiming for Stability   RNA molecules in general can usually fold into a number of energetically closely related states. At low temperatures, the difference in free energy of folding may favor an active ribozyme (or aptamer) configuration over others by only a small margin,[766] although a number of practical and theoretical studies have indicated that evolved functional RNAs tend to have more ordered and unique folded structures than random (unselected) RNAs of comparable size.[767–769] Nevertheless, relatively stable alternative folding states can be a significant problem in RNA functional analyses.[770] While thermophilic proteins are well known, RNA molecules in general are less capable of maintaining their folds at elevated temperatures. Also, less discrimination usually exists at higher temperatures (in terms of free energy of folding) between "productive" folds and alternative nonproductive folded states, doubly reducing the probability of finding thermophilic functional RNAs.[766] As an example, the efficacy of hammerhead ribozymes generally decreases beyond 50–60°C,[771,772] and even natural "thermophilic" hammerhead ribozymes lose activity by 80°C.[773] Hairpin ribozymes too cannot be subjected to temperatures in excess of 65°C without functional loss.[774]

It is interesting to note that RNA molecules artificially selected for function *in vitro* have been predicted (by thermodynamic structural algorithms) to have lower thermal stabilities than a control group of natural functional RNAs.[775] This was attributed to the specific selection processes used to isolate the artificial RNAs, leaving room for secondary improvements through selections directly targeting stability. Within the above-noted intrinsic constraints dictated by RNA chemistry itself, by *in vitro* selection procedures it has indeed been possible to significantly enhance ribozyme thermostabilities, and a correlation between thermostability (retention of structural integrity at elevated temperatures) and activity enhancement has also been noted.[776,777] Natural ribozymes use "tertiary stabilizing motifs" outside catalytically crucial regions,[733,733,778] and this observation has inspired efforts to exploit such structures for artificial ribozyme stabilization, using *in vitro* evolution[779] or specific modifications based on pre-existing structural knowledge.[780,781] A cautionary note that is reminiscent of the global properties of protein enzymes is applicable to this kind of ribozyme tinkering. Cross-linking ribozyme regions not associated with the catalytic core can sometimes alter the nature of ribozyme catalytic activity itself, as seen where the activity of a hammerhead ribozyme was deviated from cleavage toward ligation.[782] Nevertheless, hammerhead ribozymes with artificial enhancement of thermostability can retain activity at 80°C (albeit not at optimal levels).[779]

Although ribozyme stabilities can thus often be improved, they are nevertheless poorly compatible with temperatures in excess of 80°C. This might seem puzzling, since we have seen previously that ribozyme activity is a fundamental process of cellular life (as key components of ribosomes and RNase P[783]), and this certainly includes hyperthermophilic organisms that find temperatures of only 80°C somewhat on the cold side. But these examples of natural ribozymes, important though they are, do not perform their tasks in isolation. We noted earlier that the activities of natural ribozymes can be significantly modulated and augmented by protein facilitators, and both ribosomes and RNase P function as ribonucleoprotein complexes, despite the core element of RNA-mediated catalysis. With this in mind, it is interesting that although the isolated RNA components of RNase P from eubacteria are catalytically functional, most thermophilic archaea do not possess corresponding RNAs from RNAse P that have activity when free from protein.[*,785,788,789] *In vitro* experimentation has concluded that the protein subunit of RNase P from a thermophilic organism confers thermal stability to the RNase P complex, rather than thermal resistance being associated with the RNA subunit itself.[790] Based on recent observations, it has even been suggested that at least one studied hyperthermophilic organism might have its RNase P function (that of tRNA maturation) entirely supplanted by protein.[791]

These results are consistent with the supposition that RNA alone functions poorly as a catalyst (if at all) at very high temperatures, unless it is assisted by protein cofactors that can stabilize RNA tertiary structures otherwise facing thermal denaturation. On the other hand, these data do not sit well with the proposal that the origin of life itself occurred at the elevated temperatures that modern thermophilic organisms prefer. This notion hinges primarily on evidence from phylogenetic sequence analyses that the last universal common ancestor (LUCA; the root of the evolutionary tree on earth for all cellular organisms) was a thermophile.[792,793] This interpretation itself has been controversial,[794,795] but it has also been pointed out that even if the "hot-LUCA" viewpoint was indisputable, cellular thermophiles are already highly advanced biosystems, which themselves must be the result of long processes of molecular evolution. It is perfectly possible to then propose that the ancestors of the primordial thermophiles developed at much lower temperatures, with thermophily arising later as a secondary adaptation.[766,796] This view is also consistent with the observed problems of RNA function at elevated temperatures, and a "hot-start" RNA world therefore appears untenable.[766] Indeed, models for low-temperature "icy" RNA worlds have been put forward and supported.[797,798] An implicit assumption in such scenarios is that at least some local environments of the early earth were suitably cold, but the nature of the earth in such remote times has many uncertainties.

Accepting that RNA molecules have probable limits for thermophily well below that achievable with proteins, we might then ask about directed RNA evolution toward other

---

[*]Activity has been reported in one case, but weak compared to eubacteria.[784] Initial work suggested that eukaryotic RNase P also lacked *in vitro* activity through RNA alone,[785] but such RNA-mediated catalysis has been demonstrated under the correct *in vitro* conditions.[786,787]

extreme environments. As we will note again shortly, metal ions (especially Mg^{2+}) have a prominent role in catalysis for many (although not all) ribozymes, and some *in vitro* evolution experiments have aimed successfully at evolving ribozymes toward a reduced Mg^{2+} dependence.[779,799] It has also been possible to change the metal dependence of different classes of ribozymes from magnesium to calcium ions,[800,801] and in the case of the hepatitis delta virus ribozyme, a single nucleotide can determine the nature of the metal ion preference.[802] Another parameter for "robustness" in general is resistance to extremes of acidity and alkalinity, and significant improvement in ribozyme activities at both ends of the pH scale has been achieved.[803,804]

It is then clear that within limits, directed evolutionary approaches for stability have been successful for functional RNAs. But since we undertook to compare *in vitro* evolution of the latter with proteins, let us also draw further parallels between the enabling technologies involved in each case. Random mutation of sequences at the DNA level that encode RNAs, and subsequent selections for function, have obvious similarities to random mutation procedures for protein *in vitro* evolution. Since the phenotype of functional RNAs derives directly from the secondary and tertiary structures resulting from specific RNA sequences themselves, mutageneses for RNA are not concerned with the complexities of codon preferences and related issues of the protein world. Yet in our considerations of RNA *in vitro* evolution thus far, there is one aspect which has not been raised that featured loudly in the corresponding protein world. This is recombination, which surfaced repeatedly in Chapter 5 as a very effective means for plumbing new combinations within sequence space.

Recombination: Still Good for RNA?   It might seem by now that just about any feature of RNA molecules can bring forth ruminations concerning the RNA world, but the potential role of recombination within primordial RNA systems has been raised from the early days of the whole concept.[133] This issue relates directly to how diversification could have occurred in the RNA world, a continuing research topic amenable to certain experimental approaches. The frequency of RNAs within a random library that significantly interact with a defined RNA partner has thus been tested *in vitro*, and found at an encouragingly high level for RNA world scenarios.[805] Recombination between RNA molecules in modern viruses is well recognized, mediated by several mechanisms including viral RNA polymerases jumping between different transcriptional templates.[806,807] Ribozymes can be artificially engineered to act as general RNA recombinases,[808] and ribozyme-mediated recombination has been orchestrated to produce another functionally active ligase ribozyme from inactive precursor RNA segments.[809] A related aspect of RNA recombination elicited by ribozymes is the phenomenon of trans-splicing, which we will briefly discuss in Chapter 9.

Natural RNA recombination is one thing, but what is the role of recombination in the artificial evolution of RNA molecules? This question inevitably brings to mind the technique of DNA shuffling (Chapter 4), so often effective in the directed evolution of proteins. Shuffling of random mutations by homologous recombination of fragments of an amplified DNA sequence encoding a single ribozyme could be performed, analogous to that originally used with protein coding sequences.[344] But performing a conventional homologous "gene family" shuffling experiment for artificially selected ribozymes (or aptamers) is limited by the great sequence and structural diversity of such RNAs, even those within similar motif classes.[810] A different type of "recombination" between different functional RNAs employs the strategy of simply joining them to form chimeric bifunctional molecules. This is not as trivial as it sounds, since the fusion of functional RNAs can generate new cross-interactions that may mutually interfere with their original activities. A SELEX-based approach with dual selection has been used to isolate fused RNAs recapitulating both of the original functions.[810]

Homology-driven limitations on functional RNA molecular shuffling do not, of course, apply (by definition) to approaches exploiting nonhomologous recombination. RNA functional motifs have a high level of modularity, which in itself suggests that homology-independent strategies may be effective.[811] A strategy for random nonhomologous recombination between different functional nucleic acids* yielded an improved scan of sequence space (as judged by the end results) relative to error-prone PCR mutagenesis.[812] Nonhomologous fragment recombination of a DNA gene for a specific ribozyme proved to be an effective pathway toward minimizing the ribozyme sequence, and thereby defining the essential functional regions of the ribozyme.[811] This is a worthwhile exercise, since both natural and artificial ribozymes frequently have essential segments interspersed with dispensible "linker" regions.[811,813]

Folding in Two Worlds   For both proteins and functional nucleic acids, the nature of their secondary and tertiary folds is crucial in determining their molecular shapes and activities. An interesting point from Chapter 5 was that despite the diversity of protein folds, some in particular are associated with a very wide range of enzymatic functions. Our favorite in this regard was the β/α_8, or TIM-barrel fold (Fig. 5.2), which recurs in a variety of guises. Is there any precedent in the world of catalytic RNA for the reuse of a suitable framework structure for more than one catalytic task? A relatively limited amount of work in this area has been done, but this type of question formed the rationale for one study where a specific ribozyme (one of those catalyzing an aminoacylation reaction) was chosen as the starting point.[814] This RNA molecule was subjected to limited mutagenesis in an effort to evolve its aminoacylation function into a chemically distinct activity (as a polynucleotide kinase). Multiple different kinase ribozymes were in fact obtained, close in "mutational distance" from the parent molecule but with distinct folding patterns. It was further shown that the frequency of obtaining novel kinase ribozymes increased with the divergence from the parental sequence. This, together with the observation of new folds even in the mutationally proximal kinases, suggested that there was a need to "escape" from the initial parental RNA fold to acquire the selected kinase activity.[814] While protein folds (as with TIM barrels) are frequently used as scaffolds in a variety of functionally different contexts, new RNA functions appear to usually require novel folds (Fig. 6.4, top panel). It might be predicted that evolutionary pressures on functional RNAs would select for molecules with unique (or at least highly restricted) folded structures, and for natural RNAs this is supported by computational modeling.[†,768] It should be noted also that while random-sequence proteins are usually insoluble and have a low probability of assuming a stable fold, random sequence RNAs typically do fold into compact soluble structures.[767,769] Yet evolved (functionally selected) RNAs and random-sequence RNAs are not equivalent in this regard, as the latter have been found to have an increased propensity for forming more than one folded state.[769]

These observations show that the folding of proteins and RNA molecules have important areas of divergence, so let us consider some of these points from a theoretical angle. As we have seen previously, sequence and shape (structure) spaces are useful ways of conceptualizing molecular evolutionary transitions. A folded polymeric molecule of specific sequence has a neighborhood of closely related mutationally altered variants in sequence space. Each sequence has an associated folded shape, mapped onto a shape space. If neighborhood

---

*This particular report[812] studied nonhomologous recombination for improvement of DNA aptamers, which we will consider again later in this chapter.
†RNA folding analysis can be accordingly used to make predictions regarding the functional statuses of uncharacterized genomic RNAs.[815,816]

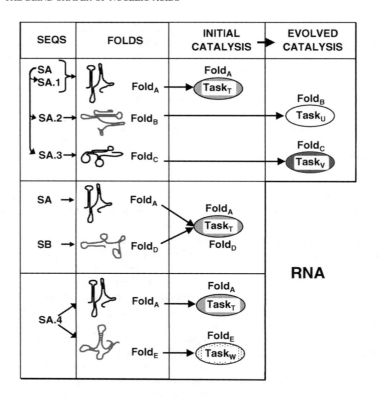

FIGURE 6.4 RNA sequences mapped onto corresponding folds and catalytic tasks; possible arrangements. Top panel: Functional RNA (ribozyme) sequences derived from parent SA and with increasing degrees of mutational distance from SA.1-SA.3, but where all have been obtained from the same evolved population. The species SA.1 has essentially the parental fold (SA.1 thus represents all neutral mutations with respect to SA), but the remainder has divergent stable predominant folds. SA and SA.1 share the original catalytic function, but the evolved SA.2 and SA.3 have novel tasks as well as novel folds. Middle panel: The fold of RNA SB is distinct from SA, but shares a catalytic function. Bottom panel: The situation where a variant of SA (SA.4) has an alternative stable fold which itself is associated with an entirely different catalytic function.

transitions in sequence space map to corresponding transitions in a shape space neighborhood,[*] then the progression can be termed "continuous."[2] If the sequence transitions map to a different shape space neighborhood, then the progression is accordingly discontinuous. A discontinuous transition can be viewed as a sudden jump from one folded structural state to another, or as an example of a "punctuated" evolutionary change on a molecular scale

---

[*]While proximity in sequence space is not hard to visualize, how does one really define "nearness" in a shape space? This has been done by considering that many points in sequence space (a local neighborhood) will map to a single fold in shape space. Or in other words, the sequence space neighborhood here is a cluster of neutral mutations that do not significantly impact upon the folded shape of the molecule. Nearness in shape space of two folds A and B can then be defined as the likelihood that fold B has a high probability of directly mutationally deriving from an "average" sequence that assumes the shape of fold A.[2]

(Chapter 2). Experimental results suggest that relatively small sequence changes in functional RNAs can result in the acquisition of very divergent folds and functions,[769,817] and these "jumps" in turn indicate that some RNA evolutionary change has a distinctly discontinuous character.[2,769] This is depicted in Fig. 6.4 for the transition between the parental species SA and variants SA.2 and SA.3.

Since functional RNA molecules can serve as both genotype and phenotype, RNA models have often been used for theoretical analyses, and the effects of neutral mutations have been an important parameter in this regard. Neutral mutations in sequence space will by definition not affect the corresponding fold in shape space, but (as we have seen in Chapter 4) neutral variation can be a springboard for evolutionary innovation. Webs of all possible neutral pathways in sequence space for a common fold are often termed *neutral networks*,[608,818] and theoretically any points on such a network (between sequences corresponding to common folds) can be mutationally traversed. Theory also predicts that neutral networks for different folds will show points of intersection.[818]

Now, since natural ribozymes and self-splicing RNAs (such as Group I introns) exist with common functional cores bearing divergent sequences,[819,820] it is inferred that RNAs can evolve as components of a theoretical neutral network. A fragment of such a network is represented by the sequence changes in Fig. 6.4 between the parental SA and SA.1 ribozymes. Different functional RNA folds in shape space could be connected via overlap of their neutral mutational networks, and at the intersection of these, duality of function might result. As noted above, experimental studies have supported the assumption that functional RNAs fold into unique and well-ordered structural states.[767,769] Yet the intersection of ribozyme neutral networks has been demonstrated with real RNA molecules.[817] In this case, two separate ribozymes with different functions and ~25% sequence identity were used to design a set of ribozymes with progressive neutral base changes, converging on a common sequence that showed dual activity. In practice, the activities of the dual ribozyme (at the "intersection point" of the neutral network) were suboptimal relative to both parental wild-type ribozymes, and thus strictly speaking not truly neutral. The phenomenon of a "bifoldable" ribozyme is also depicted in Fig. 6.4 (for the variant of the parental SA, SA.4).

We considered in Chapter 5 the notion of proteins with the same sequence occurring as different stable structural isomers (Fig. 5.5). Biological proteins almost always display a single folded functional form, although some interesting exceptions exist (intrinsically unstructured proteins, prions, and certain others as described in Chapter 5). Yet if we insist on a definition of true protein structure–function dimorphism as a single polypeptide independently giving rise to two alternative functional conformations that are stable in the absence of another protein, soluble, nonpolymeric, and with *globally* distinct folds, no known protein equivalents of bifunctional ribozymes can be found. The propensity of RNA molecules to assume alternative conformations has been termed "non-unity molecular heritability,"[799] and the existence of alternative functional conformations has implications for both natural and artificial evolution. A single sequence accordingly could be the common ancestor of diverse functional RNAs with no discernable structural identities.[817]

In both the RNA and protein universes, the question of functional convergence is pertinent, where pairs of macromolecules differ in both sequences and folds, yet converge on a common catalytic function. For proteins, it is relevant in this regard to recall the discussion of analogous enzymes in Chapter 5. For RNAs, many examples have been found with evolved ribozymes, of which the different classes of RNA ligases (noted earlier) serve as examples. A striking difference between functional proteins and RNA molecules arises at the level of their respective folding behaviors, as we have seen. In this context, it might be recalled that throughout the kingdoms of cellular life, the folding of proteins is important enough to warrant assistance from specialist proteins termed chaperones. For a small minority of natural proteins, folding cannot in practical terms be accomplished at all without the agency of folding catalysts (foldases). Do functional natural RNAs then require folding assistance?

The answer to this question is yes, RNA chaperones have indeed been characterized. A theoretical "need" for an escape route for misfolded RNAs has long been proposed, given the above-noted stability of RNA secondary structures that could maroon an otherwise-functional RNA in the wrong conformation.[821] Since the number of possible alternate conformations increases rapidly with the size of an RNA molecule, the regular and efficient folding of large cellular RNAs (especially the ribosomal 16S and 23S RNAs) seemed a difficult proposition without some form of assistance. It was not until relatively recently, though, that promising *in vitro* evidence was confirmed through *in vivo* evidence for protein-mediated RNA chaperone activity, where a self-splicing intron served as the target RNA.[822] Protein chaperones for RNA have since been demonstrated to facilitate both Groups I and II self-splicing and ribozyme activities *in vivo*.[822–824]

The evident need for protein help to assist RNA folding within intracellular environments raises an issue that (yet again) invokes the RNA world. With the discovery of the ribozyme heart of ribosomal protein synthesis, it seemed as though the chicken-and-egg paradox of the progression of the RNA to (RNA)/protein worlds was laid (so to speak) to rest. And we have spent some time to date making note of ingenious strategies used by researchers toward finding artificial ribozymes that would have been needed to constitute a viable RNA world. Yet there remain significant gaps in this enterprise, such as deriving a ribozyme RNA polymerase with truly satisfactory activity. If very large RNAs with complex secondary and tertiary structures require protein assistance for efficient folding, are we not back with the chicken and the egg all over again? For example, if large functional RNAs need proteins to fold, and protein synthesis itself requires large functional RNAs (23S and 16S ribosomal RNAs), how could the RNA world have entangled with proteins in the first place? Several answers can be proffered in response to this challenge, including noting that a low-efficiency process (as with RNA folding in the absence of protein help) is better than none at all. This is often the case with *in vitro* RNA experimentation, which can show activity but fail to go to completion due to competing RNA folded structures.[770] But a feasible model for early RNA world–protein interactions invokes the development of cooperation between functional RNA molecules and peptides with nonspecific RNA-binding activity, where the latter could serve as primordial RNA chaperones.[821] An alternative escape route would exist if RNA molecules themselves could act as RNA chaperones, and thus facilitate the folding of large RNAs in the RNA world. To date, a functional RNA that completely satisfies the definition of a chaperone has not been derived, but certainly oligonucleotides can sometimes act as "facilitators" of ribozyme activity in the sense of promoting a desired structural feature (as with, for example, an oligonucleotide used to maintain a stem region of an RNA ligase ribozyme[825]).

The general question of the coevolution of RNA and protein function is a fascinating area that should require less and less conjecture as time and science progresses. But this general topic almost inevitably raises the consideration of the relative proficiencies of these two very different types of polymers as functional molecules. Having compared some of the features of RNA and proteins in both natural and directed evolution, let us now look again at the known abilities and limitations of catalytic nucleic acids, particularly with respect to the nucleic acid alphabet.

Working with the Ribozyme Repertoire

Ever since the discovery of the catalytic potentialities of RNA, workers have sought to understand the chemistry and mechanisms that underlie these effects. Unlike the diversity of amino acid side chains in proteins, functional RNAs are limited to a four-letter alphabet (leaving aside the possibilities of secondary chemical modifications as seen in some biological RNA molecules such as tRNAs, as noted in Chapter 5). The reality is even more restricted, given that the nucleobases come in only two "sizes": purines (A and G) and pyrimidines (U and C for RNA).[821] A common initial assumption following from these inherent chemical constraints was that ribozyme activity was intimately and essentially linked with coordinated metal ions.[760,826,827] Concerted research has demonstrated that metal ions are indeed an integral part of some ribozyme catalytic strategies, including those of Groups I and II self-splicing introns.[194,826] On the other hand, divalent metal ions are apparently not essential for the activities of some small natural ribozymes (including the hairpin and hammerhead varieties).[*,829,830]

Such results suggested that despite their evident chemical limitations, the RNA nucleobases were mediating the activities of some types of ribozymes. It is one thing to propose this, and other thing to rationalize how it could occur. In particular, it was not initially obvious how RNA nucleobases could promote general acid–base catalysis, a process at the core of RNA-cleaving activities of proteins such as ribonuclease A.[194,727] We saw in Chapter 2 that the amino acid histidine features prominently in protein enzymology, and this stems from its facility as an acid–base catalyst, with its pK_a values[†] close to neutrality. On the other hand, the usual RNA nucleoside bases (cytidine, uridine, adenosine, and guanosine) have pK_a values far from neutrality, meaning that ordinarily only a low proportion of bases will be charged (and capable of promoting the proton transfers necessary for acid–base catalysis) in neutral solvent conditions.[760] Experimental evidence has nonetheless strongly pointed to the existence of RNA-mediated acid–base catalysis.[831–833] As we have just seen, the purine and pyrimidine RNA bases in isolation would seem unpromising agents for this type of activity. But in RNA molecules, they are not in isolation. An individual base in even a small functional RNA is always located within a specific sequence-dependent (and in turn a specific secondary and tertiary structure-dependent) context. And the conformation and folding of a ribozyme can evidently make all the difference between good activity and none. If a ribozyme structure can perturb the chemistry of specific nucleobases such that they become effective for acid–base catalysis, this should in theory be borne out by a deviation in pK_a values toward neutrality for the residue(s) in question. This proposal is now directly supported by experimental evidence.[834]

Nucleobases themselves have thus proven to possess previously unsuspected chemical versatility, subject to molecular context-specific factors.[760,827,833] Moreover, ribozyme catalysis shows similar proton transfers as in acid–base catalysis used by protein enzymes.[194,727,827] We saw in Chapter 2 that water itself can be a vital catalytic ingredient for proteins, and it has been proposed that some ribozymes likewise may rely on solvent molecules for catalytic purposes.[835] Then in general terms, ribozyme and protein catalyses are not alien with respect to each other, and share some important features. Yet there is no getting around the fact that the RNA alphabet, and its associated full chemical repertoire, is much more limited than that available for proteins.[194] We might recall, though (also from Chapter 2) that sometimes even proteins require help from chemical cofactors to enable or accelerate certain enzymatic

---

[*]Monovalent metal ions in these experiments could substitute for divalent ions with comparable results for ribozyme activity (as in pH/activity profiles), suggesting that the roles of each ionic environment were similar. In turn, this indicated that a specific divalent metal ion requirement was unnecessary. In the case of the hepatitis delta virus ribozyme, an initial apparent requirement for magnesium ions was later found to not be absolute, as monovalent lithium ions could partially substitute.[828] Ionic interactions may structurally stabilize such ribozymes without having a direct catalytic role.[826,827,829]

[†]The pK_a is an index of acid dissociation, and equivalent to the value of the pH when a weak acid is at a state of 50% ionization (a point of half dissociation into negatively charged ions and hydrogen ions [or strictly speaking, H_3O^+ hydronium ions]).

processes. Could not RNA molecules likewise extend the range of their functionalities by co-opting different compounds for specific catalytic tasks?

Calling for Cofactors If we are to consider cofactors in general, the first thing to recall is that we have already encountered them in this chapter, in the form of metal ions. As we have seen, an association of metal ions with a functional RNA (or a protein enzyme) is not necessarily indicative of metal-promoted catalysis, but divalent metal ions have defined roles in Groups I and II intron self-splicing,[826] and are certainly catalytic cofactors as such in this context. The folded active sites of these ribozymes can be considered as scaffolds for allowing metal ions to assist catalysis. Beyond magnesium and calcium ions, it has been possible to derive "leadzymes,"* which rely on Pb^{2+} ions to effect catalytic self-cleavage.[836,837] To cite another previously mentioned cofactor, it may be recalled that Group I introns also require the presence of guanosine nucleoside for initiating the self-splicing reaction.[750]

We also previously saw that oligonucleotides could function to stabilize ribozyme folded structures and thereby act as "chaperones" in a limited sense. In the same vein of broad definitions, such an oligonucleotide might also be thought of as a "cofactor," in the sense that ribozyme folding and activity may be compromised in the oligonucleotide's absence. Indeed, by combinations of *in vitro* selection and directed design, ribozymes have been produced whose conformations and catalysis are modulated by specific short oligonucleotides.[838,839] This kind of activity modulation is a special case of *allostery*, which we will look at in more detail shortly. Proteins too can be important "facilitators" of ribozyme activity through stabilization of RNA conformations or spatially positioning active sites in favorable manners, as we have also seen earlier. Nevertheless, these kinds of oligonucleotide or protein assistance are not classifiable as catalytic cofactors, and the latter are our present focus.

Organic cofactors (coenzymes) enabling or promoting catalysis have been used as investigative tools for mechanisms of RNA catalysis, using deliberately crippled natural ribozymes. When active site cytosine bases are chemically removed from hepatitis delta virus ribozymes (rendering a particular site in the RNA molecules into an *abasic* state), activity is lost, but can be restored by the provision in *trans* of soluble imidazole† or chemical analogs of it.[842] But are there precedents for normal natural ribozymes making use of small organic molecules as catalytic cofactors? The best (and probably most-studied) contender in this regard is a special ribozyme that is an element of a *riboswitch* found in the 5′ region of mRNA transcribed from the *glmS* gene in Gram-positive bacteria, which we will discuss later in this chapter.

It is often assumed that natural ribozymes that used organic cofactors (and thereby had extended catalytic ranges) were a feature of the RNA world. We saw in Chapter 2 that many modern protein cofactors are in fact modified nucleosides, an observation that itself is consist with an RNA world origin for these compounds. During the transition from the RNA to (RNA)/protein worlds, it is in turn postulated that such cofactor ribozymes were gradually supplanted

---

*It is an unfortunate feature of the English language that "lead" (metal) and "lead" (beckon to follow) are spelt the same. Thus, a "lead compound" (more of which in Chapter 9) can mean two very different things, although they could (rarely) coincide (where a lead (element) compound itself acts as lead compound (pointing the way for a new group of compounds). A "leadzyme,'" though, is not a "lead" onto new areas of catalysis.

†Imidazole is the chemical moiety on side chains of the amino acid histidine, which as we have seen is active in protein general acid–base catalysis. Imidazole is a five-membered heterocyclic (ring) compound with two nitrogen atoms. A conceptually related version of this kind of *trans*-complementation experiment is to modify a ribozyme active-site base itself rather than completely remove it. In two different ribozymes, replacement of an active site base with imidazole allowed the retention of function, and provided further support for the role of acid–base catalysis in ribozyme activity.[840,841]

by more efficient proteins (which often co-opted the available cofactors, or became enlisted into cofactor biosynthesis). In this view, cofactor ribozymes are largely extinct, save for relic holdouts such as the above-mentioned *glmS* riboswitch. So once again, it is up to human ingenuity to recapitulate the extended catalytic ranges of hypothetical primordial functional RNAs. In doing so, it is important to note that functional dependence of a ribozyme selected to require the presence of a small molecule does not necessarily imply in itself that the small cofactor is involved with catalysis. Often the small "cofactor" is required to engender an allosteric conformational change in the ribozyme favorable for activity, rather than directly mediating catalysis *per se*.

But for some types of catalysis, ribozyme cofactors with catalytic involvement are likely to be essential. While RNA replication is perhaps the most fundamental function to be accounted for in the RNA world (as we have seen), a fully operational RNA-based molecular ecosystem would have to possess many other functionalities, very likely including RNA-based oxido-reductases (capable of catalyzing chemical reduction and oxidation, or *redox* reactions). Therefore, artificial derivation of an oxidoreductase ribozyme is of considerable theoretical interest. Yet RNA chemistry alone is not likely to be up to the job, since even proteins, with their much more versatile subunit alphabet, use cofactors (coenzymes) for these kinds of tasks. Among these is nicotine adenine dinucleotide (NAD^+, Fig. 6.5), and its reduced derivative NADH,[*] which can participate in the electron flow needed for oxidation–reduction of enzyme substrates. If the cofactor concept is approached by the provision of NAD^+ in an RNA library, a ribozyme capable of oxidation–reduction (redox) reactions can be potentially evolved and selected. This in fact has been successfully carried out[843] (Fig. 6.5).

Assistance for ribozyme catalysis by organic cofactors is thus a realistic prospect. This question was raised in part because protein use of cofactors for catalysis has long been appreciated, but also due to the perceived additional burdens placed on the ribozyme repertoire by the limited "alphabet" of nucleobases. As with proteins, there have been some interesting and significant investigations into the RNA functional alphabet that are worth our consideration before moving on further.

Alphabetic Acrobatics   If you wish to test the effect of an available molecular alphabet on functional performance, one way to go about it is to make the said alphabet bigger or smaller. Enlarging an alphabet is undoubtedly beneficial for expanding functional potential, analogous to a mechanical engineer being presented with a fancy new toolbox that allows new tasks to be undertaken. For proteins, we have seen (Chapter 5) that alphabet expansion involves codon reassignments and generation of new pairs of orthogonal tRNAs and aminoacyl-tRNA synthetases. For nucleic acids, alphabetic enlargement carries different challenges (and we will discuss this theme later in this chapter), but what about alphabetic reductions? Studying the effects of a diminished alphabet is profitable, not only to model possible early RNA evolution but also to gain insight into minimal structures compatible with functional RNAs.

Faithful transmission of genetic information requires a high level of base-pairing fidelity, and significant *degeneracy* of base complementarity would compromise such transmission during strand copying. For example, a four-base alphabet (ABCD) where A was complementary to B, but C was complementary to B *or* D with equal affinity, could not replicate accurately. An arbitrary sequence (for example, ACDDBBA) could yield not only the complementary "strand" BDCCAAB but also BBCCCCB and others. A nucleic acid replication system based on complementarity also requires pairs of letters (A/T; C/G), and an odd number would seem to

---

[*]The reduction of NAD^+ involves a net transfer of two electrons and a proton, such that the reduced NADH form is uncharged.

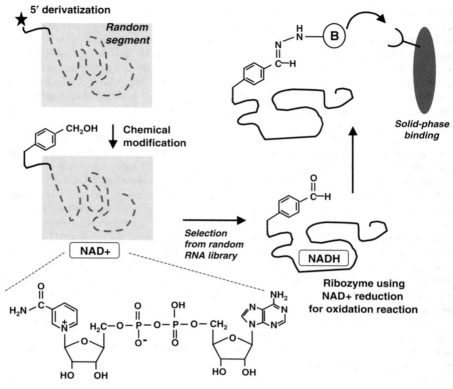

FIGURE 6.5 Chemical derivatization and selection process used for isolating NAD^+-dependent ribozyme oxidoreductase. A randomized RNA library with an introduced 5′ modification[37] is then chemically derivatized at this unique site (with a CH_2OH alcohol side chain group). In the presence of the cofactor NAD^+ (as shown), a ribozyme of the correct configuration can oxidize the alcohol to an aldehyde (CHO) group, with the concomitant reduction of NAD^+ to NADH. The chemical change in the 5′ tag allows its selection by further reaction with biotin hydrazide (B = biotin in schematic), which in turn allows selection for the chemical group and the associated ribozyme on a solid-phase biotin-binding protein (streptavidin).

force degeneracy. An ABC alphabet could have A/B complementarity, but either A or B would be required to share complementarity with C. Such a three-letter system could provide templates for replication, but without faithful information transmission. Yet this does not mean that any functional RNA strand must always have all "letters" of the nucleic acid alphabet. Obviously, single-stranded RNA molecules composed only of one base (for example, AAAAA...poly(A)), two bases (AAAUUU...), or three bases (AAAGGGUUU...) can exist, and are biologically replicable. By the nature of RNA transcription, the DNA strand that serves as a template for the transcribed RNA molecule need not have the same base composition as the "top" DNA strand that corresponds directly to the copied RNA sequence.* This means that four-base RNAs can be transcribed (depicted in Fig. 6.6), where the fourth base is used purely for

---

*This is the coding strand in the case of genes specifying proteins.

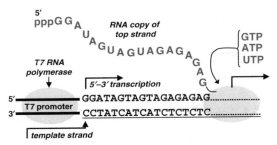

FIGURE 6.6 Depiction of transcription of a duplex DNA sequence (from a T7 promoter) with no G residues in the strand that acts as the template for RNA synthesis directed by T7 RNA polymerase. Accordingly, only three nucleoside triphosphates are required for transcription (GTP, ATP, and UTP as shown). The transcribed RNA copy of the "top" strand correspondingly has no C residues.

replication purposes and is absent from the transcribed RNA strand. Functional RNAs with only three bases can thus in principle be derived by *in vitro* evolutionary approaches (requiring repeated rounds of replication), and ligase ribozymes with only A, G, and U residues have in fact been isolated.[844] It would seem improbable, though, that if a four-base alphabet was available in the early RNA world that one of the bases would be used exclusively for replication and not for the constitution of functional molecules as well. Indeed, if cytidines are reintroduced into ribozymes exclusively composed of A, G, and U, more efficient catalysis results (due in part to improved molecular stabilization).[845]

So can the alphabetic reduction for a functional RNA be taken down to a single base pair? The answer is yes, although those successfully used in this undertaking were not among the conventional nucleobase pairs, but rather 2,6-diaminopurine* and uracil. Here, a previously obtained three-letter ribozyme ligase[845] was transformed into an inefficient but still-functional catalyst with only 2,6-diaminopurine:uracil base pairs.[846] Since a functional protein enzyme composed of only two amino acid types is very unlikely, a ribozyme of two letters stands as the logical lower limit for subunit simplicity for a biological catalyst.[846] These kinds of experiments might remind us of references made in Chapter 5 to "molecular lipograms," which correspond to biological alphabetic reductions. A linguistic lipogram continuing a nucleic acid theme and using only a single vowel might be "GRAND RNA STRAND DATA ALWAYS GRABS MANY CASH AWARDS AT HARVARD."† While linguistic lipograms would appear to have few if any practical uses, molecular lipograms, as we have seen, have very serious intent. But one thing both have in common is a reduction in range and versatility. There are limits to what you can express if the letter "a" is your sole permitted vowel, and restriction to less than the normal RNA alphabet comes analogously with a catalytic cost. The mere fact that the latter is possible, though, is a remarkable demonstration that a minimal set of biological building blocks is still combinatorially significant, revealed by the power of diversification and selection.

---

*2,6-Diaminopurine is equivalent to 2-aminoadenosine or adenosine with an additional amino group in the 2-position on the purine ring. 2,6-diaminopurine and uracil form three hydrogen bonds in base pairing, as opposed to two between adenine and uracil.[846] It was found that this purine derivative was advantageous for certain aspects of *in vitro* transcription with T7 RNA polymerase on 2-base alphabet templates.[846]

†Achievement in aptamer or ribozyme research at an Ivy League University is well-remunerated? I cannot vouch for the truth of that one, but it fills the bill as an "a-specific" monovocalic sentence, or a lipogram without e, i, o, or u. Researchers involved with DNAzymes or DNA aptamers might offer the simple rejoinder "AND DRAW AN AWARD, DNA," which is a lipogram, a monovocalic, and a palindrome.

We are not done yet with ribozymes, but this is a convenient juncture to consider some additional features of aptamers. Both, of course, fall within the broad domain of functional nucleic acids, and (as we have already seen to an extent) there are significant areas of functional interplay between them. When we have digested some more of the aptamer world, we can consider functional RNAs from a more integrated viewpoint.

ANGLING FOR APTAMERS WITH APTITUDE

Aptamers were introduced earlier in this chapter during our consideration of SELEX, and these functional nucleic acids are intimately associated with this general evolutionary selection process. Until this point, we have been accustomed to thinking of aptamers as RNA molecules, but DNA aptamers (and enzymes) also exist. The term "aptamer" has even been applied toward peptides selected from libraries, but this is less widely used, and the tag "oligonucleotide" is frequently specifically included as an integral part of aptamer definitions.[*] We will conform with the protocol that assumes aptamers are composed of RNA unless otherwise stated, although "RNA aptamer" may be appropriate in some cases to avoid confusion.

When in the form of duplexes between complementary strands, nucleic acids lose much of their propensities for assuming a wide range of distinct tertiary structures. Yet this is not to say that the helical duplex structures themselves are restricted to a single form. As well as the normal right-handed double-stranded conformation for DNA and RNA (B-DNA and A-RNA, respectively), alternative forms are known with the opposite (left) hand helical senses (Z-DNA[848,849] and Z-RNA[850,851]). DNA duplexes can also show sequence-dependent bending,[852,853] and the formation of cruciform structures.[854] While interesting and often biologically significant[854] these wrinkles on the conventional structures of nucleic acid duplexes do not offer much room for diverse molecular binding interactions.[†] On the other hand, if nucleic acid strands are maintained in isolation from their complements, a very different picture emerges. Single strands fold into secondary and tertiary structures, and out of the huge number of possible sequences even for relatively short nucleic acid polymers, some can fold into shapes with useful binding properties. These, as we have seen, are aptamers.

Usually aptamers function on a unimolecular basis, where a single folded nucleic acid molecule can bind to a target ligand, though some exceptions have been found. Both DNA and RNA strands of certain sequence compositions (especially those that are G- rich) are able to form multistrand triplex and quadruplex structures,[856,857] some of which have potentially useful ligand-binding activities.[858,859] If we took a bird's eye view of the field in general, we would see that aptamers have been selected for binding a wide variety of small and large molecular targets, far too many to discuss in detail in each case. Major classes of aptamer targets include sugar derivatives (thus giving rise to the label "RNA lectins"[860]), peptides,[861]; proteins,[862] nucleotides,[863] and many artificial compounds.

Let us take a look at an aptamer for a particular low molecular weight target molecule, to think about how single-stranded RNA molecules bind to specific ligands. Malachite green is a synthetic dye with interesting properties as a photoactivable "suicide" ligand, which affords

---

[*]Artificially selected functional peptides were known before the word "aptamer" was coined. It may seem unfortunate to dilute the meaning of a useful scientific term by encumbering it with additional referents when alternatives exist, although once a term becomes popular it usually stays. "Peptide aptamer" has been given a stricter definition by confining it to combinatorially selected peptides held within a protein scaffold,[847] but the term seems to be used more loosely than this in other contexts.

[†]Chemically modified DNA duplexes used as molecular decoys have been referred to as "aptamers," but this usage is not widespread.[855] We will say a little more about nucleic acid decoys later in this chapter.

potential biological applications. This dye has an absorption maximum in the red range (620 nm; not otherwise absorbed by cells), and upon photoactivation releases highly reactive but short-lived free radicals, which attack molecules in the immediate vicinity.[864,865] When biological molecules (including antibodies) are tagged with malachite green, treatment with laser light at 620 nm triggers this effect and destroys the dye carrier molecule.[866]

Aptamers against malachite green have been derived, and such aptamers bound with their target ligand are susceptible to specific laser-mediated destruction through the above photo-activation effect.[867] A solution (NMR) structure for the aptamer–malachite green complex itself has been determined.[868] This has the general form of a stem–loop–stem, where the conserved loop "bulge" mediates the binding of the malachite green ligand (Fig. 6.7). The stem regions are largely composed of conventional planar Watson–Crick base pairing, but non-Watson–Crick bonds are also a prominent feature of the aptamer (Fig. 6.7). Intriguingly, although the malachite green aptamer was selected only by virtue of its binding affinity for the specific dye molecule, this same aptamer actually has a measurable intrinsic catalytic effect (esterase activity) toward a molecule structurally related to the original dye.[869] The malachite green

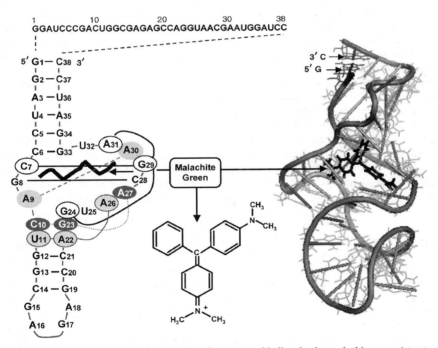

FIGURE 6.7   Sequence and solution structure of an aptamer binding the dye malachite green (structural formula of dye as shown). The primary RNA sequence is shown with a schematic of the secondary structure on the left, and representation of the tertiary solution structure of the aptamer complexed with malachite green on the right.[38] In the schematic, normal (canonical) Watson–Crick bonds are shown by horizontal black lines; base triplexes A26 U11:A22 and A27 C10:G23 are shown by gray lines/light gray circles and dotted lines/dark gray circles, respectively; and quadruple interactions G24 A31 G29:C7 are shown by black lines/white circles. A9 and A30 (stacking interaction) are shown with gray circles. For tertiary structure, phosphodiester backbone is shown (dark gray) with bases light gray; 5′ and 3′ positions as marked. Position of bound malachite green (black) as indicated. *Source*: Protein Data Bank[2] (http://www.pdb.org) 1Q8N. Tertiary structural image generated with PyMol (DeLano Scientific LLC).

aptamer then can be considered as a ribozyme as well (albeit of modest activity), a finding that raises the question as to what fraction of aptamers in general may have a cross-reacting targetable molecule in chemical space that could act as a substrate for some type of catalysis. This observation also may have ramifications for models of the frequencies with which nucleic acid catalysis could arise in the RNA world, although we should recall from our consideration of "peptide-like" catalysis in Chapter 5 that finding weak esterase activity is a relatively easy undertaking in the context of all catalytic tasks.

The general stem–loop–stem secondary structural scheme has also been observed in other aptamers, including those with specificity for flavin mononucleotide.[870,871] But this is by no means the only RNA fold associated with specific ligand binding. A variety of different stem–loop arrangements and motifs termed "pseudoknots"[*] have also been defined in the context of RNA aptamers.[870,873] Ligand recognition through the specific binding pockets of aptamers is mediated by planar stacking, hydrogen bonding, and shape complementarity.[873]

Proteins, too, are very competent for the binding of small ligands. We have already noted this in several contexts in earlier chapters, particularly in considering the organic cofactors commonly called coenzymes. These are useful for making comparisons between binding pockets of proteins for specific ligands, and comparable sites on RNA aptamers directed toward the same small molecules. Of course, these cannot be completely equivalent, due to the discrepancies between the functional alphabets in each case, and in general aptamers may not be able to attain the same degree of shape complementarity available to more-versatile proteins.[873] Yet significant parallels can be drawn between ligand-binding modes for aptamers and proteins directed against the same coenzymes or other small molecules. Structural analyses have shown that a significant number of weak interactions that collectively allow the binding of a specific ligand are shared between proteins and aptamers, including hydrogen bonding and stacking effects.[874]

While placing aptamers and proteins side by side, we may also be reminded of antibodies, since both antibodies and aptamers are recognition molecules selected by evolutionary processes. While looking at immunoglobulins in Chapter 3, we very briefly noted the concept of "induced fit," where the antibody conformation after antigen binding is altered from the unbound state. There is abundant evidence in the case of aptamers that the binding of ligand is very often accompanied by a significant conformational change in the RNA structure:

$$\text{Aptamer}_{\text{State A}} + \text{Ligand} \rightleftharpoons (\text{Aptamer}_{\text{State B}} : \text{Ligand})$$

Here "State A" and "State B" refer to the conformational states before and after ligand binding, respectively. For the "State A" in solution, loop regions of the aptamer may exist in a relatively disordered configuration until interaction with specific ligand, whereupon a stable conformational state is taken up. This (as with proteins) has been termed "induced fit"[875] or alternatively "adaptive recognition".[873] Numerous instances of this effect have been documented,[873,876,877] and the malachite green aptamer is no exception in this regard.[878] In the latter case, while the RNA aptamer is subject to "adaptive recognition", the malachite green ligand itself undergoes conformational changes in the aptamer: ligand complex.[868,878] The binding of an RNA aptamer toward its selected ligand is a very different prospect to the nonspecific binding of certain planar compounds to nucleic acids.[†] Malachite green binds very weakly to nonspecific RNA sequences.[879]

---

[*]A pseudoknot is an RNA structural motif whose simplest form consists of two helical duplex regions connected by a single-stranded loop.[872] In such an arrangement, a loop region of one stem-loop structure contributes to the formation of the stem of a second stem loop. The pseudoknot motif in general can fall within a number of different folding categories.[872] As the name implies, a pseudoknot is not a true knot in a topological sense.

[†]Aptamer binding requires a specialized interaction pocket, whereas nonspecific binders ("intercalators") can insert nonspecifically between nucleic acid base pairs. Many intercalators exist, but ethidium bromide by far is the best known.

The requirements for successful *in vitro* selection of aptamers (and functional RNAs in general) have been systematically investigated. Independent isolates of both aptamers and ribozymes driven by a specific selection pressure typically show conserved "core" regions flanked by more diverse sequences. Many of such nonconserved segments may prove to be completely dispensable, although some can be significant as spacers and stabilizers. By the nature of the spacing and stabilization roles, the specific sequences involved can often vary without necessarily affecting the activities of the core regions.[880] Given that core sequence modules are often of limited complexity, the theoretical frequency of finding functional aptamers within random RNA sequence libraries is surprisingly high,[*] consistent with the relative ease of their isolation via SELEX.[881] We noted earlier that ribozymes could be viewed as evolving over pathways of interlinking neutral networks, and that intersections of such nets could occur both in theory and practice.[817] A study with aptamers binding flavin cofactor or guanosine nucleosides indicated that neutral drift could result in a flavin-binding aptamer only a "jump" of several base changes from the transition to a guanosine-binding phenotype.[871] (This is consistent with a close approach of their respective neutral mutational networks.) Guanosine binding by aptamers is also of interest since it appears to be realizable by RNA molecules widely distributed in sequence space, and guanosine is a key cofactor for many natural and artificial functional RNAs.[871]

Aptamers are artificially selected for maximal binding *in vitro*, and in consequence no trade-off in specificity or affinity is assumed, as is the case for natural recognition molecules which are often under various constraints.[873] Nevertheless, to be useful reagents for molecular recognition, aptamers should have a demonstrably high degree of binding specificity, combined with acceptable binding affinities. Much evidence has suggested that these issues need not be limiting factors for their development and practical deployment. Aptamers can distinguish between amino acids with related side chains (arginine and citrulline[882]) or closely related small molecules differing only by a methyl group (caffeine and theophylline[†,876]). Protein-binding aptamers have also been developed that can discriminate their targets from closely related proteins.[883,884] There are, of course, limits to specificity, subject to the specific nature of the RNA-binding pocket for a particular aptamer. To use our earlier example (Fig. 6.7), the malachite green aptamer can accommodate within its binding pocket a variety of compounds related to the original dye, and such binding can still be of high affinity.[868] It has been argued that selection for high affinity will in itself push recognition molecules toward high specificity,[885] but more recent investigations suggest that aptamers with increased binding affinity do not necessarily have an increased number of ligand contacts. In such cases, stabilizing rather than ligand-directed contacts are implicated in improvements in the dissociation constants for the aptamer: ligand complex.[886] In other words, selection for aptamer affinity *per se* may not always yield improved specificity, and if specificity is the issue, selecting directly for this property is called for.[887] (Shades of "you get what you screen/select for" again?) At the same time, it should also be pointed out that extreme specificity is not always even desirable, as it may be advantageous in some cases to bind a family of targets rather than a unique structure alone. "Generalist" aptamers binding a family of related proteins have been derived, and this broadened specificity has been suggested as a potential weapon against the evolution of resistance in pharmacological applications.[888]

---

[*]Taking into account the likelihood that random RNA sequences bearing relevant core modules and flanking sequences will fold appropriately, a computational study has suggested an isoleucine-binding aptamer could be found with 50% probability in $\sim 4.10^9$ random RNAs of 100 bases in length.[881]

[†]Hydrogen bonding "pseudo-base-pairing" between nucleobase residues of the anti-theophylline aptamer and its cognate ligand is disrupted by the presence of the additional methyl group of caffeine.[876]

It was noted at the beginning of this section that the term "aptamer" has been applied toward more than just RNA. "Functional nucleic acids" then is a blanket term for any nucleic acids that possess some kind of effector function as well as the normal base-pairing capability that permits their own replication. This is now a convenient point to say a little more about the DNA counterparts of RNA aptamers.

DNA Can Do It Too

We earlier considered the roles of nucleic acid structures and strandedness in their general abilities to act as binding molecules for specific molecular targets. Given the experience of RNA aptamers and SELEX, it was logical to assess whether chemically similar single-stranded DNA molecules could fold and recognize specific ligands in a comparable manner, and the answer was in the affirmative.[889,890] Aptamers were then no longer the exclusive domain of RNA, and DNA aptamers too have proven capable of impressive feats of specificity and affinity. DNA and RNA aptamers binding the same specific ligand (adenosine monophosphate[891,892]) have been compared, and although divergent in sequence and folding, the two nucleic acid aptamers shared important ligand contacts in their respective binding pockets.[893] For any specific ligand, the number of possible binding contacts is likely to be limited, but there may nonetheless be a relatively large number of options available for folding and positioning an aptamer to achieve such contacts. As for RNA aptamers, rather than exhaustively attempting to cite a long string of examples, let us simply note that the classes of small molecules toward which DNA aptamers have been evolved for binding is broad, including amino acids, nucleosides, sugars, and a wide variety of other compounds (ftp site).

If one demands high specificity in a molecular binding interaction, an inherent requirement in this regard is stereoselectivity. Here too, DNA aptamers can deliver. For example, antipeptide DNA aptamers have been shown to discriminate between peptide enantiomers (stereoisomers of D- and L-forms), a useful property for separation purposes.[894,895] Base modifications compatible with DNA polymerases (used during replication of selected single-stranded DNAs) can extend functionality and specificity. An example of this is forthcoming with chirally selective DNA aptamers against the drug thalidomide, where thymidine is replaced by a derivatized deoxyuridine.[896]

This section has been devoted to aptamers, but before moving on, it must be noted that DNA enzymes also have proven to be selectable, and may offer advantages over ribozymes in thermal stability. At first this was surprising, because an early finding showed that ribozymes could not be converted into corresponding DNA sequences without ablation of activity.[897] Since DNA lacks the 2' hydroxyl group of RNA attributed with a role in hydrolytic catalysis, it was initially expected that DNA could have no catalytic properties. Yet the success of powerful selection for catalytic DNAs from random sequences proved this viewpoint to be wrong. And in a reciprocal manner to ribozymes, corresponding RNA sequences to "DNAzymes" no longer exhibit catalysis.[898] In other words, both RNA and DNA sequences can act as enzymes, but different folds are required for equivalent catalytic tasks, in turn requiring different samplings of nucleic acid sequence space.

We have considered functional DNA and RNA molecules in separate sections, but in the future nucleic acid analogs with altered sugars, bases, and backbones may supersede either option. This we will discuss further at the end of this chapter. Also on the subject of categories and distinctions, we noted briefly earlier that ribozymes (or catalytic nucleic acids in general) can artificially be rendered subject to allosteric control by small molecules. There are natural parallels to this, and this leads us to the topic of natural aptamers, a profitable area to explore in a little more detail.

Natural Aptamers and Riboswitches

The existence of natural RNA counterparts to artificial aptamers was postulated after the molecular recognition capabilities of RNA were experimentally demonstrated.[899] Yet while DNA aptamers remain as entirely artificial creations (as far as known), aptamers based on RNA sequences do indeed have an extensive presence in nature. First let us clarify what this actually means. A natural RNA strand functionally binding a small molecule must have evolved its sequence and resulting shape to "fit" the target ligand,* and therefore has a clear-cut parallel with an artificial aptamer. Defining natural RNAs that bind proteins as aptamers may not be quite as straightforward, since in principle both RNAs and proteins could coevolve toward optimization of their interactions. In other words, in a natural RNA–protein interaction, which has evolved to fit: the protein-binding site, the RNA shape itself, or both? If optimization of the protein–RNA interaction provides a net fitness benefit, clearly mutations in either that promote such an outcome would be selectable, provided no other function was compromised. This, of course, is distinct from an artificial *in vitro* selection for a protein-binding aptamer, where the protein is a passive framework against which the aptamer is selected and evolved. In reality, though, noncoding RNA sequences would most likely have much more "room to move" by evolutionary change than target protein sequences, especially if the latter were under folding or functional constraints. This issue aside, one can distinguish between natural RNA interactions based on straightforward base-complementarity with other nucleic acids (producing linear duplexes) and those which involve RNA folding and secondary structures. The latter "aptamer-like" interactions usually result from single RNA strands assuming a specific shape, which enables molecular recognition by a protein. It is interesting to compare early predictions that natural aptamers would prove to be widespread[900] with recent findings of extensive conserved noncoding RNAs ("lincRNAs"[901]), at least some of which may be assigned as functional aptamers encoded by genomes.

Some natural RNA–protein interactions are of particular interest, since artificial aptamers have been generated to mimic the natural target RNA structures. One such case is the RNA bacteriophage MS2, whose coat protein specifically binds to an RNA stem–loop structure in the gene for the phage replicase.[902] Structural analysis of an RNA aptamer binding to the same MS2 coat protein showed a divergent secondary structure to the natural RNA element, but comparable interfaces for molecular recognition.[903] (This convergence at the level of recognition contacts is reminiscent of comparisons between DNAzymes and ribozymes.)

Once a significant stock of sequence information regarding artificial RNA aptamers had been accumulated, it was a reasonable undertaking to use computational bioinformatic approaches to see if the ever-burgeoning genomic databases would reveal motifs suggestive of natural aptamers. Scans of this nature revealed some promising possibilities[904,905] and some motifs (such as certain artificial amino acid-binding aptamers) appeared to be significantly under-represented in natural genomes, consistent with (but far from proving) selection against spurious and potentially deleterious functional nucleic acids.[905] Subsequent database searches for aptamer motifs have been strongly influenced by the discovery of riboswitches, or natural aptamers acting as gene regulatory elements.

Here, it would be useful to first consider some genetic regulatory systems in *E. coli* based on DNA-binding proteins. Many of these act as negative controls of gene transcription (repressors), since their binding at promoter regions acts as "off-switches." In a classic example, the *lac* repressor (regulating a control unit (operon) that processes the sugar lactose) only binds to its

---

*A small molecule ligand such as a coenzyme widely used as a cofactor by many different biocatalysts would be highly constrained in its ability to selectively evolve, which would require changes in corresponding biosynthetic enzymes.

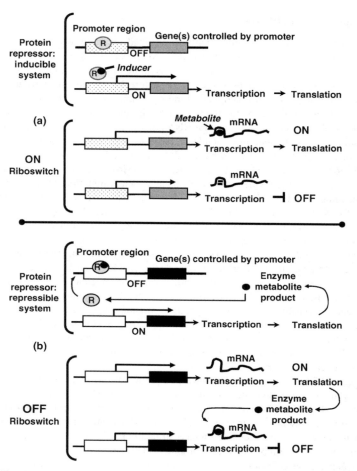

FIGURE 6.8   Comparison of protein negative control of gene expression with riboswitches. (a) Inducible system where protein repressor no longer binds to promoter region in presence of inducer. For an ON riboswitch, an mRNA requires binding by a metabolite either to allow transcription to be completed or for efficient translation. (b) Repressible system where protein repressor only binds to target promoter in presence of metabolite. Binding of a metabolite to a riboswitch analogously blocks translation of target mRNAs. OFF riboswitches can also function by blocking the continuation of transcription such that mature mRNAs cannot form.

target promoter in the *absence* of inducer (under natural circumstances, a metabolic product of lactose, as depicted in Fig. 6.8a as an "inducible" system). This ensures that the genes for lactose utilization are only switched on when lactose is available. On the other hand, it has been long known that some metabolic products of bacterial biosynthesis can repress expression of the relevant biosynthetic genes themselves. Another classic regulatory system makes this case perfectly. The tryptophan (*trp*) operon of *E. coli* can only bind to the *trp* promoter and repress transcription in the *presence* of the relevant metabolic product, where the *trp* repressor complexes with tryptophan. This establishes a negative feedback loop where high levels of tryptophan shut down unneeded continuation of the *trp* operon biosynthetic machinery (represented in general terms in Fig. 6.8b as a "repressible" system).

But of all bacterial genes subject to such feedback control, some appeared to lack analogous repressor systems as for the *trp* operon. Providing a clue to the nature of the control mechanisms in such circumstances, some relevant bacterial genes yield mRNA molecules with long 5′ untranslated regions, which were shown to be necessary for observed metabolite-repressive effects.[*] Now, control of gene expression certainly does not end once transcription has occurred, as many post-transcriptional (and indeed, post-translational) regulatory mechanisms are known, and these repression effects could in principle be accounted for by a protein which directly binds the mRNA in the 5′ regions of interest. (Recall the cases of the viral RNA-binding proteins referred to earlier, and many other cases of protein-based regulation by RNA binding are known.) Nevertheless, for the specific bacterial feedback systems in question, this hypothesis was displaced by a novel and very intriguing reality. When it was already known that RNA could fold into structures capable of binding small molecules, another possibility was suggested that turned out to be correct. Untranslated regions of some mRNA molecules (usually, but not exclusively, 5′ to the translational initiation codon) were found to specifically bind metabolite target molecules, with very significant consequences for gene expression. In the first described instance of this, the metabolite and vitamin thiamine pyrophosphate could bind to the 5′ regions of mRNAs encoding genes for thiamine biosynthesis, with consequent blocking of gene expression.[907] This effect was mediated by an allosteric change in the mRNA molecules such that they could not efficiently be recognized by ribosomes via the mRNA ribosome-binding sites.[907] Such a process constituted a novel RNA-based molecular switch, hence giving rise to the "riboswitch" label.

Numerous bacterial riboswitches[†] have since been described.[910,911] As well as modulating translation, some can directly affect transcription of nascent mRNA molecules.[912] For most riboswitches, metabolites bind to specific mRNA 5′ motifs and repress gene expression in a manner consistent with negative feedback control and an "OFF" switch (Fig. 6.8b), but some exceptions to this exist. The adenine riboswitch from *Bacillus subtilis* is an "ON" switch signal (depicted schematically in general terms in Fig. 6.8a), mediated through disruption of a transcriptional terminator in the ligand-free nascent mRNA.[913] What is the biological rationale for this? The gene under the control of this adenine riboswitch appears to encode a purine efflux pump, capable of exporting adenine and guanine from the bacterial cell. Under conditions of purine excess, increased expression of the efflux pump would be advantageous to avoid intracellular toxicity, consistent with the biological ON riboswitch in response to high adenine levels.[913] Another case of the rare ON type of riboswitches is found with a bacterial glycine degradative operon used to channel glycine as an energy source when it is present in high concentrations. This riboswitch has the additional interest of two separate tandem glycine RNA-binding sites in the 5′ untranslated regions of the relevant mRNAs. The glycine binding in this case acts cooperatively for the associated mRNA allosteric change, to ensure that the switch is only flicked to "ON" position when high glycine levels are present, as is biologically logical.[914]

There is a particularly interesting riboswitch associated with the *glmS* gene (of the Gram-positive bacterium *Bacillus subtilis*) that we have alluded to earlier, and there are major reasons for this interest. First, although it functions as an OFF switch, the *glmS* mechanism involves

---

[*]An example of this (prior to the discovery of riboswitches) is the role of lysine repression, in the 5′ untranslated region of a gene for lysine biosynthesis in the bacterium *Bacillus subtilis*.[906]

[†]It should not be thought that this form of gene regulation is restricted to prokaryotes, as riboswitches have also been identified in plants and fungi.[908,909]

mRNA cleavage by a ribozyme activated upon binding of the ligand glucosamine-6-phosphate.[911,912] And there is good evidence that the glucosamine-6-phosphate functions as a true ribozyme cofactor. Unlike "normal" riboswitches, structural studies have shown no conformational changes associated with ligand binding, indicative of a cofactor role for the ligand itself.[915,916]

Since their discovery, it has been hypothesized that riboswitches are relics from the RNA world,[907] and the diversity of their switch mechanisms is consistent with this. It has also been proposed that the scarcity of ON switches is a result of genetic necessity, where downregulation of metabolic synthetic machinery is frequently called for in the face of high metabolite levels.[913] At the same time, there is no inherent reason associated with RNA folding itself which could produce such a bias toward OFF switches.[913] While the usage of riboswitches across the bacterial world appears to be widespread,[917] they are far from universal components of gene regulatory systems. To use the *E. coli lac* operon again as an example, one could theorize (or artificially design?) a riboswitch ON system where *lac* mRNA expression was activated by the presence of lactose, which would in principle fulfill the same genetic requirements for substrate-mediated operon induction as occurs with the natural *lac* repressor/operator system. Whether or not such a system ever existed in evolutionary time, there is no evidence for its current existence. The selective pressures that determine the "choice" of regulatory systems are likely to be complex, and variable between divergent evolutionary lineages. For example, the *glmS* genetic system in *E. coli* is regulated by an entirely distinct mechanism to the ribozyme switch seen in *Bacillus* species, although the *E. coli* system still involves an RNA-based process.[918]

To conclude this introduction to natural aptamers, we can conduct a *gedanken* experiment (with a little help from a previously encountered friend) with respect to the question as to why functional RNAs are not used more frequently by nature than is evidently the case.

An Aptamerican Adventure   Lucy, ever the intrepid traveler, is again a stranger in a strange land. She has chanced upon a baffling biomolecular terrain called "Aptamerica." Here, almost all effector molecules are composed of nucleic acids, although there are some functions performed by relatively minor species called proteins. Aptamerican scientists tell her this is obviously because the modern biosphere in Aptamerica derives from a primordial "protein world," which eventually gave rise to the more efficient nucleic acid world. They explain also that owing to contingency and "frozen" systems that cannot be readily altered by natural evolution, some protein-mediated systems still persist until the present time. But Lucy is not to be bamboozled by a few lines of glib patter. "That makes no sense," she retorts, "Nucleic acids can be both effector molecules and replicators, and so could readily constitute a proto-bioworld, while this is not feasible for proteins!"

The Aptamerican authorities beg to differ, but their detailed explanations leave Lucy unconvinced. She argues with them back and forth, and eventually they come around to the topic of immune systems.

"In my world," Lucy says, "There *is* a type of RNA-based immunity for control of genetic parasites and some viruses, which we call RNAi. But high-level adaptive immunity is in the capable hands of proteins, its effector molecules." She then gives a brief overview of how the adaptive immune system works, in all its power and sublime beauty.

"Nice enough," responds one of her Aptamerican listeners. "You have some of the picture correct, but in a kind of mirror image. The effector molecules for our immune systems are entirely composed of RNA. We have shown in Aptamerica that folded RNA molecules act as universal recognition molecules for foreign pathogens, and these RNAs are produced by a Darwinian selection process. These RNA effectors are expressed from genes that are combinatorially assembled from sets of segments to give a vast range of alternatives. Moreover, specific ribozymes introduce somatic mutations into combining regions of the immune effector RNAs to further maximize diversity. When a novel foreign molecule is recognized by any of these randomly generated RNAs, the latter molecules are amplified and selected as the functional RNA effectors. It works perfectly!"

Lucy pauses and considers this. She is aware of natural aptamers such as riboswitches in her world (the "real" one), and she knows that aptamers can be functionally expressed intracellularly. What then is fundamentally wrong with the Aptamerican immune system, which seems to her rather like a natural SELEX process? Has the Aptamerican "all-RNA" world, and specifically their immune systems, somehow become evolutionarily "frozen" and unable to switch to proteins? She raises a few objections for the waiting Aptamericans, such as problems of exporting and secreting "RNA antibodies" from "ribocells" into the extracellular environment. Then she sees a related objection, and attacks:

"In the humoral immune system, each B cell can be regarded as an amplifiable recognition unit. Therefore the molecular recognition devices—antibodies in my world, at least—have to be displayed on a cell surface. And for that to occur, a hydrophobic trans-membrane segment must be used. How can aptamers answer that one!?"

This did not seem to ruffle any feathers. "We don't dispute the need for membrane-bound display, but in Aptamerica, special ribozymes attach hydrophobic lipid segments to RNA receptors, enabling them to remain membrane-anchored. Oh, and an analogous mechanism is used to attach hydrophobic leader segments to the same aptamers to facilitate their trans-membrane passage."

Remembering that instances of protein membrane anchoring by lipid-tagging in her own world, Lucy takes a different tack and hits on the issue of stability. Compared with proteins, surely RNA immune recognition molecules in extracellular fluids would be at a serious disadvantage in this regard?

"All your concerns here are quite solvable," interjects one of the Aptamerican biologists. "Natural selection and evolution are powerful forces, as well you know."

Another Aptamerican grudgingly concedes that those "ancient" proteins just might have a role in helping stabilize the Aptamerican RNA recognition molecules in some environments.

But the arguments rage on . . . and Lucy wakes, another dream within a dream. But dreams can be inspirational, and she resolves to see if it is feasible to engineer an adaptive immune system based entirely on RNA recognition molecules in her world. That way, she could see if the Aptamericans were right.

Allosteric Ribozymes and Aptamers

Riboswitches can be viewed as aptamers whose conformational changes upon ligand binding have specific functional effects on another region of the same RNA molecule. This can be generalized to the concept of a modular aptamer ligand-binding domain linked to another functional RNA domain, where the ligand binding allosterically regulates (positively or negatively) the function of the second domain. This second modulated function can be a ribozyme or another aptamer function itself.[919] We saw above that the ribozyme repertoire can be extended through the exploitation of organic cofactors for direct catalytic assistance, just as with proteins. But unlike this type of arrangement (as with the *glmS* riboswitch), allosteric ribozyme regulation inherently requires conformational changes to occur to mediate the

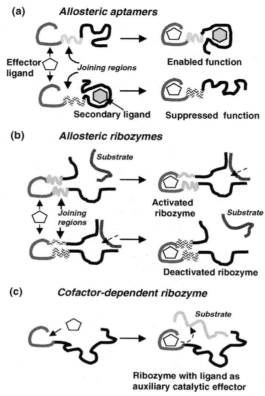

FIGURE 6.9   Distinctions between allosteric aptamers (a) and ribozymes (b, aptazymes) and ribozymes dependent on ligands as co-catalysts (c). In each case, the "effector" ligand is that which produces a conformational change in allostery, or participates in catalysis directly. This is in contrast to a secondary ligand that can exist for an allosteric aptamer. The nature of the joining regions is usually very important for obtaining the desired functionally modulating conformational changes. For (a) and (b) (allosteric aptamers and ribozymes), either a ligand-induced transition from low to high activity, or inactivation of an initially functional molecule can be produced.

functional switch. The distinctions between different classes of ligand-binding functional nucleic acids are portrayed in Fig. 6.9.

The generation of an allosteric ribozyme is at least in principle amenable to a "rational" modular approach. Ribozyme allostery can be specified in advance by choosing two pre-selected functional RNAs (the ligand-binding aptamer and the ribozyme itself) that are desired to "communicate" for regulatory purposes. It is a facile matter to link two such functional domains together; the "trick" is to ensure that the joining segment allows ligand binding to the aptamer segment to effect a conformational change that produces the sought-after functional transformation in the ribozyme domain. This has been achieved through *in vitro* selection itself for the appropriate junctional sequences.[920] Alternatively, it is possible to use "allosteric selection" to generate allostery from random sequence, by coupling a random tract with an inactivated ribozyme and selecting for restoration of ribozyme activity in the presence of a desired ligand.[921,922] It is a difficult prospect to design from scratch the interaction of a small molecule ligand with an aptamer-binding pocket, such that an allosteric effect is produced in a linked RNA domain. Yet oligonucleotides themselves can act as allosteric effectors for ribozymes, and it is a more tractable exercise to predict the conformational consequences of an interaction based on conventional base pairing between a designed oligonucleotide and a target RNA molecule. Accordingly, computational analyses of ribozymes and prospective oligonucleotide allosteric effectors has allowed the design of a variety of alternative switch configurations.[839]

Application of allostery for the regulation of aptamers (as "aptasensors") and ribozymes ("aptazymes") has mushroomed in recent times for a variety of purposes, and some more details are provided in ftp site. For the present purposes, let us concentrate on functional nucleic acid applications in the present and near future that have a general *in vivo* or ultimately clinical focus.

PAYOFFS FROM FUNCTIONAL NUCLEIC ACIDS

The direction of nucleic acid enzymes toward target RNA cleavage sites by flanking sequences is in principle a precise process, since the ribozyme or DNAzyme "arms" are *antisense* to the target RNA (usually an mRNA molecule). Provided the catalytic core of the nucleic acid enzyme is preserved, the design of such an effector DNA or RNA catalyst is then a matter of choosing the target site.[*] And in turn, this is (at least in theory) a "rational" process, although we must remember that the nucleic acid core catalytic sequences are conferred to us by either natural or artificial evolution. Nevertheless, since to a considerable degree ribozyme activities can be "rationally" routed toward specific purposes, we will relegate further discussion of this to the chapter primarily concerned with Rational Design in general (Chapter 9). Let us then see how aptamers may gainfully be employed in biological contexts. In doing so, it is pertinent to compare their pros and cons with competing recognition molecules, and most of these come from within the vast field of antibody-related technology. And as we will see in Chapter 7, additional avenues for molecular recognition now exist beyond the antibody framework itself.

Aptamer Tools: Better than Antibodies?

There are certainly some types of molecular recognition that are well suited to nucleic acid aptamers and less feasible with proteins. One of these is RNA–RNA interaction, which is of fundamental importance in biology.[923] Evolved RNA or DNA aptamers can make contacts with target RNA structural motifs by means of conventional base pairing or non-Watson–Crick base interactions, which, of course, are unavailable to antibodies.[924,925] At the same time, such simple rules of engagement for aptamer–RNA interactions are necessary but insufficient, as intricate interrelationships between folding of both nucleic acid partners is often highly significant.[923,925] While antibodies can be prepared against nucleic acids, they generally show a low level of sequence specificity, and cannot rival certain dedicated nucleic acid-binding proteins, as we considered in Chapter 3. It is probable, although not formally proven, that taken as a whole antibodies are likewise inferior to nucleic acid aptamers for RNA binding.

---

[*]Given the simplicity of the point of ribozyme or DNAzyme target cleavage, in principle any mRNA would provide many potential cleavage sites. But the specific target region on an mRNA (often eight bases in length on either side of the chosen cleavage point) must be chosen with RNA secondary structures also taken into account.

This particular kind of molecular recognition aside, we can find numerous gray areas in nature where both nucleic acid and protein-based interactions operate. For example, the regulation of gene expression has historically been regarded as the province of specific DNA- (or RNA-) binding proteins and associated protein cofactors. More recently, a hitherto hidden world of critical RNA-based regulation has been revealed, in the form of microRNAs and other short regulatory RNA molecules.[926] It is not clear how much this complex interplay between protein and RNA results from evolutionary contingency and chance, and how much is dictated by natural selection controlled by biochemical necessity and efficiency. In an analogous manner, artificial modulation of expression can be tackled at the protein level, or by artificial small interfering RNAs (to be briefly discussed in Chapter 9). Small natural regulatory RNAs ultimately rendezvous with their nucleic acid targets in *trans* by simple base complementarity,[*] but this is not the form of molecular recognition we have in mind for aptamer-like RNA interactions, as noted earlier.

Specific antibodies have been known for a very long time to bind proteins and often block their functions (as in the "neutralization" of toxins), and aptamers can also be very effective in this role. Blocking of the function of a biological effector molecule will very often be associated with impedance of a normal intermolecular interaction, either directly through an inhibitor binding to an active site (as with enzymes) or indirectly by engendering a conformational change. Aptamers are clearly capable of inhibiting protein or enzyme functions as a consequence of their specific binding.[927,928] Where the binding site coincides with that used for a natural ligand, an aptamer can be viewed as an inhibitor functioning by molecular mimicry of the natural molecule. One way of studying this is to use antibodies themselves, since a specific antibody-combining site has a ligand toward which it is directed. Aptamers can accordingly be selected which compete for the binding site of an antipeptide antibody, and thereby act as specific inhibitors of the normal binding reaction.[929]

As antigen-specific binding molecules, antibodies have an enduring association with a panoply of *in vitro* assays. There are also a great many known and potential applications for specific recognition molecules *in vivo*, and this point leads us to a prime example of two parallel streams of technological development involving antibodies and aptamers. Normally either secreted or expressed on the cell surface, antibodies in various forms have been successfully expressed intracellularly as binding proteins ("intrabodies"[930]) for designated targets. But aptamers too can be expressed such that they function in an intracellular environment, and based on specificity and other features, these "intramers" may be competitive with the antibody alternative.[931] The intracellular environment is not inherently favorable for antibodies, as they were not "designed" for that purpose, and considerable engineering of antibodies is required to overcome this difficulty. In contrast, nucleic acids are "at home" in the reducing environment of the cell, and thereby have an inherent head start.[932] There are now numerous instances of successful intracellular expression of aptamers for protein binding and dissection of signaling pathways, and other *in vivo* research topics.[932] Intramers can be functionally expressed within cultured cells growing in the laboratory, or within whole organisms by well-defined transgenic procedures. With specific protein targets, the latter type of studies have been successfully performed with the fruit fly *Drosophila*[933] and the nematode worm *Caenorhabditis elegans*, a widely used genetic workhorse.[934] These encouraging results are not to suggest that technical problems are completely overcome, but there do not seem to be fundamental theoretical objections to the widespread application of intramers.

---

[*]This involves several steps of protein-mediated processing and assistance prior to the final hybridization with target nucleic acid.

After these general considerations on some practical issues of aptamer performance, we should now have a look more specifically at some of the growing potential for aptamer clinical applications. But there will be more to say in the way of comparing aptamers, antibodies, and other recognition alternatives in Chapter 7.

Moving Aptamers Toward the Clinic

As might be expected, the key to successful application of aptamers in a therapeutic setting is their ability to bind a specific target relevant to a designated pathogenic process. The source of pathogenesis can derive from outside infectious agents or from misfiring of internal physiological pathways, which in turn determines the nature of the molecular target. As noted earlier, binding can be associated with neutralization of a target protein's function, but this is not an inevitable consequence of binding *per se*. Especially when the target is a large protein, the specific site of binding, or epitope,[*] is critically important in determining the functional outcome. It may only be necessary to block or perturb one aspect of a target protein function to achieve useful activity, including competing for a natural binding site of another protein or low molecular weight ligand. Of course, any candidate for the clinic must pass a long series of checkpoints to prove that side effects, if they exist, are acceptable in balance with the severity of the problem to be treated.

Further details of potential clinical aptamer applications are beyond our scope, but let us look a specific issue of clinical relevance. Aptamers competing for a normal nucleic acid-binding site on a protein are sometimes called "decoys," but are in effect a special case within a larger application of this term, and this has some interesting aspects that are relevant to the whole issue of molecular design. Let us then take a brief detour to examine this. Consider a protein (such as a transcription factor) that binds a short double-stranded DNA segment of defined sequence. An example of an important mammalian transcription factor is NF-κB, which was originally defined as an immune system regulatory factor, but actually controls the expression of many genes in various contexts through signal transduction pathways.[936] NF-κB binds to a limited set of duplex sequences found in promoter regions of its target genes, one of which is[†]

5′-GGGACTTTCC

3′-CCCTGAAAGG

If one artificially provides a large excess of this duplex DNA to an intracellular environment with active NF-κB, virtually all the available NF-κB binds the extraneous target and is thereby "soaked up." The artificial "decoy" nucleic acid duplexes then effectively sequester the cellular pool of NF-κB binding that sequence, and thus prevent this transcription factor from activating its target genes.[937] In fact, this idea is not recent in molecular biology. Bacterial repressors controlling specific gene expression bind to short DNA sequences known as operators, and it has long been known that an excess of operators within a bacterial cell (carried on a extrachromosomal plasmid, for example) would "mop up" available repressor and alleviate its repression. This effect is known as "repressor titration," but is in essence the same as the decoy principle.

---

[*]The specific binding site for an aptamer has been termed an "apatope,"[935] but this has not been widely adopted, since "epitope" is the general term for such interactive sites.
[†]This is one example of a family of sequences to which NFκ-B binds, since its binding specificity is degenerate. The consensus sequence corresponds to 5′-GGGRNWYYCC, where R = purine (A or G), Y = pyrimidine (C or T), and W = A or T.[936]

Now, we have seen that aptamers can be selected to compete for the interaction of nucleic acid-binding proteins with their targets, and this includes transcription factors as above. NF-κB has been thus inhibited in binding its duplex DNA target by a specific high-affinity aptamer.[938] Yet this was not a case of simply mapping duplex DNA into duplex RNA, since the aptamer sequence rendered as corresponding DNA had no such effect.[938] NF-κB is a dimer consisting of two distinct subunits, and in principle, the aptamer might be binding at a different site to the normal NF-κB region which is interactive with DNA, and causing a conformational change which subsequently affects target DNA binding. Yet structural studies have revealed that the aptamer contacts a common surface on NF-κB as for the normal DNA target duplex, and the protein interactive surface is essentially unaltered between duplex DNA and aptamer binding.[939,940] Consequently, the aptamer has in effect been selected to mimic the contacts made between the duplex DNA and its binding protein partner. In turn, this returns us to the notion of "decoy" in general, and shows that a "direct" DNA duplex decoy can be achieved in practice by a type of molecular mimicry. With this in mind, we can consider that decoys can be generated in two distinct ways. A decoy that is simply a copy of a binding site provided in excess is rationally designable, provided, of course, that one knows the target sequence of a nucleic acid binding protein of interest. On the other hand, a "mimic" decoy cannot be simply rendered from the known target sequence, and empirical *in vitro* evolution remains the most efficient way of obtaining such a nucleic acid aptamer.

Beyond the targeting of pathogenic organisms, there remain a gamut of emerging clinical applications for nucleic acid aptamers. One of the first DNA aptamers obtained was directed against thrombin, and this remains a target of interest for the generation of aptamer-based inhibitors of blood clotting pathways.[941] Numerous other human proteins have been investigated for inhibition by aptamer reagents, with ultimate therapeutic goals, including potential cancer treatments.[942,943] You will have noted the word "potential" appearing frequently in this section, which simply reflects the fact that much basic and applied research with medically relevant aptamers has not yet jumped through all the hoops required for final therapeutic approval. Now, though, there is a precedent, in the form of the FDA-approved product Macugen®, or pegaptanib.* This is a chemically modified RNA aptamer (also conjugated with polyethylene glycol) that inhibits the action of a specific isoform of vascular endothelial growth factor ($VEGF_{165}$). The latter protein is the identified culprit in the "wet" form of macular degeneration, a significant type of age-related vision loss. $VEGF_{165}$ contributes to this pathogenic effect by promoting the excessive growth of blood vessels in the eyes (angiogenesis), which results in blood leakage and vision damage.[944] An aptamer capable of binding $VEGF_{165}$ and preventing signaling through its receptor was judged to be clinically effective and safe,[945] leading to a regulatory green light from the FDA and the production and marketing of Macugen. We can be confident that this is only the first of many therapeutic aptamers to gain such status.

Finally, we should note a sort of flip side to aptamer clinical applications. An interesting and important aspect of the study of natural aptamers of pathogenic bacteria is that they can present novel targets for antibacterial drug development. This case is convincingly supported by the observation that the previously identified antibacterial agent pyrithiamine works by perturbing the normal operation of thiamine riboswitches and thereby inhibiting cell growth.[946] Chemical analogs of other normal riboswitches have been screened for activity, with promising results.[947]

---

*This name can be decoded as peg/aptan/ib, for polyethylene glycol/aptamer/inhibitor. The "ib" suffix for "inhibitor" is a widely used convention in the pharmaceutical industry, as is the "ab" suffix for monoclonal antibody-based product. Macugen® also has certain chemical base modifications designed to inhibit ribonuclease action.

The above-mentioned fact that chemical modification of Macugen is desirable to improve its nuclease resistance might draw our attention to a much more general issue. We have previously noted that the four-base nucleic acid alphabet presents real limitations in various circumstances with respect to proteins. Just as the protein alphabet can be extended (Chapter 5), what can be done in this respect in the nucleic acid domain? Along with this, there are other ways of chemically altering nucleic acids to useful ends, as we will see.

EXTENDING THE FUNCTIONAL RANGE OF NUCLEIC ACIDS

If even proteins have alphabetic limitations, then surely nucleic acids could use some additional assistance as well. There are two general approaches to this issue: the provision of cofactors, and the use of chemically modified bases that are compatible with incorporation into nucleic acids by normal polymerases. While cofactors are clearly useful for both natural and artificial molecular design, selection for a function that is directly realizable with a set nucleic acid alphabet is more efficient, and may improve the likelihood of isolating novel functions by evolutionary or rational processes in the first place. An extension of an alphabet to encompass a grander sphere of potential functions then seems a worthy (if challenging) undertaking. Though useful in its own right, the simple incorporation of chemically modified nucleotides into polymerizing nucleic acids is still dependent on conventional base-pairing interactions, and thus does not constitute a true alphabetic extension.*

And yet in the end, why should one bother with real extensions to the nucleic acid alphabet, if proteins are always going to have the edge as the mediators of molecular functions? There are two major answers to this. First, it is not clear in any case that the proteins are uniformly superior for all applications, an issue we have already encountered in the world of aptamers. But the second answer gets to the heart of the matter. The ability of nucleic acids to undergo assisted self-assembly on their own complementary templates is the underlying basis for the replication of life itself, and certainly all modern directed evolution. Artificial evolution of proteins still relies on the amplification of encoding nucleic acids, whether *in vitro* or *in vivo*. This remains true even for sophisticated *in vitro* techniques such as mRNA or ribosome display (Chapter 5). And nucleic acids, as we have now seen in some detail, can act as both effector molecules and replicators. This fact, simple in outline but subtle in its implications, is the basis of SELEX and all evolutionary *in vitro* selection of nucleic acid enzymes and aptamers. Proteins cannot be directly amplified in the same way as nucleic acids. For proteins and peptides, amplification can only be accomplished indirectly by tagging them to their encoding nucleic acids, either through physical linkages (display technologies) or through compartmentalization. In contrast, functional nucleic acids come with an inherent ready-made link between phenotype and genotype, and SELEX and related techniques fully exploit this one-to-one correspondence between function and encoded information. Increasing the size of nucleic acid alphabets then has the potential to dramatically increase the functional power of nucleic acids (perhaps to the point of rivaling proteins in many new areas), while retaining the power of self-replication. This point made, we should explore these issues in some more detail.

*Most RNA polymerases will accept and incorporate certain chemically modified nucleosides into an extending RNA chain copied from a template. Therefore, during the iterative production of RNAs from libraries for *in vitro* evolutionary experiments (as with T7 RNA polymerase), one or more of the four normal nucleoside triphosphates required for templated RNA synthesis can be replaced with equivalents with modified nucleobases, provided they are polymerase friendly[739,948] (detailed further below). This is in principle applicable toward the selection of ribozymes with chemically modified bases at their catalytic sites, which improve or enable catalysis.

Extended Alphabets

First, we should note that a true expansion of an alphabet requires that the macromolecule bearing the increased range of subunits is stably replicable, such that the full sequence information is transmissible and identifiable. Let us amplify (so to speak) the differences between proteins and nucleic acids with another thought experiment.

The Virtual Molecular Museum   Lucy is found wandering around a wondrous institution, the Virtual Molecular Museum. Here, one can call up any molecule of interest and look at an utterly convincing image of it from any vantage point. It is possible to see a protein as if you were the size of only a water molecule or smaller, allowing you to explore inner protein cavities. Or if desired, you can zoom out such that the protein molecule itself can be held in the palm of your hand and manipulated as you see fit. Even better, molecules can be pulled apart and recombined at will, as you long as you conform to basic rules of chemistry. Lucy spends much time admiring proteins, finding levels of beauty in their shapes and folds that some of us would reserve for works of van Gogh. But she is ever practical as well, and resolves to test out some ideas of hers.

She finds some proteins of small to moderate size and shrinks them to convenient dimensions for her aims. Initially she takes one by the N-terminus in one hand, and the C-terminus in the other, and stretches it out, ironing out its tertiary and secondary structures until it is in the form of a linear string of amino acids linked by peptide bonds. She lets go, and the protein snaps back like a rubber band, taking up its original shape before she can say "ribonuclease A," its name. She stretches it out again, and "tells it" to stay put in that elongated configuration. Then she brings on hand a series of other molecules she has previously defined. Each of these is "complementary" to one of the 20 usual natural amino acids, by means of an interaction that is weak compared to covalent bonds but still strong enough to be useful. For example, alanine has antialanine, tryptophan has a specific antitryptophan molecule, and so on, for all of the remaining 20. The stretched out protein then forms a template for these molecules, and she arranges each of them according to their complementary amino acid as determined by the protein sequence. She then wants to link the string of antiamino acids on a backbone framework that could also be peptide bonds, although any backbone will suffice for her purposes provided it has the right architecture to enable the appropriate complementarity and templating. She chooses the best backbone for her needs, and then uses this to link each of the antiamino acids in the sequence order specified by her unfolded stretched out protein.

As she performs this activity, she thinks of it in the light of her studies of molecular biology. She concludes that she is functioning as a virtual enzyme, which she thinks of as "Lucy-for-ase," given the standard -*ase* enzymatic suffix. And Lucy-for-ase is a very versatile enzyme indeed, because it can act in reverse, taking the linked antiamino acids and using this as a template in turn to join together regular amino acids in the same order as the original protein, through the formation of new peptide bonds. Each time, of course, to repeat the process it is necessary to separate the noncovalently joined "strands" of protein and the antiprotein reverse template. If each copy is used as a replicative template in turn, exponential growth ensues and Lucy soon retires from her catalytic role. But in the meantime, both the protein and the antiprotein strands have undergone enormous replication, and when she desires, each of the protein strands can "pop back" into their original functional shape. Maybe the antiprotein strand can "pop" into its own shape, but Lucy is not concerned with that.

Her interest lies in the general practicality of her protein replication system. To her discouragement, she finds it just does not work very well in general, for a number of reasons. For one thing, despite her best efforts, her antiamino acids do not have the desired specificity. It is hard to find 20 separate pairs of small molecules which noncovalently interact with perfect specificity, and which will function in this capacity. And not all proteins are as tractable as her ribonuclease A test case, since some refold slowly or hardly at all. When denatured *en masse*, many proteins stick to each other and interfere with the antiprotein polymerization. She ponders trying to solve some of these problems with complex chaperone systems, but instead decides on an alternative.

At first, she considered nucleic acids as very interesting from an informational point of view, but a bit of drag in functional terms. After a stint in the Virtual Molecular Museum of folding single-stranded RNA and DNA molecules, though, she starts to change her mind. Those "boring" nucleic acids turn out to be far more versatile functionally than she had thought, and furthermore come with their own natural replication systems. She could then shelve her ambitious protein template replication plans, retire as Lucy-for-ase, and let natural enzymes do all the hard work. But she still remembers the functional power of proteins, and then she decides on a type of compromise. What if she retains the main features of nucleic acids, such as their phosphodiester backbone and ribose moiety, but extends the base pair alphabet? She spends time adding new pairs of complementary bases, using additional interactions such as hydrophobicity as well as hydrogen bonding. Some of her more interesting additions in this regard (which deviate the most from the natural base repertoire) work poorly with natural polymerases. Rather than revert to Lucy-for-ase, she takes the polymerases themselves and tinkers with them until they do work as desired. Then she sets up systems for selecting new functional extended-alphabet nucleic acids. It seems as though she will be in the Museum for some time to come.

Let us now consider the basis of nucleic acid alphabet extension in more explicit terms, to illustrate the difference between chemical modification of bases that is compatible with existing base pairing, and the generation of entirely novel base pairs.

During *in vitro* evolution experiments, selected RNA libraries are amplified at the DNA level (after reverse transcription) by PCR. Consider an arbitrary RNA sequence 5'-GAUUCUAG-3', as a part of a longer ribozyme. After reverse transcription into the complementary DNA strand and PCR amplification, the original RNAs (including our model oligomeric sequence here) can be "read-off" by RNA polymerase if each amplified DNA duplex is equipped with a suitable 5' promoter (as in the classic SELEX approach; Fig. 6.2). Now, during this transcription of the RNA library, one or more of the four nucleoside triphosphate substrates can be chemically altered provided they remain compatible with the RNA polymerase and modified bases can pair (by hydrogen bonding) with their complementary bases. If, for example, the nucleoside triphosphates provided for RNA synthesis were exclusively ATP, CTP, GTP, and *UTP (where *U is a chemically altered uracil base), then obviously the above model sequence becomes 5'-GA*U*UC*UAG-3'. If this modification enables the ribozyme function, this modified RNA (out of all the other modified RNAs in the library) can be physically selected. To identify this sequence, it is necessary to amplify it. Since the *U base (necessarily) still can base pair with A residues, after reverse transcription, PCR and cloning the original selected RNA can be identified by its corresponding DNA sequence copy GATTCTAG. If *all* uridines in the original RNA library were modified, this DNA sequence nonetheless still tells us that 5'-GA*U*UC*UAG-3' is part of the original ribozyme of interest. Both what if less than 100% of the uridines in the original RNA library were altered? A selected RNA sequence might be 5'-GA*UUCUAG-3', and the pattern of modification/nonmodification might be crucial to activity. If we try to reverse transcribe and PCR amplify this, however, we still end up with the DNA copy GATTCTAG. The reverse transcriptase in the first step cannot distinguish between a template with U or *U; equivalent to a *degeneracy* of recognition, as noted earlier in the context of ribozyme alphabet reductions. Consequently, A is used as the complementary base for either U or *U templates, and the information specifying the original modification pattern is lost.

Though there is a substantial size difference in the protein versus nucleic acid alphabets (20 to 4, respectively, in the majority of circumstances), the most fundamental difference is nucleic acid base-pair complementarity. Certainly, protein structures are often stabilized by charge and other interactions between amino acid residue side chains, but there is no parallel in proteins with the specific hydrogen bonding between nucleobase pairs.[†] Extensions of the protein alphabet with one or more unnatural amino acids can thus be made, as we have seen in Chapter 5. An extension of a nucleic acid alphabet, on the other hand, requires a pair of new complementary "letters," if faithful replication of the super-alphabet is to be maintained. So the above example of chemical modification and selection with the *U analog is not a stable alphabetic change. For this to be the case, we would need another base alteration (^U) which does *not* base pair with conventional adenines, but only with another modified base ^A. The novel ^A base should likewise be faithful to ^U only. To enable RNA identification after amplification, ribo-(^A:^U) base pairs should also be compatible as corresponding deoxynucleotides for incorporation into DNA. This would constitute a two-base alphabet extension, since one could replicate and identify any sequence combination of A, U, G, C, ^A, and ^U. This novel alphabet could be reduced as well, to (for example) ^A, ^U, G, and C.

The notion of nucleic acid alphabet expansion is not new, and goes back at least two decades. It was known that the natural Watson–Crick base pairs could accommodate alternative hydrogen-bonding arrangements, and this was used to design chemically distinct pairs of pyrimidine–purine analogs that together constitute a novel base pair.[949] After chemical synthesis, these modified bases allowed formation of stable duplexes, and as a primer-template system a novel base pair were accepted by RNA or DNA polymerases.[949] Numerous additional candidate novel base pairs have since been investigated.[950,951] While many of these use hydrogen bonding as the basis of the required complementarity, it has also been found that hydrophobic stacking interactions can also substitute in this regard.[952–954]

Structural studies using chemically synthesized duplexes have found that the DNA backbone is tolerant of bulky unnatural base pairs, with "stretching" of the normal double helix,[955] but to be useful in most biotechnological contexts, novel bases (as nucleoside triphosphates) must be compatible with DNA and RNA polymerases. But if a hammer cannot bang in a nail, instead of throwing away the nail, perhaps one can make a better hammer. With modern directed evolution of proteins, it is entirely feasible to seek to redesign polymerases such that they gain the ability to use unnatural nucleoside triphosphates as substrates. Phage display systems for the evolution and selection of novel polymerase activities have been established, and applied toward the isolation of polymerases that can incorporate unnatural nucleoside triphosphates into nucleic acids.[956,957] The success of this is clearly dependent on the chemical nature of the modified base pairs, and one step in the right direction is the isolation of polymerases that can tolerate a novel base pair at the 3′ end of a primer template designed for enzymatic extension.[958]

Leaving aside current technical challenges, how far can the addition of new base pairs go? Can we end up with a replicable template system with sufficient "letters" to rival proteins? Unfortunately, there are very likely to be significant trade-offs as one increases the numbers of base pairs in a nucleic acid template-directed replicable system beyond a certain point. This question can also be framed as to why the natural alphabet for DNA consists of only two base pairs.[959] In natural organisms, an increased alphabet will come with significant metabolic costs, in terms of the need to acquire extra synthetic and transport machinery for the additional bases. Strains on replicative fidelity (and repair systems) are also likely to increase as an alphabet grows, since to "fit" the general architecture of duplex DNA, certain chemical features of the

---

[†]If this was the case in proteins, 10 pairs of specifically interactive residues would exist. Under this set of rules, radically different folding would result, probably too restrictive to be useful for biological purposes.

different bases will tend to overlap, increasing the probability of polymerase misincorporations. But the sophistication and copying fidelity of nucleic acid replicative polymerization has increased over evolutionary time, and some modeling studies have indicated that with high-fidelity replication, three base-pair (six letter) nucleic acid alphabets have a competitive advantage over two base-pair systems.[*,960] Beyond this level, though, the overall fitness scheme inevitably declines.[960] Still, factors that might prohibit the competitiveness of extended alphabets under the unforgiving eye of natural evolution do not necessarily preclude its use under the controlled conditions of evolution in the laboratory. The above consideration of polymerase engineering or evolution toward accepting novel substrates is a case in point in this regard. If one "universal" polymerase for both the natural bases and a gamut of alternatives is not possible, replication or transcription of a hybrid nucleic acid might require the joint cooperation of multiple types of engineered polymerases with different specificities. Such an arrangement might never evolve naturally for a variety of reasons, but could readily be envisaged in principle in the laboratory. In effect, many fitness costs unacceptable to natural organisms can be "worn" by laboratory experimentalists.

Since the extension of nucleic acid alphabets is still an early stage undertaking, it is not clear what the functional "letter limit" will be, but a point of diminishing returns will surely be reached at some level. Nevertheless, nucleic acids with extended alphabets based on complementarity could theoretically acquire a highly extended range of functions, especially in the field of catalysis. In an ideal situation, such extended molecules would acquire all the functionality of proteins but retain the ability of nucleic acids to be replicated by base pairing on complementary templates.

It has been suggested that one limitation of normal nucleic acids in functional roles is their difficulty in forming deep hydrophobic pockets excluding water, which could act as a significant impediment toward certain types of catalysis.[739] In principle, the artificial provision of unnatural base pairs, with hydrophobic rather than hydrogen-bonding complementarity, could help address such a shortcoming (as schematically depicted in Fig. 6.10). There remains a large window of opportunity for the development of functional DNA and RNA aptamers and catalysts with extended alphabets of this type. Though useful, the application of unnatural base pairs for site-specific labeling purposes[961,962] would seem as the metaphoric tip of the iceberg in this regard. As well as directly exploiting expanded aptamer functionality, additional unnatural base pairs also afford the possibility of creating entirely novel codons, for which orthogonal tRNAs charged with unnatural amino acids could be evolved.[951] If we recall the section dealing with unnatural codons and unnatural amino acid insertion from Chapter 5, we will note that most work until the present (especially *in vivo*) has used "spare" stop codons; provision of novel codons totally unused by the host organism background would greatly facilitate further progress in this field. Thus, base-pair expansion and unnatural amino acid insertion can develop in parallel.

Altered Backbones

The base sequence of nucleic acids can carry information, whether by encoding proteins or functional RNAs, or regulatory signals for their expression. Specific sequences of bases also are important for single-stranded nucleic acid folding, but the phosphodiester/sugar backbone is

---

[*]With the assumption of low-fidelity replication in modeling, four-letter alphabets were indeed found to be optimal. These findings were also cited as evidence in support of the contention that the four-letter/two base-pair nucleic acid alphabet is a "frozen" relic of the RNA World, since in primordial times, copying is not likely to have had a high level of fidelity.[960]

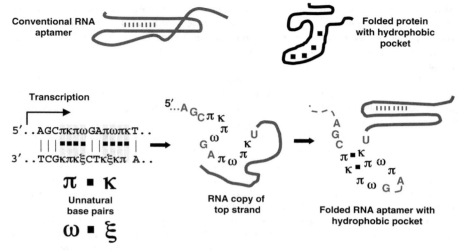

FIGURE 6.10 Depiction of an aptamer incorporating two unnatural base pairs (κ/π; ω/ξ) whose complementarity is based on hydrophobic interactions (indicated as small black rectangles). Transcription of a DNA duplex including the unnatural bases (with appropriate complementarity) into corresponding RNA enables folding of an "unnatural" aptamer bearing a pocket with hydrophobic character formed through its specific base-pair hydrophobic interactions, as shown.

not a passive spectator in determining the functional properties of such nucleic acids. This is a major reason why it is not appropriate to forget about this backbone, which enables nucleic acid polymerization and is amenable to chemical manipulation in many ways. And as in so many other circumstances, such studies have ramifications in the fields of both pure science and applied technology.

The "pure" side to this kind of work clearly enters at the level of nucleic acid structures themselves, with an aim toward an improved understanding of what structural features are necessary and sufficient for genetic information transmission. Such studies also either explicitly or implicitly have an impact on our understanding of how these molecules originally arose, and what their precursors might have been. Since it is generally accepted that the hypothetical RNA world itself must have had its antecedents,[744] it is logical to assume that simpler polymer templates must have arisen first. The challenge then is to define in precise chemical terms what such structures may have been, and some interesting variants on the broad nucleic acid theme have been derived. Going down in complexity from the ribose pentose (five carbon) sugar, it has been possible to synthesize nucleic acid equivalents with a phosphodiester backbone containing the 4-carbon sugar α-L-threose[963] (thus termed threose nucleic acids or TNAs). Simpler polymers using glycerol or other small compounds replacing the sugar moiety have also been considered as candidate progenitors of modern nucleic acids.[964,965]

Another interesting nucleic acid analog takes the form of *peptide nucleic acid* (PNA), where the entire phosphodiester/ribose backbone is replaced with a repeating peptide structure.[*],[966] PNA strands can form very stable duplexes based on base complementarity

---

[*]Strictly speaking, PNAs have a "pseudopeptide" polyamide backbone. A protein is a polymer of α-amino acids ($NH_2-CHR-COOH$) polymerized through $HN-C=O$ peptide linkages, with (normally) 20 different possible "R" groups attached to the α-carbon atom. A PNA then is not simply a protein analog where the R groups are nucleobases. The geometry of PNAs is important for allowing base-pair recognition with conventional nucleic acids.

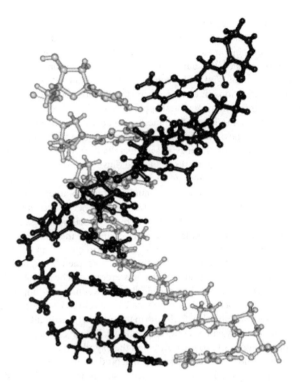

FIGURE 6.11   Structure of a DNA (light gray)–PNA (dark gray) duplex (solution structure by NMR[39]).
Source: Protein Data Bank[2] (http://www.pdb.org) 1PDT. Image generated with Protein Workshop.[5]

with each other[967] or with conventional nucleic acids[968] (Fig. 6.11). Considerable interest has focused on the possible relevance of PNAs (or an analogous molecule) to the origin of life,[969] in part due to the ability of PNAs to act as heterogeneous templates for the ligation of nucleic acids[970] or vice versa.[971] Also, the observation of complementary PNA–PNA duplex formation has shown that phosphodiester/ribose (or deoxyribose) backbones are not the only way to obtain a DNA-like "double helix" structure.[967]

Many applications for PNAs have been envisaged and investigated.[966] The great stability of PNA-DNA or PNA–RNA duplexes[972] is relevant in this regard, as PNAs can invade DNA double helices and form triplex structures, affording an opportunity for sequence specific targeting and manipulation of gene expression.[973,974] Some of these applications have been restricted by cell delivery limitations for PNAs, though experimental avenues for circumventing such problems have been identified.[975] PNAs (and other backbone-modified nucleic acids) have also shown promise for antisense-mediated inhibition of gene expression.[*,966]

To be useful in any cellular environment, and above all *in vivo*, nucleic acid molecules as therapeutic or sensing agents must possess reasonable stability. And most of the time, under these conditions normal DNAs and RNAs do not. They are very susceptible to ubiquitous nucleases, and RNA is also much more chemically labile than DNA.

---

[*]Antisense inhibition strategies are still in use but face competition from other technologies, including ribozymes and RNAi, as we have noted earlier and will continue in Chapter 9.

A major motivation for wide-ranging tinkering with nucleic acids has been to render them resistant to chemical and nuclease-mediated degradation. The unnatural status of the backbones of PNAs and other nucleic acid analogs renders them "foreign" to most nuclease enzymes, and thereby confers high degrees of resistance and survival prolongation for the above-mentioned applications. Aptamers synthesized with such alterations can be accordingly generated. Such chemical backbone modifications indeed result in stabilization, but a quite distinct and highly effective option also exists for many potential applications. This is based on certain properties of *chirality*, which we discussed in Chapter 5. Nucleic acids are chiral molecules through their ribose sugar moieties, which are in the D-chiral conformation in common with other naturally used sugar molecules. Since molecular recognition events are usually highly specific for the chiral status of participating molecules, normal nucleases recognize "inverted" L-nucleic acids very poorly.

So how could this be used to generate highly nuclease-resistant aptamers? Consider the situation where one has selected a normal (D-nucleic acid) aptamer against (for example) a normal L-peptide. It is quite feasible to synthesize an L-isomer of this aptamer with identical sequence but inverted chirality, but it will no longer bind the original target. The L-aptamer will specifically bind the unnatural peptide D-isomer, owing to the nature of "mirror image" molecules. In molecular binding interactions, mirror image chiral molecules possess a reciprocity of functions, which simply means that inverting the chiral status of each binding partner will preserve the interaction. If the L-stereoisomer of molecule A binds the D-stereoisomer of ligand B, then reciprocal binding between the D- and L-stereoisomers of A and B, respectively, will occur. This effect can be exploited toward obtaining complex binding molecules of opposite chirality to the norm that would be difficult to produce otherwise. For aptamers, chiral reciprocity can be applied as follows: First evolve a normal aptamer against a target with the opposite chirality to the norm, and then synthesize a reverse chirality aptamer with the same sequence. The latter molecule will then bind the natural target of normal chirality (Fig. 6.12). Such aptamers have been termed "spiegelmers,"* from the German for "mirror."[977,978] Though the spiegelmer is necessarily prepared synthetically,† the vital information specifying the correct sequence is thus obtained by means of a normal SELEX-based aptameric selection process. As would be expected, both RNA and DNA spiegelmers can be prepared.[977,978,980]

The idea of making complex molecules and systems with inverted chirality is an old one,[981] and has been a common theme in science fiction. (An "inverted" human could not eat food containing normal biomolecules, but would require nutrients of opposite chirality to survive.) Certainly, the reciprocity effect applies to proteins, and has been exploited in phage display to find D-peptides binding biomolecules of normal chirality.[982] Direct chemical synthesis has also been used to prepare chiral enzymes that likewise have specificities toward substrates of swapped chirality themselves.[981] Since spiegelmers are possible, then "spiegelzymes" (that is, L-ribozymes) should likewise be feasible, where substrates and products have unnatural chiralities. This has considerable potential for useful application in synthetic chemistry (where chirality is crucial and often a difficult factor to contend with), and has been verified for the important Diels–Alder reaction,[983] to which we have previously referred.

---

*Upon reflection, this is another apt term. As stereoisomers (enantiomers) of natural single-stranded nucleic acids, spiegelmers should not be confused with the left-handed variants of nucleic acid duplexes (Z-DNA and Z-RNA) that we discussed earlier. L-DNA molecules can form duplexes with their L-DNA complementary strands, which themselves are of left-handed helicity.[976] Another way of making this point about helical handedness is to note that PNA single strands are themselves nonchiral, but acquire chirality (optical activity) upon forming complementary duplexes, since two possible helical directions can be formed.[967]

†Although natural D-nucleic acids may have a theoretical stability advantage favoring their ancient adoption by living systems,[979] this does not preclude the artificial preparation of L-nucleic acid stereoisomers.

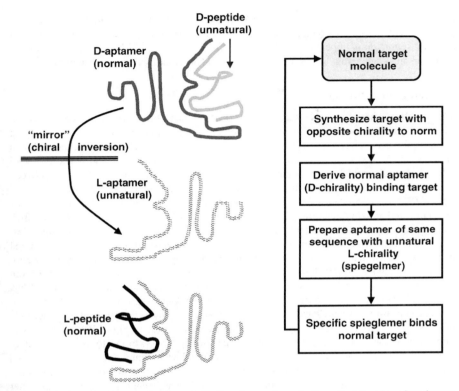

FIGURE 6.12   Spiegelmers and reciprocal chirality, shown schematically on the left and as a flowchart on the right. A natural chiral target molecule serves as the starting point for synthesis of an isomeric molecule with reversed chirality. An aptamer binding this "reverse target" is then obtained by standard SELEX procedures. Chiral reciprocity means that an aptamer identical except for reversed chirality will now bind the original target as desired. The L-aptamer ("spiegelmer") has the desired recognition specificity combined with high resistance to nucleases.

Any nucleic acid with a modified backbone is in principle amenable to the generation of aptamers, provided such molecules are compatible with available DNA and RNA polymerases.[*] It is encouraging in this regard that polymerases compatible for TNAs have been identified,[984,985] opening the door for the generation of functional TNA aptamers. And just as we have seen above with modified bases, the potential exists for artificial polymerase evolution toward toleration of novel backbones that are otherwise refractory. Furthermore, there is no reason why the use of altered nucleobases (as in the preceding section) and backbones cannot be combined, and steps in this direction have already been taken for PNAs.[986]

Through such strategies, the world of functional nucleic acids may undergo a marked extension whose boundaries remain to be seen. Many practical ramifications of this work are

---

[*]If one possessed a complete *in vitro* replication and amplification system of opposite chirality to normal biology (L-nucleic acids and D-polymerases), it would certainly be feasible to directly obtain L-nucleic acid aptamers (spiegelmers) against targets of normal biological chirality. This would enable the direct generation of nuclease-resistant aptamers binding normal molecules of interest, but is currently limited by the difficulties and costs of creating the "opposite" replication system.

expected,[987] which can only amplify the "tip of the iceberg" principle noted earlier for altered base pairs alone. Yet despite their potential for extensive chemical tinkering, functional nucleic acids in a natural setting do not operate in isolation. We may be wise to recall this during the artificial evolution and engineering of DNA and RNA molecules, as the next section briefly relates.

The Best of Both Worlds?

We have already noted that it is not accurate to refer to modern* biosystems as the protein world, since RNA has functional roles in all cellular organisms as well as acting as an informational intermediary between genomic DNA and protein. The entanglement between nucleic acids and proteins operates at several different levels, including RNA catalysis, and hence the label "RNA–protein" world is more suitable,[721] as noted earlier. The close relationship between catalytic RNAs and protein "facilitators" in Group I self-splicing introns,[724] ribosomes,[725] and RNase P[988] are significant cases in point. We have also seen that certain functional RNA molecules and proteins are intimately entangled through the agency of RNA chaperones, which are important for the efficient adoption of the correct conformations of specific RNA molecules.

What does this have to say about the practical utilities of evolved and engineered functional nucleic acids? Basically, since in many spheres (especially catalysis) proteins have superior properties, it comes down to the old aphorism, "If you can't beat 'em, join 'em!" Or at least join forces with them. Molecular design should accordingly take inspiration not just from the RNA and protein world in isolation, but from the integrated RNA–protein world as well. One way to approach this is to make nucleic acids more "protein like," which we considered from the angle of novel base pairs for expanding the nucleic acid alphabet (as in Fig. 6.10). Another is to directly combine proteins and nucleic acids (specifically RNAs) in the form of ribonucleoprotein complexes in an analogous manner to those found with natural ribozymes, from which much has been learned. It should be noted that proteins engaged in promoting or modulating ribozyme activities include those providing chaperone-like assistance. Exemplifying the latter case, chaperone-associated enhancement of hammerhead ribozyme activity has long been known.[989,990] On the other hand, specific proteins are intimately associated with the activities of many natural ribozymes in a more direct fashion, and those within the RNase P ribonucleoprotein complex have received intensive study. RNase P-associated proteins thus modulate RNA substrate (pre-tRNA) recognition, stabilize the catalytic RNA conformation, and modulate metal ion binding.[988] Interestingly, RNase P proteins also increase the versatility of the catalytic RNA component of RNase P in evolving toward altered substrate recognition, as shown by *in vitro* evolutionary experiments.[991,992,993]

It is clear from such natural lessons that proteins have much to offer for the extension of functional RNA activities. Early work in this area showed that the activity of a hairpin ribozyme was compatible with artificial incorporation of a recognition site for a foreign RNA-binding protein, and complexing with the protein itself.[994] The evolution of functional ribonucleotide complexes may have evolved by the "takeover" of certain RNA–RNA interactions by RNA-binding proteins, and this has been successfully modeled in natural ribozymes by judicious insertion of specific foreign RNA-binding motifs for known RNA-interactive proteins.[995,996] Such work sets the stage for the development of ribonucleoprotein ribozymes with a variety of possible additional protein-based *in situ* activities or regulatory opportunities.[997] We have seen, though, that there is extensive evidence that nucleic acids themselves can function as molecular recognition elements for regulatory purposes. Consequently, while proteins can certainly mediate molecular switching events, DNA or RNA molecules greatly extend the "recognition

---

*"Modern" as within the last three billion years or so.

toolbox" in this regard, and can signal the binding of specific proteins or peptides by allosteric conformational changes.

Engineering or evolving a sizable functional ribonucleoprotein is performed most efficiently if the major functions of the RNA and protein components are taken into account. A broad breakdown of ribonucleoproteins into three categories has been proposed[998]: those whose structures are RNA-derived, those with largely protein-derived structures, and those where the RNA acts as a flexible scaffold upon which multiple proteins can be associated. In the latter category, there is no specific structure for the overall ribonucleoprotein, and the prototypical example of this type of RNA–protein complex is the eukaryotic telomerase enzyme,[999] which maintains the repeated structures at the ends of chromosomes (telomeres). This kind of flexible nucleic acid scaffold may find artificial application in the construction of synthetic higher order macromolecular complexes.

Only a few decades ago, the entire theme of this chapter was an unmined lode of knowledge, the extent of whose riches were unsuspected even when the existence of ribozymes was first revealed. Yet as the world of functional nucleic acids has come under the spotlight, so has the field of protein-based molecular recognition advanced dramatically, and certainly a majority of the recognition-based topics raised in this chapter are not the exclusive preserve of nucleic acids. So it is appropriate that we should look at the development of protein-based molecular recognition in more detail. This is the role of the succeeding chapter, where antibodies inevitably have a high profile.

7 Evolving and Creating Recognition Molecules

OVERVIEW

If we subtract molecular interactions and recognition from our thinking about functional molecules of value to humans, there is little left to deal with. Yes, we could think of exceptions, such as sensor molecules responding to photons of an appropriate wavelength, or to some other stimulus that does not require a signal from another specific molecule or ionic group. Almost always, though, a recognition event of some kind is the basis of a useful molecular function, ranging across the huge gamut of all catalytic, small molecule–protein, protein–protein, protein–nucleic acid, small molecule–nucleic acid, and internucleic acid interactions. While we have already seen many examples of these, there are still numerous areas we have not yet uncovered. We have, of course, already taken note of antibodies and the Darwinian processes by which they are naturally derived and seen that nucleic acid-based recognition can rival that of proteins in many respects. Yet antibodies (and many variants of them) remain the prime exemplars of "made to order" molecular recognition units for the majority of applications at present. We should then spend a little more time looking at this general field, and indeed many aspects of these technologies are of great interest in their own right. Also, it is important to present more detail on the world of antibody-based recognition, if only to better compare it with the increasing range of alternatives. Aptamers are a significant subset of these, but they are certainly not the only ones. First, there are several alternate protein frameworks for variable combining sites that have been mooted as competitive with (if not superior to) antibodies in certain aspects. Then, there is the completely nonbiological approach of molecular imprinting, whose full potentialities remain to be plumbed. Of course, we do not have to insist on a monolithic one-size-fits-all solution to molecular recognition, and various "alternatives" may find different specialist niches where they excel.

Where does this chapter stand in relation to our overarching themes of empirical versus rational molecular design? While the earlier chapters were intended to highlight natural evolutionary processes and "blind" molecular acquisition, many levels of rational input were often involved. As one example among many, sophisticated chemistry can be applied toward the development of unnatural nucleic acids, which in turn then be used for *in vitro* evolutionary selection of novel functions. Natural or artificial evolutionary methods have also provided springboards for higher levels of design that are rational in their application. In some circumstances, this process has continued to the point where the rational and empirical strategies are often closely intermingled, and the history of antibody technology development is a good case in this regard. The interplay between empirical and rational design, which we

Searching for Molecular Solutions: Empirical Discovery and Its Future. By Ian S. Dunn
Copyright © 2010 John Wiley & Sons, Inc.

have noted already, will thus be even more prominent in this chapter. Initially, let us revisit the power of the immune system as a paradigm of empirical selection for human application.

THE IMMUNE SYSTEM IN A BOTTLE AND ON A LEASH

There is still a great deal to be learnt about the amazing vertebrate adaptive immune system, whose full complexity (even as presently understood) could only be hinted at in Chapter 3. But the salient underlying principles of both the antibody and T cell systems are understood, and a central recurring message is diversification, selection, and amplification. Before this knowledge was gleaned, it was not possible to think of "taming" the immune system, even in principle, for the generation of antibodies remained utterly mysterious. (The nature of the T cell receptor was not even defined until less than 30 years ago.) This ignorance did not, of course, prevent the wide-scale application of antibodies for many decades. For this, animals (including horses, sheep, and goats) were "black boxes" where antigens were the input and antibodies the output, with the intervening processes shrouded in uncertainty. Long before the understanding of antibody generation mechanisms, much empirical information was gained as to the best way of arranging the input (adjuvants and the nature of antigen presentation) and output (for example, optimization of times for obtaining sera) in order to maximize the amounts and specificity of the desired antibody.

From this early stage, the current picture is completely different, thanks to great strides in fundamental knowledge and steady advances in both cellular and molecular technologies. Ultimately, a complete and self-contained "artificial immune system"[1000] is desired to maximize control over the generation of immune recognition molecules. Although the reality still falls short of an ideal system, much has been achieved. The T cell receptor has been developed for certain useful applications, chiefly for the augmentation or generation of specific cytotoxicity against cellular targets (especially tumors; ftp site). But by far the biggest effort in the harnessing of the mammalian immune system has been focused on antibodies, and our focus will be in this area too.

Decades of antibody technology development around the world has resulted in impressive progress in a variety of systems, but this superficial diversity is underpinned by a relatively limited number of core advances. Among these, hybridoma (monoclonal antibody) and display systems feature as being of the highest importance for the generation and selection of immunoglobulins directed against predesignated antigens. Taking antibodies all the way from the identification of a suitable target molecule to high-level commercial production is a long road, especially if the antibody product has a clinical role. It should not be surprising that many different technologies are required to enable such scaled-up production to become a reality, and details of many such procedures are beyond the scope of this book. Our interest is primarily concerned with the processes whereby immunoglobulins with useful combining specificities are obtained, the nature of artificial antibody-based recognition molecules derived through the immune system, and certain aspects of the deployment of such recognition molecules. It will be most logical to introduce antibody technology by considering the pathways for artificially manipulating the evolution and selection of desired immunoglobulin combining sites. And the broadest demarcation here is *in vivo* and *in vitro*.

New Wrinkles on the Old Way to Antibodies

The time-honored *in vivo* way of making antibodies allows natural immune systems to do the work of generating novel combining specificities of human utility. Yet even an absolute reliance on natural immunological diversification and selection mechanisms does not preclude performing useful manipulations *after* such processes have resulted in the appearance of B cells with the appropriate rearranged and somatically mutated immunoglobulin genes.

Because a natural immune response (even toward an antigen of restricted size) will involve the activation of multiple independent B cell clones, naturally formed antibodies are termed "polyclonal." A major innovation in antibody technology (since 1975) has been the invention of *monoclonal* antibodies, where antibody-producing cells derived *in vivo* are clonally immortalized during *in vitro* culture.[1001] This is based on the ability to fuse antibody-producing cells (B cell-derived plasma cells[*]) with special immortal myeloma cell lines that allow selection of cellular fusions[†] (hybridomas). Each resulting hybridoma clone secretes large amounts of antibodies with the same variable region combining site as expressed by the original antibody-producing fusion partner cell. Since appropriately immunized animals can act as donors for the antibody-producing cells, selected hybridoma clones can have "predefined specificity."[1001] By this process, a plasma cell secreting the correct antibody is in essence plucked from the *in vivo* milieu, immortalized by cell fusion and cloned. Natural GOD and selection *in vivo*, combined with cellular immortalization and screening for desired antibody specificities, are thereby fruitfully combined for the generation of antibodies of monoclonal origin *in vitro*.

These monospecific immunoglobulins have long been highly useful reagents for many diagnostic and research applications *in vitro*. But in the minds of many researchers working with monoclonal antibodies, the image of antibodies as "magic bullets" for clinical applications was compelling, and in the early days of this technology hopes were very high.[1002] The chief limitation and disappointment for these aspirations derived from the requirement for murine cells for the initial monoclonal antibody technology. Here, the problem was not mice themselves, but the simple fact that to a human, a mouse protein is much more foreign than a corresponding human protein. In turn, any internal treatment of patients with mouse immunoglobulins generates robust anti-immunoglobulin responses,[1003] which tend to block potential therapeutic action and directly trigger adverse effects. As a classic self/nonself issue, any other mammalian immunoglobulin but human would be expected to elicit similar effects to proteins from mice, and the replacement of murine monoclonals with a human equivalent was therefore called for.

In terms of general hybridoma technologies, two broad strategies here can be contemplated. One could take monoclonal antibodies conveniently generated through the standard murine hybridoma technology and alter them to become less immunogenic in humans. The alternative is to establish a fully human monoclonal antibody system. It might be thought that the latter is the obvious route to take, but technical difficulties have limited its application until relatively recent times. Fusion of human lymphocytes with mouse myeloma cells to form xenogeneic hybridomas allowed monoclonal human antibody production,[1004] but such hybrids were of low durability. Fully human hybridomas were shown to be feasible,[1005,1006] yet still with unsatisfactory stability. For human systems, a problem also exists at the stage of antibody evolution and selection. Mice can be immunized with any antigens for the raising of antibodies of desired specificities, but this is not usually an ethical option in humans. In principle, the antibody selection itself could be performed using an *in vitro* cellular system that mimics the entire *in vivo* process (*ex vivo* immunization), but this is rarely a practical option, especially for

---

[*]Activated B cells with mature immunoglobulin antigen receptors differentiate into "plasma cells," which secrete large amounts of the corresponding soluble immunoglobulin, rather than express it on the cell surface.[258]

[†]A myeloma (or plasmacytoma) is a tumor derived from antibody-producing plasma cells. Myeloma cell lines used for generating hybridomas have mutations such that they die in special culture media. This defect is overcome in hybridomas by genetic complementation from the fusion partner antibody-producing cell, and thus hybridomas can be positively selected in the special medium. Georges Köhler and Cesar Milstein were awarded a Nobel Prize in 1984 for ushering in the age of monoclonal antibodies.

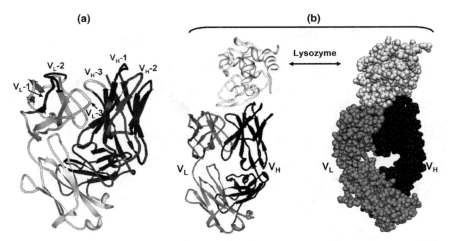

FIGURE 7.1 Structures of antibodies, represented by Fab fragments (monovalent antigen-binding fragments of whole immunoglobulins; Fig. 7.2) from a monoclonal antibody (D44.1) against hen egg lysozyme.[40] (a) Isolated Fab fragment light chains (light gray) and heavy chains (dark gray), with the complementarity-determining regions (CDRs) as shown (V_H-1 = heavy chain CDR1, etc.). (b) Same antibody Fab fragment complexed with its antigen lysozyme (same shading scheme for heavy and light chains; lysozyme white), on left, ribbon backbone image, on right, space-filling view. *Source*: Protein Data Bank[2] (http://www.pdb.org) 1MLB (a); 1MLC (b). Images generated with Protein Workshop[5] (ribbon diagrams) and PyMol (space filling; DeLano Scientific).

the generation of high-affinity secondary antibodies. In any case, an *in vitro* option is indeed available, at the level of selection directly from antibody libraries, as we will explore in more detail shortly. Before doing so, let us look again at the first alternative above, whose aim is the transformation of murine antibodies into a nonimmunogenic state for humans. Then, we should consider recent solutions to the human hybridoma problem.

As we noted in Chapter 3, the antigen-binding function of immunoglobulins is centered on the hypervariable complementarity-determining regions in the light and heavy chain variable regions. While the remainder of an immunoglobulin molecule has important biological functions, it can be dispensed with as far as antigen recognition goes. Fig. 7.1 shows the structure of antibody directed against the antibacterial protein lysozyme, (alone and complexed with the lysozyme antigen), with the complementarity-determining regions (CDRs) indicated. So, parts of a mouse immunoglobulin can be replaced with human equivalents without destroying its ability to specifically bind its original antigen target. This process has been termed "humanization" and has gone through a series of stages from initial replacement of the constant regions to more sophisticated "grafting" of mouse CDRs for a desired antigenic specificity onto a human immunoglobulin background (Fig. 7.2).[1007]

These strategies largely, but not entirely, eliminate the antimouse immune response in an antibody recipient. The reason the dampening of the unwanted immune recognition of such therapeutic antibodies is not complete lies in the foreign nature of the complementarity-determining regions themselves.* We can recall from Chapter 3 that anti-idiotypic responses

---

*It must also be recalled (from Chapter 3) that antibody generation against most antigens requires T cell help, and T cell epitopes can be identified in immunoglobulins.[1008] Removal of potential T cell epitopes from monoclonal antibodies is another therapeutic option in conjunction with humanization.[1009]

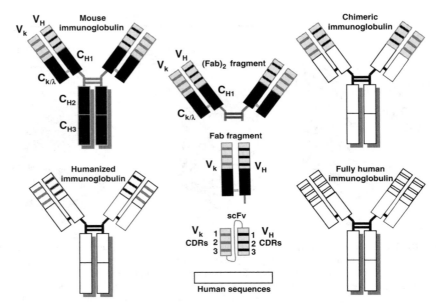

FIGURE 7.2 Schematic depictions of an IgG-type immunoglobulin with a V_k light chain, and certain fragments and altered derivatives of it. A mouse monoclonal antibody can be rendered decreasingly immunogenic by successively replacing its constant regions (chimeric immunoglobulin) and then variable region framework sequences (retaining the original selected CDRs) with appropriate human sequences ("humanization"). Finally, a fully human immunoglobulin selected for the same combining specificity lacks any mouse sequences. The structures of divalent antigen-binding (Fab)$_2$ fragment, monovalent Fab fragment, and single-chain Fv fragments are also schematically depicted. Note that the latter is connected by a flexible linker from the C-terminus of one of the V regions to the N-terminus of the other, such that the two V regions can align in the correct manner. Straight lines between heavy and light chains represent disulfide bonds.

can be mounted against somatically generated regions of immunoglobulin variable regions that are not registered as "self" by a host organism. Responses against the mouse-derived CDRs are thus predictable, do occur, and may be impossible to remove absolutely. It is feasible, though, to reduce this effect by further dissecting CDRs in terms of the key residues making contact with antigen. With such knowledge (based on detailed structural information), "abbreviated" CDRs can be generated or ultimately broken down into the "specificity-determining residues" that constitute the minimal sequence requirements for antigen recognition.[1010]

These advances aside, the aim of practical production of fully human monoclonal antibodies has been best addressed through advances in transgenic technologies. If large genomic segments representing the entire human light and heavy chain germline gene repertoires were transferred to the genome of another mammal (and the endogenous immunoglobulin locus removed or inactivated), in principle human antibodies could be produced by the host animal through the usual selection and affinity-maturation processes.[*] This has been indeed accomplished by means of immunoglobulin transgenesis of mice[1012] and cattle,[1013] using artificial

---

[*]Note that this is quite distinguishable from transgenic systems for expression of a *single* preselected rearranged and somatically mutated immunogloblin. The latter kind of production has certain advantages over bacterial systems, and an example is the expression of human antibodies in chicken egg whites.[1011]

chromosome transfer in the latter case. A "Xeno-mouse" equipped with a human immunoglobulin repertoire can then be used as a donor of antibody-secreting cells for the creation of monoclonal antibodies in the conventional manner, with the happy difference that the antibodies will be human, even though produced by murine cells. This has led to the regulatory approval in 2006 of the first fully human monoclonal antibody derived from a mouse transgenic system.[*,1012,1014]

The great strength of monoclonal antibodies is the precise definition of their antigenic targets, but monospecificity in an antibody therapeutic can also be a weakness. Both tumor and parasite targets are adept at losing or changing antigens by which they are flagged by the immune system, and loss of a target antigen will render useless antibodies exclusively binding such a target.[1015] On the other hand, complex targets composed of many proteins and other potentially antigenic molecules offer multiple epitopes for immune attack, and a polyclonal response will therefore not fail due to loss of any single epitope. Large animals transgenic for human immunoglobulin loci offer the promise of the direct production of human polyclonal antibodies in response to any desired antigenic stimulus.[1013,1016] Another approach for polyclonal production with human antibodies is to use *in vivo* antibody specificities from humans themselves. Although (as noted above) the use of human immune cells raises ethical issues, many individuals possess useful high-affinity antibodies against pathogenic or other targets of interest through natural immunization. Cloning and expressing panels of such antibodies (against specific complex targets) in mammalian cultured cells allow an artificial polyclonal to be harnessed for therapeutic application.[1017]

We have already paid heed in Chapter 3 to the "affinity ceiling" observed with naturally formed antibodies, and noted that artificial directed evolution *in vitro* has been able to far exceed apparent natural limits on antibody-binding affinities.[281,1018] This kind of augmentation of the affinity of antibodies already matured *in vivo* has also been attainable by means of rational design strategies.[†,1019] But *in vitro* selection for antibodies with desirable binding properties does not require naturally matured antibodies as a starting point, which leads us on to further thoughts regarding the power of *in vitro* evolution and selection in an immunoglobulin context.

Antibodies and Libraries

The advent of recombinant DNA technologies allowed the characterization of antibody genes and an understanding of the means for their somatic diversification. An *in vitro* library of diversified heavy and light chain genes in principle could then be used as an "immune system in a bottle" for the expression and selection of antibodies with a desired antigenic specificity. For early work toward this goal, the polymerase chain reaction (PCR) was a key enabling technology. Both heavy and light chain mRNAs have relatively conserved regions that flank the hypervariable CDRs, allowing common sets of primers to amplify whole arrays of naturally diversified immunoglobulin variable region (V) genes. At this point we should consider that harvesting and expressing a library of V regions allows the artificial *in vitro* selection of antibodies, where V-gene diversification is produced entirely through the normal natural channels. But in addition to this, the generation of V-gene diversity itself can also be artificially manipulated *in vitro*, as we will come to shortly. As we have also seen with other *in vitro* empirical approaches, including SELEX (Chapter 6), we can distinguish "pure" selection from a library in a "single pot" experiment from true evolutionary progression involving multiple

---

*This antibody, "panitumumab," has specificity against epidermal growth factor receptor, for therapy of colon cancer.
†This may be cited as another example of interplay between empirical and rational protein design, where natural empirical selection provides the initial antibodies, and rational design does the rest.

rounds of diversification and selection toward a molecular optimum. Both "pure" selection and *bona fide* directed evolution are applicable to antibody library technology *in vitro*.

Initial work showed that it was feasible to express amplified V-gene libraries as Fab fragments (monovalent antigen-binding fragments, as in Fig. 7.2) in *E. coli* using a bacteriophage vector.[1020] In these pioneering experiments, Fab fragments were expressed within plaques of the phage (lambda) vector, requiring tedious screening methodologies, yet successful in demonstrating that individual antibody specificities could be isolated from the library.[1020] The next major technological development, following closely on the heels of this 1989 study, enabled antibody technology to really come into its own. It was found possible to display assembled Fab fragments on the surface of filamentous phage, either via the gene III[1021] or via gene VIII products[1022] (refer back to Chapter 4 for more details on these phage display systems). This linkage of phenotype and genotype allowed rapid selection rather than screening for a desired antibody specificity, and enormously facilitated the efficacy of the technology as a whole. We might recall from Chapter 3 that a B cell with a surface antibody directed against a specific antigen will be selected for proliferation (almost always) via T cell help (Fig. 3.2). In the case of phage antibody libraries, the selection is via binding under the aegis of the experimenter, and proliferation simply corresponds to the growth of the selected phage in its bacterial host. Libraries of greater than 10^{10} separate clones have been generated and widely used.[1023]

Although isolated heavy chain variable regions can sometimes show appreciable antigen recognition,[1024] usually high-affinity antigen binding requires interactive residues from both heavy and light chains that are supplied by the use of Fab fragments[1020] (Fig. 7.2). A simpler alternative to the latter antibody "pieces" exists in the form of *single chain variable region fragment* (scFv) molecules, which have been extensively used for phage display libraries. Here variable regions of light and heavy chains are covalently joined by means of a flexible linker segment (shown schematically in Fig. 7.2 and structurally in Fig. 7.3). While it is very convenient for an antigen-binding function to exist in the form of a single contiguous polypeptide, for some applications Fab fragments are nonetheless preferred owing to their tendency to avoid forming dimers, which scFv fragments can often manifest.[1025,1026] A compromise position with single-chain Fab fragments (scFab) has recently been introduced.[1027]

Antibody libraries whose diversities derive from recapitulation of natural heavy and light chain repertoires can be broadly divided into two categories. A library can be derived from donor animals that have been actively immunized against antigen(s) of interest, or one can alternatively use nonimmunized or "naïve" sources for library construction. Expressed germline V-gene libraries that have not been actively biased and shaped by antigen exposure have a much higher theoretical chance for containing specificities against self,* or antigens of low immunogenicity or toxicity.[1028] Naïve antibody libraries are typically prepared with heavy and light chains from IgM$^+$ B cells (bearing surface immunoglobulin M) that have not undergone further maturation (including class-switching to other immunoglobulin isotypes). Although the IgM compartment is certainly the right place to look for truly naïve antibodies (which bear germline (albeit rearranged) sequences), in fact it is naïve to imagine that all antibodies in this population have such characteristics. Although somatic hypermutation and affinity maturation is associated with class-switching (and therefore a feature of matured isotypes such as IgG and

---

*This point raises interesting issues of self/nonself discrimination and self-tolerance, some aspects of which were mentioned in Chapter 3. Although deletion of B cell specificities against highly expressed self antigens occurs during B cell maturation[258], some self-reactive B cells can survive into maturity. Most of these will not be activatable *in vivo* owing to the absence of T cell help, where T cells against the corresponding antigen are thymically deleted or suppressed by regulatory cells.[258]

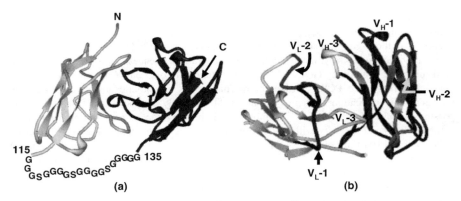

FIGURE 7.3 Structure of a specific single chain Fv construct.[41] (a) The light chain variable region segment (light gray, N-terminus (N) as shown) ends at residue 115 and is linked to the N-terminus of the heavy chain (dark gray) segment at position 135 via a serine–glycine linker $G_3(SGGGG)_4G$ as depicted (as an unstructured segment, the linker is not directly visible in X-ray diffraction images). The position of the C-terminus of the whole structure is as shown (C). (b) The same structure looking down the antigen-binding region, with CDRs for light and heavy chains as shown. *Source*: Protein Data Bank[2] (http://www.pdb.org) 2GJJ. Images generated with Protein Workshop[5] with serine–glycine linker superimposed in the left panel.

IgA classes), a distinct compartment of memory IgM cells in humans bears somatic mutations.[1029]

At an early stage of antibody library development, it was considered unlikely that a high-affinity antibody could be isolated from a library in a single "hit." By analogy with the natural adaptive immune system's secondary affinity maturation of antibody specificities by somatic hypermutation, it was anticipated that comparable mutational "tuning" of primary *in vitro* antibody isolates would be routinely required.[1030] Certainly there is a place for the *in vitro* correlate of affinity maturation in antibody technology, but repeated studies have shown the feasibility of "pulling out" high affinity antibodies from libraries directly, even from naïve libraries.[1028,1031] Interestingly, and consistent with what one might predict, the affinity of antibodies selected from libraries is proportional to library size.[1023] In other words, the chances of finding an antibody clone with high-affinity binding for an antigen of interest increase with the size of the antibody library.

A converse strategy to naïve libraries is the use of "immune libraries" where the donor V genes are obtained from animals deliberately immunized with antigen(s) of interest, or humans known to be naturally exposed to pathogens or other antigenic sources of relevance. If one can find an enormous range of binding specificities in a naïve antibody library of sufficient size, what then is the advantage of using a library with a predetermined antigenic bias? If an experimenter's primary interest is a specific pathogenic organism, an immune library will provide a focus in the desired direction. But beyond this, antibodies generated *in vivo* may reveal the scope of pathogen epitopes targeted by naturally formed antibodies,[1032] or uncover special (and clinically relevant) antigenic structures that might easily escape detection from probing a naïve library. An example in this regard is HIV, where an epitope on the viral protein gp120 is induced through interaction with certain host proteins,[1033] discovered using libraries from humans exposed to this virus. Immune libraries can also be effective tools for studying the nature of autoimmune antibody responses.[1034]

Aside from the natural processes for antibody GOD, antibody engineers have been very good at introducing antibody diversity themselves within antibody libraries *in vitro* (Fig. 7.4).

The V-gene diversity can be introduced on a completely synthetic basis, with a wide range of options to choose from. These can involve complete random mutagenesis of all or specific CDRs,[1023] or the targeting of sites known to be hotspots for somatic hypermutation.[1035] As well as mutational changes, varying the lengths of CDR coding sequences in libraries has also been used for artificial antibody GOD,[1036,1037] and mutations throughout V regions can be introduced by error-prone PCR, considered in Chapter 4.[337] Synthetic and natural V-region mutagenic diversification strategies are not at all mutually exclusive and can have many possible permutations.[1023,1038,1039] Combinatorial shuffling can be superimposed on initial natural or artificial diversity within V regions themselves ("intrachain," Fig. 7.4), either by PCR-mediated shuffling of diverse CDRs[1040] or through conventional DNA shuffling using panels of related V regions.[1041,1042] Also, since both heavy and light chains contribute to antigen recognition, binding site diversity can also be generated by *chain shuffling* (Fig. 7.4), equivalent to combinatorial reassortment of large collections of V_H and V_L regions.[1043]

In the context of antibody libraries, another factor is of considerable practical importance. Thus far, we have spoken of the CDRs without reference to the relatively conserved framework regions that are interspersed between them. Of course, the V-region β-sheet structural framework (Chapter 3) is essential to configure and display the CDRs for antigen binding itself (as exemplified by Figs 7.1 and 7.2). Although these V-region heavy and light chain "scaffolds"

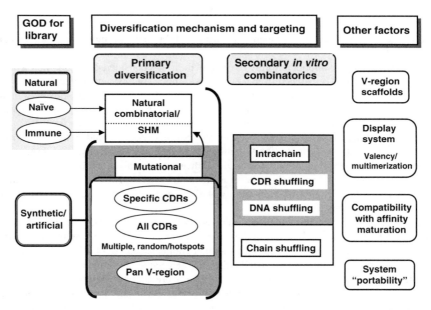

FIGURE 7.4 Features of antibody libraries, including the source of generation of diversity (GOD), how it is applied, and other relevant factors. The "diversification mechanism and targeting" heading has been subdivided into the primary source of diversification (whether natural, using combinatorial rearrangements and somatic hypermutation [SHM]), as opposed to additional ("secondary") combinatorial diversification procedures that can be subsequently applied *in vitro*. (Note, though, that DNA shuffling approaches can combine additional mutagenesis with combinatorial shuffling, as outlined in Chapter 4). The curved arrow indicates that the use of natural and synthetic mutational diversity can be combined as desired. "Other Factors" (as shown) can influence the efficiency and utility of the library at several different levels. "Portability" refers to the ease with which a library can be converted into another display or expression format.

have much less variability than the CDRs, the scaffold regions show sequence differences that assist their grouping into a number of separate V-gene families.[1044,1045] Libraries whose diversity derives from natural sources have often used the available natural variable region scaffolds, but as always the antibody engineer need not be constrained by this and can be as "unnatural" as desired. From a biotechnological viewpoint at least, it is clear that not all antibody scaffolds are created equal, with some germline families showing significant inferiority in terms of expression levels, solubility, and folding.[1046] On the basis of performance and frequencies of natural utilization, rational choices of consensus scaffold family sequences can be assigned for expression of libraries in a particular host (such as *E. coli*).[1047] DNA shuffling of human framework regions has been used to select for antigen-binding retention in the context of murine CDRs, as an alternate humanization approach.[1048,1049] Yet for consistency and control over the behavior of selected antibodies from libraries, it is often desirable to limit the range of variable region scaffolds used. Using synthetic or semisynthetic diversification (Fig. 7.4), it has been found possible to isolate high-affinity specific antibodies with libraries based on a single chosen scaffold.[1037,1050]

All of the antibody library features described in Fig. 7.4 are directed toward not only obtaining a desired antibody specificity, but also ensuring that such an antibody has a useful binding affinity. Previously, we have noted that in natural biosystems, the optimal affinity for any protein–ligand interaction will evolve in response to numerous selective pressures, and often will be much less than an absolute theoretical maximum. (Hence, the *in vivo* antibody "affinity ceiling" and its surpassing in the laboratory, which we have revisited earlier in this chapter.) Very often, there is indeed a direct correlation between antibody affinity and practical benefit for human application, but even in this artificial domain, the "benefit" component in an affinity-benefit relationship can reach a plateau before the attainment of final affinity maxima. It has been pointed out, for example, that antitumor antibodies of very high affinity may have inferior tumor-penetrating characteristics.[1051,1052] Even so, antibody affinity is a very important yardstick for most developmental purposes. We noted above the comparisons between the two-stage natural evolutionary acquisition of high affinity antibodies and their counterparts derived from *in vitro* libraries through human agency. In the latter case, well-documented "single-pot" isolations of high affinity antibodies exist, even from naïve libraries as we observed earlier. While acknowledging that the direct isolation of high-affinity antibodies from naïve libraries is possible, an important theoretical issue is its generalizability. In turn, this raises the question of the practical significance of artificial affinity maturation, as opposed to the well-known importance of the *in vivo* process. A simple answer is that promising primary antibody isolates from libraries often need or benefit from further manipulation of their affinities for antigen,[1023] and *in vitro* evolutionary methods are a productive route toward this goal.

Some further considerations (or at least revisitation) of antibody diversification seems warranted at this point. If we think about the theoretical extent of antibody library diversity, we soon run up against the familiar numbers problem. Even if we restrict our calculations to the third complementarity-determining region (CDR3) only,[*] we can readily deduce that it is impossible to create a single library bearing *all possible* CDR3 combinatorial sequence variants for a $V_H:V_L$ *in vitro* library, if 10^{11} clones constitute an approximate practicable library upper size limit.[†] Yet these kinds of calculations are quite misleading, since this CDR3 maximal-diversity

---

[*]This is not an entirely unreasonable restriction, given the general importance of CDR3 for antigen recognition, especially as contributed by the heavy chain.[269] On the other hand, contributions of other CDRs for both chains cannot be ignored, especially when deliberately diversified *in vitro*.

[†]Although the length can vary, if we take V_H and V_L CDR3 tracts as 7 and 9 amino acid residues, respectively,[1053] then all possible sequences are $20^7 \times 20^9$ (assuming 20 amino acid alphabet variation at each CDR3 position, in conjunction with chain shuffling combinatorics), or about 6.6×10^{20}.

figure is already orders of magnitude above upper-level estimates of *in vivo* antibody diversity of $\sim 10^{16}$ (Chapter 3). Natural combinatorics, junctional diversity, and somatic hypermutation do not produce a "library" of all possible CDR sequence combinations. Recall from Chapter 3 that the number of circulating lymphocytes in an animal with an adaptive immune system constitutes an upper "library" boundary at any one time, which is far below the theoretical maximal antibody diversity *in vivo*. Many sequences *in vivo* also are "forbidden" for reasons of self-reactivity, and certain sequences may be incompatible with protein folding or expression.

In Chapter 3 we also considered the concept of shape space, and the deduction that a mere 10^8 discrete antibody-binding specificities could provide "coverage" of this hypothetical construct.[124,276] Or in other words, the "shape space" model proposed that 100 million antibodies would suffice for all possible structures that might be encountered. Aside from the overly simplistic view of the shape space model of the realities of antigen-antibody interactions,[278] how does the prediction of 10^8 specificities measure up in the real world? While experimental results with real *in vitro* libraries suggest that 10^8 clones cannot routinely yield high-affinity antibodies to a chosen antigen,[1023] they do indicate that larger libraries of still relatively modest sizes (on the order of 10^{10}) have generalizable utility. ("Modest" is used here in comparison to the monstrous levels for complete CDR diversification obtained through blind calculations as above.) We should remember too, especially in the context of naïve libraries, that germline antibodies are believed to have inherently greater conformational flexibility than affinity-matured antibodies *in vivo* (as discussed in Chapter 3). With increased flexibility comes an increased propensity for cross-reactivity, but with the accompanying benefit of an increased likelihood for primary "hits" during challenge with a foreign antigen. "Polyspecificity" of antibodies (and T cell receptors) may be an inherent feature of the primary immune repertoire.[1054] Primary antibody candidates can then be "tuned" to higher levels of affinity and specificity through affinity maturation, the second tier of natural adaptive immunity. The power of the latter 2-stage evolutionary approach is illustrated by the ability of 10^8 B cells in the murine *in vivo* immune system (at any one time) to at least equal the specificity responses of naïve *in vitro* antibody libraries up to three orders of magnitude larger.[1055]

Clear-cut analogies can be made between the natural processes of antibody combining site refinement and *in vitro* honing of antibody affinities. Although the latter artificial process has been targeted to specific CDR sites based on rational considerations or previous information concerning mutational hotspots,[261,1056] empirical *in vitro* display-based methods have been widely used. Moreover, there is evidence that CDRs, and even an additional conserved loop within a V_H framework region, tolerate more diversity from a purely structural point of view than natural antibodies would suggest.[1057] In other words, room for improvement over the natural repertoire is available by unnatural tinkering with immunoglobulin V-region CDR loops. For a completely empirical approach to affinity optimization, however, targeting of CDRs can be by-passed in favor of global V-region mutagenesis combined with suitable display and selection procedures, with the same end result of affinity improvement.[1058] With respect to the selection system for affinity adjustment, ribosome display is efficient and may prove advantageous over traditional phage display.[1058,1059] Likewise, mRNA display is a valuable alternative for such purposes.[1060] Evolutionary tuning of antibody affinities can eventually deliver "ultrapotent" antibodies of potential therapeutic value in some circumstances.[1061]

For *in vitro* molecular libraries of any description, the approach to diversification is a key issue, as we have seen repeatedly for directed evolution in general. Itself rapidly evolving over the last two decades, the scope of artificial antibody *in vitro* diversification is so large that for our purposes only a brief overview has been possible. Given this technological diversity, once again the question arises for whether there is a clear-cut winner in the approach to library antibody GOD. General comparisons based on reports emerging from the field as a whole suggest that diversification (whether natural or synthetic) in multiple CDRs is an effective strategy for antibody leads with good affinities.[1023] Ruminating about optimal strategies for antibody diversification might have the side effect of causing us to wonder about the use of antibodies *per se*. Even before one chooses the best way to make an antibody, should not an observer step back to the more fundamental question as to whether an antibody-based strategy is the best way to go in the first place? After all, in the last chapter we spent some time looking at the world of aptamers and their considerable promise. There is indeed a lot more to say about potential antibody alternatives, and this we will come to a little later in this chapter. Before this, though, there are still some important aspects of antibodies and antibody-related proteins that we should examine more closely.

The GOD of Small Things: Minimizing Antibodies

A trend toward streamlining antibody proteins can be perceived from early days of their structural characterization, from divalent $(Fab)_2$ to monovalent Fab fragments, and then isolated variable region segments (Fig. 7.2). As noted, there are certain conveniences attached to the manipulation of smaller (and especially single-chain) molecules, but the size of normal immunoglobulins was a major motivating factor toward "minimization" in its own right. Bivalent antibody molecules such as IgG have molecular weights of $\sim$150 kDa, and certain potential epitopes on proteins can be excluded as recognition elements simply on the basis of structural inaccessibility. Eliminating regions of immunoglobulins irrelevant for antigen binding also reduce the chances of unwanted superfluous side effects. So for many applications (not least of which are those in the therapeutic domain), a protein with the antigen recognition properties of an antibody but greatly reduced size is highly desirable.

Various small antibody fragments have been made, proceeding from the scFv fragments as noted in Fig. 7.2 (further detailed in ftp site). But for the most interesting developments, we can ask the seemingly bizarre question, "what do sharks and camels have in common?" There are actually plenty of things, if one descends several biological levels, but let us make the question more meaningful and specific by attaching the phrase "in an immunobiotechnological context" at the end. As it turns out, both of these animals, widely separated in evolutionary terms, have evolved functional single-domain antibodies based on isolated heavy chain variable regions that do not associate with light chains. In both cases, these function as homodimers with accompanying C-terminal constant regions of different lengths (two C-region domains in the case of camels and related mammals,[1062] and five for the shark[1063]). The isolated monovalent heavy chain domains from these kinds of antigen receptors for both animals are only $\sim$12–15 kDa in size,[1064] less than 10% of a full bivalent immunoglobulin.

Cartilaginous sharks and related fish have immune systems with recognizable antibodies, T cell receptors and major histocompatibility complex (MHC) proteins.[1065] Yet in addition to more conventional heavy/light chain immunoglobulins, sharks have the additional intriguing homodimeric recognition molecule referred to above, which has been termed IgNAR.[1066] This signifies "immunoglobulin new antigen receptor," given that this molecule is divergent from conventional immunoglobulins (equally related to immunoglobulins as for T cell receptors) yet is still firmly within the immunoglobulin family. Shark IgNAR variable regions have an evolutionary deletion that almost eliminates the CDR2 domain[1063,1067,1068] and partially accounts for their small size.

And what of camels in the world of single-domain antigen receptors? We should more accurately use the term "camelids," since these receptors are not only restricted to camels *per se* (irrespective of hump number) but also extended to other closely related members of the camel family such as llamas.[1062] The camelid heavy chain homodimers are derived from immunoglobulin isotypes IgG2 and IgG3, both of which have deletions of the CH1 constant region domains* that enable their dimerization and secretion.[1069] The antigen-binding variable regions of camelid heavy chain homodimers have been termed VHH domains† and show divergence from shark IgNAR in having three recognizable CDRs. Yet in common with IgNAR, camelid VHH domains exhibit increased sequence variability outside CDR loops.[1062] Both shark and camel single-domain receptors have conformational similarities in CDR1 suggestive of common functional requirements for single-domain antibodies.[1063] In particular, though, IgNAR and VHH proteins both have unusually long and complex CDR3 loops.[1063,1067] The numerous parallels between shark IgNAR and camelid VHH molecules have been collectively cited as a true case of evolutionary convergence of function.[1070]

But the most striking parallel between IgNAR and VHH comes at the level of their modes of antigen recognition, and this is also one of their most interesting and potentially useful features. Broadly speaking, as we briefly noted in Chapter 3, conventional immunoglobulin recognition of antigens operates through the formation of pockets, grooves, or extended contact surfaces,[258] which can be viewed as a "concave" recognition mode. This in turn is a generalizable limitation on the range of molecular recognition for normal immunoglobulins, since it can restrict certain topological features of macromolecules from access by antigen-recognizing CDRs. Normal antibodies show variation in CDR3 length, but within limits that in fact are significantly exceeded by both the special shark and camel single-domain recognition molecules. The extended CDR3 regions of IgNAR and VHH are able to enter clefts and pockets in protein targets that are inaccessible to conventional immunoglobulins. Thus, CDR3 regions of VHH domains can penetrate into the active site of an enzyme (carbonic anhydrase) and act as a competitive inhibitor of catalysis, which has not been observed with normal antibodies against the same target.[1071] The long extended CDR3 loop constitutes the major recognition VHH structural element, and selection for specific high-affinity VHH-binding molecules tends to favor CDR3 penetration of clefts in target proteins, in order to maximize the VHH-antigen interface surface area.[1072] In contradistinction to conventional immunoglobulins, this type of binding has been termed a "convex" recognition mode.[1071,1072]

Penetration of clefts in proteins is also important for IgNAR recognition, and this is shown in Fig. 7.5 for the recognition of a target protein (the enzyme lysozyme), along with analogous VHH recognition of the same antigen. Many (although not all) VHH molecules prepared against lysozyme compete with low molecular weight chemical inhibitors for binding to the lysozyme active site.[1072] It is also instructive to compare the lysozyme recognition of both IgNAR and VHH (Fig. 7.5) with lysozyme binding by a conventional antibody (Fig. 7.1), where an extended surface is bound, and the active site cleft is not involved.

The evolutionary implications of the shark and camelid single-domain antibodies are interesting, and we will put these recognition molecules in a broader context later in this chapter. But there is another capability of antibodies of particular interest that we have previously noted in passing, and this falls at the intersection of recognition and catalysis.

Antibodies as Catalysts

In the context of the directed evolution of enzymes in Chapter 5, it was necessary to refer to the relatively recent finding that antibodies, the quintessential binding and recognition molecules, can also function as enzymes. We also briefly considered the intriguing area of the generality of antibody catalysis and comparisons of antibody-enzymes with their "normal" biological

---

*In contrast, camel IgG1 is produced by conventional heavy/light chain assembly.
†Also known as "Nanobodies®" by a commercial interest (Ablynx).

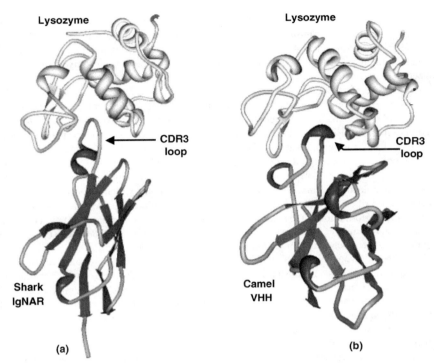

Lysozyme

Lysozyme

CDR3
loop

CDR3
loop

Shark
IgNAR

Camel
VHH

(a)

(b)

FIGURE 7.5 Comparison of crystal structures of the antigen recognition domains of shark IgNAR[42] (a) and camel (dromedary) VHH molecules.[43] (b) α-helices and β-strands dark gray; remainder (including variable loops) light gray, both complexed with a common target antigen, lysozyme (white). *Source*: Protein Data Bank[2] (http://www.pdb.org). Shark IgNAR: 2I25; Camel VHH: 1ZVY. Images generated with Protein Workshop.[5]

counterparts. Given the earlier introduction to the idea of antibodies performing enzymatic tasks, this section is the appropriate stage for examining the development of this area in a little more detail.

For many decades, similarities between enzymes and antibodies have been noted through the abilities of both types of proteins to recognize small molecules with specificity and affinity, and also the general patterns of their association–dissociation kinetics.[1073] Even so, early tests with antibodies raised against labile ester compounds (readily attacked by esterases) did not find specific antibody-mediated enhancement of the rate of hydrolysis[1074] and seemed to place antibodies and enzymes into separate broad functional realms. Later experiments with polyclonal antibodies[*] against cofactor-substrate catalytic intermediates[1073] showed a small antibody-associated reaction rate increase with the targeted substrate *in vitro*. In a subsequent study (made possible by the advent of hybridoma technology), monoclonal antibodies directed against a hapten moiety of an ester compound[1077] showed modest substrate hydrolysis rate elevation above the spontaneous background. By 1986, additional technical advances convincingly confirmed the possibility of antibody catalysis. As noted in Chapter 5, rational design of

---

[*]From initial experience, it was believed that catalytic antibodies could be not derived as a polylclonal population, but later polyclonal successes in this regard suggested that the design of the immunizing enzymatic reaction analog was the most important limiting factor.[1075,1076]

stable chemical analogs to transition-state intermediates for specific enzymatic reactions allowed the generation of monoclonal antibodies with true catalytic activities.[1078,1079] By using other transition-state intermediate analogs tailored to different types of enzymatic activities, numerous catalytic antibodies (or "abzymes"[1079]) have been derived since these pioneering experiments.[572]

In the above applications, the transition-state analogs serve as low molecular weight immunogens* (haptens) for antibody selection and maturation. The combining sites of the resulting antibodies are then predicted to recognize and bind true transition states for the reaction of interest.† Binding-induced stabilization of otherwise extremely short-lived chemical intermediates can dramatically accelerate catalytic rates, but this can occur in different ways. In some cases, the antibody combining site can act as an "entropic trap" and constrain the intermediate(s) into a conformation directly conducive to reactivity.[572] This is particularly evident for bimolecular events such as the Diels–Alder reaction,†† which we encountered in the last chapter in a ribozyme context. If the antibody provides a binding pocket that is a "snug" fit for a specific conformation of the reactants (Fig. 7.6), molecular movements that could result in alternative stereoisomer formation are minimized.[1082,1083] Catalytic antibodies can thus offer a high degree of stereochemical precision over reaction pathways.

But antibodies would be very limited in their catalytic ranges if they were solely restricted to a "grab and hold" process, for the participation of amino acid residues proximal to the binding site is critical for many cases of antibody-mediated catalysis. Directly screening a panel of candidate monoclonal antibody-secreting hybridomas for the appropriate enzyme activity will accordingly require antibodies with both transition state binding and suitably placed residues for interaction with the substrate, in order to obtain a positive signal in the first instance. Clever hapten design, though, can make a substantial difference to the prospects for success, by in effect pushing the resulting immune response toward selection of antibody combining sites that favor the desired catalytic effects. Completely reproducing all the features of a transient intermediate in a stable molecule for immunization is not feasible, and thus the catalytic antibody designer must focus on the features of the haptenic analog molecule which are most salient for catalysis.[572] The "instructions" to the immune system by hapten design are then reflected in the nature of the antibody combining site, and in turn the efficacy of the resulting catalysis when the antibody is presented with the appropriate substrate *in vitro*. One way of exploiting hapten design has been termed "bait and switch," and is based on the placement within an immunizing hapten of charged groups that are intended to result in complementary charge from an amino acid residue at a desired position within the antibody-binding site relative to the hapten itself. Thus, a hapten with a judiciously placed positively charged group can promote the selection of an antibody with a negatively charged carboxylate group in its specific combining site,** which is efficacious for catalytic events involving proton transfer.[1085–1087]

---

*As noted in Chapter 3, in order to obtain a satisfactory response against a foreign low molecular weight compound, it is necessary to conjugate it to a large macromolecular carrier molecule, usually a protein.

†It could also be predicted in advance that the original immunizing haptens should be effective inhibitors of the *in vitro* reaction catalyzed by the antibody, since of course the antibodies are initially selected for hapten-binding capability. This was quickly borne out by the initial studies.[1078,1079]

††It is particularly striking that it has been possible to generate catalytic antibodies for reactions where natural enzymes are poorly represented or uncharacterized.[1080] For a considerable time, it was considered that no natural precedents existed for enzymes catalyzing the important Diels–Alder reaction. More recently, likely natural "Diels–Alderases" have been identified in the biosynthesis of certain fungal secondary metabolites.[1081]

**The interpretation that an interactive ion pair is formed (by the introduced positive charge in the hapten and the negative charge from an amino acid side chain carboxylate group of the antibody) has received structural confirmation.[1084]

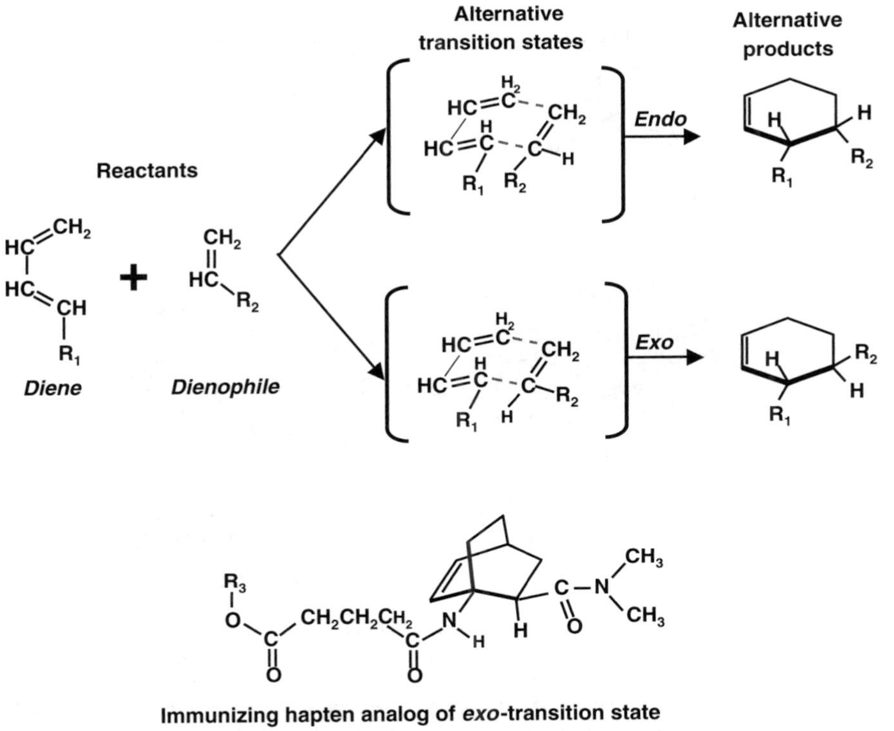

Immunizing hapten analog of *exo*-transition state

FIGURE 7.6 Transition states for alternative stereoisomers of a Diels–Alder reaction, and a specific hapten analog of the *exo*-transition state for selecting antibody combining sites favoring this product.[44,45] The bicyclic ring structure of the hapten stabilizes its structural conformation as a transition state mimic. R_1 and R_2 = molecular groups not directly involved in reaction; R_3 = chemical group for conjugating onto a carrier protein, or methyl group if used as an inhibitor of the antibody-catalyzed reaction.

A very distinct but no less interesting approach for the generation of catalytic antibodies has been to exploit the anti-idiotypic responses of the immune system. To appreciate this, we should hark back to Chapter 3 and its section on antibody idiotypes and their implications for immune networks and recognition. The central premise for the present purposes is that a primary antibody against an immunizing antigen will itself engender an immune response through the novel epitopes arising from the CDRs of its variable regions, which are "nonself" in the eyes of the immune system. The antigen-binding site second-round (or anti-idiotype) antibody may then recapitulate the original antigen as an "internal image." So, if the immunizing protein was a (nonself) enzyme, it could be speculated that if the active site of the enzyme constituted a recognition epitope, then an anti-idiotypic antibody might recapitulate the structure of the active site itself and in turn its activity (as portrayed in Fig. 7.7). This proposal has received experimental support.[1088–1090] As we also noted in Chapter 3, sequence and structural analysis of anti-idiotypic "internal images" has shown that the molecular mimicry is topological rather that at the linear sequence level. This applies equally to anti-idiotypic antibodies recapitulating enzyme active sites, and topological mimicry may therefore be insufficient to replicate enzymatic activity unless key residues for promoting catalysis are preserved in the antibody active site "image."[1089] By its nature, this general approach is limited through the ability of an

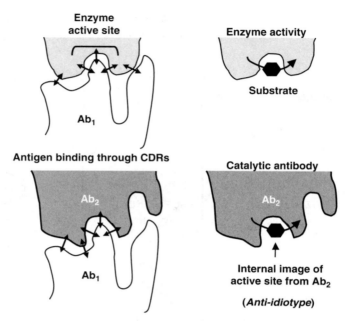

FIGURE 7.7   Depiction of generation of catalytic antibodies by means of anti-idiotypic response to an enzyme active site. If a primary antibody response (Ab$_1$) against an enzyme recognizes the active site, an anti-idiotypic response against the first antibody (Ab$_2$) may recapitulate the shape and activity of the original active site pocket. Small double-headed arrows depict contacts at antibody-binding sites; similar but not identical for the original enzyme and Ab$_1$ as for Ab$_1$ and the anti-idiotype antibody Ab$_2$.

antibody to recognize an enzyme active site, which in many cases will be in a relatively inaccessible cleft within the enzyme molecule. (This is potentially less of a problem for small single-domain antibodies considered above, which might therefore have potential applicability for the anti-idiotypic catalytic antibody strategy.)

Catalytic antibodies are an excellent example of the productive interplay between very sophisticated rational design and empirical selection procedures. We have seen that high levels of understanding of specific reaction chemistries are needed for effective haptenic design, but natural evolutionary selection of the immune system is required for the isolation of a protein with a binding site for the immunization hapten within which catalysis can occur. This is not to say, of course, that rational (or "semirational") improvements cannot in principle be further applied after an initial catalytic antibody is isolated, by directed mutageneses within the binding site region.[572] Likewise, systematic studies of the relative importance of specific residues have been very informative for the general understanding of antibody catalysis.[572,1091]

But thus far, in terms of the empirical selection side of the process by which "abzymes" are acquired, we have been focusing on *in vivo* immunization. From much previous discussion on antibody selection from libraries, it should not be a surprise to find that *in vitro* display-based evolutionary selection processes are applicable to antibody catalysis. It is accordingly possible to use such *in vitro* approaches to further evolve or optimize a catalytic antibody previously obtained by conventional *in vivo* immunization.[1092–1094] The more fundamental strategy of primary *in vitro* selection of antibody catalytic combining sites themselves is also possible.[1095,1096] We have seen that the *in vitro* generation of novel ribozymes requires the devising of ingenious selection strategies (Chapter 6), and this dictate is no less important for obtaining novel catalytic antibody specificities from artificial libraries. An inherent problem is coupling the genotype of a library member (as easily readable information) to the product created by the encoded active enzyme, and this can be overcome by coupling-linkage strategies, or *in vitro* compartmentalization (Chapter 5). A strategy eminently applicable to catalytic antibodies has been "covalent trapping," where a specially designed substrate in the solid-phase forms a reactive structure only after appropriate enzymatic action, which rapidly forms a covalent linkage to the enzyme responsible.[1096] In this manner, within libraries displayed on phage, specific candidate antibodies can be "pulled out" by virtue of their desired catalytic properties.

Usually, catalytic antibodies generated by conventional *in vivo* immunoglobulin selection and maturation processes are screened *in vitro* for candidates with the appropriate activities. It is possible, though, to select for suitable antibodies within bacterial cells, if the required antibody activity can by-pass a mutation necessitating a nutritional supplement for growth. Bacteria with such mutational lesions (auxotrophs) can resume growth in minimal media (lacking the supplement) if an antibody expressed in the cells can catalyze either the formation of the missing nutrient itself, or a precursor compound that the bacterium can then utilize for nutrient synthesis. Expression in *E. coli* of antibody libraries from mice immunized with a rationally designed hapten thus permitted selection of specific antibodies that enabled auxotrophic growth.[*,578] Such candidate antibodies could (as expected) also perform the required enzymatic activity *in vitro*.

We have seen that nucleic acid enzymes, with their reduced alphabets relative to proteins, are usually less proficient than their polypeptide counterparts, and likely to be more limited in the range of chemical reactions that they can catalyze.[194] Yet for diverse activities, even protein enzymes require extension of their capabilities by co-opting metal ions or organic coenzymes as cofactors (as noted in Chapter 2). It would logically follow then that antibodies in the role of enzymes likewise would benefit from the adoption of cofactors for catalytic assistance. This has been approached in several ways. A hapten corresponding to a complex of a target ligand and cofactor can be used for immunization, aimed at generating antibodies whose recognition and catalysis of the ligand can be accommodated by simultaneous binding of the desired cofactor. This has been accomplished with metal ion complexes with peptides, for generating specific peptide cleavage mediated by antibody.[1097] Immunization with protein conjugates of pyridoxal-5'-phosphate (a derivative of vitamin B_6) has also been used to generate catalytic antibodies utilizing this cofactor.[1098] Pre-existing antibodies can be also engineered for the acquisition of metal-binding sites of catalytic potential,[1099] or antibodies can be selected for the binding of specific metals as desired from *in vitro* libraries.[1100] Predetermined metal-binding sites (usually at a specific complementarity-determining region) can then prime an antibody for further mutation and selection at other CDRs toward the evolution of metal-assisted catalysis.[1100]

Many applications of catalytic antibodies have been contemplated. In the ADEPT procedure (antibody-dependent prodrug therapy), a "prodrug" (an enzymatically activatable drug precursor) is converted to active form at a desired *in vivo* site by means of an antibody-enzyme conjugate.[1101,1102] While abzymes are not competitive with their natural enzyme counterparts in terms of overall catalytic efficiencies, for some applications, including ADEPT, a reduced

---

*This study[578] is of interest from the point of view of the particular enzymatic activity studied. The complemented bacterial host defect was the pyridine synthetic pathway enzyme orotidine 5'-phosphate decarboxylase, which is the most efficient natural biocatalyst known[513] (as noted in Chapter 5). The immunization hapten was designed for antibodies catalyzing the related (but nonidentical) salvage pathway reaction of the decarboxylation of orotic acid, which enables by-passing of the mutational defect and selection for activity. Although very good by the standards of catalytic antibodies, the best selected orotate decarboxylase catalytic antibody is surpassed in efficiency by natural orotidine 5'-phosphate decarboxylase by a factor of 10^8.[513]

turnover rate may be acceptable[1103] and possibly even desirable to reduce excessive diffusion and side-toxicity of the activated drug. Bifunctional antibodies can be engineered for recognition of both an *in vivo* target (usually on the surface of a tumor cell) and a prodrug for catalysis toward cytotoxic activation.[1104] Antibody catalysis has also been developed with an aim toward the neutralization of addictive drugs, as a potential weapon against HIV, and for degradation the amyloid fibrils associated with Alzheimer's disease.[1105]

In Chapter 5 we considered factors limiting the use of catalytic antibodies, despite their virtues in the development of novel catalytic processes. Chief among these has been the growing appreciation of the functional role of enzyme regions beyond the active site for the realization of high-level catalytic efficiencies.[517,518] One of the reasons that antibodies are relatively primitive catalysts compared with natural enzymes thus derives from the restriction of catalytic antibody design to the combining site alone. It has been pointed out that natural selection for enzyme activity over megayears and immune selection for binding *in vivo* over short time periods are themselves not likely to yield products of comparable efficiency.[572] Further options must be considered, and we will do this in the broader context of alternatives to antibodies in general, and then return to the issues raised by catalysis.

ALTERNATIVES TO ANTIBODIES

To begin this section, we should first look back to Chapter 3 to recall the extended discussion there concerning antibody limitations, as viewed as "holes" in the universe of all possible antibody ligands. A significant issue raised was the probable suboptimality of antibodies for some molecular recognition tasks, and the issue of specific binding by antibodies to long DNA sequences was considered as a case in point. Yet when we are considering what antibodies* or mooted alternatives can and cannot do from a completely pragmatic perspective, we must consider not only the essential recognition function, but other important properties as well. These include (but are not necessarily limited to) molecular size, solubility, immunogenicity, ease of intracellular expression or transport, and stability to host or environmental factors ranging from degradative enzymes to thermal challenge.

These considerations aside, thoughts of functional RNAs (and other nucleic acids) capable of impressive feats of molecular recognition will naturally remind us that we have already noted that aptamers are a major area of potential competition with antibodies. Since we have already looked at aptamers in some detail, most of this section will deal with other antibody alternatives, although at the end we will consider all of the diverse molecular recognition technologies on a side-by-side basis, as far as that is possible. In any case, it is not necessary to find a universal framework solution for all applications, and the challenge becomes the rational identification of the best macromolecular recognition system for the job at hand. Superimposed on all these considerations is the ever-pragmatic issue of development and production costs. Another important issue in the real world that can influence developmental decisions in biotechnology is intellectual property rights, which can sometimes (as with monoclonal antibody technologies[1106]) become a complex minefield for commercial enterprises to negotiate. An unrestricted alternative technology can then potentially become an attractive prospect for such purely "artificial" reasons. At the end of this chapter we should look at this broad area again, but for the time being let us cast our gaze upon where protein-based recognition alternatives are headed.

---

*In this broad context, "antibodies" include conventional immunoglobulins and all natural or artificial variants.

Other Protein Scaffolds for Molecular Recognition

Going back to first principles, it should be realized that solutions to some molecular recognition problems can be found without recourse to antibodies if a "dedicated" natural-binding protein for a desired target is already known to exist, or if another natural protein can be readily altered to gain the required specificity. This is so because many targets for diverse applications are themselves part of natural molecular recognition events that can be exploited. In fact, the very concept of "biosystems" is predicated on a complex network of inter-molecular interactions in which all biological molecules participate. So, for virtually any given target biomolecule, in principle another naturally interactive molecule exists that might be adapted into a useful molecular binding tool for the required purposes. Natural molecular interactions have long been used for analytic applications. A case in point in this regard is the "far-Western" blotting technique,[*] where an immobilized sample on a filter membrane is probed with a labeled natural ligand for a protein found within the sample. (The protein of interest is previously separated from others in the sample by electrophoresis, before transferral to the filter.)[1107]

In practice, adaptation of a natural partner molecule is feasible in only a minority of cases, since a great many biological interactions lack the affinity or specificity to be suitable for biotechnology. But certain receptor–ligand systems have indeed been applied as direct alternatives to antibodies. A good example of this comes from therapies for the affliction of rheumatoid arthritis, which is known to be associated with pathologic production of the inflammatory cytokine mediator tumor necrosis factor-α (TNF-α).[1108] Two monoclonal antibodies with therapeutic regulatory approval (infliximab and adalimumab[1052,1109]) are directed against this cytokine. In fact, an analogous but nonantibody approach for rheumatoid arthritis is also in clinical use, by means of an adaptation of a cellular receptor for TNF-α into a soluble molecule, given the generic name of etanercept.[1110] (The C-terminus of etanercept is in fact derived from an immunoglobulin constant domain, but the essential point for our purposes is that its recognition of TNF-α does not involve an antibody combining site.) Having a choice of therapies (all directed against the same mediator) may not be redundant, since it appears that differential clinical responses can be obtained with the three options above. In other words, patients may be refractory to one blocker of TNF-α, but respond to another[1111].

From a theoretical point of view, a ligand–receptor interaction can be blocked equally well by targeting either interactive partner. Thus, blocking of specific TNF-α receptors would be an alternate pathway toward reducing cell signaling involving TNF-α in either soluble or cell-bound forms.[†] This approach of *receptor antagonism* has in fact resulted in the approval of additional nonantibody therapies for rheumatoid arthritis, although against different inflammatory immunological pathways than TNF-α. One of these is mediated by the cytokine interleukin-1β, whose native activity is controlled by a natural interleukin-1 receptor antagonist (IL-1ra).[1113] A recombinant form of this protein goes by the generic name of "anakinra," for therapy of rheumatoid arthritis.[1114]

---

[*]We noted in a footnote of Chapter 5, the origin of the nomenclature for "Western" blotting as arising from Southern blotting (for DNA by nucleic acid hybridization). So, the conventional Western blot, where immobilized proteins within a biological sample are probed with an antibody against a specific protein of interest, has been taken "far" by substituting antibodies with a naturally interactive ligand against a protein of interest. It is nevertheless perfectly possible to continue to use antibodies as the *detection* system of a far-Western, if a suitably labeled antibody is available against an epitope of the probe ligand, provided such an epitope does not clash with the site of interaction of the probe with the immobilized protein of interest. In such circumstances a detection "sandwich" is formed of labeled antibody:ligand probe:immobilized protein.

[†]Both TNF-α and one of its two distinct receptors exist in different isoforms, leading to complex signaling patterns with significant functional ramifications. The more selective that an anticytokine or antireceptor agent can be, the more likely that unwanted side effects (which have certainly been seen with anti-TNF-α agents used to date[1112]) will be reduced.

Making use of natural-binding proteins necessarily must be done on a case-by-case basis. Sometimes "ready-made" solutions exist, as with anakinra, but this will be the exception rather than the rule in general. The use of pre-existing natural-binding proteins might prompt an analogy with the innate immune system itself. It is fast, but it cannot deal with all contingencies. For that, we need an adaptive immune system, and complete replacement of antibodies requires a versatile molecular substitute for antibody diversification and selection *in vitro*. Such an alternative scaffold would necessarily have regions of diversity that can be "pasted" onto a conserved framework, analogous to antibody complementarity-determining regions.

Finding and Designing Proteins for Diversification   If we want to generalize protein formats for molecular recognition, we need to define what the basic requirements are. And it would seem foolish to ignore the natural lessons provided by antibody and T cell receptor recognition, even if the ultimate goal is to exceed their capabilities. As we have seen repeatedly, the immunoglobulin family solution to generalizable molecule recognition is a defined β-sheet framework supporting three hypervariable loops, the CDRs. With this formula, small ligands can be recognized within a CDR pocket, or the CDR loops can make contacts with a large protein target over an extended recognition surface. This is not to say, of course, that quite distinct artificial approaches should not be investigated, only that the natural antibody-type recognition solution remains a "gold-standard" for purposes of comparison.

In the broadest sense, the simplest approach toward novel biological recognition molecules has been the generation of peptide libraries using a variety of *in vitro* display formats, and specific peptide binding of protein targets was shown from early demonstrations of such technologies.[424–426] One could picture a variable peptide borne on a carrier protein as representing a "mono-CDR" (Fig. 7.8). Such peptide variable segments can be displayed as free sequences either at the N- or C-termini of a carrier, or in a conformationally constrained manner (Fig. 7.8). The latter choice can be an advantage for stabilizing useful peptide conformations and increasing their chances of selection, and may be achieved either by the introduction of disulfide bonds[1115] or by inserting a random peptide loop into a small defined peptide framework.[1116,1117]

In some cases, it is possible to display peptide (or small protein) sequences at very high densities on phage surfaces, as with the filamentous tubes of phage fd/M13,[1118,1119] or the tail tube of icosahedral phage.[1120] An interesting aspect of such high-density display in filamentous phage is the observation of "emergent" binding phenomena, where target interaction appears to require a higher order structure deriving from the surface-modified phage itself, rather than from individual displayed peptide sequences. These have been termed "landscape" phage and suggested as possible antibody alternatives in some circumstances.[1121] It seems clear, though, that high-level display of a single peptide segment will show limitations in the types of ligands that can be bound, especially with respect to small haptenic targets. More complex approaches are then called for.

Surely the most ambitious way of obtaining a desired recognition molecule is to attack the problem at the level of *de novo* protein design, and we saw in Chapter 5 that with the most powerful techniques for *in vitro* evolution, nonbiological-binding proteins for specific ligands have been derived. In this regard, we might be reminded of *in vitro* evolutionary selections for RNA or DNA aptamers, which in many cases are *de novo* processes in the sense that no preconditions are set upon the random sequence libraries from which candidate binders are extracted. Nevertheless, given the size of the protein alphabet, *de novo* selection for protein recognition molecules remains an impractical goal in most circumstances at present. In fact, it is also largely unnecessary, owing to the wide variety of natural protein folds that can be contemplated as potential alternative antibody scaffolds. Since on the order of 50 types of

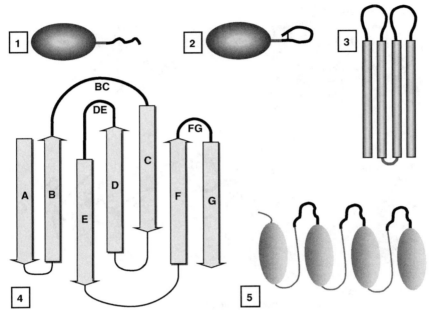

FIGURE 7.8   Schematic depiction of some examples of protein scaffolds for variable peptide segments (thick black lines). (1) Carrier protein with unconstrained single variable peptide; (2) as for 1 but with constrained peptide sequences; (3) two variable peptide loops on 4-helix bundle; (4) three variable loops on a β-sheet scaffold with seven β strands (A–G) defining six loops, top three as shown between strands B and C, D and E, and F and G. (5) Peptide variable loops on a repeat unit scaffold.

these have already been investigated,[1122,1123] we are prompted to concentrate on several key cases in order to illustrate the most salient points.

If a peptide library is freely appended to a common carrier protein, the nature of the carrier will generally be unimportant provided the peptide segment does not interact with the carrier itself and is maintained in an accessible state (often via a flexible linker sequence). The carrier itself should also have suitable physical properties such as solubility, moderate size, and reasonable stability. Topological constraint of a peptide sequence necessarily requires provision of a structural means for fixing the peptide sequence in effect as a single variable loop (as depicted in Fig. 7.8), and when we move beyond a single segment of peptide variability, the nature of the protein framework obviously becomes of paramount importance. While the unstructured nature of protein loop regions renders them highly tolerant of sequence diversity, it does not completely preclude the use of specific secondary structural regions, within defined limits. As a case in point of the latter, specific solvent-exposed residues within α-helical segments of the Z domain of staphylococcal protein A (with a three-helix bundle configuration; Fig. 7.9) have been used for diversification and selection experiments to yield derivatives with novel binding properties. Owing to their lack of disulfides and small sizes and stabilities, such "affibodies"[1124,1125] have been considered for diverse applications. Affibodies can themselves be displayed on certain bacterial cell surfaces.[1126]

Numerous scaffolds for variable peptide loops (many along the general lines of Fig. 7.8) have been used for randomization and *in vitro* selection for specific binding functions.[1123] Loops formed by natural four-helix bundle proteins[1127] or loops interspersed between specific

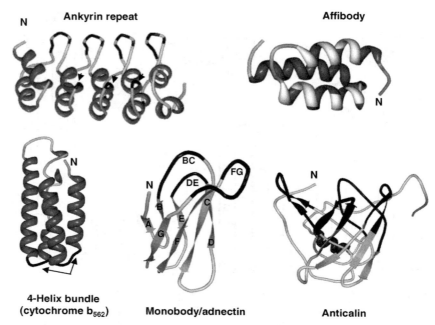

FIGURE 7.9   Structures of some alternative scaffolds for molecular recognition. Turns and coils light gray, β-strands gray, α-helices dark gray. N = N-termini. *Sources*: Protein Data Bank[2] (http://www.pdb.org): Ankyrin repeat[46] (1SVX), randomized regions in loops are shown as black segments, and in α-helices indicated by arrows; affibody[47] (1H0T), randomized regions of α-helices are shown in white; four-helix bundle cytochrome b_{562} structure[48] (1M6T), randomized loops are shown in black and with arrows[49]; Monobody/adnectin (1FNA),[50] β-strands A–G and loops are shown (compare with schematic of Fig. 7.8), randomized loops (black) as for[51]; anticalin (1T0V), randomized regions in black.[52,53] Images generated with Protein Workshop.[5]

repeat motifs[1128,1129] are among those that have been exploited in this manner (Fig. 7.9). As noted earlier, the lessons of antibodies have often been used as a guideline for seeking other analogous (but evolutionarily unrelated) scaffolds of superior physical properties. Immunoglobulin-like folds are in fact commonly found in protein databases,[1130] giving many potential choices for alternate scaffolds, but a specific example of this is of particular note. Fibronectin is a protein associated with the extracellular matrix and cell adhesion, and consists of a number of types of repeat domains. One of these, the tenth Type III domain, mediates cell adhesion and has an immunoglobulin-like character.[1130,1131] In addition, it is thermostable, soluble, and lacks disulfides.[1132,1133] Derivatives of this specific fibronectin domain used as molecular recognition scaffolds have been termed "monobodies,"[1134] but are also referred to as trinectins or adnectins.[1123] The β-strands of this domain define a total of six loops (Fig. 7.9), all but one of which can tolerate sequence diversity without overall structural destabilization.[1135] Three of these tolerant loops can be viewed as analogous to immunoglobulin CDRs (Fig. 7.9), and randomization and selection within such loop regions has been pursued *in vitro* for the isolation of novel ligand-binding specificities.[1133,1134,1136] Another scaffold to note is derived from a class of proteins called lipocalins, which possess a funnel-like β-barrel structure connected by four diversifiable loop regions.[1123,1137] The flexible loop regions of engineered lipocalin derivatives ("anticalins," Fig. 7.9) can offer variable binding cavities that may service a broad range of potential ligands, including small molecules that may be incompatible with other scaffolds.[1123,1138]

The above recognition platforms are based on finding a specific high-affinity binder recognizing a designated ligand or a single epitope on a macromolecular target. Yet a majority of biological targets (usually proteins) will in fact possess multiple potentially targetable epitopes, and this can be exploited for recognition purposes. Consider a recognition scaffold that has allowed the isolation (from an *in vitro* library of high diversity) of multiple binders toward separate epitopes of the same target protein. If these distinct binders (all with the same structural framework) are then linked together, the overall binding strength, or avidity,[*] will be dramatically increased. This scenario has been accomplished with a specific small novel scaffold, the "A-domain,"[1139] which is a common repeated motif in many receptor proteins. Analysis of the natural diversity of these domains allowed designation of residues for randomization, with selection for binding carried out in a phage display system. Binding domains identified in initial rounds were coupled with additional randomized domains in successive selection rounds until high-affinity combinations were isolated.[1139] These "avimers" performed as successful target inhibitors in animal models[1139] and have generated considerable commercial interest.[1140]

The "avimer" strategy of increasing the net avidity toward a target by multifunctional binding can be in principle generalized to any target bearing multiple discrete epitopes. Often, though, a target may bear more than one structural copy of the same epitope, as with polymeric antigens or homomultimeric proteins. In such cases, simply rendering a binding molecule with a valency greater than one will effectively increase target avidity, which is a natural strategy of immunoglobulins (especially IgM, as noted in Chapter 3). Other natural homomultimeric proteins might then be investigated as potential multivalent scaffolds for artificial molecular recognition systems. This has been done with a trimeric small human protein, tetranectin.[1141] Each tetranectin subunit consists of an N-terminal dimerization domain and a C-terminal C-type lectin domain bearing five loops[1142] compatible with high levels of diversification.[1141] Interestingly, the C-type lectin fold is widely dispersed in all kingdoms of life, and a bacteriophage example has been found to exhibit massive natural sequence diversity. A receptor-binding protein of a phage for the bacterium *Bordetella* has a C-lectin fold domain that has been found to support the existence of theoretically 10^{13} variants.[†,275]

As our short inspection of this field has indicated, any limitations on its overall growth do not stem from a shortage of putative recognition frameworks. It has therefore been suggested that the early phase of general scaffold exploration needs to enter a secondary stage of consolidation such that the best-performing alternatives can be chosen for further refinement.[1123] Certainly a high degree of commercial interest exists in the field as a whole,[1140] no doubt related to the great commercial successes of monoclonal antibodies.[1109] There is a multiplicity of both proven and potential clinical targets, and much competition between alternate protein products all ultimately hinging on specific molecular recognition.[1143] To provide one example, competition appears to be particularly intense in the area of inhibition of TNF-α, the cytokine we noted above, which is associated with certain autoimmune and inflammatory conditions, and which is already a target for commercial antibody and modified receptor products.[1144] It is interesting in this regard that *in vitro* mRNA display has been used for selection of high affinity monobodies/adnectins (Fig. 7.9) binding TNF-α.[1136] Also, the tetranectin scaffold has been used to isolate

---

[*]Here, we should recall our consideration of avidity versus affinity from Chapter 3.
[†]This protein enables phage host determination and entry, and its high propensity for variation permits rapid adaptation of the phage to mutation in the host cell surface receptor protein.

binders of TNF-α,[1141] of particular interest in this case since both tetranectins[1145] and TNF-α[1146] are trimeric, allowing for an associated boosting of avidity.

To date, the search for antibody substitutes has unsurprisingly been restricted to molecular recognition alone (binding). But we have seen earlier in this chapter that recognition is also a prerequisite for a second level of activity, catalysis. What are the prospects for the development of novel enzymes by an "antibody-like" route using alternative frameworks? We can remember that the fundamental requirement for successful antibody catalysis is selection for recognition and binding of immunizing transition state analog haptens of appropriate design. An alternative scaffold that is well-suited for binding large protein epitopes would not necessarily be also able to efficiently form suitable recognition pockets to accommodate small molecular haptenic targets. And any scaffold deficient in hapten recognition would be an unlikely source for the *in vitro* selection of catalytic derivatives.* As we noted above, not all recognition scaffolds are functionally equivalent, and only a subset (possibly including the anticalin motif[1138]) may be candidates for the binding of low molecular weight ligands. Another important requirement we have seen for many catalytic processes is the presence of suitable sets of residues† at the scaffold's binding site, but this is likely to be amenable to rational engineering in combination with empirical selection. Ultimately, it may be possible to search for enzyme activity through *in vitro* selection for specific molecular recognition processes in a variety of protein structural scaffolds beyond the antibody β-sandwich/loop framework, which must surely increase the prospects for improving novel enzyme catalytic power and range.

As well as their primary roles of recognizing and binding antigens, immunoglobulins also possess "effector" functions mainly mediated through their constant regions (Chapter 3). An artificial-binding protein would thus be unlikely to possess any comparable effector functions unless steps were taken to rationally incorporate them. Certain immunoglobulin variable regions even possess additional "bonus features" beyond molecular recognition *per se*, including sequences mediating dimerization, the binding of other specific proteins, and cell penetration.[1147] But just as the peptide segments mediating such "superantibody" activities can be identified and grafted into normal V regions,[1147] the responsible peptide sequences should be in principle transferable to at least a subset of the range of antibody alternatives if desired.

All of the range of scaffold choices rely on rational structural information for the choice of sites for randomization, and the powerful empirical evolutionary selections that we looked at in the last two chapters. We could also be reminded at this point that efforts to go beyond the existing constraints of antibodies are echoed in widespread research striving toward altered protein and nucleic acid function in general. Yet there is another technology quite relevant to antibodies and molecular recognition that takes a very distinct tack, and this we should now uncover.

Molecular Imprinting: Making its Mark

The generation of antibodies *in vivo* and selection for function *in vitro* with any molecular scaffold (whether antibody-based or not) share a common principle of action: a candidate binding protein matching the desired target can be found if the library of variant receptor shapes

---

*Of course, unlike the classical pathway toward the generation of catalytic antibodies, primary selection for recognition of haptenic transition state analogs with a nonimmunoglobulin scaffold would necessarily be performed *in vitro*.
†Among these are serine–histidine catalytic dyads for protease activity.

is large enough. Diversify, identify, and amplify—a commonplace principle now, it was not always so. Its origin goes back to the initial conceptualization of the immune system in terms of the clonal selection theory, where a pre-existing binding protein (an antibody on a B cell surface) is in effect amplified by engagement with a specific antigen target. We considered this in Chapter 3, and also noted in passing some competing notions regarding the nature of the immune system, which became discredited as evidence in favor of clonal selection mounted. Central among these superseded proposals was the idea that an antigen could actively mold a pluripotent antibody into a specific shape with binding complementarity for the foreign antigen molecule. In this old hypothesis, rather than relying on preformation of appropriate antibody receptors, it is the antigen itself that directly provides the information resulting in the formation of a suitable binding site. The latter scenario and clonal selection have accordingly been cast as "instructive" versus "selective" models.[12]

It is somewhat ironic then that a technology potentially competing with antibodies specifically uses an "instructionist" mode. This nonbiological process is termed *molecular imprinting* and is based on template-directed polymerization. The general idea itself is not new, going back to the 1930s and 1940s using silica gels and small organic molecular templates, based on a simple "footprint" model (Fig. 7.10a).[1148,1149] Even more ironically, the most famous proponent of the instructionist hypothesis for antibody formation, Linus Pauling, claimed in the 1940s to have created "artificial antibodies,"[1150] by what appears now to be a variation on the molecular imprinting theme. In these experiments, proteins (including γ-globulin serum fractions containing antibodies, and also others such as bovine serum albumin) were denatured and allowed to refold in the presence of low molecular weight target compounds. It was claimed that certain treatments along these lines resulted in protein preparations that could differentially bind the antigens to which they had been exposed.[1150] Possibly such treatments (which included incubations of "several days" at elevated temperatures) can result in irreversible conformational changes that "fix" a binding site around the compounds of interest. Regardless of this, time has of course proven that "instructive" processes are not at all how antibodies are really formed, and the *in vitro* "antibody imprinting" observations are of little practical value. Nevertheless, the sum of early work in general within molecular imprinting covers two broad concepts: the formation of a binding "footprint" around a target molecule or the refolding of a preformed polymer around a similar target. (These are depicted schematically in Fig. 7.10.)

While the refolding of a preformed polymer to form an imprint is a theoretical possibility, a realistic problem is its stability. Proteins, for example, may under special conditions themselves serve as imprintable polymers, but this is not of practical importance under physiological circumstances. The flexibility and dynamics of protein structures inevitably mean that an unusual binding pocket induced by another molecule will be highly unstable under normal conditions in the absence of the inducing molecule itself. But interestingly, if we specifically go beyond conditions in the normal aqueous cellular environment, these dictates do not necessarily apply. There is considerable interest in adapting enzymes toward function in organic solvents (noted in passing in Chapter 5), and in nonaqueous (anhydrous) media protein flexibility is drastically reduced.[1151] As a consequence of this, a ligand-induced conformational imprint that is rapidly lost in water can be retained as a "memory" in anhydrous conditions.[1151,1152] (This need not involve full denaturation of the protein, as transient conformational variants selected by the ligand can then be "fixed" by the conformational rigidity imposed by nonaqueous media.) Likewise, conformational variants at enzyme active sites can alter substrate specificities if these too are fixed under the same anhydrous conditions.[1153,1154]

As we have seen, the above "bio-imprinting" processes are not generally relevant for ligand binding under normal cellular circumstances, and most current interest in molecular imprinting

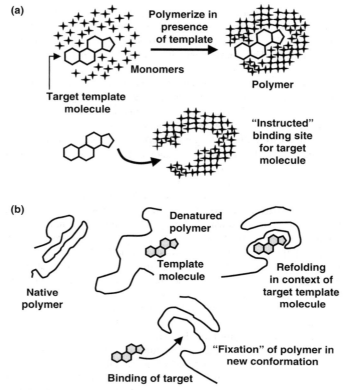

FIGURE 7.10 Simplistic depictions of processes of molecular imprinting in its early implementation. (a) As a "footprint" effect. A target molecule functions as a template around which monomers of an appropriate chemical nature can polymerize (forming a matrix if cross-links can form directly, or if a chemical cross-linker is added). Removal of the template results in a "mold" that is complementary to the original target and can then act as a specific binding site for the same molecule. (b) Instead of template-directed polymerization, this depicts template-directed folding, or fixation of a conformational state that is normally short-lived.

involves template-directed polymerization rather than template-directed refolding or conformational selection.* But before continuing, we should note that "imprinting" has diverse meanings in biology in general, all radically different from each other. Say it to one person and they might think of Konrad Lorenz being followed by young geese (behavioral imprinting) while many others would think exclusively of the genomic/genetic imprinting that we briefly alluded to in Chapter 2. For the remainder of this section, "molecular imprinting" will refer only to template-directed polymerization effects. Clarity of nomenclature is not helped by the occasional use of "molecular imprinting" in reference to genomic imprinting.[1156,1157] Still, confusion is avoidable through the context of the term's usage, provided these two domains of imprinting remain separated.

Molecular imprinting can generate a template-directed "molecular cast," which can then (at least in theory) act as a binding site for the original molecule. But the simplistic "footprint"

---

*Some work has been done with "conformational imprinting," where conformations of a modified polypeptide which bind a ligand are fixed on a solid surface.[1155]

model as in Fig. 7.10 is an inadequate description of the process, since the chemistries of both the polymerized matrix and the target template ligand determine how the polymer becomes imprinted by the template. (Without attention to this, a polymer such as silica gel can only accommodate a limited range of template types.[1149]) A rational approach is to control the mechanism by which the template is coupled with the polymerization process, which creates a specific cavity occupied by the template molecule. Since the early 1970s and beyond, two general strategies to template-directed polymerization have arisen, where the target template is rendered suitably active by either reversible covalent bonding[1158] or noncovalent interactions[1159,1160] (depicted in Fig. 7.11 for the covalent process). A number of variants of these approaches, including some that combine covalent and noncovalent interactions, have been described,[1161,1162] although in general terms, the noncovalent strategy may be the most versatile.[1160] Is then molecular imprinting a fully rational process, in comparison with the evolutionary strategies used to obtain antibodies and other binding proteins either *in vitro* or *in vivo*?

Certainly, molecular imprinting does not qualify as "molecular evolution," although it belongs in this chapter by way of its direct parallels with conventional binding molecules. The technology itself still remains largely empirical in terms of the choices of monomers,

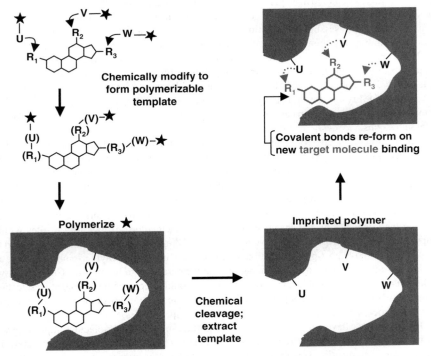

FIGURE 7.11 Molecular imprinting using covalent bonding with the template, where the chosen template is covalently modified with appropriate reactive groups (stars denote chemical groups (containing a C=C double bond) suitable for copolymerization with monomer and cross-linker used to obtain the polymerized matrix. Groups in brackets denote chemical residues remaining following specific derivatization. The corresponding noncovalent process is conceptually related, except that the reactive groups self-assemble onto the target template via noncovalent interactions.

cross-linkers, and polymerization conditions,[1163] and indeed empirical combinatorial approaches can be used to establish optimal chemical formulations for imprinting specific templates.[1164] Nevertheless, it can be expected that in time most if not all of these factors will become fully optimized.[1165] Yet by the strictest definition, molecular imprinting still fails to qualify as a completely rational solution to a designated recognition problem, since the precise structure of the recognition element is not a defined single molecular configuration predicted in advance. Of course the monomeric and cross-linker *composition* of an imprinted binding pocket is initially defined, but not the precise spatial distribution of all polymerized subunits once the binding pocket is formed. Because of this, individual pockets in a single preparation of an imprinted polymer have nonidentical binding properties, and overall affinity measurements must necessarily take this heterogeneity into account.[1166,1167] Molecularly imprinted polymers (commonly referred to as MIPs) are therefore always "polyclonal,"[1149] at least with current technologies.

These kinds of considerations need not detract from the practical utilities of imprinted polymers. Many *in vitro* applications for this technology exist where the detection and quantification of small target molecules is concerned, ranging from analytical and preparative processes to environmental molecular sensing.[1149,1159] Imprinted polymers with defined specificities have many advantages over virtually any biorecognition molecules in the general area of stability, whether to heat, pH, salt concentrations, or any other relevant factor. The ability of some imprinted polymers to function effectively in aqueous environments has been a limitation in some cases, but can be rationally improved.[1168] Imprinted polymers have achieved averaged binding constants in the submicromolar to nanomolar range, less than the best antibodies but certainly competitive with them.[1169,1170] If antibodies indeed become routinely replaceable by specific molecularly imprinted polymers (at least *in vitro*), it seems that the term "plastibody"[1171] coined for the latter might become widespread. Certainly, imprinted polymers have shown utility *in vitro* for analytical purposes in "pseudoimmunoassays."[1172,1173] Imprinting also holds promise in certain areas of drug discovery,[1174,1175] delivery, and controlled-release systems.[1176,1177]

Another important area for applied molecular imprinting is preparative purification based on an affinity chromatography step, where the necessary recognition is supplied by a specifically imprinted polymer. One potential drawback to this activity is the heterogeneous nature of imprinted polymers, the extent of which is another variable itself. After binding of a target ligand to an imprinted polymer, an excessive binding affinity range will result in a broad "tailing" effect as the ligand is eluted from the specific matrix.[1149] Some studies have addressed the problem of binding heterogeneity by the selective "poisoning" (inactivation) of low affinity sites within imprinted polymers in order to improve overall net binding affinity.[1149] We might be tempted to compare this kind of post-polymerization modification of imprinted polymer-binding sites as an analog of antibody affinity maturation, but thus far the parallel is rather inexact. Culling of low-affinity binding pockets within an imprinted polymer could be considered as a negative selection operating within an aggregate population of binders, whereas true affinity maturation is a positive selection process at the clonal level of single-binding specificities.

But thus far, we have not yet uncovered an important subfield with molecular imprinting and its applications. Since we have regularly been associating molecular recognition with catalysis, it will come as no surprise if catalysis rears its pretty head in the imprinting context as well. And not unpredictably, if plastibodies are the molecular imprinting analogs of antibodies, then for enzymes we have . . . "plastizymes,"[1171] although this term too has a way to go before assuming universality of usage. We could compare molecularly imprinted catalysts with the generation of catalytic antibodies. In both cases, the processes are established in highly rational manners, requiring sophisticated chemical knowledge, and both use similar design principles with chemical analogs of transition state intermediates. For catalytic antibodies, if catalysis requires additional chemical assistance beyond a simple entropic trap mechanism (as in our above survey of abzymes), then it must be supplied by reactive amino acid side chains in the active site. As such, it is limited to the 20 natural amino acids, unless the selection for the catalytic antibodies is performed in a system using an unnatural extension of the conventional amino acid alphabet (Chapter 5). In contrast, any transition state analog and catalytic cofactor can in principle be coimprinted into a polymeric reaction pocket, as long as noncovalent association exists between the cofactor and the analog compound.[1178,1179] This is depicted in Fig. 7.12, showing that post-polymerization removal of the transition state analog leaves a catalytically functional binding pocket. By manipulating the transition state analog, the coimprinted catalytic ligand groups, or the monomer, it is possible to vary functional group positioning within the active site for tuning of catalysis.[1178,1179] Metal ions, so useful in innumerable catalytic contexts, can also be imprinted by various strategies.[1149] To date catalysis by imprinted polymers cannot challenge natural enzymes, but has risen to efficiency levels as good as (or even better than) many catalytic antibodies.[1179,1180]

A nonenzymatic application of molecular imprinting that nonetheless has some similarities to catalysis is the use of imprinted polymer cavities as "molecular reaction vessels." This is an interesting approach to organic synthesis where imprinting is performed with an initial template molecule, leaving a cavity that can be used to direct chemical reaction between two other smaller molecules. The shape of the imprinted cavity constricts the reaction pathway toward a

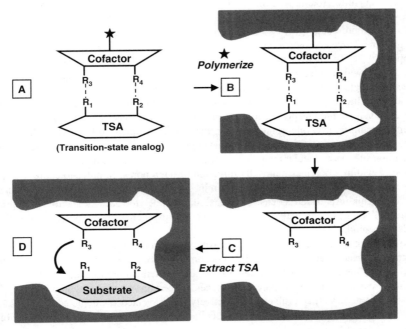

FIGURE 7.12 Schematic showing generation of an imprinted polymer with a transition state intermediate analog (TSA) for a specific reaction in noncovalent association with an enzymatic cofactor (dotted lines) that also has a functional group available for copolymerization (black star). The cofactor then becomes incorporated into polymer cavities that function as catalytic sites for the appropriate substrate when the transition state analog is extracted.

specific product that is a mimic of the original template shape. Some natural enzyme binding pockets can also be used in an analogous fashion, for site-directed synthesis of enzymatic inhibitors.[1181] This process has been dubbed "anti-idiotypic" imprinting by analogy with anti-idiotypic immune networks.[*,1160,1183] As practiced, this is not a catalytic process, since reactants required to form the template mimic are added sequentially, and no turnover is involved (products are extracted from the matrix after a single round of reaction).[1181]

Imprinting of Macromolecules   If molecular imprinting was forever confined to small molecules, it would still find numerous applications, but it could never be considered as a rival for antibodies in general. To rise to the latter level, imprinting must demonstrate that it can specifically discriminate proteins and other biological macromolecules. At this point, we should consider certain important distinctions between antibody recognition and molecular imprints. Although antibody combining sites are highly versatile, there is an upper limit to the size of a molecular entity that can be *directly* contacted by a single V_H/V_L recognition unit.[†] Of course antibodies can bind to relatively very large structures such as multimeric proteins, viruses, and whole cells, but this proceeds through recognition of discrete accessible sites termed epitopes, as we have previously seen. Surface area maxima associated with target recognition necessarily apply to all alternative protein- and nucleic acid-based recognition molecules. In contrast, recognition through polymeric molecular imprinting is in principle free of these restraints. By the nature of the imprinting process, formation of a specific cavity (Fig. 7.11) is not strongly size-limited, and imprinting can thus in principle simultaneously recognize an entire protein. And there is no need to stop there. Polymeric imprints have been made of whole viruses and even whole yeast or bacterial cells.[1187–1189]

But for our present purposes we can stay on the scale of individual macromolecules, starting with proteins. Three alternatives are acknowledged for protein imprinting: "bulk," surface, and epitope.[1163,1190] The first of these, as the name implies, refers to whole-protein imprinting (Fig. 7.13a), which creates challenges at several different levels. Compared to simple low molecular weight templates, even a small protein offers many more opportunities for spurious interactions with the polymer matrix that can reduce binding specificity. The inherent problem of heterogeneity of binding pocket affinity tends to become likewise increased. Also, for "bulk" protein imprinting, it is important to enable the protein template to be extractable from the matrix post-polymerization, and to allow other copies of the same target to access the imprinted recognition site. As molecular templates grow in size, the more difficult it becomes for them to enter sites within a polymer matrix by diffusion through pores. In order to address this, the degree of polymer cross-linking (which determines pore size and rigidity) must be manipulated, but as cross-linking and matrix density decrease, in general so too does the stability of imprinted polymers. A range of alternative polymers and cross-linkers have been investigated with this problem in mind.[1190]

So although molecular imprinting can easily vault over antibody interactions in terms of recognition surface area, this facility does not come without practical penalties that need to be coped with. One answer to the problems of bulk protein imprinting has been to restrict imprinting cavity formation to polymer surfaces, by certain ingenious technical steps.[1190] In this way, the accessibility problem is effectively circumvented (as depicted in Fig. 7.13b). A subset of this type of surface imprinting is the third broad protein imprinting technology

---

[*]In general terms, this kind of process falls into the domain of "receptor-accelerated synthesis,"[1182] and we will revisit this area in Chapter 8.

[†]Contact surfaces of antibodies with proteins can approach 1000 square angstrom units ($\mathring{A}^2$; $1\,\mathring{A} = 10^{-10}$ meters or 0.1 nanometers)[1184,1185]. Even a moderate-sized globular (roughly spherical) protein of molecular weight 50,000 Da with a radius of $\sim$25 Å,[1186] thus has a surface area of $\sim$7800 $\mathring{A}^2$, much in excess of the antibody surface area contact range.

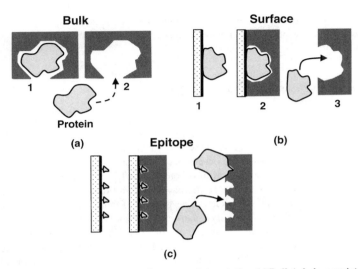

FIGURE 7.13 Depictions of major approaches to protein imprinting. (a) Bulk (whole-protein) procedure. (This employs the noncovalent imprinting method, not shown here for simplicity.) (b) Surface imprinting, where protein template targets are immobilized on a solid support before the polymerization step. Removal of the support and extracting the protein results in a surface-imprinted polymer. (c) Epitope imprinting, which is a subset of (b) using relatively small defined protein regions in the form of peptides. The final surface-imprinted polymer can bind not only the original peptide but also proteins that display the same epitope in accessible form.

where only small protein regions are functionally imprinted as relatively small peptides (Fig. 7.13c). If these are accessible on the target protein, they can act as epitopes allowing binding of the whole protein to the surface-imprinted polymer.[1191,1192] It should be clear that the epitope imprinting approach has the most direct parallels to the binding of proteins by antibodies or other recognition molecules.

A general conclusion of this brief overview is that protein imprinting has overcome many of its early difficulties, but still needs to develop further to seriously challenge antibodies. In this regard, though, we should note that for *in vitro* and industrial applications trade-offs exist between recognition affinity and the stability of the recognition molecule. In some circumstances a molecularly imprinted polymer of great strength and stability with reasonable affinity for a specific protein might therefore be chosen in preference to an antibody with much higher affinity (as we will consider further in the next section).

Before moving on from this highly dynamic field, we could return for a moment to the instructionist versus selective dichotomy with which we began this section, to note that the hard-and-fast distinction between the two blurs a little when we look at recent advances in molecular recognition. This is not to suggest that the principles of diversification and selection for the generation of antibodies are under challenge; the instructionist school for immunoglobulin recognition surrendered many decades ago. The point to be made is that although selected and amplified by a Darwinian process, many structural studies have confirmed that conformational changes in antibodies resulting from antigen binding are far from rare[264] (as noted in Chapter 3). We have also seen that for nucleic acid aptamers, conformational alterations upon ligand binding are the rule rather than the exception (Chapter 6). At this level at least, we can see a ghost of the instructionist theory coming back to haunt the orthodoxy of the selective viewpoint. And for molecular imprinting, the instructionist approach is alive and thriving. But let us now attempt to integrate imprinting within the "big picture" of molecular recognition.

COMPARING RECOGNITION ALTERNATIVES

Confronted with a multiplicity of alternate pathways to molecular recognition, it falls upon us to try and identify which are the best choices for a given task. This is a somewhat daunting task in its own right, because such decisions are necessarily made against a background of constantly changing competing technologies. It is necessary to define the most important variables that contribute to recognition molecule utilities from a broad background. These are listed in Fig. 7.14, along with the general range of alternative recognition options. The environment in which a recognition molecule is expected to perform is a fundamental issue, and this can be divided into the fields of *in vivo* activity versus everything else.

Clinical or veterinary applications require profound attention to factors that are irrelevant or of greatly diminished *in vitro* (Fig. 7.14). Any such reagent must be of demonstrable safety in terms of direct toxicity or from spurious (and deleterious) recognition of extraneous biomolecules. This is an aspect of specificity, which is of course a primary issue for any molecular recognition, accordingly included within the "general" variables. This is also implicit within *in vivo* applications, but "cross-reactions" can go beyond binding specificity *per se*. The reason for this is the parsimony of usage of biomolecules in complex organisms. A protein used for one purpose in one cell lineage may have a different (if overlapping) role in another differentiation lineage within the same organism. As a result, even an exogenous binding protein of perfect specificity for one target may engender complex effects *in vivo* that may be difficult to predict purely from *in vitro* results in a limited number of assay types. *In vitro* applications in the form of assays using whole cells (of any origin) or complex mixtures of proteins (such as transcription/translation systems) must also pay heed to this in principle, but it is less likely to be a problem than with a whole animal or human being. We have previously encountered another issue of great significance for clinical applications of recognition molecules, that of the immunogenicity of the novel therapeutic binding molecule. Any protein-based reagent that is foreign to the host will be immunogenic to some degree, although not necessarily strongly so. (For example, camelid VHH domains appear to have relatively low immunogenicity compared to whole foreign antibodies, attributed to their resistance to aggregation.[1062]) And we have also seen that antibody idiotypes themselves will provoke a response as "nonself," although this is usually a much less significant challenge than the full-blown responses to foreign nonhumanized antibodies. Small nucleic acid aptamers, on the other hand, are unlikely to result in the generation of an immune response that could limit their *in vivo* efficacies.

Some recognition variables (such as specificity and affinity) are particularly important for distinguishing one molecular recognition alternative from another. A rough-scale ranking is shown in Fig. 7.15 of some recognition options, grouping different alternatives into classes that themselves have much internal diversity. Further explanation of this is needed, since it is intended to account for likely near-term developments as well as options already available. In this vein (for example), it can be seen at a quick glance that all of the choices except molecularly imprinted polymers (MIPs) are given an equally high rating for affinity and specificity. This is intended to convey the message that aptamers, antibodies, other small V regions, and artificial protein scaffolds all have the potential to attain approximately equivalent high-level performances against a given target by these criteria. On the other hand, at the present time it does not

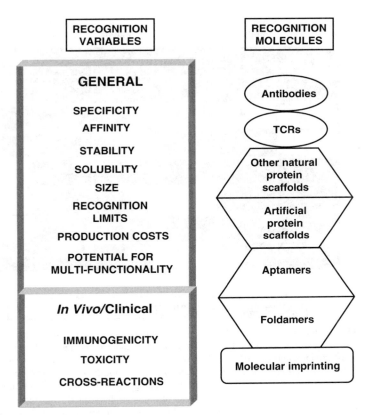

FIGURE 7.14   The overall range of recognition molecule alternatives and the factors relevant to their relative efficacies. TCRs: T cell receptors; "Other natural protein scaffolds": natural protein frameworks with regions tolerant of variation that can be adapted for broad recognition purposes.

seem that imprinting can equal the best-performing protein or nucleic acid recognition, although as always it is risky to make emphatic predictions in most areas of this general field. Molecular imprinting as currently practiced also is inferior when size and solubility are important.

When we move on to the issue of stability, though, the tables are turned in favor of MIPs (Fig. 7.15). All protein alternatives are here grouped at the same approximate level, on the assumption that rational design or empirical evolutionary approaches will be able to improve stabilities when it is initially suboptimal. (Although that is not to say that all protein-based alternatives will be equally amenable to this.) While nucleic acid aptamers are broadly less stable to chemical or physical challenges,* their resistance to nucleases (highly relevant for *in vivo* applications) can be dramatically increased by a number of alternatives we saw in Chapter 6, including the ingenious spiegelmers. Different environments place different physical demands

---

*By the nature of their respective chemistries, RNA aptamers are less chemically stable than DNA aptamers, but both are nonetheless susceptible to thermal denaturation that will destroy their specific recognition properties. Proteins in general may be more competitive in terms of thermostability, but (as noted in Chapter 6) protein cofactors can in some circumstances be used either naturally or artificially to help stabilize functional nucleic acids.

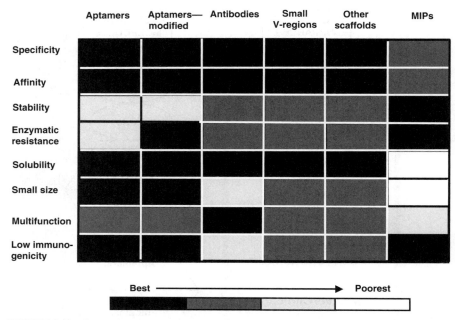

FIGURE 7.15 Contrasting strengths and weaknesses of different molecular recognition technologies, as broad guidelines. "Aptamers-modified" refers to chemically modified aptamers such as spiegelmers. "Small V regions" refer to camelid VHH, shark IgNAR, or other small immunoglobulin-derived V regions. "Other scaffolds" denote all other protein scaffold alternatives, obviously a very diverse collection and only groupable here in the most general terms. Relative rankings (shadings) refer to levels of a variable quality (specificity, etc.) that can reasonably be attained with present technologies for best representatives of each class of binding molecules. For example, the best examples of all protein and nucleic acid classes are approximately equal in affinity and specificity, but the best MIPs (molecularly imprinted polymers) are at present less potent in these areas. Note that the order of rankings of each variable also corresponds to most desirable status (for example, small size and low immunogenicity). "Stability" denotes physical stability (temperature, pH, etc.); most relevant to *in vitro* applications. "Multifunction" indicates the potential of each recognition approach for the generation of single molecular entities with multiple binding specificities.

on recognition molecules, and one such example briefly alluded to previously is the adaptation of antibodies toward the intracellular environment (intrabodies). By the same token, intracellular aptamers (intramers) have an inherent "head-start" in terms of their abilities to fold and function intracellularly. As in many other situations, it is not clear whether either options will in the end show an overall competitive advantage, or whether it will come down to choice on a case-by-case basis.

When we are thinking in general about any drugs for biological use, molecular size is a very important consideration. In fact, the topic of size is a major theme within the next chapter. For the present purposes, we can note that molecular weight and dimensions are significant factors in determining the efficiency of drugs in general, related primarily to drug diffusion rates and accessibility to target sites. And the battery of alternatives of Fig. 7.15 span a wide range in this regard. Aptamers have a clear lead in this aspect, especially when compared with conventional immunoglobulins.[941] Aptamers may be on the order of 20 bases (molecular weight ~6860), but a normal antibody molecule such as immunoglobulin G has a molecular weight approximately

22 times larger (150,000 Da). While a useful guide here, molecular weight alone is insufficient, as molecular folding and final shape is very important in determining the behavior of molecules in diffusing through nanopores. As an example, the molecular volumes assumed by the folds of specific aptamers are poorly predicted purely by molecular weights alone, as aptamers of disparate lengths in primary sequence can have comparable folded molecular dimensions.[941] This issue aside, the dimensions of all artificial aptamers are indeed small compared to conventional immunoglobulins, but this advantage is considerably diminished by the advent of the single domain antibodies referred to earlier. Camelid VHH or shark IgNAR domains[1064] are thus only on the order of twice the molecular weights of small aptamers. While still significant, a size-related advantage of an aptamer over a potentially competing small V domain of comparable specificity would need to be clearly demonstrated.

We have noted in earlier contexts that the "gold standard" of molecular recognition, antibodies, possess a range of other "effector" functions as well as the recognition and binding of molecular targets. This is the "multifunctionality" referred to in Figs 7.14 and 7.15. Antibodies here have been singled out (Fig. 7.15) owing to their natural effector function assets, although many factors impinge upon such secondary functions *in vivo*, rendering prediction of optima difficult.[*,1195] But in any case, initial advantages of whole antibodies in this sphere can rapidly change. It is thus in principle unlikely that all the other options should continue to be less capable than antibodies for adaptation for multifunctionality. Certainly, there is abundant favorable experience with the modularity of protein functional domains in general, and we might recall that work with artificial riboswitches has shown that distinct aptameric functions can be usefully linked together in the same molecules (provided suitable "communication" segments are used). In contrast, imprinted polymers would not seem easily amenable to functional modularity, at least in the same sense of linking together preformed functional domains. Yet it is well established that noncovalently bonded templates are imprintable (as with the general noncovalent approach, and as applied to catalysis, Fig. 7.12), so multifunctional options in this sense certainly exist for MIPs.

Molecularly imprinted polymers in general show the greatest contrasts between some variables and others (Fig. 7.15), and thereby raise the issue of trade-offs. Where stability in a designated environment is of paramount concern, it might result in the choice of a recognition molecule of adequate if less than perfect affinity and specificity. An issue not directly related to recognition performance in itself, yet of immense significance in the real world of large-scale manufacture, is of course the cost of production of any of the noted alternatives, as included in Fig. 7.14. Complications linked to this are whether post-translational modifications are required for optimal function, including natural glycosylation,[1195] or any form of artificial derivatization. Production costs are also artificial to the extent that the motivation for generating improvements in manufacturing efficiencies is driven by factors such as the overall size of the target market and degree of commercial competition. The market picture for natural or artificial recognition molecules can be drastically altered if rival low molecular weight inhibitors emerge that can tackle a clinical problem from a different angle. In Chapter 8 we will look at some aspects of this general issue in more detail.

One can make evolutionary analogies between competing technologies. If two alternative technologies with comparable functions emerge at roughly similar times, during their often prolonged developmental phases, it can seem that little communication proceeds between enthusiasts in either camp. Sometimes one of the molecular alternatives has clear superiority

---

*Some natural modifications may actually be detrimental to the performance of antibodies as therapeutics. Prevention of antibody fucosylation (but not glycosylation with other sugars) has been found to improve antibody-dependent cellular cytotoxicity, one means by which antibodies can mediate their desired therapeutic effects.[1193,1194]

over the other in one particular niche, and vice versa for another application (as in Fig. 7.15). If so, representatives of both should eventually find a home in the marketplace. As noted earlier, intellectual property rights themselves can be a driver toward molecular innovations, but alternatives have to be performed at least as well as the current "gold standards" in specific applications. On the other hand, if widely recognized universal advantages of one technology over another exist, theoretically rationality should eventually prevail, with the "fittest" alternative being "selected." (In practice, this might require waiting for the winner to come off-patent before some commercial alternatives fall by the wayside.) Yet we noted previously (in the context of rheumatoid arthritis and TNF-α) that in a clinical setting the complexity of responses between different patients tends to favor the retention of multiple alternative products as therapeutics.

In any case, meaningful differences between recognition alternatives may be subtle and require accurate side-by-side comparisons. These can be time-consuming, expensive, and a distraction from other research and developmental goals. It may prove difficult to obtain both types of recognition reagents with the comparable affinities and specificities required to make valid inferences regarding relative efficacies. Ultimately, as molecular recognition technologies make the transition into practical realization in the marketplace, economic realities also become a factor in "selecting" one techno-competitor from another. Production costs for continuous manufacture of a reagent are of obvious importance. In this respect aptamers may have an edge, as they are almost always short enough for quite feasible chemical synthesis (especially DNA aptamers), unlike even "mini" protein antibodies.

Another cost-related issue is the relative ease of rapidly producing new recognition molecules on demand. From this point of view, it is interesting to note that automated methods for aptamer selection (or auto-SELEX of sorts) have been developed[1196,1197] and on-going refinement of such processes can be expected. High-throughput generation of new aptamer specificities may prove more cost-effective than traditional monoclonal antibody techniques, and if so will become a significant additional factor in the balance sheet between the relative practical values of the two technologies. At least we can end this rumination on a positive note, on the theme that biotechnology would appear to be much less susceptible to the "locked-in" phenomenon so frequently invoked for technology in general by the QWERTY type format (as noted in Chapter 2), and also used as a metaphor for "frozen accidents" of evolution. But given the complexities and demands of diverse applications, there is likely to be a place for most of the alternatives in Fig. 7.15, given their areas of specialist performance (Fig. 7.16). It seems highly improbable that a global best-possible single type of recognition "scaffold for all seasons" will emerge.

This in turn raises a general issue not included within Fig. 7.15, which we might designate as "Recognition Limits." Such a question returns us to fundamental aspects of molecular recognition in the broadest sense and imposes the question of limitations on a particular recognition framework, which are inherent in its design. Where such intrinsic limitations are encountered, a radically different design solution may be called for. We have noted some aspects of this previously, such as the comparison between antibodies and DNA-binding proteins in Chapter 3. Apparent limitations of some scaffolds may succumb to continuing advances in the fine details of their design, and some convergence of capabilities may therefore be expected for the future. Yet from the present perspective, it is clear that some recognition approaches are better suited to some recognition problems than others, and this perspective is represented in Fig. 7.16.

We have also seen previously that a variety of scaffolds for catalytic purposes are possible, and decision based on the broad alternatives on offer (most of the options of Fig. 7.16) may be made on an increasingly rational basis as knowledge accumulates. But before thinking about

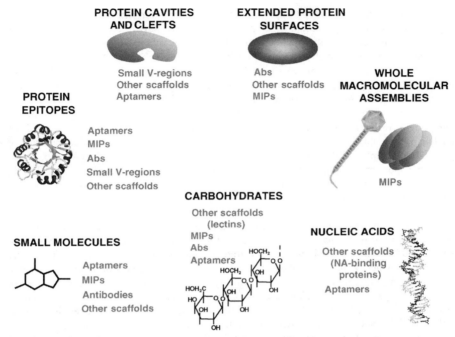

FIGURE 7.16   Recognition of diverse targets by different molecular strategies. A range of targets is shown (not to scale; from small individual molecules to large macromolecular assemblies such as high molecular weight protein multimers and viruses) with the associated recognition molecules in gray text. For "whole macromolecular assemblies," the recognition event refers to simultaneous binding of the entire unit which is feasible with molecular imprinting. Abbreviations: Abs = antibodies; MIPs = molecularly imprinted polymer; NA = nucleic acid.

such catalytic generalities, we should glance at an area of technological progress that will strongly impact upon the future of artificial molecular recognition in general. Although there are quite basic differences between them at a chemical level, both functional proteins and nucleic acids assume specific molecular configurations associated with their activities. Folding in a specific and ordered manner can then be considered as a fundamental property of functional biological macromolecules. But why stop there? We have already seen (in this and the preceding chapters) that both proteins and nucleic acids are being extensively modified in highly unnatural ways, with the end results increasingly diverging from the original biomolecules that inspired them. Inevitably, a point is reached where it is no longer useful or appropriate to use the "bio" prefix for certain functional artificial macromolecules. These in principle could be composed of any set of chemical subunits, as long as they allow the macromolecule to fold into a specific ordered conformation, and this is the theme for the next section.

GENERALIZING TO FOLDAMERS

It has long been known that proteins are amazing molecules with hugely diverse functionality, and more recently nucleic acids too have revealed their functional capabilities. The exploitation of protein and nucleic acid design (separately or in combination) in turn enables hugely diverse applications, whether by rational or empirical means. We have previously thought about the hyperastronomical size of protein sequence space, and one of the most salient points was the tiny size of the numbers of all biological proteins in relation to the size of the sequence space itself. This number was even more striking if the diversity of natural proteins was broken down into specific folding categories. Nucleic acid sequence space is much smaller, if we restrict the functional size range to approximately <200 bases, but is still of enormous magnitude. This, of course, is why evolutionary selection is required, if rational design cannot be applied toward obtaining a desired function. Yet true molecular maestros would not be limited to peptide or nucleic acid combinatorics, but would extend their domain to include *any* folded molecule of potential use. Implicit in the word "fold" in this context is the use of noncovalent interactions for the stable and reproducible adoption of a specific three-dimensional molecular shape. Within relatively recent times, it has become clear that certain nonbiological molecules composed of specific subunits can fold in stable and ordered ways, and these artificial forms have been termed *foldamers*.[1198-1200]

A folded macromolecule by definition has a high degree of order, but of course not all foldamers will be functionally relevant as agents for molecular recognition. In this chapter we have included enzymatic binding of substrate within the broad umbrella of "molecular recognition" where appropriate. We can therefore consider that aspect of molecular function that is encompassed by catalysis, and previously we have thought about the limitations of catalysis wrought by both nucleic acids (in Chapter 6) and proteins (in Chapters 2 and 5). Especially when one begins with nucleic acids and proteins with limited alphabets and without cofactors, it is possible to perceive a hierarchy of abilities as one progresses from 2-base ribozymes upward. The greater catalytic versatility of proteins then is at the next level, followed by artificially modified proteins.

When considering the development of catalysis in broad terms, the efficiency of the best protein enzymes still is far in advance of the rest of the pack. And we spent some time in Chapter 5 looking at means for further improving enzymes using a variety of artificial evolutionary approaches. But a special category of proteins, antibodies, can themselves approach catalytic goals from a different type of evolutionary selection to that shown with natural enzymes. While abzymes themselves also fall short of matching the full power of enzymes fine-tuned and honed by natural evolution, as we have seen earlier in this chapter they offer a very useful strategy for finding prototypical catalytic binding pockets for reactions absent or under-represented in nature. This is entirely made possible by the power of evolutionary selective procedures, whether *in vivo* or *in vitro*. Yet certain catalytic tasks might still remain beyond the reaches of even extensively modified folded biomolecules, at least at an efficient level of performance. Or the tasks themselves may be realizable, but not under the desired environmental conditions. We can picture this scenario simply by considering the maximal practicable thermal limits of protein-based catalysis, and then asking that some other catalyst steps in to extend the limits of the possible beyond such a boundary. Hypothetically, the universe of all possible foldamers could plug any such remaining gaps in catalytic tasks or catalytic performance unattainable through nucleic acids, proteins, or their modified derivatives. This is depicted in Fig. 7.17, and analogous extensions of versatility could be proposed for other desirable functions as well.

So in the long run, investment in foldamer research would seem "the way to go," and the field is very young. Only since the mid-1990s has "foldamer" emerged as a label for artificial oligomeric structures with sequence-specific folding patterns.[1199] Depending on the strictness of one's definition, though, modern foldamer research as such could be regarded as having older antecedents in the controlled folding of certain rigid polymer chains.[1200,1201] In any case, there has been a great upsurge of interest in this field, no doubt in part owing to its inherent

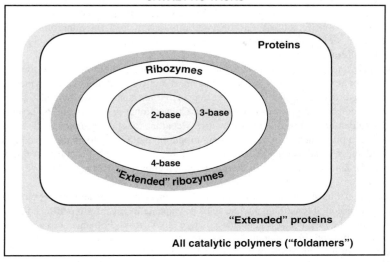

FIGURE 7.17 Probable relative limits of different polymers for the universal set of catalytic tasks. Ribozymes (or DNA enzymes) with limited alphabets are shown as being a smaller set enclosed within four-letter ribozymes. Ribozymes with extended alphabets (equivalent to those using cofactors) have in turn a greater catalytic range than natural ribozymes. Proteins subsume ribozymes in this regard, but modified proteins with extended alphabets are shown with superior catalytic prowess. More hypothetically, some catalytic tasks may be outside the range of even modified proteins, but not necessarily beyond completely artificial folded polymers ("foldamers").

fascination, combined with the underlying challenge of ultimately by-passing nature at a very fundamental level. Foldamers can be broadly subdivided into those that take a lead from nature ("bioinspired") and those that are wholly artificial ("abiotic").[1198] The former are usually based on a wide array of chemical variants of peptides ("peptoids").[1198] Full realization of the more radical abiotic foldamer approach would produce a completely distinct set of monomers as a novel molecular alphabet, which could polymerize into specific sequences capable of protein-like folding. Of course, for this to occur, strict attention would have to be paid to many chemical properties of the monomers, to provide a distribution of charge, hydophilicity, hydrophobicity, and other properties driving the formation of stable folds as in proteins themselves. (Certainly not just any combination of monomers would permit protein-like folding.)

Given our broad division of molecular discovery into empirical/evolutionary and rational strategies, wherein do foldamers belong? For any foldamers, systematic analyses of the relationships between subunit chemistry, sequence, and structure lay the foundation for increasingly sophisticated rational design. "Bioinspired" foldamers could be viewed as a case of boot-strapping upon the information provided by billions of years of the blind process of natural evolution. Information obtained from analysis of natural biocatalysts may feed back into artificial functional foldamer design. For example, we have noted that nucleic acids super-ficially do not appear good contenders for certain enzymatic activities, owing to the limitations of their 2-basepair alphabets and the associated nucleobase chemistries. But a very interesting (and perhaps highly significant point in general) is the extent to which the apparent limitations of the chemical properties of ribozyme subunits may be altered by virtue of subtle sequence and (in turn) structure-dependent effects, which modulate specific subunit properties. (This effect was noted earlier for the deviation of pK_a values in catalytically important ribozyme nucleobases.[834]) Obviously this effect will be dependent on the specific chemistries of the foldamer in question, but it does raise the possibility that an apparently unpromising foldamer alphabet could yield polymeric sequences with useful functions as a consequence of local chemical environments in the folded state.

An immediate difference between foldamers and folded biomolecules is the ability to replicate and amplify the latter. Nucleic acids are directly amplifiable with polymerases, and proteins are indirectly amplifiable if they are tagged in some manner with their encoding nucleic acids (as we considered in detail in Chapter 5). Conventional *in vitro* evolution is thus not an option for foldamers.* Most foldamer structures (let alone functions) cannot yet be specified in advance based on sequence alone, although rational predictions of folds for some peptidomimetic foldamers is advancing (Chapter 9). For the present, the generation of functional foldamers for a novel function will therefore remain empirical. But given their generalizable promise, we will return to the foldamer field briefly in the final chapter.

This section thus serves as an introduction to foldamers in general, and as a conclusion to this chapter. Discussion of functional nucleic acids has repeatedly raised compelling and intriguing questions into the origin of life, and the putative RNA world. Foldamer research inevitably makes one go beyond this into a more fundamental "why" question. For what reason should a limited set of folded molecules be used for life on earth, and could alternatives have been used?[1198] This is a classic conundrum of chance versus necessity, but advances in both functional nucleic acid and foldamer research may allow the formulation of convincing answers to these questions.

A foldamer cannot by definition be classed as a "small molecule," since it must be composed of subunits that themselves qualify for this label. Yet small molecules dominate as useful drugs for many cogent reasons, notwithstanding the recent commercial successes of monoclonal antibodies and the continuing promise of other (folded) protein and nucleic acid-based therapeutics. We should therefore spend time looking at small molecules, especially from the stance of their rational versus empirical discoveries. These themes are covered in the chapter to come.

---

*In this area, another indirect tagging approach has potential, where DNA "tags" are used to specify the order of foldamer subunits, and this will be considered further in Chapter 8.

8 Molecules Small and Large

OVERVIEW

For much of human history, when a problem could be overcome with a molecular solution, the problem was of a medical or veterinary nature. And in almost all such cases, the responsible molecules were small in relation to biological macromolecules such as proteins, nucleic acids, and complex polysaccharides. Of course, as we noted in earlier chapters, empirically making use of a natural source of beneficial molecules within a complex mixture is far removed from actually understanding the physical reality involved. The latter has been achieved only in recent times (historically speaking), but defining the molecular structure of the active constituents of a natural extract does not in itself render the drug discovery process less empirical. But as much of this book seeks to highlight, empirical discovery itself is attaining increasing levels of sophistication. Some of these advances are thoroughly applicable to the identification of new small molecules of potential human value.

The other side of the coin in terms of drug acquisition is rational design, and nowhere else than in the area of small therapeutic molecules have rational approaches had such an impact. This we will consider in Chapter 9, but for the present purposes, we will focus on discovery pathways that are fundamentally of an empirical nature. Again, we define an empirical process as one using a screening or selection process to identify a candidate molecule, and where the exact structure of such a molecule is not specified or predictable in advance. Yet it is very important to consider a fact that we have also stressed previously, to the effect that many rational approaches have an empirical underpinning that enables them in the first place.

Before moving into such areas, it would seem appropriate to first of all consider the "smallness" of molecules from the most basic of viewpoints. While the use of polyclonal antibodies in some therapeutic contexts goes back many decades, it is only in very recent times that artificially engineered proteins and other large molecules have become serious clinical contenders. Indeed, while monoclonal antibodies are referred to as drugs in popular media outlets and elsewhere, in the minds of many involved with pharmacology, the term "drug" inevitably conjures up a mental image of a small(ish) molecular species. Another important issue arising in this context is that of "druggability," or the capacity of a biological molecule to act as a target for drug action, and we will also consider this further within this chapter. Still, this does not yet begin to answer the question, "Why do small molecules have such predominance in the world of traditional drug therapies?" Let us now look at that seemingly naïve query a little further.

Searching for Molecular Solutions: Empirical Discovery and Its Future. By Ian S. Dunn
Copyright © 2010 John Wiley & Sons, Inc.

AS SMALL AS POSSIBLE, BUT NO SMALLER

Making broad scientific or technological pronouncements as aphorisms can be catchy, but dangerous for the long term. (Remember the demise of the premise, "All enzymes are proteins, but not all proteins are enzymes.".) With this caveat in mind, look at the statement "For a specific role, the smallest drug alternative of equivalent functional efficacy will always be the logical choice." Is this necessarily true? If we take into account all possible scenarios and include all relevant factors, we would have to say no, since cost-effectiveness and other factors arise to muddy such simplistic waters. And yet there is much to be said for such a generalization, especially if we are lumping all agents from aspirin to immunoglobulins under the same banner of "drug," as long as they can act in some kind of therapeutic role. Before elaborating further, let us first look for a better, though succinct, guideline here. Famous sayings are often paraphrased (or parasitized if you will) out of their original contexts to make a point. The title of this section has duly taken a sentence attributed to Einstein, "Make things as simple as possible, but no simpler," and made a "small" alteration. This is intended to convey the message that low molecular weight has many advantages in therapeutic applications, as long as no other scientific or practical function is compromised. It also echoes Einstein's statement, in a more context-defined manner. But we still need to explore the matter of how molecular size relates to drug efficacy.

To a pharmacologist, singling out size as an issue might seem inappropriate, since although molecular weight is undeniably a factor in identifying "drug-like" compounds, it is certainly not the only one. We will extend this point later, but the apparent size fixation of this chapter is justified by the striking dichotomy in molecular weights between all the protein and nucleic acid effector molecules we have considered to date, and the large pantheon of much smaller molecules that play critical roles in cellular activities. So before looking at the artificial exploitation of small molecules, let us first focus on some interesting aspects of natural molecules that qualify for the "small" label, especially as they relate to drugs of small molecular size for human use.

Small and Natural, Small and Useful

Low molecular weight secondary metabolites from a variety of organisms have long been of great importance in pharmacology, but many small molecules have essential functions in the life of cells. Biological small molecules have been compared with a missing link in the classic flowchart of the "central dogma" of biology,[1202] where encoded information in DNA is transcribed into mRNA and thence translated into protein. Enzymes create small biomolecules, which not only act as cofactors and regulators of other enzymes but also serve as modulators of other noncatalytic proteins and nucleic acids[*] as well. And the versatility of natural small molecules can be mirrored by their useful artificial counterparts or drugs. All small molecules are a subset of OMspace, but is this "small" sector itself infinitely large? It cannot be infinite, but there is no clear consensus on a realistic size of "low molecular weight chemical space." (Variation in estimates from 10^{13} to 10^{180} is indicative of this.[1205,1206]) Before we embark on a tour of low molecular weight therapeutic compounds, let us then take a closer look at some aspects of the world of natural low molecular weight compounds.

---

[*]We have seen examples of small molecule interaction with RNA in riboswitches (Chapter 7). Small molecule interactions with DNA are commonly indirect via a protein transcription factor (as, for example, in the case of hormone nuclear receptors[1203]). But direct interactions are certainly known, as with a range of antibiotics that can bind to DNA molecules by intercalation or other means.[1204]

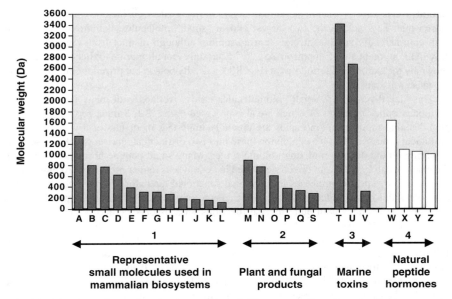

FIGURE 8.1 Comparison of molecular weights of (1) a range of small molecules important in mammalian (and many other) biosystems, excluding encoded peptides, proteins, and nucleic acids; (2) certain plant and fungal products; (3) certain marine toxins; and (4) representative small peptide hormones derived from encoded precursor polypeptides. Specific examples (molecular weights): Within group (1): A = cyanocobalamin (vitamin B_{12}, 1355.4), B = acetyl-CoA (enzyme cofactor, 809.5), C = thyroxine (modified amino acid hormone, 776.9), D = ferro-heme (oxygen-carrying group in hemoglobin, 626.6), E = cholesterol (membrane structural lipid and hormone precursor, 386.7), F = glutathione, reduced form (redox/detoxification peptide, enzymatically synthesized, 307.3), G = arachidonic acid (important lipid, 304.5), H = thiamine (vitamin B_1, 265.4), I = glucose (simple metabolic sugar, 180.2), J = ascorbic acid (vitamin C, 176.1), K = dopamine (neurotransmitter, 153.2), L = niacin (vitamin B_3, 123.1). Within group (2): M = chlorophyll A (photosynthetic cofactor, 893.5), N = digoxin (cardiac glycoside, 780.9), O = reserpine (tranquilizer/antipsychotic, 608.7), P = pyrethrin II (insecticide, 372.5), Q = penicillin G (antibiotic, 334.4), S = morphine (analgesic, 285.3). Within group (3): T = maitotoxin (from marine single-celled dinoflagellate, 3422), U = palytoxin (from a marine coelenterate ((sea anemone-like animal), 2674), V = tetrodotoxin (from puffer fish, 319.2). Within group (4): W = somatostatin-14 (growth hormone regulator, 1637.9), X = Arg-vasopressin (nonapeptide vasoconstrictor/antidiuretic, 1084.2), Y = angiotensin II (nonapeptide vasoconstrictor, 1046.2), Z = oxytocin (nonapeptide causing contraction of smooth uterine muscle, 1007.2). All molecular weights from PubChem-compound except for maitotoxin (T) and palytoxin (U).[54]

Oddly enough, while the biological molecules in question are indeed smaller on average than most functional proteins and nucleic acids, to label them by size alone could sometimes prove misleading. "Small molecules" in biology themselves vary substantially in molecular weight, such that on occasion one is forced to distinguish between "small and large small molecules."[*,1202] An indication of the range in molecular weights of natural small molecules is shown in Fig. 8.1, where the representatives were chosen chiefly to indicate the molecular weight span. Many important small molecules are <300 Da in molecular weight, but some

---

[*] "Large small molecules" still has meaning for us, just as Little Bighorn is not really a place of uncertain size. But perhaps a better term for biological small molecules is called for.

natural products are an order of magnitude larger. Marine toxins were included in Fig. 8.1 for the reason that they include the two largest known "small" molecules, maitotoxin (associated with commercially significant ciguatera poisoning, although distinct from ciguatoxin) and palytoxin (a potent shellfish poison).[1207,1208] (Certainly, not all marine toxins are so large, as shown by the example of tetrodotoxin (Fig. 8.1), a deadly poison but prized in minute doses in Japanese *fugu* cuisine.)

   This size diversity of "small" biomolecules easily overlaps with many natural peptide hormones, some examples of which are also provided in Fig. 8.1. Simple peptides, modified peptides, and modified amino acids are crucially important mediators of a wide variety of biological processes. But we can divide these into two distinct categories, which in itself has a bearing on the definition of small biomolecules. Many small peptide hormones (including the examples in Fig. 8.1) are directly encoded in genomes as longer precursor proteins, which are cleaved by specific proteases into the final functional mediators. Perhaps the best-noted example of this is the product of the pro-opiomelanocortin gene, which is differentially cleaved into multiple peptides with quite diverse functions, in a tissue-specific manner.[1209] In fact, there is accumulating evidence that in addition to cleavage of "dedicated" hormone precursors, many proteins with defined functions donate "cryptic" peptides with entirely different roles, following specific cleavage events.[1210] This may even involve abundant proteins such as albumins.[1211] Irrespective of such interesting observations, all these peptides derive from encoded information directly in the genome; their coding sequences are embedded within the genes for their polypeptide precursors. Yet certainly not all bioactive peptides are derived in this manner. While all biological proteins are translated from mRNA on ribosomes, there are many known instances of the synthesis of nonribosomal peptides by specific enzymes. One such example is the important and ubiquitous peptide glutathione, included within the "small molecules" of Fig. 8.1.

   Despite the size overlap, peptide hormones arising from processed precursors are usually distinguished from other small biomolecules. A key feature distinguishing these two categories is then not size *per se*, or their chemical natures, but whether they are directly encoded within genomes, as opposed to arising indirectly via the agency of genomically encoded enzymes. This is a useful distinction, but still insufficient to pin down small biomolecules as a separate group, since enzymatic synthesis is also responsible for large carbohydrate-derived molecules (polysaccharides such as cellulose, starch, and glycogen) that are also not directly encoded in genomes, but inappropriate within the set of small biomolecules as in Fig. 8.1. Such sizable carbohydrates are derived from a limited set of building blocks, and are thus a class of enzymatically synthesized biopolymers. So another feature of our (small) biomolecules as a group is that they are of a nonpolymeric (or oligomeric) nature. In other words, the structure of such biomolecules cannot be defined simply as a specific sequence of simpler monomeric constituents.[*]

   To qualify as a "small biomolecule" by the above definition, a molecule should therefore satisfy three criteria: it is not directly encoded by a genome, it is synthesized by specific enzymes, and it is nonpolymeric. This description is not linked to the chemical nature of the molecule in question, and therefore includes nonribosomal peptides or certain nonpolymeric modified carbohydrates. "Smallness" does not directly enter into this picture as framed in this way, and consequently some relatively low molecular weight biological molecules, such as peptide hormones (Fig. 8.1), are excluded. Yet most of the "small biomolecules" embraced by

---

[*]Note that this does not exclude the possibility that certain chemical motifs are repeated within a small biomolecule. Maitotoxin thus can be described a polyether,[1207] but not as a polymer (or oligomer), since its polyether motifs are not repeated as regular subunits (ftp site).

this definition are indeed small (<1000 Da) compared with proteins and nucleic acids. Both evolutionary and functional factors are likely to drive this. First, let us consider that molecules that are not synthesized by an informational template or translational adaptor system (as with nucleic acids and proteins) require an increasing number of specific enzymatic steps as they grow in size and complexity, and thus synthesis of very large biomolecules with unique structures comes with rapidly escalating energetic costs. In evolutionary terms, the selective advantage of a large enzymatically synthesized biomolecule must outweigh the fitness loss due to the metabolic investment in the molecule's synthesis. Eventually, a size and complexity ceiling will theoretically be reached beyond which it is not evolutionarily profitable to proceed. (Such a biological boundary is, of course, quite distinguishable from what is *chemically* possible.) This limit will vary between different organisms in different environments, and will depend on both the specific enzymatic steps involved[*] and the fitness benefits conferred by the synthesized molecule. So an absolute limit is not easy to ascertain, and we can only look to current natural precedents (Fig. 8.1).

Evolutionary considerations then place an indirect size and complexity limitation upon the range of small nonpolymeric biomolecules synthesized by enzymes. But the functions of these molecules are obviously critically important as well, and this can be more directly linked with size. The key issue here is target accessibility, where molecules beyond a certain size range may be unable to act as effective ligands for their natural targets *in vivo*. Relevant to the issue of target : drug ligand interaction is the fact that molecular passive diffusion rates are related to molecular size and shape (in general, the larger the molecule, the slower the diffusion rate). But there are evolutionary issues here as well. For example, we have noted earlier that many enzyme cofactors are believed to have arisen as modified nucleosides during the RNA World stage of molecular evolution. From this viewpoint, protein enzymes using such small molecules as catalytic cofactors arose after the cofactors themselves, and the co-opting of coenzymes and evolution of appropriate enzyme-binding pockets then necessarily post-dated the RNA world itself.

Another aspect of accessibility is the ability of small molecules to cross cell membranes. The compartmentalization of biosystems into defined spatial boundaries is regarded as a fundamental step in the origin of cells as we know them,[136] and we have previously seen (Chapter 4) how artificial compartmentalization can be a valuable aid for *in vitro* evolutionary studies. By definition, compartmentalization requires partitioning of a system from the rest of its environment, and restriction of access by exterior molecules. In an aqueous environment, an effective partitioning structure, or membrane, must involve hydrophobic interactions to prevent unrestricted diffusion of outside solutes, and forbid escape of the constituents of the organized system itself. The classic simple membrane lipid bilayer is characterized by amphipathic molecules that stack their hydrophobic segments and expose their hydrophilic regions to form a structure that impedes free diffusion of hydrophilic molecules. In a spherical form, such a bilayer thus defines and partitions an inner and outer aqueous environment. But just as *in vitro* compartmentalization requires effective means for "communicating" with the interior of droplet compartments, membrane-bounded biosystems must regulate molecular traffic with the exterior environment for general nutrients and monitoring of environmental conditions. Small molecules with a hydrophobic character can often passively diffuse across phospholipid membranes,[†] but this is not possible for very hydrophilic molecules, at least at a rate that is

---

[*]The evolutionary costs involved would be reduced if one or more of the enzymatic steps were also used for other cellular functions.

[†]It is interesting to note that even where passive membrane diffusion of lipophilic molecules (those with hydrophobic character) is theoretically possible, at least in some instances specific trans-membrane import mechanisms have evolved, as found in relatively recent times with steriod hormones.[1212]

useful for metabolic purposes. It is then an inevitable consequence of biosystem compartmentalization that specific mechanisms for the cross-membrane transport of hydrophilic molecules must be established, and this is so for all cellular life. (This is also the case for charged inorganic ions, for which many trans-membrane protein structures have evolved for the purpose of forming channels for ionic access.[*])

Having considered these features of natural small molecules, let us now briefly compare analogous aspects of the use of small molecules in therapeutic applications. These drugs can, of course, themselves be natural products that prove to be beneficial for human use, and completely artificial small molecules with "drug-like" character and human utility are also to be included. Generic advantages of smallness in the domain of molecular therapeutics not surprisingly mirror the above points raised concerning natural biomolecules of drug-like size (roughly <1000 Da). Accessibility in its general sense is again of paramount importance. The mode of action of many drugs operates through inhibition of the activity of a target enzyme, and for this to occur at all, the chemical therapeutic agent must have access to the active site of the enzyme. We saw in the previous chapter that extended protein loops such as camelid antibody CDR3 can in some cases penetrate enzyme active sites or clefts and act as inhibitors. Merely in terms of access alone, though, even these versatile recognition systems cannot rival a specifically tailored low molecular weight inhibitor. Proteins exploited for therapeutic use in general remain "macrodrugs" compared to conventional small drug compounds. This is contrasted schematically in Fig. 8.2.

Many conventional drugs thus require small size as an integral part of their modes of action, either for compatibility with their molecular target sites or for compatibility with host processes that affect their ability to access such targets. In this vein, drugs can be directly modeled after natural mediators, for improving desirable pharmacological properties. Among many examples in this regard we can find the important area of peptidomimetics (compounds designed to mimic the functions of specific peptides), which will be noted again in Chapter 9. Molecular size and complexity can also be a pragmatic factor for production costs. Obviously, an analog of a compound with acceptable pharmacological returns compared to a more expensive parental molecule is a preferable alternative. But as a generality, this is not as straightforward as it sounds, because the source material for synthetic starting-points is important. Despite the daunting complexity of many natural product compounds, most have been successfully synthesized by highly ingenious organic chemists,[1214] and with continuing chemical advances it is extremely unlikely that any such molecules will remain intractable for synthesis in the long term.[†] Yet success with laboratory synthesis, often requiring a very large number of steps, is a very different prospect to feasible industrial-scale production of a drug. An example of this is the complex vitamin B_{12} molecule (cyanocobalamin, size noted in Fig. 8.1), whose complete chemical synthesis was achieved decades ago, but which is far more easily and cheaply obtained from bacterial sources.[1217]

In addition to this, harvesting of natural sources can provide useful molecular frameworks for subsequent chemical modifications, borne out by the large number of such semisynthetic compounds in current use. Semisynthetic modification of natural products is not in itself

---

[*]Although phospholipid membranes are impermeable to ionic passage, fatty acid bilayers modeled as protocell membranes permit transit of magnesium ions and nucleotides.[1213]

[†]The current record-holder for nonpolymeric natural product molecular size noted earlier, maitotoxin (Fig. 8.1), has been partially synthesized,[1208] and complete synthesis can be expected to follow. A major problem restricting syntheses historically was the complexity arising from increasing numbers of stereocenters (atoms within a compound which bear an asymmetric arrangement of different chemical bonds; that is to say, atoms whose chemical attachments would yield a stereoisomer if rearranged[1215]). Advances in organic chemistry in the last few decades have resulted in syntheses of compounds with large numbers of stereocenters.[1216]

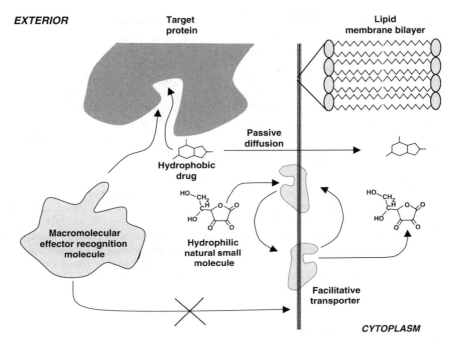

FIGURE 8.2   Contrasts between low molecular weight drug and macromolecular effector molecules. Small-molecule inhibitors can more readily access enzyme active sites or protein clefts than most protein loops. Large protein therapeutics are usually denied cell entry, but low molecular weight compounds with hydrophobic character can passively diffuse across biomembranes. While small hydrophilic molecules cannot readily do this, cells provide a variety of transporters that enable intracellular access. A facilitative transporter (as shown schematically in this case to depict a conformational change in the transporter following substrate binding) assists the passage of specific solutes across the membrane barrier with a favorable concentration gradient (higher external concentration of the solute than intracellular). Transporters acting against such a gradient (concentrators) require an additional energy source.[55] The particular example here depicts the facilitative transport of dehydroascorbic acid (oxidized form of vitamin C), which is a substrate for the GLUT1 glucose transporter.[56]

an unnatural activity, in the sense that many organisms can partially synthesize cofactors from core scaffolds derived from microbial sources. This statement can be illustrated again the case of vitamin B_{12} used above, for although cyanocobalamin indeed functions as a vitamin in the sense of providing the required nutritional supplementation, it does not act directly as an enzymatic cofactor. Rather, it first requires enzymatic conversion to the derivatives methylcobalamin or adenosylcobalamin, to form active coenzymes.[1218] Thus, humans and many other organisms use the complex cobalamin core created by bacteria to obtain the final coenzyme through a "semisynthetic" enzymatic operation. In any case, the above observations demonstrate that size and molecular complexity can be relatively insignificant issues for drug development and deployment if a natural source is available, where the required complex synthesis is wholly or partially performed by cultivatable organisms.

Why then are many complex compounds much more readily prepared in microbes than by chemical synthesis? The more complex the molecule of interest, the greater the need for each step to operate with the highest possible efficiency and specificity, especially with respect to the correct stereochemistry. The most efficient syntheses require the right tools for the job, and efficient, complex catalysts can be the only answer. The latter, of course, are represented by natural biosynthetic enzymes, and enzymes found within the microbial world are consummate masters of biosynthesis.

Yet in the present age we are not confined to purely natural biocatalysts alone. We have previously spent considerable time looking at the intense interest in both reshaping natural enzymes for improvement and generating new enzymatic catalytic activities. For the complete synthesis of complex natural products, it is common to find that a multiplicity of enzymatic steps are required. (To return to the vitamin B_{12} example once more, synthesis of the cobalamin core from its precursor molecule uroporphyrinogen III (itself requiring multistage synthesis) requires about 20 discrete enzymatic steps, depending on the specific pathway involved.[1219]) Engineering or evolving altered or novel biosynthetic products then is inherently a challenge at the level of biosystems rather than discrete enzymes, and this is considered within Chapter 9. But simply because macromolecular catalysts offer so many efficiency advantages over conventional chemistry, the future development of complex artificial small molecules intersects with the future manipulation of the large molecules that can be used to generate them.

So from small to large, and back again, under the lens of our size-wise molecular focus. But as already noted, there are a number of other very important qualities that render a molecule drug-like. Both rational drug design and empirical drug searches benefit from increased information in this regard, so a short tour of this big field is in order.

HOW TO FIND NEW DRUGS

Titles in the scientific literature that begin with "How" can often be unsatisfying if they overstate their objectives for eye-catching effect. A paper or book entitled "How x works" may draw attention, depending on one's interest in x, but it tends to convey the message that the full and complete story will be presented as an open-and-shut case, when this is rarely (if ever) permitted by the current state of knowledge in the particular field involved. (Clearly, this depends on the scope of the x topic, and the level of explanation one is eagerly hoping for.) Is this section heading in the same "x-asperating" direction? It may, if it was taken as attempting to provide a level of detail which more properly can be found in textbooks of pharmacology, but this is not the intention. This section aims at providing a brief background lead-in to the ensuing discussion of some significant features of classical and modern drug discovery, especially as filtered through the twin themes of empirical versus rational approaches. The main aspect of this is a consideration of what constitutes "drug-like" character in a molecule, and a very broad-brush overview of some aspects of useful drug targets. Rational design as a central topic awaits the beginning of Chapter 9, so here we concentrate on low molecular weight drug discovery with an empirical foundation at heart.

Know Your NME

Bringing an entirely new drug for human use into the world is not a cheap undertaking. (If the above "How to. . ." title suffered a single i-to-u vowel mutation, its meaning would be radically altered, but not its relevance to the entire field of drug discovery. Such a "fund-amental" change would make the question easier to answer though: simply find a backer with very deep pockets, if your pockets are not already so well endowed.) With this in mind, it is understandable that a great many clinical trials feature drugs (with previously studied human pharmacokinetics) in possible new therapeutic applications. In numerous additional cases, chemical analogs of a known drug are assessed with the hope of preserving established characteristics of the parental drug's mode of action, while achieving improvements in specific areas of pharmacological performance. But recently identified drug targets will often require an innovative small molecular solution, or a "New Molecular Entity" (NME) in official parlance, referring to previously unapproved drug molecules. While the potential reward for an approved NME "blockbuster" drug are enormous, so to is the uncertainty as to whether the hopeful candidate drug will be able to jump through all the metaphorical hoops leading to the final green light. It has been observed that NMEs with novel mechanisms of action are more likely to fail late-stage (Phase III) clinical trials,[1220] adding a high level of risk to an already costly process.

There are many complexities involved in bringing drugs toward regulatory approval, and given the required investment in time and money, there is a perceived need to identify candidate drugs "developability."[1221] Some key aspects of this have been identified and commonly referred to by the acronym ADME, for absorption, distribution, metabolism, and excretion; often the tag of "toxicity" is thrown in as well to yield ADMET.[*,1222,1223] Another frequent term that overlaps with some of these areas is "bioavailability," which can be considered a measure of how well a drug moves from its site of administration (usually orally) into the systemic circulation and toward its intended target. Not surprisingly, all of the ADMET properties are themselves strongly influenced by the nature of drugs at the molecular level, as well as the physiological responses of the host organism. For the present purposes we will be more interested in some fundamental aspects of the molecular nature of drug-like compounds, and certain very general considerations regarding proteins as drug targets.

Drug design, drug screening, and the nature of drug targets necessarily operate in a linked fashion. Obviously, one cannot even begin to attempt rational drug design without information concerning the relevant specific target, and screening can be greatly streamlined if the field of possible drug candidates can be narrowed down from a purely random collection of molecules. And such winnowing of non-contenders in the drug stakes depends on a level of understanding of what might characterize a useful drug at the molecular level. Researchers have sought to pin down the essential properties of "drug likeness" in a variety of ways, including the enumeration of common drug molecular functional groups and scaffolds, and statistical analyses of drug clusterings in chemical space.[1224] A widely known set of guidelines for drug identification has been defined, usually termed "the rule of five" or Lipinski's rule, in honor of its formulator.[1225,1226] Among these properties we indeed find size, but other critical factors as well. This rule can be stated along the following lines: A drug candidate is likely to perform poorly for absorption or permeability if it has >5 hydrogen bond donors (hydrogen atoms bonded to oxygen or nitrogen atoms), > 10 total nitrogen plus oxygen atoms (that can act as hydrogen bond acceptors), the molecular weight is >500 Da, and the calculated log P value[†] is >5. (The "five" aspect of the rule arises from each numerical property being a multiple of 5.) Candidate molecules violating more than one of these specifications are then designated as having a low probability of succeeding as viable drugs, according to the rule. The rule of five and related dictates for assessing "drug likeness" have been judged within the pharmaceutical industry as very useful guidelines for the general evaluation of prospective new molecular entities.[1224,1227]

---

[*]As an acronym, ADMET could rearrange to "TAMED," perhaps appropriate for a drug that finally passes all hurdles leading to regulatory approval.

[†]Log P is an index of lipophilicity/hydrophobicity, and is defined for a compound as log ([solubility in octane]/[solubility in pure water]). Thus, the higher the log P value, the higher the lipophilicity. A very hydrophilic molecule such as glucose will have a negative log P value (-2.3 in this case), while compounds with hydrophobic character will be preferentially soluble in octane and show a positive value (for example, 8.2 for cholesterol).

Yet at the same time, it has been cogently argued that it is a mistake to take these rules as absolutes.[1227,1228] An implicit aspect of the rule-of-five is its focus on drugs that are orally bioavailable, a factor of considerable practical and economic importance. Nevertheless, many highly effective therapeutics requiring administration by alternative routes would be rejected in the first instance if the rule was applied too slavishly. Also, it is very notable that most natural products used as drugs do not conform to the rule's dictates,[1228] and obviously "macrodrugs" such as monoclonal humanized antibodies are also in violation of the same precepts. We noted above that some drugs can utilize natural transporter systems for their uptake, and those within this category too are poor rule-of-five conformers. Continued refinement of drug-likeness guidelines into higher order "rules of N-multiples" can proceed and no doubt be useful, but a complete division of chemical space into drug-like and nondrug sectors may not be absolutely attainable.[1229] Seemingly in support of the latter assertion, surveys of FDA-approved drugs show that only slightly more than half are compliant with the rule-of-five and orally bioavailable,[1230] although statistics of this nature are continually changing over time.

Another way of looking at drug efficacy from the viewpoint of drug chemistry is to consider common features shared by drugs active on the same or related protein receptor targets. This is the basis of the concept of a "pharmacophore,"* definable as a particular collective spatial chemical arrangement common to all drugs of a class that interact with the same receptor target, and trigger or block its response.[1232,1233] Alternatively, one can express this as the minimal structural requirements a drug molecule must exhibit to preserve a specific functional activity. A pharmacophore as such is an abstraction and does not directly correspond to a particular compound, but rather defines the qualities of "drug likeness" in a very specific manner for drugs aimed at a narrowly defined target. Structure/activity relationship analyses and computational molecular superpositions (overlays) among sets of active compounds have been used to identify essential pharmacophore features.[1234] Pharmacophore description is of practical significance for the pharmaceutical industry in the areas of patents and protection of intellectual property,[1231] but more fundamentally, spatial and chemical definition of pharmacophores can lead to the useful computational screening of candidate drugs for activity. This in turn heavily impacts on rational design (Chapter 9) and also the prescreening of collections of compounds for chemical libraries, as we will explore a little further below. But before doing so, we should look at the other side of the drug discovery equation, which is the question of the amenability of targets to drug action in the first place. Intimately related to this is the process of identifying the most suitable targets for drug development.

Aiming in the Right Direction

If we think about distinct human biomolecules, is it possible to put a figure on the number of possible drug targets in existence? While even in the recent past this question was quite imponderable, with the sequencing and annotation of the human genome it might now appear that at least an estimate of the answer could be forthcoming. Yet things are not so simple in practice, for a number of reasons. Some of these are concerned with levels of genomic complexity and control mechanisms that have only recently begun to be recognized. Traditionally, proteins serve as drug targets, but a variety of nucleic acids and nucleoproteins are also of considerable pharmaceutical potential. But even if we restrict ourselves to the proteome as a drug target, we must factor in differential splicing, editing, and many different post-translational

---

*This term was coined early in the twentieth century by Paul Ehrlich (whom we have encountered several times previously) as "a molecular framework that carries (phoros) the essential features responsible for a drug's (pharmacon) biological activity," which remains close to the spirit of the modern definition.[1231]

modifications that considerably increase the numbers of protein forms over the raw human gene count of $\sim$20,000. Yet adjusting for these complexities still does not hand over even a rough target tally. By the nature of their cellular functional roles, some proteins will be inherently more interesting than others as future drug targets, and most notably, not all proteins are equally susceptible to drug action. This feature of "druggability"[*] is therefore inextricably intertwined with the identification of truly useful drug targets, and only a subset of the encoded proteome is currently amenable to modulation with conventional low molecular weight chemical agents. By definition, the function of a "druggable" protein must be drug modifiable, and a corollary of this is that such a protein must interact with specific drugs with significant affinities (less than $10 \, \mu M$[1236]).

Druggable Struggles What then renders a protein target druggable? In the majority of cases, it is the possession of a fold motif (generally a pocket, cavity or cleft) that is amenable to interaction with low molecular weight "drug-like" compounds.[1237,1238] Not all protein pockets are necessarily accommodating for drug binding,[1237] and a protein lacking a suitable binding site will functionally be scored as poorly druggable. Surveys of known drug targets encoded within the human genome have indicated quite a restricted number of different proteins (on the order of 200[1230]). Estimates of the druggable total by extrapolating to the whole genome have been approximately 1000,[1230,1239] necessarily a tentative figure requiring continual updating. But even if doubled, this figure represents less than a tenth of the genome's protein coding capacity, even excluding the above-mentioned complications of differential splicing and processing. Such "ballpark" drug target estimates are also complicated by some potential targets accommodating more than one drug (especially large and multimeric proteins).

It may be recalled that we have previously briefly noted technologies for deriving intracellular antibodies and aptamers (Chapters 6 and 8). It is hoped that these "intrabodies" and "intramers" could be very effective in targeting otherwise "undruggable" proteins.[1240] Conventional drugs perform poorly when a designated target protein does not present a suitable binding pocket for activity modulation, but only offers extended contact surfaces. Therapeutic disruption of specific protein–protein interactions is likewise often a difficult prospect for low molecular weight drugs. As we have seen, though, these situations are no problem for larger protein- or nucleic acid-based recognition molecules, and these are accordingly attractive in many therapeutic contexts, especially when their sizes are reduced as much as possible. Yet there are exceptions to such a generalization, where small molecules have been derived for seemingly difficult protein-binding applications. One avenue toward this end is the use of small peptides that correspond to specific epitopes within part of a larger protein–protein interaction surface. Such peptides in isolation can then interfere specifically with the protein–protein recognition as desired. As we noted earlier, though, normal peptides have severe limitations through their poor ability to cross cell membranes. Yet in some cases, small molecules unrelated to peptides can also elicit protein interaction disruption. Some protein domains that mediate protein–protein interactions are widespread and do offer potentially druggable sites for therapeutic drug disruption of protein dimer or multimer formation. This need not necessarily occur directly at the protein contact surface, if drug binding to a protein pocket elsewhere results in an allosteric conformational change and subsequently indirectly modulates the normal protein–protein interaction.[1241]

---

[*]Sometimes rendered with one "g" as "drugable." While spelling pedantry is unbecoming, to some of us "druggable" looks more correct through comparison with other words. For example, a "huggable druggist" (a hypothetical lovable pharmacist) would look odd with single g's. A related (but less frequent) term is "targetability,"[1235] which is a largely self-explanatory.

A take-away general conclusion is that many circumstances of protein–protein interactions formerly regarded as "undruggable" are more accommodating than formerly realized.[1242] What does this portend for the future of druggability of protein–protein interactions in general? Increasing success against hitherto poorly druggable targets has in part been achieved through improved definition of specific potential drug-binding subsites, and increasing chemical design sophistication. Yet advances in chemical library design and high-throughput screening (as below) have also been productive. Does this mean in the end that "undruggability" will become a thing of the past? Clearly, protein target access for drug action varies widely on a case-by-case basis, as certain permissive interactions are likely to be a relatively easy subset of the total. An extended flat contact surface without any usable drug-binding pockets remains poorly druggable. Again, size comes into the picture, where a larger recognition molecule has increased opportunity to form stabilizing contacts over a wide contact surface than a small molecule (and therefore more chance of acting as a successful inhibitor of the target). There is no inherent chemical restriction on the nature of such a "large recognition molecule" (it need not be a protein or nucleic acid in principle), but generation of such larger molecules, and the identification of those with desirable binding properties, is more readily achieved with proteins and nucleic acids within current technological limits.[*]

Overcoming druggability barriers in any area is obviously a good thing, but we should keep in mind that a protein–protein interaction of interest will usually perform as a component of a larger biological system, which might offer entirely different target opportunities toward attaining an equivalent functional end. Complex systems may also offer multiple targets for therapeutically beneficial modulation. As an example, the inhibition of transcription has many potential pathways of attack for small molecules, not only at the protein or DNA-binding levels but also through RNA, where suitable mRNA structures exist. In this regard we might be reminded of the therapeutic potential of prokaryotic riboswitches as antibacterial targets (Chapter 6), and viral RNA structures involved with gene expression regulation are also applicable to intervention with low molecular weight drugs.[1244] Druggability and optimal target choice therefore should not be considered in isolation from each other.

High-throughput Put Through High Loops   Much more could be said about the issue of target suitability, but for present purposes we should return to the issue of small molecule libraries, a diverse topic encompassed by the theme of high-throughput screening. Historically, evaluation of chemical functions in biology was a linear (and laborious) process where candidates were tested separately, one by one. This was done just over a century ago by Paul Ehrlich, one of the main pioneers of chemotherapy, for the screening of drugs against the syphilis treponeme[1245] (as mentioned in the Introduction). As a labor-intensive and time-consuming undertaking, this screening often required near-heroic levels of dedication in the face of repeated failures, until successful candidates were isolated. So modern chemical libraries are most definitely not screened on a linear basis, but by means of massive parallel processing, or simultaneous evaluation of large numbers of library members at one step. For this reason, high-throughput screening and chemical libraries are intimately linked technologies.[1246]

By definition, high-throughput involves rapid assessment of a large number of alternatives from a large collection, tested against a target of some kind (that can range from a single molecule to a complex system such as a mammalian cell). The "alternatives" in this arrangement are members of a library, but the physical nature of a molecular library has no specific boundaries. We have previously considered libraries of mutated proteins and nucleic acids, but there is no

---

[*]But fragment-based approaches with small molecules such as molecular Tethering[1243] offer promising alternatives for deriving binding molecules for difficult protein targets; considered in more detail later in this chapter.

reason for limiting the definition to these biological molecules. Various technological advances in recent times have converged to render feasible the generation of large collections of diverse small organic molecules, which then become chemical, rather than biological, libraries. Of course, a major difference exists between the biological and chemical library groups. Nucleic acids can be directly amplified, and proteins can be indirectly amplified through linkage with nucleic acids that encode them. This is not possible as such with small nonpeptide molecules (although later in this chapter, we will consider "tagging" approaches that endeavor to capture nucleic acid amplification for small molecule identification). Screening of chemical libraries must accordingly be performed in a systematic and necessarily high-throughput manner.

We first noted high-throughput approaches in Chapter 1, and some more technical details regarding the implementation of this type of screening are provided in ftp site. High-throughput screening technologies are relevant in a great many areas of scientific and technological investigation. As an example, consider once again molecular imprinting, as raised in Chapter 7. Here, it was noted that a basic issue in the production of molecularly imprinted polymers is the choice of monomers and cross-linking agents, and linear manual trials of different conditions for imprinting optimization is superseded by automated high-throughput combinatorial approaches.[1164] Libraries of polymer combinations imprinted with the same target template molecule can thus be systematically evaluated for choosing the imprinting conditions which elicit the best binding.[1164,1247]

While acknowledging the importance of the screening process, our focus will lie more directly with small molecule libraries themselves, and the parallels that can be made between such libraries and other forms of empirical molecular discovery. The optimal composition and assembly of libraries of chemical compounds requires a much higher level of detailed knowledge and analysis than might meet the eye at first inspection, and their importance and continuing potential for drug discovery has warranted this kind of focus. The nature and origin of these libraries of small molecules are thus what we should turn to now.

CHEMICAL LIBRARIES AND THE GOD OF SMALL MOLECULES

Collections of chemical compounds are not at all a recent development,[1248] although beyond two decades ago they were not usually referred to as "libraries." We could define a chemical library purely as a collection of small molecules that are systematically screened for some drug-like or other useful property, independent of the screening process itself. Nevertheless, high-throughput screening in itself is a liberating factor in allowing the practical size of the library to substantially increase. The other major development in relatively recent chemical library technology is the design and rationale of the library chemical diversity. Let us then focus first on this issue of chemical library content.

Generalities and the Natural Approach

The design, preparation, analysis, and optimization of chemical libraries have become an extensive field in its own right, reflecting the many levels and types of chemical diversity that can be implemented. Like protein space, small-molecule chemical space is far too enormous to be encompassed within any physical library of practical significance. Even designing a single library of maximal diversity and universal relevance is an unlikely prospect, given the diverse areas where chemical libraries can be applied. What types of such libraries have then been generated, and what underlying principles determine library assembly? Two philosophies of chemical library generation could be reduced to: (a) "Let's rummage through all organic compounds we have at our disposal, whether natural or artificial, and pick out the ones with the best prospects. We have some good ideas about what renders a molecule drug-like, and we can therefore focus on the most likely contenders" and (b) "Let's screen for completely novel small molecules using special *combinatorial* chemical syntheses. We can do this by choosing sets of functional building blocks and linking them together in different sizes and sequences, just as for the artificial chemical synthesis of peptide libraries. Alternatively, we can take specific chemical scaffolds as starting points and "decorate" them with a variety of combinations of functional groups."

Before elaborating on these approaches, we should note that both are limited by a familiar nemesis: the game of finding the ceiling of numerical practicality for large libraries of any type. For all libraries, the "practicality" factor is entwined with the screening process, but this is even more fundamental with chemical libraries. Without high-throughput, the feasibility of screening the latter plummets, and reduces them merely to uninformative chemical collections.

We have earlier looked at some of the properties of molecules considered to render them drug-like (such as the rule-of-five), and these can, of course, be applied toward the evaluation of pre-existing compounds for library inclusion. A more restricted set of criteria has been refined for attempting to pinpoint the quality of "lead likeness," or compounds that are predicted to show promise as high-quality leads for drug development.[1249,1250] It was also noted previously, though, that natural products with useful pharmacological properties often fit poorly with drug likeness, as narrowly defined by such rules.[1251] Indeed, systematic comparative studies have shown that synthetic drugs and most artificial chemical libraries have marked differences to natural products by a variety of criteria.[1252] Of particular note, natural products tend to have more chiral centers (atoms with substituents that can exist as alternative stereoisomers), differences in the average numbers of N, O, S, and halogen atoms, and more nonaromatic fused-ring polycyclic structures.[1252,1253] Aggregate molecular properties for compounds (molecular "descriptors") can be compared by a mathematical approach termed principal component analysis, through which property distributions demarcating drugs and natural products have been derived.[1252,1254]

Natural products as a whole are structurally both diverse and complex.[1255] Molecular complexity in general terms is associated with drug value, although only up to a point.[1256–1258] An inherent requirement for useful drug action *in vivo* is an acceptable level of specificity of action. This requires a minimal number of specific contacts between the drug and its primary target molecule, in turn dictating that a molecule below a certain threshold level of complexity is unlikely to have good drug-like qualities. But beyond "moderate" complexity, the chance of finding a specific binding interaction for a random complex ligand falls,[1256] and pharmacokinetic problems tend to increase.[1257] Given the perceived benefits of natural products with respect to diversity and sources of drug leads, libraries of archived natural products have been assembled (Fig. 8.3a).[1259,1260] Natural product libraries have been commercially offered as crude extracts or nominally pure,[1259,1261] with some tens of thousands of compounds included. Yet for a period beginning in the 1990s, screening of natural products took a back seat to synthetic combinatorial libraries.[1262] To some extent, reluctance to use natural product libraries was associated with certain technical limitations (including the average increased complexity of natural compounds), but these potential drawbacks have been mitigated with recent advances.[1262] Still, some natural products do present problems of developmental and scale-up costs as drug leads. Where natural products are obtained from third-world primary sources but developed elsewhere, complex intellectual property issues can also arise.[1263]

But for assembly of a chemical library, one can take heed of these caveats while "cherry picking" any desirable compounds from all available sources, including archived synthetic "hand crafted" compounds as well (Fig. 8.3a). Chemical compound archives are maintained by

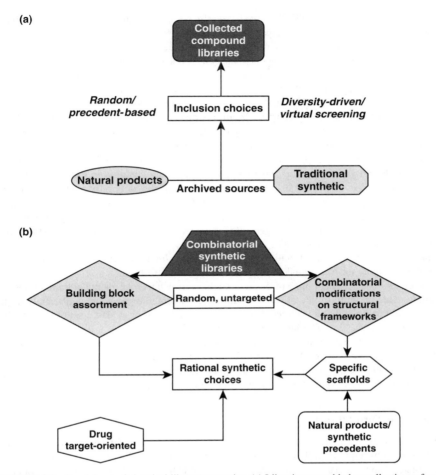

FIGURE 8.3   Breakdown of chemical library categories. (a) Libraries assembled as collections of pre-existing compounds, where the choice itself can be untargeted (random) or influenced by factors as shown. (b) Two major aspects of combinatorial synthetic libraries: sequential couplings with specific building blocks and combinatorial variation on chemical scaffolds. Here also, the combinatorial chemistry can be random or rationally designed for diversification purposes and tailoring to specific drug targets.

certain companies and organizations (for example, the Novartis Institute[1264,1265]), and companies specializing in compound library development have emerged relatively recently.[1266–1268] Numerical constraints on library sizes enforce the need to define criteria for compound inclusion, and the vastness of chemical space dictates that this requirement will always exist. In any case, the "shelf" of low molecular weight compounds available for purchase is already large (on the order ten million[1269]), indicating a need for practical library selectivity even if this was the only set of molecules under consideration. To cover a gamut of structural possibilities within chemical space, one reasonable objective is to maximize chemical diversity and eliminate redundancy. Diversity itself correlates with the degree of library molecular dissimilarity, as might be logically expected.[1269]

With the kinds of numbers involved, clearly computational screening of assembled candidate library members is an imperative. To do this, some kind of measure for diversity/dissimilarity is required, to move away from what is otherwise a subjective assessment.[1255] The above-mentioned principal component analysis is one way to tackle this,[1254] and a variety of other methods using physicochemical and molecular topological yardsticks have been used[1205,1206,1255] although without a universally accepted consensus approach yet emerging.[1269] It is important to note the different levels where assessment of diversity can apply. Most attempts to quantitatively characterize library diversity are based in the end on structural criteria of one form or another, but a highly relevant consideration is diversity at the level of function. This very practical aspect of library quality relates to the utility of "universal" libraries for yielding multiple hits for entirely different functional screens. A library with superior functional diversity will therefore have more widespread applicability in drug screening for multiple targets, an obviously desirable feature that is unfortunately only practically gauged through screening experience.[1254] Although one can define diversity through specific molecular structural descriptors, these may not always bear on what is most biologically meaningful. Nevertheless, structural diversity in general is expected to show a correlation at least to some degree with the practical read-out of functional diversity.[1254] Notwithstanding ongoing issues in its definition, measurement, and implementation, maximal possible structural diversity is therefore a highly desirable goal for chemical library optimization.

But we noted at the beginning of this section that a truly universal chemical library is a hard thing to come up with. In lieu of universal libraries for screening, one can deliberately try to design a library such that it is directed against a specific target, or at least a set of targets with common characteristics. The latter feature suggests that for focusing a library for targeting common protein structural features, one should think globally rather than the local attributes of a single specific protein. And in turn, this inevitably directs us to modern genomics and proteomics, and their derivative science for small molecular discovery of "chemogenomics" (ftp site). But there is yet more to say about chemical libraries themselves. . ..

Make Your Own Library

Many of the above issues regarding analysis of diversity and drug target-oriented library generation apply equally well to libraries of synthetic compounds (Fig. 8.3b) as to those derived from natural sources. The obvious difference, the synthetic procedures themselves, ultimately derive from solid-phase techniques with peptides dating from the early 1960s,[1270] and the underlying importance of solid-phase technology for synthetic chemical libraries has been stressed.[1182,1254] Peptide synthesis can be used as a good example of combinatorial library generation, which can be done in high density format.[1271] The idea of generating synthetic libraries of peptides might immediately remind us of biological recombinant peptide libraries produced via display technologies (Chapter 5), and indeed synthetic methods are hard pressed to rival the numbers of individual peptide variants in a library that can be obtained by the best *in vitro* display methods (such as mRNA display). Yet all synthetic libraries have their own inherent advantages. By definition, synthesis is not subject to biological constraints that can (for example) limit the representation of certain peptide sequences from some display libraries. Synthetic peptide libraries can thus include all manner of unnatural amino acids, including D-amino acid stereoisomers. It is also possible to use peptide leads generated through display technologies for optimization through systematic synthetic approaches.[1272]

In vitro peptide synthesis is a combinatorial process using solid-phase supports and sequential addition of new amino acid building blocks, and this principle can be extended to any chemical set which can be manipulated in an analogous manner (Fig. 8.4). Yet this

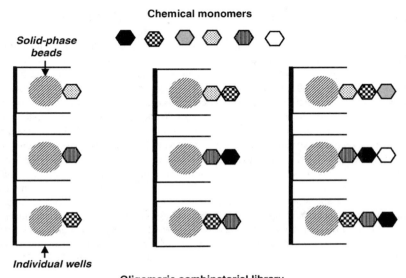

Oligomeric combinatorial library

FIGURE 8.4 Basic combinatorial synthesis with sequence of serially added monomers (as with amino acids for solid-phase peptides) determined by spatial position.

comparison between synthesis and biological display raises an important issue for all combinatorial chemical libraries. We have seen a repeated message with display technologies in the all-important linkage of phenotype and genotype, but this cannot occur in the same manner with completely synthetic chemicals, even if they are originally of biological origin (as with peptides). A solution is to encode information into the synthetic process such that a library member can be identified after its isolation by binding to a target. Various chemical "tags" can achieve this, and by virtue of its importance this topic has its own section that follows later in this chapter. A simpler approach is to use *spatial encoding*,[1273] where the identity of a library member is specified by its spatial position in a grid or array system.* Simple in principle, this notion entails screening each library candidate in a separate identifiable spatial compartment such as a well of plate, which clearly carries escalating demands as library size increases. High-throughput automation, though, has met such challenges up to a point.

Combinatorial chemistry and chemical libraries, of course, go far beyond peptides, indeed in principle bounded only by the vast chemical space of all small molecules.[1274,1275] While as we noted earlier, the number of compounds encompassed by this Space is at best enormous and at worst hyperastronomical,† the reality is slightly more comforting. Analogously with the vast biologically nonapplicable areas of protein space, a huge fraction of the set of all small molecules has a very low probability of ever becoming drugs. Moreover, a battery of rational pathways for winnowing candidates for library inclusion (as we touched on above) promises to further narrow the field. Application of available synthetic routes can accelerate the progression from initial "hits" toward *bona fide* lead compounds. Also, flexible synthetic

---

*This kind of encoding was the basis of the metaphorical treatment of library screening in Chapter 1.

†As an example, molecules of only 11 atoms and composed only of carbon, nitrogen, oxygen, and fluorine have been computationally modeled as forming $\sim2.6.10^7$ compounds (with over 10^8 stereoisomers).[1276] Only $\sim0.2\%$ of these molecules were present in 2007 databases.[1276]

options hasten the acquisition of data concerning structure–activity relationships for candidate molecules.[1254]

Enthusiastic investigations of combinatorial chemical libraries in the 1990s tended toward the "bigger the better" principle, but owing to the size of chemical space, a large purely random collection may be inferior to a much smaller compound set with predetermined desirable qualities. For any particular chemical diversification, "coverage" of as much of the maximal diversity as possible is desirable, but practical concerns often intervene between a theoretical desirable library size and what is realistically attainable and screenable. An understanding of the trade-offs between library diversity and size can show when a point of diminishing returns is reached. As the potential combinatorial complexity of a library increases, so too does the library size required for coverage.[1205] A more effective principle for chemical library design and assembly can be phrased as "quality over quantity," leading to a more recent trend toward smaller, but more effectively-designed libraries.[1263,1277,1278] Indeed, rational considerations have featured more and more heavily in the design of synthetic libraries.[1279,1280] Computer-generated virtual screening features heavily in this process, taking prototype libraries through a series of checkpoints until the actual library for real-world synthesis is virtually selected.[1279] We will return to the area of virtual screening as a facet of "rational design" in Chapter 9, but for now let us look at some direct issues of chemistry involved in the synthesis of chemical libraries, at least in broad terms.

A well-trodden path for combinatorial chemistry is the "mix-and-split" (or "split-and-pool") process,[1254,1281] which introduces increasing chemical diversity after each successive round (Fig. 8.5). In effect, this kind of approach combinatorially shuffles chemical building blocks, which encourages an analogy to be drawn between it and genetic recombination.[1281] The building blocks in such combinatorial syntheses need not be a small set of monomers (as with amino acids and peptide synthesis), but can be composed of a wide range of different chemical groups which serve as a potentially rich source of diversification. Conventional combinatorial chemistry aims for diversity, but when it is focused on library synthesis based on a specific scaffold (as in Fig. 8.5) then it is clearly sampling a restricted volume of chemical space.[1255] On the other hand, pragmatic observations show that small substitutional changes on a molecular framework can make a large activity difference. Molecules with a common scaffold have thus been found to have only ~30% chance of sharing a functional activity.[1282]

In any case, not all molecular scaffolds are created equal. For example, take the scaffolds offered by natural products, whose advantages were noted above. In particular, it is believed that a major reason for the frequent success of natural products is their inherent evolutionary linkage with the set of protein folds observed in the biosphere. As a result, natural molecules selected by evolution for binding to protein motifs will tend to have special features promoting such interactions.[1283] For example, molecular volumes of most natural products are within the volume size range found within protein cavities.[1284] The gift of evolutionarily "prescreened" scaffolds can in principle be combined with artificial chemical diversification, and many groups have done just that.[1253,1263,1285,1286] Combinatorial synthetic manipulations upon judiciously chosen natural product scaffolds have also been perceived as a route for circumventing some of the technical and commercial problems associated with native-compound natural products themselves.[1263] An activity allied with investigations of natural product-like libraries is the systematic classification of known natural product scaffolds,[1284] and global structure–activity relationship analyses of large sets of them.[1287]

The utility of the core scaffold ring structures within natural products is relevant to the pharmaceutical concept of "privileged structures" within important sets of drug ligands.[1288,1289] This stems from observations that some molecular core structures are found in ligands against multiple protein targets encoded by broad gene families.[1290–1292] For this

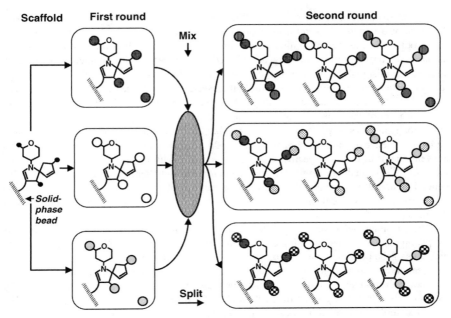

FIGURE 8.5 "Mix-and-split" approach to combinatorial library generation from a starting framework scaffold structure on solid-phase beads. A small library can be produced after even the first round of derivatization, and successively diversified in succeeding rounds. Synthesis begins on a predetermined structural scaffold with a limited number of reactive groups (black dots). In any one round, each separate compartment undergoes reaction with a specific chemical reagent (dots in right-hand corner of each capsule in the diagram) to produce a derivative that is capable of further reactivity. After a reaction round, mixing the products and redistributing them into new reaction compartments allows progressive diversification as indicated. The process can be repeated beyond the second round as many times as desired. Note that each reaction solid-phase bead carries a unique chemical species; identification of functionally selected compounds requires a means for encoding the synthetic combinatorial information and a means for reading it out after selection (as in the "tagging" section later in this chapter).

to occur, a logical inference (supported by investigative evidence) holds that there must be common structures in "privileged" ligand-binding sites among (otherwise disparate) protein family members.[1293,1294] Given the importance of the G-protein coupled receptor family to pharmacology (as noted in Chapter 3), these receptor proteins have been extensively investigated in the light of the privileged structure concept.[1293–1295] The notion of privileged pharmacological "masterkeys" for target families[1296] is an intriguing and productive line of investigation, subject to its correct interpretation. Perhaps reminiscent of overdedication to the rule-of-five (as referred to earlier), a cautionary note has been drawn against slavish application of the privileged structure paradigm in new areas, without adequate verification of the specific association of a putative privileged structure with a target family.[1297] (Showing a high frequency of a particular "privileged" structure in association with a related set of targets would not in itself rule out the possibility that such a structure's utility stemmed from more general drug-like properties.)

But exclusive dedication to consensus natural product scaffolds (whether deemed privileged or not) would logically exclude the possibility of discovering radically new compound structural motifs. Specific scaffold libraries can be seen as hypothesis-based, proceeding logically from the proposal that a specific scaffold is an excellent framework to generate a useful compound. In contrast, a library where scaffolds were included in diversification as well would therefore be a *hypothesis-generating* approach,[1255] since it does not rely on preconceptions at the outset. In this regard, multiscaffold libraries strive toward an ideal of a completely unbiased set that can empirically seek new pathways for molecular development. A systematic approach to this goal has been termed diversity-oriented synthesis,[1281] which enables change (diversification) in starting-point scaffolds themselves through the specific synthetic steps used.[1182,1298] As well as "skeletal" multiplicity, diversity-oriented synthesis must also engineer diversification of the repertoire of side-chain and linker building blocks, and pay attention to stereochemical variety.[1298] (With respect to the latter, we might recall that a considerably higher prevalence of chiral centers in natural product compounds is found compared with most conventional drugs*.) The specific chemistries involved with this process are obviously key factors,[1301,1302] and their continual improvement remains a challenge for the future. Within this general area, multicomponent reactions (as the name implies, reactions involving more than two participating molecules) are recognized as being of special value for the purposes of diversification.[1281,1301,1303]

Often the component members of chemical libraries are referred to as "elements," which is simply another shorthand for "single chemical entity" or its verbal equivalent. But just as the elements of the periodic table are composed of atoms that are not at all indissoluble entities, so to can most "small molecules" be considered as amalgams of different chemical moieties. This consideration leads us to an important and relatively recent development in searching for useful drugs, which at the same time as it looks "downward" at subunits of small molecules, must also look "upward" at proposed drug targets.

Frag-mental Inspiration

When challenged by complexity, a time-honored approach is to seek ways of breaking down a multifaceted entity into smaller and more easily managed units. If one seeks to apply this in the molecular domain, the relativity of "smallness" soon becomes an issue, and at the beginning of this chapter we saw that a "small molecule" can cover a substantial range of molecular weights. Within this size span, we find drug molecules, and it is usually possible to represent these small molecules as derivatives of still smaller compounds, or "fragments." A complete drug molecule can also be seen as the result of merging or linking multiple fragments.[1304] Conversely, many drugs themselves are structurally encompassed as fragments of other drugs of larger molecular weight.[1305] From the point of view of targets, the sites of binding of useful drugs can be analyzed and subdivided into local areas of molecular interaction, which in turn can be related to interactions mediated by molecular "fragmental" components of the drug of interest. Molecular discovery by fragment analysis is consequently intimately related with target binding itself. But before looking in more detail at how targets feature here, we should spend a little time dealing with the fragment concept itself. In so doing, this chapter's focus on small molecules is "downsized" even more, perhaps to its logical end point.

How can one define a "molecular fragment"? The notion of such fragments might bring to mind the previously used submolecular concept of "scaffolds" or "skeletons." The latter refers to a core structure that is a central feature of a molecule preserved through a variety of side-chain alterations, such as the classic common skeleton of steroidal drugs and hormones. Pharmacological molecular fragments themselves can possess ring systems, and the steroidal ring system can certainly exist as fragments of larger and more complex molecules. While a

---

*And in fact, libraries using "natural product-like" scaffolds (noted above) can also be developed using the principles of diversity-oriented synthesis.[1286-1300]

fragment of anything would seem to be only definable in relation to the whole from which it is derived, in practical terms drug fragments are compounds necessarily on the low end of the small molecule size range (as in Fig. 8.1), and which collectively share certain characteristics. This follows from observations that drugs that have passed all screening hurdles and entered clinical use tend to adhere to a range of properties, as we have seen with the rule-of-five dictum earlier. (Although we have also made note that such "rules" are at best just a set of useful guidelines, and many drugs, especially natural products, escape the net that they cast.) But for fragments, perhaps the situation is somewhat simpler, as molecular size for one thing is an obvious factor. By direct parallel with the rule of five, the "rule-of-three" has been proffered,[1306] which of course stipulates a size boundary (≤ 300 Da), and also values of ≤ 3 for $\log P$ and numbers of hydrogen bond donors.[*] An implicit feature of this "fragment rule" is therefore not only mere constituency within a larger molecule but also continuing functionality. Drug molecules can be "deconstructed" by systematically removing substituent groups, but simpler progenitor molecules must be identified as retaining functional activity toward a specific target. Reiteration of this process logically leads to the identification of minimal fragment elements that possess a measurable trace of the original activity above background levels.[1307]

While bioactivity of fragments is a useful guide (or at least a pointer toward a real drug lead), an inherent expectation is that fragment functional levels will be low compared to complete drug entities. Thus if the functional read-out is binding affinity toward a protein target, affinities will be anticipated to dramatically increase as fragments are merged toward becoming a useful drug molecule. (Remarkably, an almost linear relationship between molecular weights and binding affinity toward target has been observed as lead compounds are "deconstructed" into progressively smaller fragments[1307].) Yet if two small molecules (considered as fragments) are to be merged together to form a larger molecule combining their relevant functional features, an implicit assumption is that the properties of the entire drug molecule are collectively localized to its functional groups, and not an "emergent" property of the molecule as a whole. If such functional additivity applies, then fragments of a larger molecule will retain at least some of the original capability of forming noncovalent interactions with a target molecule. Of course, depending on their positions within a "merged" molecule, functional groups themselves may interact with each other and change the pattern of target binding. Also, by the very nature of chemical transformations, there are exceptions to the simple additivity notion if it is taken too far beyond the level of discrete functional groups. A basic example would be the gain or loss of aromaticity[†] in unsaturated ring systems as they change in size, showing that the fragmentalization principle can break down if applied indiscriminately.

But the overall usefulness of merging or linking of fragment molecules reminds us that molecular combinatorics can operate on a number of different levels. Much of the previous chapters dealing with directed evolution of proteins and nucleic acids are essentially devoted to ways of combinatorially diversifying strings of amino acids or nucleosides for screening or selecting novel activities. This, of course, is combinatorics at the level of polymers with defined alphabets, and can be contrasted with combinatorics operating through building up libraries of small molecules (as in Fig. 8.5). While the combinatorial libraries we have already made note

---

[*]Refer back to the earlier discussion of the rule-of-five for more detail on the meaning of these parameters. Often a cut-off molecular weight boundary of ~ 250 Da is cited as an upper boundary of fragment size,[1243,1307] making the rule-of-three a bit less precise in this regard.

[†]An "aromatic" chemical compound does not indicate anything odoriferous, but refers to a special property of certain rings containing double bonds. As in the prototype compound, benzene, "π orbital" electrons particpating in bonding are delocalized around the ring structure, resulting in very different chemical properties to that expected from conventional double bonds. The structure of benzene and the development of aromaticity theory are milestones in the history of chemistry.

of could be considered examples of "fragmentology" in the sense of progressively assembling small molecules from smaller parts, the difference with true fragment-based approaches lies at the level of screening. The members of conventional combinatorial libraries are subjected to high-throughput screening for direct drug leads, while fragment-based screening initially looks for weaker target binders with libraries of smaller fragments.[1308] Fragment analyses also serve to simplify analysis of ligand binding by computational methods, and as an aid to virtual screening,[1308] considered further in Chapter 9.

But a major reason for the perceived benefits of fragment-based approaches is the fundamental nature of chemical space itself. By definition, fragments on average are composed of fewer atoms than "complete" small molecules. It is also inevitable that fragment combinatorics (combining fragments in many different arrangements) means that there are hugely more "complete" small molecules than there are fragments in chemical space.[1243,1308,1309] In other words, fragments occupy a smaller region of chemical space, while representing a useful level of diversity. A fragment library at the same time would theoretically embody a greater degree of potential secondary diversity than molecules of larger mass, through chemical "shuffling" of fragment constituents. Also, since modeling indicates a negative correlation between compound complexity and frequency of target hits during random screens,[1256] fragment libraries should yield a significantly higher hit rate during screening than conventional chemical libraries. This prediction carries obvious practical ramifications, and has been supported by experimental analyses.[1310]

Thus, the evaluation of fragment libraries constitutes a partial answer to the conundrum of the size and diversity of chemical space. At the same time, considerations of fragment combinatorics and hit rates would suggest that fragment libraries of relatively modest size would be productive for screening, in comparison with those used for conventional chemical compound high-throughput. Yet for practical implementation, all this is predicated upon the assumption that low-affinity binding of targets by fragments is amenable to dissecting *structure–activity relationships** in a reproducible manner, a supposition that has been vindicated by experience.[1308] Low-affinity interactions may nonetheless require high working concentrations of fragments, and often escape detection by conventional molecular screening assays.[1309,1311] Relevant data can nevertheless be obtained through direct structural approaches such as nuclear magnetic resonance ("SAR by NMR") or X-ray crystallography.[1304,1309,1312] In such cases, the library sizes are necessarily very restricted, but gaining a foothold on a target with a molecular fragment allows extensive optimization to commence. This can operate by extending the fragment molecule through systematically "building" or "growing" it in size,[1313] with concomitant screening for increased target binding or functional properties (Fig. 8.6). If two discrete fragments bind proximally to each other with respect to a specific target site,[†] then "merging" these compounds (as referred to above) has the prospect of summing their binding interactions in a resulting small molecule lead (also depicted in Fig. 8.6).[††] It is here that the potential of fragment combinatorics comes to the fore, since combinatorial assortment of a relatively small fragment library theoretically rivals much larger traditional compound libraries.[1243] In contrast to the "building up" of fragment hits, a significant fraction of conventional chemical library hits actually require a decrease in molecular mass for

---

*Structure–activity relationships (SARs) define correspondences between the structure of a compound and its functional results in relation to its target. This basic avenue for drug optimization is most useful when quantifiable (hence QSAR) and leads to rational drug design strategies (Chapter 9).

†When structure-based methods are used (such as NMR or X-ray crystallography, as referred to above), screening can identify fragment ligands binding to defined structurally adjacent sites in the target molecule.[1312]

††An area of molecular discovery that overlaps with the concept of fragment merging is termed "'Tethering[1243],'" and is detailed further in ftp site.

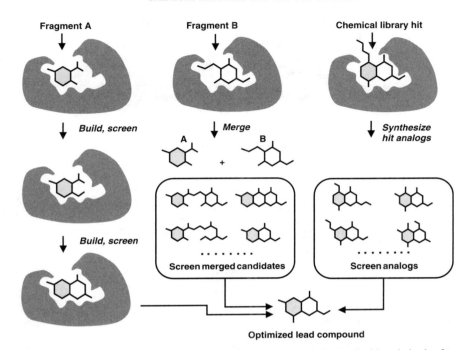

FIGURE 8.6 Fragment optimization by building versus merging, and compared to hit optimization from a conventional chemical library. Detection of target binding from within a fragment library allows candidate fragments to be optimized by extending their structures ("building" or "growing"). If multiple fragments binding the same target site are defined, then fragments can be "merged" as depicted to approach the optimized ideal. In a conventional chemical library, an initial hit provides a structure for optimization by screening a sublibrary of chemical analogs. Combinatorial assortment of fragment library members (such as A and B) allows relatively small libraries (compared to conventional compound collections) to attain high levels of binding diversity.

optimization (also portrayed in Fig. 8.6).[1256,1314] Even a very large and highly diverse conventional small-molecule library (with average molecular weight >300 Da) is unlikely to furnish primary hits that are fully compatible with the spatial arrangement of a specific binding site within a randomly-chosen target. Both positive or negative adjustments of molecular sizes of conventional hits during the optimization process are then not surprising. On the other hand, fragments of <250 Da are far more likely to be encompassed within a macromolecular target site, and require upward building of molecular mass for affinity improvement. The optimization of a conventional chemical library hit of greater than usual size or complexity may in fact benefit from the deconstruction of the molecule into fragments of more manageable dimensions.[1315]

Fragment combinatorics is inherently guided by the intended target used in successive screenings. Let us now look at some powerful approaches where this general concept can be harnessed more directly.

Self-Assembling Molecular Combinatorics   A macromolecular target by its very nature provides the sites that serve as recognition points for the specific binding of small molecules. The molecular surface terrain of targets can therefore serve as a means for assembling fragments into small molecules with useful net interactive properties. If library members

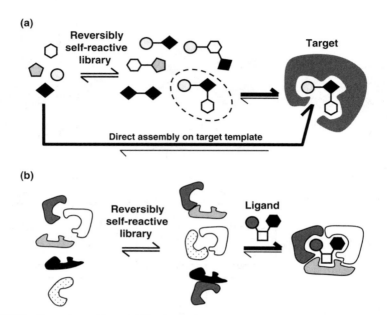

FIGURE 8.7   Dynamic combinatorial libraries. (a) Target-mediated shifting of equilibria in a reversibly self-reactive chemical library, either by stabilizing a species formed in solution or directly within the target binding site ("virtual" combinatorics[57]). (b) Ligand-mediated shifting of equilibria in an analogous manner.

can be brought into spatial proximity on a target structure, and their chemical reactivities are complementary, there is the prospect of enhancement of reaction rates through the large increase in their effective concentrations with respect to each other. Here, we might also reflect on its conceptual overlap with some instances of antibody-mediated catalysis and "anti-idiotypic" molecular imprinting (Chapter 7). In the context of imprinting, it was also noted that enzyme active sites can in some cases be used as nanoreaction vessels.[1181,1316] Such processes fall under the general ambit of receptor-accelerated synthesis, where irreversible chemical reactions between library building blocks are used.[1182] But reaction *reversibility* and molecular interchange can be highly significant in combinatorial chemistry with a molecular fragment library. In such circumstances, the presence of a macromolecular target has the effect of perturbing a chemical equilibrium in favor of a specific combination of library elements that has affinity for a target site (Fig. 8.7).

A chemical library can generate enhanced diversity beyond its starting composition if it can undergo reversible interactions among its members, which in addition to noncovalent bonding also includes the formation of reversible covalent bonds and molecular conformational changes.[1317] The addition of a target molecule can then "select" for the best combination of small molecules (or fragments, depending on one's defining criteria) that self-assemble within a target binding cavity. The target-mediated stabilization of a specific combination of self-reactive library members out of a large number of possibilities then forces the redistribution of the interchange equilibrium in favor of the specifically bound set. This kind of scenario is the basis of *dynamic combinatorial libraries*,* as depicted in Fig. 8.7a. Upon attaining a perfect

---

*Dynamic combinatorial libraries are encompassed within the term *receptor-assisted combinatorial chemistry*, which also includes the above-noted receptor-accelerated synthesis strategy.[1182]

thermodynamic equilibrium under defined conditions, the components of a chemical library with reversible self-reactivity will reach constant steady-state levels. This equilibrium is perturbed by a target binding a specific product and stabilizing it, since the steady-state concentration of the free product falls and thereby thermodynamically drives formation of more of the product from interacting monomeric chemical units. In this manner, the target template in effect amplifies a specific tightly binding product from within the interactive chemical units of a dynamic combinatorial library.[*,1317,1319] This can also be viewed as a net flow from more weakly binding combinatorial variants toward more effective binders of target.[1320]

But as well as using dynamic combinatorial libraries to stabilize and thereby identify specific combinations of chemical building blocks, the logical converse also applies. The same conceptual background is equally applicable to the isolation of novel artificial receptors (as depicted in Fig. 8.7b) where the ligand becomes the focus for multiple library building blocks that assemble around it to form a cage or artificial receptor pocket. This has been achieved by exploiting the reversibility of disulfide bond formation for generating dynamic libraries of disulfide-bonded macrocycles, whose combinatorial equilibria were perturbed by addition of small-molecule ligands. Tight-binding macrocyclic "hosts"[†] were amplified through this process at the expense of weaker binders.[1321] In another example, complex artificial receptors with interlinked ring structures binding the neurotransmitter acetylcholine have been assembled and identified from dynamic libraries of small peptide-related building blocks.[1322]

Dynamic combinatorial assembly processes have generated terms that themselves are reminiscent of molecular imprinting (Chapter 7), through some superficial similarities. Thus, target-mediated combinatorial building block assembly has been called "casting," and ligand-directed receptor assembly has been likened to "molding."[1323,1324] The ability of a target to "select" the "fittest" tight binders from a diverse library population has also led to parallels being drawn between dynamic combinatorial library application and directed *in vitro* evolution.[1325] Certain problems with this evolutionary analogy have been noted, in that under some experimental conditions, amplification of the strongest binders does not necessarily outcompete those with more mediocre characteristics. Happily, this shortcoming has been successfully addressed by close attention to the correct molar concentration of target template.[1326]

An obvious issue arising from applying targets as fragment assemblers is the identification of the resulting combinatoric products. The target-mediated self-assembly of the appropriate fragments itself serves to amplify the appropriate configuration of fragments far in excess of random levels, facilitating direct chemical analyses. But biological amplification systems can also be brought to bear in such circumstances. One such approach of particular interest is to provide each library element with a nucleic acid "tag" that enables its ready identification and amplification, and target-directed self-assembling chemical libraries have been constructed with this principle in mind[1327] (extended in ftp site). The concept of informational encoding is very general and directs us to the next section.

CODES FOR SMALL MOLECULES

Within the vastness of imaginary OMspace and mathematically modeled chemical molecular spaces, solutions to a huge panoply of human needs exist, if only they can be found. Finding a valid "lead" of any sort can be seen as a massive leap in the right direction, as finding

---

[*]In a variation termed "pseudo-dynamic" combinatorial chemistry, the target-mediated selection is combined with cycles of chemical destruction of unbound species, further enhancing the amplification effect.[1318]

[†]The chemical terminology of "guest" (small ligand) and "host" (larger molecule or polymeric structure in which the guest "resides") is frequently used in the context of dynamic combinatorial libraries.

a metaphorical very tiny needle in a very, very large haystack. In earlier chapters, we saw that protein space too is hyperastronomically large, but from this potential space an impressive array of designs has been yielded through the blind agency of evolutionary selection. Proteins come naturally packaged with a molecular encoding scheme mediated through nucleic acids, and this feature is exploitable for *in vitro* evolution, as we have seen in some detail previously. The general concept of an amplifiable informational molecular tag is a very powerful one, which can be taken beyond its original evolutionary context into many new artificial domains. This theme has its own specific foci, but all are very relevant to modern molecular discovery with an empirical evolutionary underpinning, and have areas of intersection.

Tag Team Technology: Encoded Combinatorial Chemistry

A novel molecular function is only useful if the responsible molecule can be identified and synthesized, and library selections or screenings are pointless unless this elementary requirement can be fulfilled. Consider two categories of libraries with arbitrary designations of Types I and II. In Type I, all library members are defined in advance, as with a large collection of small-molecule compounds of known structures, whether natural products or artificially derived. A Type II library is combinatorially derived from smaller subunits, which can be chemical functional groups (as in Fig. 8.5) or biological nucleic acids or proteins (as we have considered in previous chapters). With the Type I library, it is only a matter of computational bookkeeping to ensure that the exact constitution of the collection is known at the outset, but a combinatorial Type II library is not necessarily so precisely defined. To make this clear, let us again consider peptide libraries, since these can be derived either biologically (as with phage display, for example) or by chemical means, but are of a clear combinatorial nature in each case.

Combinatorial peptide libraries of limited size can be synthesized such that all sequence variations are predetermined. In other words, by successive syntheses at spatially defined sites, a combinatorial library can be built up in a "parallel synthesis" that is just as precisely defined as a Type I collection above, and is "pre-encoded" through the chosen synthetic strategy itself. This can correspond to a series of short overlapping peptides for "scanning" a protein sequence,[1271] which has been very useful for defining immunological epitopes.[1328] Pre-encoded combinatorial libraries in general are thus defined in both location and chemical identity during the course of synthesis and after its completion.[1273] There are various ways of implementing a pre-encoded library synthesis. A useful strategy has been placing solid-phase reaction resins (on which library building blocks are assembled) into permeable containers with pores retaining the resin but permitting reactant entry (these are commonly termed "teabags," which encapsulates their operating principle into a single evocative word). Pre-encoding of such containers can use a color-based system with manual sorting[1329] or machine-readable radiofrequency tagging with an automated sorting operation[1330,1331] (detailed further in ftp site).

But compare the general notion of pre-encoding with the combinatorics involved with a biological phage peptide display library, where a short tract of codons is randomized and fused in-frame with the gene encoding the protein directly functioning as the display vehicle. This results in a population of phage bearing surface proteins that vary at the short randomized site—or a peptide library, in other words. Now, unlike the defined combinatorial synthetic peptide library above, the contents of this kind of library can only be defined statistically. For example, if the random tract was four codons in length, then 20^4 peptide combinations could result, or approximately 1.6×10^5. Libraries of 1000 unique members obviously could not possibly contain all sequences, and the chance of finding any specific sequence would be low. Take the library size up to 10^9, though, and the picture is reversed. Thus, theoretically if the library size is large enough, the probability of finding a specific sequence approaches 1.0, but in reality various biological constraints render some sequences "forbidden" or under-represented. (And most real peptide libraries use randomized tracts longer than the four residues of this example.) Short of exhaustively sequencing every library member, the precise compositions of such libraries are consequently indeterminate, but this is not a crucial matter for their deployment. As we have seen repeatedly already (Chapter 4), an *in vitro* selection for display peptide binding coupled with genotype–phenotype linkage allows identification of functionally significant clones in the library. In such circumstances, the selectable phenotype carries an *informational tag* in the form of the easily amplifiable and readable genotype.

Does this mean that a synthetic combinatorial peptide library must always be completely predetermined, as opposed to the randomization strategy possible with biological peptide libraries? Not at all. Both biological libraries of random encoded peptides and random artificial combinatorial libraries must have some means for the identification of hits isolated during their respective selection or screening processes. There are numerous ways this can be achieved for combinatorial libraries, some of which have parallels with their biological counterparts. First, let us recall the classic mode cited above of encoding of predetermined "Type I" chemical libraries for screening, that of spatial positioning. In this kind of scheme, there is an assigned correspondence between the chemical identity of a library member and a spatial coordinate in a screening matrix (such as a multiwell plate). Obviously, this is not possible in a fully randomized combinatorial library where the identity of any specific member is not known until it is pulled out and characterized.

Combinatorial chemical libraries generated by mix-and-split strategies (Fig. 8.5) are amenable to analysis by a "proceeding backward" approach termed recursive deconvolution[1332] (ftp site). But deconvolution methods, spatial encoding and pre-encoding/direct sorting become increasingly cumbersome for combinatorial library evaluation as library size increases toward very high levels. Diverse combinatorial libraries that are not addressed in these ways require some other way to identify hit compounds found during screens. By the nature of the process, combinatorial mix-and-split syntheses on beads result in "one bead–one compound" libraries, and if the amount of compound on each bead is high enough, the bead itself can serve as a reference stock of compound for direct analysis as well as screening. In contrast to biological libraries where amplification of a selected nucleic acid (or indirectly its encoded protein) is possible, this kind of bead strategy in effect carries along a preamplification of the molecule of interest by means of the initial synthesis itself (Fig. 8.8).

Since beads for these applications are very large on a molecular scale (>50 microns (1 micron $= 10^{-6}$ meters), where a typical rod-shaped bacterium is 1–2 microns in length), screening rather than selection processes are appropriate. Millions of beads can be screened with appropriate assays, and positive candidates physically isolated by microscopic manipulation.[1333] The compound of interest is released from the bead and identified by an appropriate technology (microsequencing for peptides; or mass spectrometry for other organic combinatorial compounds). Screening can also be performed by means of an iterated sib-selection procedure using controlled partial release of compound from the bead surface, retaining sufficient for final identification purposes.[1334]

If combinatorial chemical library members are scored as hits and identified directly, the only "informational code" involved is the chemical structure of the compounds of interest themselves.[1335] But this raises the question as to how far this can be taken, which relates to the analytical limits of mass spectrometry, generally the most sensitive practical approach available. Could binding of a protein target by small molecules be evaluated by treating the protein with a chemical library, removing unbound material, and directly identifying the bound compounds? In fact, it is not practical to attempt this with an entire large library by screening

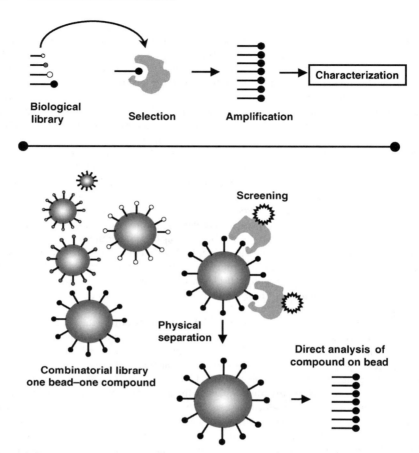

FIGURE 8.8 Biological library selection and amplification (top panel) versus screening a "one bead–one compound" combinatorial library with direct analysis of the specific bead compound after its physical isolation (bottom panel; each structure on each bead represents an assembled combinatorial library element).

library members separately, but pools of several hundred molecules at a time have been indeed been successfully investigated with protein targets, by size exclusion of low molecular weight unbound compounds and direct mass spectrometric analysis* of the protein-bound fraction. Whole libraries can thus be "run past" specific protein targets in this manner, by subdivision into pools of appropriate sizes. This "SpeedScreen" technology[1336,1337] and conceptually equivalent approaches using nuclear magnetic resonance[1338,1339] might seem to resemble a selection process more than screening (Chapter 1), since irrelevant material is removed from the target in an analogous manner to binding-based selection in a phage display experiment. But, of course, the crucial difference is the inability of bound small molecules to be amplified and reapplied to the target in an iterative manner, as is routinely done for isolating biologically displayed binding molecules. This "nonamplifiability" is directly relevant to the pool size limitation of

---

*This process is also of interest for demonstrating the power of advances in technologies for enabling new screening options, since it is wholly dependent on the ability of mass spectrometry to distinguish the bound library members.

the "SpeedScreen" technologies and their ilk. Even if the analytic sensitivity is taken to the single-molecule level, the problem of background noise must be contended with. In a display-binding experiment, the relative magnitude of the nonspecific background is progressively reduced through iterated binding cycles, but cannot be mitigated in this manner when nonamplifiable binding molecules are used.

Despite progress with direct chemical identification of library members comprising positively scored screening hits, this meritorious aim thus runs up against limitations of practical pool sizes, sensitivity, and costs. Also, and very significantly, direct chemical analysis of combinatorial library elements rises in complexity and difficulty as the number of chemical building-block combinations increase in parallel with library size. This consideration brings us back to the contrast with biological libraries using nucleic acid-encoded informational tags, by which this section is introduced. Since the advent of combinatorial chemistry itself, researchers have investigated a variety of ways whereby specific combinatorial compounds can themselves be chemically encoded. As opposed to the "pre-encoding" strategies we looked at earlier, chemical informational tags that encode a specific structure formed during combinatorial library synthesis must be created in parallel *in situ*, via "in-process" tagging.[1273] This is necessary to enable the progressive encoding of each specific combinatorial chemical coupling during the iterative synthetic steps of library creation. An essential and obvious feature of a chemical tagging system is that it must allow for rapid and easy decoding, much more readily than the direct chemical analysis of the encoded library member itself. This must be stressed since chemical tags in themselves do not offer the biological advantage of replication, in common with the combinatorial library compounds that they encode.

A variety of molecular tagging schemes can be devised, and have been implemented.[1273,1340] In this area, a straightforward concept is the maintenance of a linkage between encoding molecules and the combinatorial library product (reminiscent of biological display genotype–phenotype linkage, but again distinct through the inability of the tag to be replicatively amplified. Both in principle and practice, this can be via a covalent link between coding tag and the combinatorial compound, or through a co-bound solid-phase entity such as the mix-and-split beads referred earlier (depicted in Fig. 8.9). In the latter case, specialized beads

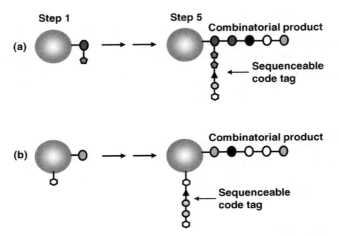

FIGURE 8.9 Depictions of sequenceable encoding tags for combinatorial synthesis from cosynthesis, either coupled to the library element (A) or independently to the bead matrix (B).

have been designed with spatially segregated regions for chemical tagging, to minimize potential interference of the tag with functional screening of library compounds.[1341,1342] Tag linkage directly with library molecules has the advantage of maintaining "tracking" with encoded compounds after release from beads, but can potentially interfere with functional activity.[1343] Another important practical consideration is whether the chemistry of tag coupling is "orthogonal"* to that required for the progressive library chemical combinatorics (having mutually exclusive chemical requirements).[1344] Ideally, a tag and library building block will be coupled specifically under the same conditions.

To report the structure of a library element, a coding tag must define both the nature of a specific chemical building block, and its sequence in a progressive combinatorial addition process. One way to define a sequentially built combinatorial molecule is to encode each chemical component by a specific tag, and encode the combinatorial sequence itself by using a directly corresponding string of covalently joined tag molecules (Fig. 8.9). Of course, the utility of this presupposes that obtaining the sequence of the encoding tag string is easier than direct analysis of the combinatorial library member itself. Peptide tag sequencing has been used for such encoding purposes from early days of combinatorial chemistry,[1345,1346] and despite certain drawbacks has been useful for the development of tag/library compatible chemistries,[1341,1342] facilitated by advances in sensitive sequencing technologies.[1347,1348]

But it is not necessary to encode a sequence directly by another sequence with one-to-one coding correspondence. Codes with discrete sets of molecular tags can be designed to specify both the chemical nature of a combinatorial unit and its position in a synthetic progression. One could simply define separate tags for each specific building block and each corresponding synthetic step number. But if there are N building blocks and M synthetic steps, one would require NM tags to unambiguously report a specific combinatorial sequence, and very soon the numbers of tags becomes prohibitive. But other numerical coding strategies can keep the tag numbers to a minimum. An example of this was described during the relatively early stage of combinatorial chemistry development[1343] using discrete tags for defining both specific combinatorial blocks and the corresponding step numbers by means of a binary code, greatly reducing the number of tags required (extended in ftp site).

Many other molecular tagging systems have been devised, including the use of differential isotopic labeling as means for increasing tag diversity within the same chemical format.[1349] However their encoded information is embodied, the limitations of chemical tags are closely associated with the sensitivity of measurements by which their informational content can be read out. But in biological systems, we have already seen abundant evidence for the great utility of nucleic acids for carrying information in an amplifiable manner. It was suggested in the early 1990s that nucleic acid "codons" could also serve as encoding elements for combinatorial chemistry,[1350] and it will come as no surprise to find that many workers have sought to exploit this principle for such applications, as we should now take the time to examine further.

Marrying Chemistry and Biology with Tags

As we have seen, a series of distinguishable monomeric compounds can be used as identifying molecular tags if they are sequenceable, as has been shown with peptides.[1345,1346] Yet the power of the now-familiar "genotype encoding" in biological protein or peptide display libraries

---

*Compare this chemical use of "orthogonal" with its analogous application for specific pairs of tRNAs and aminoacyl-tRNA synthases, as in Chapter 5. Nonorthogonal chemistries require the combinatorial building and tagging to be performed separately, thus significantly increasing the overall time scale for the process, an obvious drawback.[1344]

comes from the crucial point that, unlike polypeptides, the DNA itself can be readily replicated. Whether *in vivo* (in a bacterial plasmid or phage vector) or *in vitro* (by polymerase chain reaction), a single DNA molecule can be amplified into billions of copies, to the point where it can be easily sequenced and further manipulated. To co-opt nucleic acids as informational tags for general combinatorial application, "codons" must be chosen to encode each combinatorial unit of the library of interest, as well as addressing technical problems involved with cosynthetic nucleic acid and library chemistries. Initially, this concept was realized with a synthetic peptide library itself where seven amino acids were encoded by chosen dinucleotides.[1351] As this early study demonstrated, the choice of nucleic acid coding sequence blocks need not have any relationship to biological codons, and coding block length can be adjusted according to the numbers of library combinatorial elements involved. On the other hand, such encoding is not entirely arbitrary if applied optimally, as design considerations should aim for minimization of potential artifacts during the PCR-mediated amplification step.[1350] As long as a physical linkage is maintained between library elements and their encoding nucleic acids, this strategy is directly analogous to the genotype–phenotype linkage of biological display. A generalized depiction of nucleic acid-encoded combinatorial synthesis is given in Fig. 8.10.

If the encoding DNAs of Fig. 8.10 were covered with protein, and the small molecules to which they are informationally assigned were conjugated to the protein coat surface, a situation analogous to phage display would arise. Of course, if the genotype–phenotype linkage is broken, *in vitro* selection and evolution is thwarted. Yet if a phage carries an assigned "genotype" by which a small molecule is arbitrarily encoded, the viral protein surface can be duly modified with the corresponding compound to create a "chemical display" phenotype linked to an informational code. A single-phage population can thus be bequeathed with a unique genotypic sequence assigned to one type of chemical modification produced on its surface *in vitro*. If a sizable number of such phage (each with a specific genotypic code and

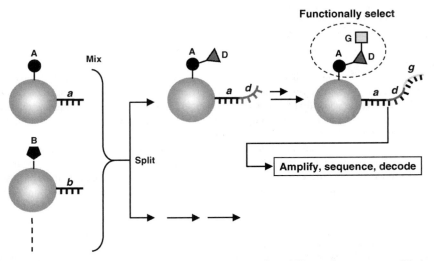

FIGURE 8.10   Depiction of simple nucleic acid-based encoding of library elements on a solid-phase bead. Library elements are shown in upper case letters; DNA "codons" (arbitrarily four-base here) in lower case. Only one strand of a limited mix-and-split synthesis (as in Fig. 8.5) is shown.

a corresponding distinct surface compound) are pooled together, a chemical display phage library has been created.[1352,1353] Owing to the protection of the informational phage DNA from nucleolytic degradation by protein coat packaging, such small-molecule display libraries are potentially useful for *in vivo* applications involving identifying and tracking compounds modulating cellular uptake.[1353]

When we think about the well-established picture of biological display, it is clear that while nucleic acids are intimately linked with both their own replication and their encoded protein-mediated functional phenotypes, these two properties are distinguishable. In other words, it is one thing to replicate DNA or RNA that encodes protein sequence information, and another thing to express the information into a physical protein, and these processes are mechanistically distinct.* It should be clear that in schemes such as Fig. 8.10, the tags function only as replicable information carriers, and do not feed back into the system by directly specifying the synthesis of the combinatorial chemical entities that they encode. By way of further analogy, nonsense mutations in a protein coding sequence result in termination of synthesis, but "mutations" in a simple DNA code as in Fig. 8.10 have no effect on combinatorial synthesis, only altering the accuracy of the informational read-out. Can DNA-based encoding of chemical libraries go beyond replicable informational tags, and at least approximate the other side of the biological roles of genes? This question has a direct bearing on how efficiently nucleic acid-encoded chemistry can mimic biological directed evolution. There is much to note in this area, which has seen rapid progress in recent times.

In fact, beyond simply encoding library members and their sequence arrangements, DNA tags can actually direct combinatorial synthesis itself, mediated through base-pair hybridization. This can operate through the sorting of pools in mix-and-split syntheses, essentially involving hybridization-directed "routing" of specific sequence tags as desired, with concomitant DNA encoding of progressive combinatorial couplings of specific compounds. As depicted in Fig. 8.11, this process has been termed "DNA display."[†,1355–1357] The ability to accurately perform the routing by hybridization is essential for the success of this approach, and is a significant achievement in its own right.[1355] Efficient routing itself serves as an enabling technology for the DNA-encoded directional sorting and simultaneous chemical subunit encoding. The physical enactment of the link between the designated codons and their encoded compounds has been termed "chemical translation."[1356] The DNA display principles have been validated with peptide, "peptoid" (peptide analog), and synthetic ligand combinatorial libraries.[1356–1358]

Nucleic acid tags on small molecules can also serve as a means for actually promoting chemical reactions themselves. To look at this further, we must reconsider some aspects of chemical reactivity. The rates of bimolecular chemical reactions are greatly subject to reactant concentrations, and under natural biological circumstances, usually reactants are too dilute to undergo significant product formation by themselves, and thus require catalytic assistance.[1359] Sometimes, the mechanism of protein biocatalysts involves simply the sequestration of reactants in a cavity of the appropriate shape, an "entropic trap" that promotes chemical interaction whose rate otherwise would be vanishingly slow. We have seen examples of this in Chapter 7 with catalytic antibodies for the Diels–Alder reaction, selected for binding an analog

*Of course, for aptamers and ribozymes the informational content and functional units coincide (Chapter 7). We should recall also that beyond the RNA world, the replication of organisms is dependent on the entanglement of nucleic acids and proteins, since proteins require nucleic acids for their encoding and expression, and nucleic acids require the proteins that they encode for their own replication.
†This should not be confused with a biological DNA display system for genotype–phenotype linkage,[1354] analogous to mRNA display (ftp site).

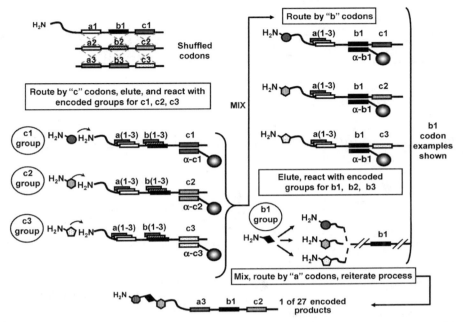

FIGURE 8.11 "DNA display" hybridization-mediated routing and DNA encoding of combinatorial libraries. From top left, a series of DNA strands is designed with constant regions interspersed with regions assigned as unique "codons" for small-molecule library elements, and a 5′ end with an amino group for derivatization. If the codons are shuffled, 27 possible combinations result (a(1-3)|b(1-3)|c(1-3)). When the "c" set of codons is sorted by hybridization (through sequences complementary (antisense) to each "c" codon, designated α-c1, α-c2, and α-c3), corresponding chemical groups can be attached for which each c1, c2, and c3 codon acts as an encoding tag. As in a typical mix-and-split synthesis, the first round of reactants are mixed and then subjected to routing for the next set of codons b1-b3 (On the right; only shown in detail for b1). The corresponding encoded chemical groups are then coupled, and the cycle reiterated. One example of 27 distinct combinatorial products is indicated at bottom.

of the transition-state intermediate for the interaction of two molecular participants in the reaction. Some nucleic acid catalysts can also act in an analogous manner (Chapter 6).[194] The molecular "induction of proximity" in biological systems in general* is very significant for signaling and as an evolutionary pathway.[1360] But consider alone the notion of enforced spatial proximity of small-molecule chemical reactants, without recourse to specialized binding pockets in macromolecules. A simplistic mental image emerges of some mediator (a Maxwell demon with nothing better to do?) grabbing two possibly reluctant participants and holding them close together until some bonding interaction occurs. Should this in itself be sufficient to accelerate chemical reaction rates, then straightforward nucleic acid hybridization might serve as a real-world "grabber," provided chemical reactants were conjugated to the complementary strands forming a duplex.

Confining a pair of reactants within a small volume through their cotethering to a DNA duplex effectively is theoretically equivalent to performing the same reaction with isolated

---

*We also noted earlier in this chapter that target-mediated fragment covalent assembly is another effect related to "enforced proximity" in dynamic combinatorial libraries.

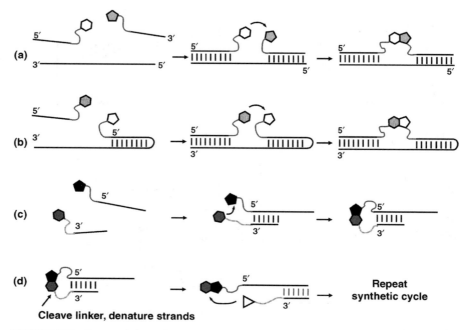

FIGURE 8.12   Representations of DNA-templated synthesis with different architectures. (a) Reactants conjugated at 3' and 5' ends and brought into proximity by a shared template; (b) Similar to (a) but with a hairpin loop for the 5' conjugate; (c) End-structural proximity, where one of the reactants is coupled via a cleavable linker. Upon linker cleavage and strand denaturation, another cycle of template-directed synthesis can take place, using a different hybridization site within the original template with a new coupled group (d).

components at far higher concentrations. And in fact there is extensive evidence that this can indeed occur under the right conditions, studied under the general term of *DNA-templated synthesis*.[1359] To clarify this, consider two single-stranded DNA molecules each bearing different appended chemical moieties with potential reactivity with each other. If the two DNA strands share an adjacent region of complementarity, then hybridization between them may greatly increase the effective concentrations of the reactants with respect to each other, provided the resulting duplex has a suitable structure. There are various alternative arrangements by which this can be brought about, using different template architectures as shown in Fig. 8.12. The nature of the DNA template system is not completely arbitrary, as intramolecular base pairing within one DNA strand bearing a reactant can impair inter-molecular duplex formation and adversely effect reactivity.[1361] On the other hand, a degree of secondary structure within single-stranded intervening tracts (as in (a) and (b) initial duplexes of Fig. 8.12) can promote activity, presumably by reducing the average spatial separation of reactants.[1361]

Before proceeding further, we should back up a little and note that while the schematic reactions portrayed in Fig. 8.12 depend on the specific chemistries of the small-molecule participants, they need have no chemical relationship with the nucleic acid backbone or nucleobases of the DNAs to which they are attached. This is significant because DNA-templated synthesis can also refer to the ligation-mediated assembly of nucleic acid analogs with altered backbones or unnatural base pairing on a conventional DNA template. Examples of this were examined in Chapter 6, as with peptide nucleic acid assembly on DNA templates (and vice versa). DNA-templated synthesis in a general sense is thus decades old, but the systematic evaluation of DNA templating as a more diverse synthetic tool is of relatively recent origin.[1359] A variety of reactions have been achieved, in itself suggesting that it is not necessary for reactants to align in precise orientations as with nucleobase pairing, and arguing for a generalizable effect consistent with profound effective increases in local reactant concentrations.[1362] Also, the observed dependence of successful reactions on sequence complementarity in participating duplexes (as in Fig. 8.12) is predicted through the requirement for stable hybridization, to enable DNA template-mediated juxtaposition of the reactants.[1362] The versatility of DNA-templated synthesis has been further extended through multistep reactions programmed by reagents with distinct DNA tags ("codons"), a process which is facilitated by cleavable linker chemistries,[*,1359,1363] as depicted in Fig. 8.12d.

Although many other technical issues exist with respect to DNA-templated synthesis, let us now consider its impact on molecular discovery. In common with the DNA display process described above, DNA-templated synthesis has potential for the *in vitro* selection of function from combinatorial libraries. Multistep syntheses are particularly useful in this regard, as depicted in a simplified reaction scheme for 27 possible triplet assortments of chemical monomers, where formation of multiply derivatized products are encoded by their template strands (Fig. 8.13). Another facet of DNA-templated synthesis very relevant to molecular discovery lies in its ability to reveal new chemical reactions themselves. To achieve this, powerful selections can be established using the convenient biotin/streptavidin system, reminiscent of pathways used *in vitro* for deriving functional ribozymes or DNAzymes (as described in Chapter 6), where the transfer of a biotin moiety is the essence of the selective process. Conceptually related schemes have been successful for reaction discovery by DNA-templated synthesis.[1364]

Some aspects of DNA-templated synthetic discovery of both functional molecules and chemical reactions might prompt comparisons with enzymatic catalysis by proteins or nucleic acid enzymes. Clearly, the reactions accelerated by DNA templating are very limited compared with the diversity and efficiency of biocatalysis. As implicit in searches for reactions through DNA templated pathways,[1364] reagents can be encouraged to react if effectively brought into high concentrations with respect to each other through their appended complementary DNA tags, but only if they are chemically capable of this though their intrinsic properties. In other words, with a simple DNA-templated synthesis, the nucleic acid provides only specificity of spatial proximity, and nothing more. This is not to deny the possibility, though, that DNA-templated reactions cannot be further extended by coapplication of specific catalytic DNA sequences. Irrespective of this, it is interesting to note that there is a case to made for the significance of syntheses templated by nucleic acids in the molecular evolution of translational adaptor processes during the origin of life.[1365]

Advances in the encoding of chemical combinatorial diversification will certainly take place and further extend existing options. These considerations aside, an overall conclusion is that it is meaningful to speak of artificial *in vitro* chemical evolution, especially in the context of nucleic acid encoding mechanisms. Let us briefly extend this comparison, which has cropped up in fragmentary form previously, but has not been spelt out clearly. This is relevant to the screening of chemical libraries and directly pertains to the overarching theme of this book.

---

*It is also possible due to observations that DNA-templated synthesis is still efficient even if the conjugated reactants are separated through displacement of the region of hybridization along the template strand.[1362,1363]

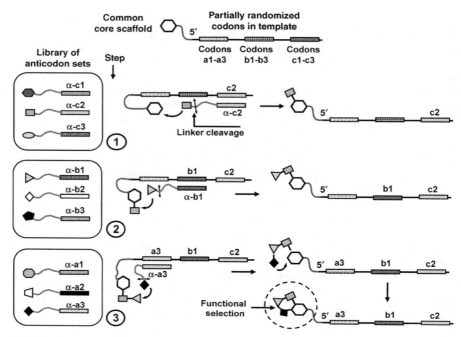

FIGURE 8.13   Multistep DNA-templated synthesis for selection of novel molecular functions. In this example, three "codon" positions exist in a set of templates, with three distinguishable alternatives for each position (during synthesis, appropriate codon positions are randomized to give a partially degenerate sequence population providing specific matches for each "anticodon"; shown as "α-a1," etc.). The latter are necessarily grouped in sets for each position and reaction stage, where each is conjugated to its designated functional group according to the predetermined codon assignments. At each stage, DNA-templated synthesis occurs specifically via codon:anticodon complementarity, with linker cleavage to enable the subsequent synthetic steps. In the final step, conditions favoring a cyclization reaction are introduced, followed by functional selection, whereupon the specific template sequence reveals the relevant functional groups. (Here one specific case out of a library of 27 possible products is depicted.)

AN EVOLUTIONARY ANALOGY

At this point, we have seen plenty of evidence that the planning, assembly and synthesis of chemical libraries is a rational process of the highest order. The likelihood that library screening will yield nonartifactual primary hits, and then useful drug leads, can thus be strongly influenced by knowledge-based decisions on the specific content of the library. A library that is rationally biased in a drug-like direction is therefore clearly of higher value than a randomly picked assortment of compounds. A library whose members are specifically tailored toward a family of proteins (or even a specific protein) may be even more efficacious in yielding potentially useful molecules. But we inevitably return to a point made earlier: isolating a candidate compound from a library is still an empirical process irrespective of the level of rational input into the library composition. True, such advanced libraries are at the higher end of a spectrum of empirical approaches, but they do not cross the line into full rational design.[*] But the

---

[*]See Chapters 1 and 9 for further discussion of similar issues. In the present context, "library" implies a sizable (and often very large) number of members.

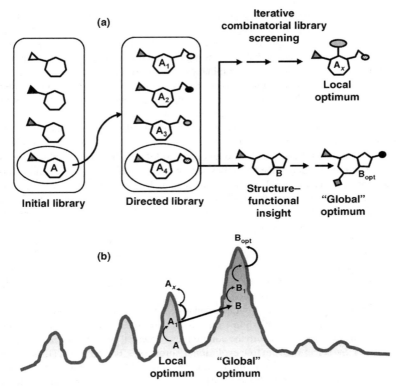

FIGURE 8.14   Evolutionary parallels in applications of chemical libraries. (a) Functional selection of compounds from unbiased libraries can then lead to their optimization, reaching optimal forms based on the relevant chemical scaffolds. Such an optimum (A_x) may nevertheless be only the best possible solution based on simple substitutions on the existing scaffolds, and inferior to alternatives using different scaffolds. But information from structure–function studies with initial library hits (combined with information regarding the target itself) can result in a "jump" to a new scaffold (B) which in turn can be optimized (to B_{opt}) by a screening library based on this scaffold itself. (b) The progressions of (a) depicted as a fitness landscape.

continuing interplay between rational input and empirical output is nonetheless to the combined benefit of the success of combinatorial chemical libraries.[1366] Since the need to screen chemical libraries in the first place is fundamentally based on knowledge limitations,[1275] it might be presumed that as knowledge advances, chemical library screening requirements will progressively change in tandem.

Some processes with an underlying empirical basis can have a distinct evolutionary character. When guided by continual empirical trial and feedback from experimental results, drug discovery based on synthesis of analogs of known biological mediators has itself been likened to a Darwinian system* (Fig. 8.14). Let us think about this a little further to see how much it is justified, especially with respect to modern combinatorial chemistry and molecular

---

*This comparison has been made by Sir James Black,[1367] a major figure of twentieth century drug development.

discovery. Already we have seen the generation of diversity (GOD) by artificial means in chemical libraries, and in previous chapters GOD has been highlighted as the raw material for both natural and *in vitro* selection. We must be careful to note, though, that functional evaluation of chemical libraries by conventional high-throughput analysis is in fact a screening process, and not selection as such (recall the rumination on this distinction in Chapter 1). Small molecules are not directly amplifiable, and therefore direct selection for them cannot be applied. Yet nucleic acid "tagging" approaches allow an amplification step which serves to indirectly identify functionally significant small molecules, as we have seen.

But let us leave aside this issue for a moment, and consider the next stage of obtaining "hits" from a primary chemical library screen, and the subsequent identification of molecular leads for further development. Leads can be thought of as prototypes, but to obtain this status in the first place, they must exhibit some of the desired functional properties of an ideal compound. Although a primary lead is usually suboptimal as a "molecular solution," it furnishes a core structure from which variant "mutants"(chemical analogs of the lead compound) can be synthesized and tested. Fragment-based screening and dynamic combinatorial libraries can also be viewed through an evolutionary prism, as already briefly noted. When small molecules are deconstructed into minimal functional fragments that can be combinatorially assorted, the parallel with evolutionary recombination and shuffling becomes more poignant. In any case, the reiterative process of screening and subsequent synthesis of next-generation candidates may go through multiple rounds of evaluation with increasing refinements toward optimization, based on structure–function information.

The ability to amplify nucleic tags appended to chemical library members allows indirect selection for small molecules binding to a designated target, since a DNA or RNA tag provides the information linked with the compound of interest. But here we must once again distinguish *in vitro* selection from true evolutionary processes. For example, DNA-templated synthesis can be used to devise increasingly complex *in vitro* library formation based on many "codons" (as in Fig. 8.13 in simplified form), but selection of a functional molecule from a single cycle of this process is not evolutionary *per se*. Yet all nucleic acid-based encoding schemes can be used for chemical evolution in the sense that each round of selection and amplification is logically based on information obtained from the previous cycle. Thus, with the "chemical phage display" outlined earlier in this section, a small molecule on the phage surface could bind to a target cell and be identified through replication of the phage genome in which the "bar code" for the relevant compound is embedded. This small molecule in turn can serve as the basis for the generation of chemically related analogs, which in are successively given a DNA code and used to prepare the next-generational set of surface-modified chemical display phage.

Combinatorial chemistry based on DNA display or DNA-templated synthesis approach evolutionary models further still, since the arrangement of codons and their temporal "expression" directs the production of libraries themselves. This enters at the level of diversification, the GOD that we have frequently encountered as an indispensable requirement for any evolutionary process, natural or artificial. To consider current limitations on DNA-encoded chemical diversification, it is instructive to compare it with natural genetic equivalents. Although at any one time within in a cell, all tRNAs charged with their cognate amino acids are present, the progression of protein synthesis from $5'$ to $3'$ on an mRNA molecule is accurately controlled by the ribosome (as noted briefly in Chapter 5). No such directional control is available (at present) for combinatorial syntheses directed by hybridization, whether DNA display or DNA-templated synthesis (as depicted in Fig. 8.15). We have previously seen on more than one occasion the power of DNA shuffling, and a lack of directional control in a combinatorial scheme is a significant limitation that prevents the unfettered shuffling of

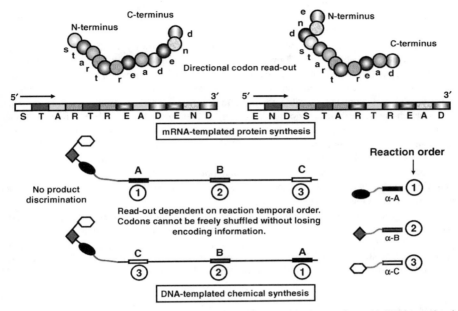

FIGURE 8.15 Comparing biological protein synthesis on an mRNA template with DNA-templated synthesis. An important distinction is the directional read-out of the biological translational process, shown by respective codons (capital letters) and amino acid residues (lower case; single-letter code). For DNA-templated synthesis (and also DNA display; Fig. 8.11), the order of combinatorial progression must be controlled by codon position in relation to reaction order. In this example, it is thus not possible to distinguish codon order ABC from CBA (circled numbers indicate order of synthetic cycles with compounds coupled to codon-complementary strands (α-A, denotes antisense sequence to A codon, etc.). In the first cycle of this example, the "α-A" reagent can both anneal and promote synthesis equally well from either position, and likewise for the "α-C" reagent in the third cycle.

chemical library "codons."* With DNA-templated synthesis or DNA display, a set of codons for different chemical groups can be only shuffled at a specific position designated for a corresponding set of reactions, as depicted in Figs 8.11 and 8.13.

While in some respects the evolutionary analogy with chemical library screening and subsequent drug development is imperfect, there are some parallels in terms of diversification and cumulative variation. As we have seen, the evolutionary parallel is closer with nucleic acid tagging than simple chemical library screening, since functional selections are involved at each step. But as with artificial evolution in general, *in vitro* chemical evolution can deviate from the blind evolutionary algorithm when new information comes to hand. The evolutionary functional development of small molecules can "change course" through the accumulation of empirically obtained knowledge, which lays the groundwork for subsequent rational decisions. This distinction is also depicted in Fig. 8.14. We have previously noted that some complex

---

*If codon "A" is present in one position, the same codon cannot be used in spatially distinct position, since it will not be distinguished through base complementarity (as in Fig. 8.15). But if desired, different codons for the same compound could be designed to give positional information. For example, a codon order "A-A-C" in Fig. 8.15 could be encoded by A3/A2/C codons, where A3 and A2 have distinct sequences but encode the same compound as for the original A codon itself. However, the positional specificity of such a system still obviates free codon shuffling among all positions, since positional information is lost when codons are freely interchanged along a DNA strand.

biomolecules or biosystems (especially those involved with fundamental processes) may become evolutionarily "frozen," or at least reach a position from which is difficult to move to an alternative configuration. Compounds found by chemical library functional screening can be subsequently "evolved" toward optimal forms by screening biased libraries based upon substitutions on the relevant chemical scaffolds (Fig. 8.14a). But these improved forms may not necessarily represent global optima, and might remain "frozen" at a local peak of "fitness" (Fig. 8.14b) if it were not possible to move to a superior scaffold requiring more radical chemical changes. Of course, empirical screening of chemical libraries is not as blind as this. Functional screening itself allows the accumulation of vital structure–function relationship data, which in effect permits jumps across valleys in a "chemical fitness landscape" (Fig. 8.14b). Such iterative rounds of library screening and chemical design are termed "sequential" or "informative library" screening.[1368–1370] Rational options thus constantly emerge as a result of empirical functional data, emphasizing again the constant feedback between these two pillars of molecular development.

Finally, when considering evolution and drug development in the same breath, we should remember the usefulness of libraries based on natural product scaffolds, which themselves are the outcomes of natural evolutionary processes. It is precisely this feature that gives such scaffolds a head start over purely random compound collections, and thereby provides a way to indirectly exploit biological evolution for the directed exploration of chemical space.

FROM SMALL TO LARGE AND BACK AGAIN

In terms of molecular discovery, to think small you also should think big. To efficiently identify optimal small molecule drugs, the big molecules that they target need to be identified and characterized. To be sure, a phenotype-based screen can be performed where candidate lead molecules are identified based only on the property evaluated, such as cellular proliferation. But this empirical lead-in itself is a foothold toward the identification of the targets with which the candidate leads interact.

The opening sentence of this section is also true when read as a qualified vice versa: when thinking big one should also think small, at least on a regular basis. To interpret this, in current biology "thinking big" most often refers to scanning entire genomes for functional properties or functional sites of interest. But insights from genomic analyses (and other high levels of analysis that derive from genomics) can feed back into the identification of useful small molecules through the agency of target discovery. A detailed understanding of metabolic pathways, both within an organism and during its evolutionary development, is a boon to the identification and improvement of small molecule natural products, and this too requires high-level analyses.

But now we have completed a triad of proteins, nucleic acids, and small molecules under the lens of discovery that is largely empirically based. It is then time to think back to Chapter 1 and move into rational design, and its impact in all these areas and more.

9   An Empirical–Rational Loop Forwards

OVERVIEW

Rational design, the highest level of the three ways to molecular discovery outlined in Chapter 1, has frequently popped up in other preceding chapters as a counterpoint to empirical pathways toward useful molecules. In this chapter, the intention is to return to some of the fundamental issues of rational molecular design for comparison with empirical alternatives. To this end, a short survey of some important rational approaches in the design of both small and large molecules is provided to portray the current flavor of the field, although as packed into one chapter, it is necessarily only a superficial tour of an immense and growing body of scientific enterprise.

APPROACHING RATIONAL DESIGN

Beginning in Chapter 1, and at several previous junctures, we have seen that "rationality" *per se* is certainly applicable to what usually would be called an empirical approach to molecular discovery, if such a strategy is the best and most productive available with existing technologies. And we have also observed that between pure empiricism and true rational design can be found a range of "semirational" strategies. The incremental appearance of rational design during the development of molecular solutions to functional problems is the "gray area" referred to in Chapter 1, and this we will amplify a little here, before considering some general pathways to the rational creation of useful molecules. These fall within the general areas of small molecule design, the engineering of proteins, the design of certain types of useful nucleic acids, and the global analysis and control of biological systems.

Scope of Rational Design

As noted in Chapter 1, a rational design definition is satisfied when a functional lead molecule with desired properties is predicted before its actual physical realization through processing of structural information applicable to the project at hand. Rational design can be viewed as the logical applied end point of knowledge acquisition and technological development. Even so, each molecular design project is a unique case in the sense that structural knowledge of the target must be obtained, but this itself becomes increasingly streamlined as the underlying technologies progress. For example, the time lag between identification of a protein sequence

Searching for Molecular Solutions: Empirical Discovery and Its Future. By Ian S. Dunn
Copyright © 2010 John Wiley & Sons, Inc.

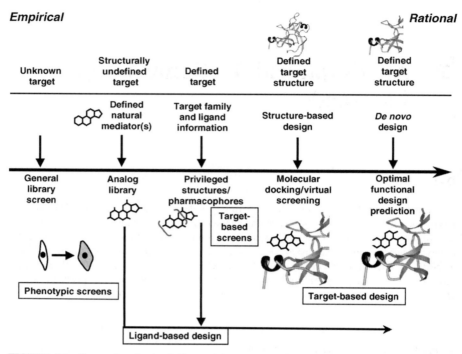

FIGURE 9.1 Progression from wholly empirical (the left-most category) through "semirational" approaches toward "strong" rational design at the far right. Though this is based on protein–small molecule design, analogous pathways apply for any class of molecular discovery. The major left–right arrow shows increasing target information; ligand information and ligand-based design are shown below major arrow. When a specific target is known, but not necessarily its structure, target-based screens can occur.

of interest and determination of its three-dimensional structure has increasingly narrowed, to the point where both acquisition and analysis of structural information have become amenable to automation and high-throughput.[1371,1372] Knowledge and technological abilities determine the interplay between empirical and rational approaches, and within this range (Fig. 9.1) we find the area of so-called "semirational" design. In Chapter 1, it was briefly noted that design fulfilling the above "rational" definition is itself not monolithic in its scope, and these themes will be explored further here. To do this, let us start with a more detailed look at a specific case of a successful semirational molecular discovery as the lead-in to "strong" rational design.

The chosen example is the development of histamine H2 antagonists, leading to identification and marketing of cimetidine (Fig. 9.2). The underlying assumption here is that *syntopic* antagonists* that bind at the same site as normal mediators will mimic the natural ligand, and that the natural structure of the mediator is a logical place to begin a search for pharmacologically active analogs.[1367] The progressive screening of such analogs is itself subject to feedback from experimental results, at least in theory directing the experimental pathway toward the desired ends. Though this process is clearly sensible, and with an established track record, it falls short of true rational design by our earlier definition. In the development

---

*Antagonists of this type also bind competitively, in that they compete for the same site as the natural ligand.

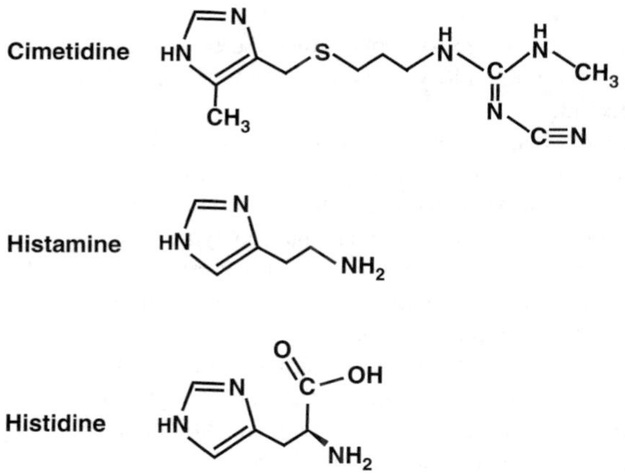

FIGURE 9.2 Comparisons of structures of cimetidine, histamine, and histidine (the amino acid from which histamine is derived). Cimetidine was the first histamine H2 receptor antagonist, effective in controlling gastric acid secretion and treating ulcers. (Drugs blocking the effects of histamine in allergic effects block H1 receptors and are ineffective toward H2 receptors.) Cimetidine was also the first "billion dollar blockbuster" drug.[58]

of H2 receptor histamine antagonists, the initial screening leading to cimetidine was performed with knowledge of the structure of histamine and its gastric physiological effects (among other activities), and the failure of this specific function to be inhibited by classical antiallergic antihistamines. This led to the hypothesis that a second receptor* (H2) mediated the effect of histamine on gastric secretion, distinct from the H1 receptor mediating allergic effects.[1367]

The second-receptor hypothesis was strongly supported by the synthesis of histamine analogs with selective ligand activity, leading to the eventual identification of H2 receptor antagonists, including cimetidine.[1367] Yet limitations on this approach are apparent from a number of observations. Antihistamines for allergy treatment also act as competitive antagonists toward their cognate (histamine H1) receptor,[1375] but are diverse and chemically distinct from histamine.[1367,1373,1375] Later-generation histamine H2 receptor antagonists (such as ranitidine and famotidine) have improved efficacies and selectivities,[1376] but lack the imidazole ring present in histamine and cimetidine (Fig. 9.2). Therefore, although with hindsight the syntopic strategy was the best one to pursue with the knowledge base at the time, success could not in principle be guaranteed at the outset, nor could it immediately deliver the best possible outcomes.

But an increasing database of diverse ligand-binding properties toward a target receptor, combined with sophisticated computational analyses, can in effect lead to rational design of improved ligands even if the target structure remains unknown. (We referred to this in Chapter 1, and this "ligand-based" *de novo* design strategy will be noted again later.) Such an exercise would pass the rational design definition if it predicted a new successful lead compound, but this returns us to the original issue of the "strength" of rational design itself. The essential point is that a molecular function can be defined in terms of its primary effects on a specific target molecule, but matters rarely stop there. For an *in vitro* application where the target molecule is

---

*Four types of histamine receptor (H1–H4) are now known.[1373] The H2 G-protein coupled receptor was not identified and cloned until 1991.[1374]

an isolated system, it is sufficient to use the target as the basis for rational design of a ligand with the desired properties. But for most applications, the target itself operates as an embedded component of highly complex biosystems, and strong rational design should be capable of taking this into account.

As an example of this, strong rational design must also cope with the increasingly recognized complexity of receptor signaling pathways. If we look at the ubiquitous G-protein coupled receptors once more (which includes histamine receptors), it has become apparent that the same receptor molecule can be involved in multiple types of signaling events, which can be modulated by allostery, oligomerization or level of expression.[1377,1378] Biological precedents also exist for differential receptor signaling based on ligand affinities. In the specific example of histamine H1 receptors, different artificial small-molecule ligands can elicit activation of distinct signaling processes.[1379] Molecular design is indeed put on a demanding level if it is to accommodate and rationally predict such subtleties of receptor performance.

Figure 9.1 portrays an escalation of design sophistication for a small-molecule drug, ending when *de novo* design for a receptor ligand is possible. This progression is necessarily knowledge-based, but the required "knowledge" for rationally generating a *clinically* effective drug extends far beyond the primary target itself. Rationality in drug design for clinical uses can enter into the picture at different levels, if we consider that a process that successfully moves from laboratory to clinical trials to marketplace is the most rational and that excessively narrow target-based drug screening can be counterproductive. From this higher level view, even the most sophisticated rational design of a drug for a specific target is not completely rational if the complex systems (the whole genetically diverse organisms) in which the drug must operate are not taken into account. This amounts to "proceeding logically from a false premise," where the latter misapprehension is the supposition that biosystem factors can be ignored during drug development. Of course, this is easy to state, but accommodating it on a wholly rational basis without any empirical input is another matter. (It is essentially asking for *de novo* prediction of all possible drug interactions and side effects in a complex multicellular organism.)

We can again use one of the histamine receptor classes to illustrate some of these issues. The first generation of drugs antagonizing H1 receptors for histamine was effective in reducing allergic symptoms, but often had highly undesirable sedative sign effects attributed to passage of the blood–brain barrier and modulation of neural H1 receptors,[1380] although the extent of these problems is very variable between individuals. Second-generation antihistamines target the same receptor, but have greatly reduced tendencies to induce sleepiness. While these more recent series of drugs have improved H1 selectivity, the reduction in their neural side effects largely results from altered molecular properties unfavorable to penetrating the blood–brain barrier.[1380] On the other side of the coin, deliberate targeting of certain neural H1 receptors (especially in the hypothalamus) has potential for the therapeutic regulation of the sleep–wake cycle and obesity.*,[1382] In order to design the most rational H1 antagonists, a full knowledge of histamine biology and all its receptors is therefore required. And in turn, the highest level of rational design should encompass all of these factors, which span the panoply of issues for drug design in general. Thus far, we have been basing the rational design concept on the derivation of useful small molecules, but any molecule can be considered in the same light.

A definition for "strong" rational design thus hinges upon the "desired functional properties" component of the general rational definition used in Chapter 1. Such properties can be defined purely at the level of modulation of a molecular target, or (at the strongest level) extended to

---

*The relatively recently described histamine H4 receptor is also a promising neural target, as well as in certain immunomodulatory and anti-inflammatory roles.[1373] The H4 receptor was discovered through genomic sequence searches through its homologies with other histamine receptors.[1381]

include all effects that a novel molecule will induce in entire complex organisms. Strong rational design by this definition might seem like a "big ask." Before thinking about how far existing technologies have come, and whether they are up to the complete job, we should first take a brief tour of some rational design specifics at different molecular levels, with a few examples thrown in along the way. This will set the stage for the final chapter and thinking about how far strong rational design can be taken.

PATHWAYS TO RATIONAL DESIGN

In preceding chapters, we have looked at molecular discovery (largely or wholly based on empirical approaches) in separate broad categories of proteins, nucleic acids, and small molecules, although there is significant overlap between each. Our relatively brief survey of rational design will also feature these categories, but with some additional background included.

To begin this overview, let us start at the very most basic level...

Quantum Chemistry and Rational Design

Stable chemical bonds form as an inherent consequence of the electronic structures of atoms, and this is formulated through quantum theory.[*] By any criteria, this towering intellectual achievement has been extremely successful in both explaining and predicting phenomena operating on atomic and subatomic scales, even though many of its ramifications are hard to reconcile with everyday macroscopic experience as espoused by classical physics.[†] Quantum mechanical analyses also explain the formation and structure of molecules through molecular orbital theory, which is supported by a vast body of experimentation. Although classical physics and chemistry serve as good approximations for describing many aspects of the behavior of molecules (and especially the macromolecules commonly encountered in biology), quantum effects cannot be ignored in some circumstances. In this respect, it is not at all necessary to invoke highly speculative proposals regarding the role of quantum processes in the origin of life (as noted in Chapter 2) or consciousness.[1383] It is in the enzymatic realm that the possibility of more widespread biological quantum effects has been proposed and supporting evidence proffered.[1384–1386] In this regard, a specific effect cited as potentially catalytically significant is the phenomenon of quantum tunneling, where hydrogen atoms can make a discontinuous jump across energy barriers prohibited by classical physics.[††] Nevertheless, the extent to which quantum tunneling is fundamental to catalysis is controversial.[1387] Should it prove to be important, it would be inferred that evolutionary selection promotes enzymatic protein folds where its occurrence is favored.

---

[*]The underpinning of this is Schrödinger's wave equation, where atomic electrons are analyzed as wave functions that describe their collective quantum states.

[†]Among such well-known effects appearing to confront "commonsense" notions of existence are wave-particle duality, the uncertainty principle, quantum decoherence, and quantum entanglement. One of the key people in the development of this science in the first half of the twentieth century, Niels Bohr, is famously attributed with the quote. "Anyone who is not shocked by quantum mechanics has not understood it." Nevertheless, understanding and application of quantum principles are fundamental in much modern electronics.

[††]Unlike larger atoms, hydrogen atoms are small enough to be subject to quantum effects, including wave-particle duality. A quantum-scale particle is described by the Schrödinger wave function as a nonlocalized distribution, such that its position in space is only defined by probability. This wave-like property allows hydrogen atoms to "tunnel" (discontinuously jump) across otherwise insurmountable physical barriers.[1387]

But regardless of the direct role of quantum phenomena in the activities of large biological molecules, the existence and chemistries of these and all other molecules ultimately arise from the electronic properties of matter, which requires quantum mechanics to accurately interpret. Ideally, then, all chemistry would be done *ab initio*, based on fundamental quantum principles. All chemical design could then be made by such means *in silico*, and proposed reaction pathways tested computationally before anyone needs to venture into a laboratory. Surely this would take rational design to the most fundamental level, and much effort has been devoted toward its realization. The basic problem comes down to an exponential increase in mathematical complexity with molecular size,[1388,1389] which then translates into the limitations of currently available computation. Directly analyzing molecular systems above the size of small molecules by pure quantum mechanical approaches rapidly becomes prohibitive, but fortunately some useful compromises can be made. Strategies allowing satisfactory approximations of quantum wave equations have been developed, which can feed into classical molecular mechanical approaches.[1389,1390] Large proteins remain formidable challenges for pure quantum analyses, but for the majority of purposes, only a localized region of interest (or active site) of a protein can be focused on as another reasonable (and computationally tractable) approximation.[1389,1391]

We have previously referred to the enigma of accounting for the biological "entanglement" of proteins and nucleic acids during the origin of life. A very different meaning of entanglement is also seen with quantum phenomena, where interacting particles remain coupled in their quantum properties even when subsequently separated by sizable distances, until measurement takes place. Molecular modeling through quantum chemistry may become, in another sense, "entangled" with a quantum technology at the level of information processing. We have noted that computation becomes a limiting factor in quantum chemistry, and advances in this field in recent times have been directly linked with the onward march of computational power and efficiency.[1391] Although "Moore's Law" of computational development has held true for decades, conventional computing is expected to eventually hit constraints imposed by the laws of physics.* Yet quantum effects may themselves circumvent this in the form of *quantum computing*.† As the information needed to describe the energy levels of a molecular system grows, quantum mechanical calculation times increase exponentially with classical computation but only polynomially (according to a polynomial function) with quantum computing.[1388] Although the potential of quantum computing may have been oversold for certain applications,[1393] it has considerable promise for accelerating accurate quantum mechanical analyses of macromolecules and thereby contributing to rational design. We will explore this theme a little more in a later section.

While quantum mechanical analyses of biological molecules are of obvious relevance to rational design, and will surely steadily grow in importance, modeling based on classical thermodynamic properties of molecules is still predominant. Let us now take a look at applying rational knowledge to small molecule design.

Small Molecules the Rational Way

The current identification of useful small molecules, usually synonymous with drug discovery, is a field where empirical and rational designs are heavily interwoven. Much of this we have

---

*This may take longer than initially predicted if new circuitry innovations pay off as hoped.[1392]

†Quantum computing exploits the ability of quantum systems to exist in superposed states until measurement. While a conventional informational bit must exist as 0 or 1, a quantum bit ("qubit") can exist in a superposition of both states, allowing the potential processing of vast amounts of data.

already encountered in Chapter 8, with respect to rational approaches toward the optimization of chemical library diversity and design, and has been depicted above in Fig. 9.1. Most work understandably concerns drug design based upon analyzing and modeling interactions of drug molecules with their direct targets (usually proteins), but an increasingly important area is also the rational prediction of drug metabolism and toxicity.[1394,1395] Cytochrome P450 isoforms are crucial in this regard, and structural modeling based on drug interactions with these enzymes is advancing.

But to continue with this general sketch of small-molecular rational design, let us look at some of the tools of the trade. As a start, we can briefly return to the concept of pharmacophores noted in Chapter 8, as the structural "common denominator" of the function of drugs active against the same targets. This approach can progress even without knowledge of the target receptor site structure and can extend to lead optimization[1396] (Fig. 9.1). An assigned pharmacophore can be used for ligand-based screening (by computationally searching chemical databases for compounds with matching properties to serve as new drug leads[1232,1397]) or as a starting-point scaffold for synthetic design.[1398]

As our knowledge-based process, it is logical that rational design of small molecules should correlate information concerning their biological activities and their corresponding structural features, using measurable physicochemical properties. As noted in Chapter 8, applied study of such *quantitative structure–activity relationships* (widely known by the acronym QSAR[1399,1400]) is an important field within drug discovery and can lead to rational predictions for the consequences of specific structural changes on the activity of a drug. Molecular structure–activity relationship studies *per se* are not at all recent innovations (even traceable back to the mid-nineteenth century), but QSAR itself has developed since the 1960s.[1398] To implement QSAR, the most relevant quantifiable molecular features ("descriptors") need to be identified, and these can include quantum mechanical functions referred to earlier. Other descriptors include shape, thermodynamic, hydrophobic, and electrostatic properties.[1399–1402] In its simplest "one-dimensional" form, QSAR treats a drug ligand as a unitary entity for ascribing a value for descriptors, but increasingly sophisticated multidimensional QSAR variants have emerged.[1403] Three-dimensional QSAR (3D-QSAR) relates variation in biological activity of compounds (such as binding affinity to target) with their three-dimensional shapes[1399] and can be assisted by pharmacophore-based information.[1396]

For higher dimensional QSAR, we need to consider an important aspect of molecules both small and large, which is their conformational flexibility. At normal temperatures in solution, a small molecule may have a favored conformation, and also constantly interchange with low-energy conformational alternatives. Yet when a small molecule binds as a ligand to a macromolecular target, a specific conformation will "fit" into the receptor-binding pocket. The predominant solution structure of a flexible molecule may show little if any change upon target binding, or it may profoundly differ. Protein conformational changes following ligand binding have been well known for an even longer time and are a significant factor determining the affinity of protein–ligand interactions.[1404] As with conformational changes in ligands themselves, the extent of ligand-associated conformational alterations varies from none to large between different protein–ligand systems.[1405]

Where such ligand-induced changes in proteins are significant, they fall under the heading of "induced fit," which we have noted previously in the context of antigen–antibody (Chapter 3) and aptamer–ligand interactions (Chapter 6). Higher dimensional QSAR then must take molecular flexibility into account, but potential structural variability of molecules does not end there. In aqueous solution, acidic and basic functional groups equilibrate between protonated and nonprotonated forms, as determined by their pK_a values. Also, some molecules can exist as alternative structural isomers that can rapidly interconvert with each other, a phenomenon termed *tautomerization*. Again, under specified conditions usually one form will predominate at equilibrium, but this is important information to incorporate into structure-activity modeling. Four-dimensional QSAR accordingly represents ligands as collections (ensembles) of different conformations, protonated states, tautomers, and other relevant variable molecular properties. The next step up (five-dimensional QSAR) takes protein flexibility within the binding pocket into account as well, and therefore models induced-fit events. Another six-dimensional stage of comprehensiveness for QSAR includes solvation effects as well as the previous levels of modeling.[1403]

We can move from the upper end of QSAR modeling to consider molecular modeling in general, a key feature of rational design. Manual modeling has long been important in molecular biology, as exemplified by the cardboard cut-out models for base-pair hydrogen bonding used by Watson and Crick in unraveling the structure of DNA.[1406] But clearly doing this kind of activity by hand has serious limitations, and computational modeling soon gained prominence, leading to the formation of numerous commercial enterprises dedicated to modeling software.[1407] In view of the earlier brief discussion of quantum mechanics, it is important to note that nonquantum-based computational approaches based on molecular mechanics (derived from classical principles or combinations of both approaches) are important for modeling of large molecular systems.[1390,1391] Nevertheless, a significant sector of the industry concerned with molecular modeling produces software for quantum-based computational analyses.[1407]

The generation of computational molecular models is a necessary prerequisite for the next theme in rational design, which is "docking" and virtual library screening. In fact, virtual screening in general can apply toward the right end of the empirical–rational spectrum of Fig. 9.1 as "nondeterministic" rational design (as outlined in Chapter 1), if the computational process models both target and ligand and directly produces leads as a result. But it is also applicable within in the intermediate "semirational" stages between the ends of this spectrum, as it can have a very useful role in narrowing down the scope of an empirically screened library (as noted in Chapter 8). We have already noted an example of this at the beginning of this section, in reference to the virtual screening of chemical databases for hits based on established pharmacophores.[1232,1397] While demonstrably useful, the pharmacophore concept breaks down in certain specific instances. An underlying assumption, that structurally similar drugs will bind to structurally similar target sites, has been supported by extensive protein/small molecule structural database searches, although exceptions constitute a significant fraction of the total.[1408] Sorting biological ligands based on ligand shape similarities results in functionally meaningful groupings and can reveal hitherto unanticipated drug–target interactions.[1409] But more explicitly, the pharmacophore concept would hold that compounds binding the same target site should share an identifiable pharmacophoric motif, even if they are superficially chemically distinct. Unfortunately, experimental evidence shows that this is not generalizable. A definitive case in point in this regard concerns compounds that target microtubules (ubiquitous components of the cytoskeleton, intracellular protein support, and transport structures). One such drug is the useful antitumor drug taxol, but other microtubule-binding drugs are known, and their respective structures were initially used in attempts to deduce a common pharmacophore. Yet once 3D structures were solved for taxol and another micro-tubule-binding drug (epothilone) in complex with their protein targets, it was revealed that their binding modes were quite distinct, and therefore a common pharmacophore could not exist.[1410]

This vignette also serves to emphasize why basing a drug search on a postulated pharma-cophore motif in the absence of target structural information remains in the semirational realm (Fig. 9.1). In principle, any pharmacophore-based virtual chemical library screen for a target of interest might identify a lead compound with superior functional properties, but in the absence of target structural information, this would necessarily involve a significant element of chance.

A screen that is structurally blind in terms of the target might in principle overlook an alternative-binding mode at the same target site involving structurally distinct compounds, and this alternative-binding mode could prove the best pathway toward the desired ultimate molecular solution.

Virtual Virtues   In Chapter 1 the case was made that if computational screening methods (based on information regarding the target structure and other background knowledge) can entirely return an accurate lead molecule as a design solution, then this falls solidly within the definition of rational design. Any computer-generated "virtual" world and any virtual screening strategies can only exist as accurate surrogates of the real world they seek to recapitulate if the input information is accurate. Virtual modeling (and any other type of modeling, be it molecules or weather patterns) can be a good approximation to reality, but such models stand or fall on the quality of the relevant background knowledge. A good example is the elucidation of the structure of DNA (as already referred to above in a modeling sense). The DNA nucleobases that obviously are crucial to the hydrogen-bonding stabilization of the double helix can exist in different tautomeric forms (recall the above-mentioned importance of tautomers, incorporated into 4D-QSAR). Identification of the standard DNA structure required precise information with respect to base structures, and the use of incorrect base tautomers stymied initial modeling attempts until this error was pointed out.[1411]

All virtual strategies fall within the general ambit of computer-aided drug discovery (CADD), sometimes given an extra D at the end for development or design,* but virtual screening itself can be broadly divided into two categories.[1413] We have already noted *ligand-based* approaches, as in the above considerations of pharmacophore-based virtual searches. And as previously noted in Chapter 8, ligand-based virtual screening can also rationally reduce the field for actual high-throughput screening of chemical libraries.[1414,1415] The second broad area of virtual screening is *receptor-based*, which encompasses molecular "docking" and virtual library screening, and by definition requires high-level structural information regarding the target (which we will assume to be protein for the rest of this section). The ligand-based/receptor-based division is not mutually exclusive, though, since a combination of approaches can be effective. One can streamline virtual library screening by initially performing pharmacophore-based and other computational "filters" to remove unpromising background[1398] and pharmacophore information can act as a guide for the design of docking algorithms.[1416]

Molecular docking itself is the computer-based prediction of a ligand structure within a target-binding site,[1417] taking into account the specific bound ligand configuration (which includes, as we have seen, its rotational conformation, tautomerization, and other factors) and its spatial orientation in the target site context. Evidence has been produced confirming that docking success is indeed affected by the conformational representation of the receptor (by comparing modeled receptor sites with "apo" (unbound) and "holo" (bound) conformations[1418,1419]). We have seen earlier that ligand conformations are often also significant. Where ligand conformational changes occur, they are rarely in the lowest possible energetic configuration,[1420] showing the need to model a diversity of conformational states (in other words, it is insufficient to calculate lowest energy state as the "correct" ligand conformation). And for docking success, another important factor to take into account for the necessary virtual modeling is the role of water, since the aqueous solvent is more frequently involved in protein-ligand binding than not, and water molecules often mediate contacts between protein

---

*This naturally tends to bring to mind the well-known CAD acronym for general computer-aided (or assisted) design. CAD and CADD have both arisen to streamline the development of new technology-based products.[1412]

and specific ligand.[1421] Docking screens are assessed and controlled by using known protein–ligand pairs in comparison with libraries of "decoy" noninteractive compounds, although ensuring that the latter control set are true nonbinders is important.[1421]

The basics of docking include the computational approaches used to represent protein and ligand (including allowance for molecular flexibility), the search methods, and the scoring/ranking of the screened virtual protein–ligand interactions.[1417] This can be computationally rendered as a combination of a search algorithm and a scoring function.[1422] The latter addresses the question, "After a virtual docking operation, how does one assess and quantitate the effectiveness of the ligand fit?" Such an assessment of course corresponds to affinity in real protein–ligand interactions. Scoring virtual affinity assessments, and ranking a large series of virtual ligands in order of priority for this property, is thus a very significant area within the field of docking development.[1415,1417] A number of alternative docking programs are available, with regular refinements incorporated into updated versions.[1422] Among several alternatives, some of these programs use evolutionary search algorithms (Chapter 4) to deal with alternative ligand conformations in the binding site.[1415,1417]

The closest computational equivalent of high-throughput screening is high-throughput docking[1423] or efficient virtual library screening. Initially, work in this area yielded unimpressive results attributed to the use of rigid receptor models, but increased computational power has allowed the use of programs that incorporate molecular flexibility and induced-fit protein–ligand interactions.[1415,1424] High-throughput docking is becoming increasingly important for the generation of drug leads, including those directed against the important target classes of nuclear receptors[1425,1426] and kinases.[1427] In some cases the acquisition of adequate target structural knowledge has been a bottleneck for implementation of rational design by virtual screening. It has been notoriously difficult to obtain high-quality structural data for membrane proteins, especially those with multiple transmembrane domains, of which the major drug target family GPCRs (G-protein-coupled receptors; noted in Chapter 3) are a prime example. For a considerable time, the only available GPCR structure was that of the visual signal transducer rhodopsin,[1428] and this was used to "rationally" model other GPCR structures. But the more recent report of the β-adrenergic GPCR structure[1429] has allowed accurate *in silico* docking studies to be performed with this receptor.[1430] This advancement in knowledge, combined with increasing confidence in the validity of computational GPCR protein models in general,[1431] suggests that this kind of bottleneck will not impede the progress of the field for very long.

Short of pinpointing specific leads, both ligand-based and receptor-based virtual screening methods can be used for the generation of focused (or "smart") libraries for subsequent high-throughput screening, as noted in Chapter 8.[1279] Virtual screening in general has also shown the ability to uncover "unexpected" drug hits; those that a human expert would be unlikely to propose when provided with the same starting target and known ligand starting information.[1421] Virtual screening by docking procedures can be turned on its head to give rise to "inverse docking." In this situation, an "orphan" ligand is screened against a protein structural database in the hope of finding an interactive and potentially druggable protein match. To do this, libraries of predesignated protein druggable active sites are test-docked and scored with the ligand(s) of interest, and this has been successful with a library of over 2000 members.[1432] Inverse docking has been proposed as a pathway toward "reverse pharmacognosy," where natural products of structural interest but unknown target proteins are screened *in silico* for protein partners.[1433] Full development of this type of virtual screening also has the potential to become one of the major boosters to the development of novel natural products.

Continuing improvements in design underscored by rational structural principles and virtual modeling are inevitable, but as yet we have not considered cases of small molecule design which pass stringent rational design criteria purely from first principles. Let us now briefly take a look at the lofty goal of design without any empirical "leg-up."

The De Novo Ultimate? At the end of the empirical–rational scale of Fig. 9.1, we find "*de novo*" rational design, where a molecule is built "to order" from the ground up, with only knowledge of the target structure and/or its binding ligands provided. The definition of this small-molecule creation process parallels that of rational design in general, in that a complete (high-affinity) ligand is an unlikely outcome in lieu of a lead prototype.[1434] With a defined target-binding structure, one approach (receptor-based design) uses computational chemical fragment fitting, which can progress by successively linking fragments within a virtual pocket, or by placing them in parallel. (This will inevitably prompt comparisons with real fragment screening approaches, as outlined in Chapter 8.) Alternatively, potential spatial positions for ligand atoms are mapped as a grid, and suitable ligand patterns matched for best-fit characteristics. In any strategies that build ligands in a defined site, the range of combinatorial possibilities becomes huge and impossible to comprehensively evaluate. This necessitates decisions for winnowing the potential ligand range, such as accepting only specific types of noncovalent interactions, using pre-existing pharmacophore information if available and imposing stringent steric constraints.[1434] *De novo* design can still be applied in the absence of accurate target binding structures by means of "ligand-based" strategies,[1434,1435] which use patterns of binding by known ligands to develop pharmacophore models and a model for the receptor site itself ("pseudoreceptor"). The latter receptor model in turn can then be used in receptor-based design strategies, albeit with less certainty than when a complete physical structure is available.

De novo design is ideally a deterministic process, but certainly nondeterministic algorithms frequently feature in its current computational implementation. Indeed, heuristic optimization techniques in general offer potential escapes from local fitness optima toward improved (if not literally universally optimal) solutions. Evolutionary algorithms and particle swarm optimization approaches (as outlined in Chapter 4) have been productively applied in this regard and have the potential to locate new "activity islands" in chemical space (clusters of compounds with shared features and functional activities).[367] A design approach (BREED[1436]) has been developed that assesses feasible fragment combinations of known ligands for a target structure, and thereby acts as a kind of virtual chemical recombination system, automating the common pharmacological practice of ligand fragment joining. Algorithms that are based on iterated cycles of ligand virtual synthesis and screening for convergence toward an optimum are collectively referred to as "adaptive design" *de novo* strategies.[367] A practical issue with these approaches in general is excessive complexity of delivered solutions, especially in terms of the numbers of stereocenters present. Accordingly, filtering of this kind of molecular complexity in favor of simpler compromises is a useful adjunct design strategy.[1437]

Restriction of an optimization process to a specific small-molecular chemical scaffold ignores potentially very superior solutions elsewhere in chemical space. Searching for novel scaffolds, or *scaffold-hopping*, is thus an important aspect of *de novo* design. This can be compared with a search for superior chemical mimics of known ligands, and a prime example of this is the important area of design for peptide mimicry (peptidomimetics),[1438] which can also be undertaken with a fragment-based strategy.[1439] But as generalizable chemical entities, "scaffolds" have not been precisely defined, leading to a level of subjectivity as to what exactly constitutes a "hop" from one scaffold to another.[1440] For a real scaffold transition, clearly significant changes are required in the small-molecule skeleton, but functional properties of the original compound must be retained. Scaffold-hopping thus seeks to generate chemically distinct *isosteres* (size and shape equivalents) of the parental molecule that can orient in the target binding site in a similar manner. In this undertaking, one can look beyond the bounds of carbon-based chemistry, and isosteres incorporating silicon are a profitable region of chemical space for ongoing exploration.[1441,1442]

Many success stories in *de novo* design have been chalked up,[367,1434,1435] and the area will predictably burgeon in the near future. But there is much room for improvement, particularly in accounting for conformational flexibilities of both receptor pockets and ligand in design modeling.[367] The great majority of such receptor structural targets for *de novo* design are contributed by proteins, so at this point we should take a glance at rational design and structural predictions of proteins themselves. In Chapter 5 we saw that directed evolutionary approaches are an effective route toward altered protein function and that *de novo* proteins could be derived by selection from large libraries or through randomization with simple patterning constraints. Owing to the vast size of protein sequence space, though, the wholly empirical *de novo* creation of proteins is feasible only in a limited set of cases. To what extent can rational design extend and overcome this limitation?

Proteins by the Book

In order to rationally design whole proteins or segments thereof, structural information is essential. This is true whether the project is "top–down" (starting from a pre-existing structure) or "bottom–up" (designing the protein target from first principles). An aim to rationally modify a given protein obviously is stymied if no structurally relevant data is in hand, but *de novo* protein design is no less reliant on structural knowledge. Success in this case is dependent on the ability to predict protein folding, or the relationship between primary amino acid sequence and the final folded three-dimensional shape. And algorithms for folding prediction could never be developed without feedback between models and real protein folding results, as revealed through structural analyses. The "gold-standard" method for protein three-dimensional structural characterization has long been the X-ray crystallography, but structural determinations by nuclear magnetic resonance (NMR) have increased in utility. The latter have the advantage of probing the native structure on solution, but are still not feasible for most large and multisubunit proteins, although much progress is being made.[1443]

Yet not all proteins are tractable for these approaches, at least in their native states with conventional crystallization methods. As noted earlier, membrane-bound proteins have historically been difficult to structurally characterize, and these are a major subsection of all drug targets (especially the G-protein-coupled receptors [GPCRs]). Ingenious solutions to these problems have nevertheless been devised. Mutants of parental membrane proteins sometimes allow crystallization while retaining overall structure, and certain wild-type membrane proteins have been coaxed into suitable crystals by means of solutions bearing phospholipids. In fact, systematic rational screens for defining either crystallization-tractable mutants[1444] or effective solution conditions for crystallization[1445] promise to enhance the efficiency of structural studies in general. The recent solving of some GPCR structures (referred to earlier) has been achieved by means of advances in both crystallization and X-ray diffraction technologies.[1446] In addition, other techniques for obtaining protein structures in cases resistant to conventional methods exist and are rapidly developing.*

Natural protein structural analyses are encompassed by *structural genomics*, the ultimate goal of which is nothing less than the definition of the three-dimensional configurations of all components of complete proteomes, particularly that of humans.[1449] For this to be accomplished in a reasonable time, much automation of the data acquisition exercise is called for,

---

*"Site directed spin-labeling" is an important example, applicable to both soluble and membrane proteins.[1447,1448]

rendering structural characterization into a high-throughput process.[1372,1450] Computational analysis of X-ray diffraction data to derive structural information has been shown to be amenable to complete automation.[1371,1451]

If protein structural predictions from primary sequence alone were highly reliable, physical structural determinations would take on only a confirmatory role. This situation does not (yet) exist, but advances in protein structural knowledge and protein folding prediction proceed in parallel and will ultimately converge. Although we have looked at folding in general terms previously in Chapter 5, another glance at progress in rational folding prediction and design is in order.

Getting Folding Right Folding of proteins into specific and functional shapes can be considered as a special outcome of the conformational flexibility of both polypeptide backbones and side chains. If the bond angles between all atoms in polypeptides were always predetermined and invariable, then each polymeric peptide sequence would behave as a rigid and elongated structure and protein-based life would be impossible. But given conformational mobility, complex interactions between amino acid residues in a polypeptide occur in a sequence-dependent manner, and for a small subset of all possible sequences,* condensation into a globular protein fold results. The folding of almost all proteins† corresponds to a pathway toward a global minimum for the thermodynamic quantity termed free energy, as applied to the conformations of their α-carbon backbones and amino acid side chains.[1452,1453] If this can be computationally calculated, then the native folded state of the protein is in principle revealed.

Yet it has been long recognized that proteins themselves could not possibly "compute" their lowest energy folding states by randomly sampling every possible conformational configuration, as the time this would require could exceed the lifetime of the universe. (Natural proteins in fact fold very rapidly, on the order of milliseconds or even microseconds in some cases.) Even a small bias away from a totally random search can have a dramatic effect,[1454] and this logically must be mediated through specific types of molecular interactions. Exclusion of hydrophobic residues from solvent by packing into a hydrophobic core, and patterns of hydrogen-bonding interactions are the most important (though not the only) factors in determining folding toward the global energy minimum state.[1455] Specific amino acid residue sequences can modulate folding by locally affecting backbone configurations or through nonlocal packing interactions. The folding pathway (the mechanistic course taken by polypeptide configurations between the completely unfolded and specific native folded states) reflects changes in the net free energy, and these transitions can be modeled as "energy landscapes." At any instant during the folding progression, free energy is a function of the total number of backbone and side chain conformational alternatives, which decrease continuously during folding. This leads in turn to the conceptualization of a "folding funnel" as the corresponding landscape shape, resulting from the existence of a very large number of nonnative states with high free energy channeled down ultimately into a unique low-energy native folded configuration.[1454,1456] A smooth funnel landscape corresponds to a rapid and unique folding pathway, whereas the existence of many local energy minima in a "rugged" funnel indicates folding heterogeneity (Fig. 9.3). (Trapping of a folding polypeptide in such a local minimum constitutes a misfolding event.)

---

*In Chapters 3 and 6, we noted the vastness of protein sequence space in comparison with the numbers of natural biological proteins.

†One type of exception appears to be the substrates for steric foldases, which we encountered in Chapter 5.

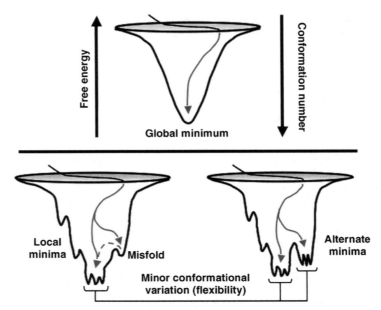

FIGURE 9.3   Folding funnel energy landscapes for folding pathways. The main message is that although proteins have many high-energy non-native states, folding converges toward one low-energy native configuration.[59] At the top, a smooth pathway is depicted for rapid folding toward a single global minimum free energy state. More rugged funnels have local minima that represent misfolds (failure to attain global optimum) if the polypeptide is trapped in such a state. But the "global minimum" can be a set of closely similar alternating conformational states corresponding to native-state molecular flexibility. On the bottom right is depicted the rare circumstance of alternative near-global minima, resulting from translational pausing[60] or other intervention during a folding process.

For the study of protein folding, the measurement of folding rates and determination of folding intermediates is important.[1455,1457] During folding, nucleation motifs (often referred to as "foldons"[1458]) form and interact highly cooperatively. Experimentation has suggested that some proteins exhibit apparent heterogeneity of folding pathways (as in the rugged funnels of Fig. 9.3), all leading to the same global energy minimum, but this effect can be equally well attributed to a single pathway with "optional" offshoots into local misfolded minima.[1459] The rates of folding for different proteins can vary over many orders of magnitude, but a simplifying relationship between the native folded state (topology) and folding rate has been observed. This is based on "contact order"; a quantity defined as the average sequence distance between amino acid residues in physical contact in the native protein, as a ratio with the number of residues in the protein. (Or in other words, measuring the average distribution of residue interactions in the protein sequence in relation to their locality.) Proteins with low contact order (a higher degree of local interactions) consistently fold faster than those with contact orders of a higher magnitude.[1460] Other experimental results, based on studying folding rates of specific mutations in proteins of interest, have suggested that small proteins with similar folds (though diverse in primary sequences) fold via similar pathways.[1460,1461] Though true in many instances, notable exceptions to this correlation have been noted.[1462]

The correlation between average locality of interactions in the native state and folding rates is consistent with the current general mechanistic understanding of the protein "folding code."

In general terms, it seems clear that proteins escape the conundrum of exploring all possible configurations by rapid formation of independent local structures (foldons) followed by cooperative assembly of substructures.[1455,1463] But the most convincing demonstration of understanding the thermodynamics of protein folding is prediction, and this we should now briefly look at.

Making Rational Predictions   A definitive solution to the "protein folding problem" would permit the confident assignment of a protein structure when one is supplied only with the primary amino acid sequence and no other information. This has long been regarded as a task of supreme difficulty, sometimes simply relegated to the "too hard basket," given that computational theory seems quite unfavorable to a "ground up" approach. Just as proteins themselves cannot sample all possible configurations in order to "decide" how to fold, a serious practical issue immediately emerges if one contemplates a global computational analysis and comparison of all possible conformational combinations. These balloon into vast numbers even for small proteins and rapidly threaten intractability. Computational problems are classifiable by their complexities, and the feasibility of designing algorithms for solving them in a reasonable period of time. Problems that are solvable in "polynomial time" (P) are computationally tractable, since the time required for an algorithm to process such a problem ("a problem in P") grows as a polynomial function (nonexponentially) as problem complexity increases. A special class termed "NP-complete*" is "hard" and not believed to be solvable in polynomial time, although this is not formally proven. Demonstrating that a problem falls inside the NP-complete class thus declares that an algorithm that can deliver a precise solution to the problem in reasonable time is an unlikely proposition, at least as presently understood. Calculating protein folds from a fundamental "ground-up" level by exhaustively evaluating all possible conformational combinations to compute the global free energy minimum has been shown to be an NP-complete problem, and therefore intractable for delivering a precise answer in reasonable time.[1464,1465]

But this seemingly pessimistic result is not at all the end of the story, as there are practical routes for finding good approximations to the problem. First, a productive way to predict the structures of novel proteins is to seek evolutionarily related proteins with known sequence as an initial guide, a process termed template-based or comparative modeling.[1466] This approach is of major significance in structural genomics,[1449] and its applicability has itself steadily grown as genomics and structural information has provided more and more evolutionary templates upon which to build models. Nevertheless, the direct comparative approach is not applicable if no known homologous protein models exist. In the latter case, "free modeling" must be applied, and this field too has seen very significant advances in recent times. The most basic approach to "free" or *ab initio* modeling is from physical calculations of all interactions occurring among the polypeptide and solvent in order to simulate the process of folding itself *in silico*. But as noted, the "NP-hardness" problem necessitates simplification of the total search space that can be undertaken. A very useful strategy (applicable in diverse routes toward protein folding prediction, and protein design also) lies in the observation that the rotational conformations of polypeptide side chains do not vary continuously but fall within a number of "preferred" specifically definable configurations termed *rotamers* (for rotational isomers), and collections of these for each type of residue are termed "rotamer libraries."[1467] But for simulations of even small proteins over short folding times, additional simplifications involving grouping of certain atoms into single centers of mass are required, along with extensive computational processing time.[1468,1469]

---

*Here NP denotes "nondeterministic polynomial time."

The most successful avenues for folding predictions of larger proteins can be collectively considered as knowledge-based "fragment recombination" methods, based on the evidence that a protein backbone configuration can be accurately modeled by stringing together appropriate short segments from other structurally characterized proteins.[1469,1470] One such successful computational embodiment of this principle, the Rosetta prediction program,[1471] uses fragments (3–9 residues in length; approximately 25 separate choices for each segment) from known proteins in the Protein Data Bank structural repository, which are inserted into the target (structurally unknown) amino acid sequence. This is based on matching between donor and target sequences, and also matching the known donor secondary structure with the predicted local secondary structure of the target.* A "Monte Carlo" nondeterministic computational method (which relies on a random-sampling approach) is then used to assemble fragment combinations into structures, based on optimization for free energies of hydrophobic and hydrogen-bonding interactions, in combination with other knowledge-based input regarding polypeptide interactions.[1471,1472] A conformational sampling of fragments whose assembly generates the lowest energy level corresponds to a coarse-grained folded state, followed by further Monte Carlo-based refinement (including rotamer optimizations) that can take the modeling to atomic-level resolution. Reasonable approximations for energy functions work because of the substantial gap between the low-energy folded state and the multitude of higher energy conformations in non-native states.[1471]

The observed utility of fragment-based assembly approaches for structural prediction has interesting ramifications for the structures of proteins in general. It might be thought that totally novel folds absent from existing databases would be poorly modeled by fragment assembly, but novel folds themselves tend to be reducible to smaller motifs shared with other proteins.[1469] Viewed from this perspective, "protein fold space" becomes continuous rather than representing discrete folding states.[82] Moreover, it has been suggested that peptide fragments with appropriate properties ("antecedent domain segments") were important in protein domain evolution.[1473,1474] In any case, fragment assembly methods such as Rosetta have been successful at predicting the structures of hitherto novel fold categories.[1475,1476]

The success of fragment assembly methods has led to hybrid approaches that add on folding simulation steps.[1469] But, as is often said, "the proof is in the pudding" and the best available rational prediction methods should themselves be readily testable in a systematic and rational manner. To that worthy end, since 1994 a series of experiments has been organized in the form of protein prediction challenges to the relevant worldwide scientific community, termed CASP (Critical Assessment of Structural Prediction),†,[1478] which has separate sections for comparative and *ab initio* modeling approaches. The first such experiments highlighted many predictive inadequacies, but after a decade of successive CASP rounds, "blind" prediction to atomic resolution of certain small protein structures from their sequences was possible.[1479] Continued refinement of structural prediction can be anticipated and monitored in ongoing CASP challenges.††,[1481]

And yet not all CASP targets have been predictable to such a high level of accuracy, indicating the need for further progress.[1482] The status of the protein folding problem as framed in structural prediction unsurprisingly depends on the size and specific structural features of the target polypeptides. One relevant factor in the degree of predictive difficulty is the level of

---

*Relatively simple algorithms for local secondary structural predictions have long been in existence, but this itself is far from a prediction of tertiary structure.

†Proteins for the CASP community are typically in the "soon to be released" category (those whose structures have been physically determined, but not yet publically reported). An analogous series of community experiments in protein–protein docking modeling and prediction is called CAPRI (Critical Assessment of Prediction of Interactions).[1477]

††The eighth CASP round (CASP8) was conducted and analyzed in 2008.[1480]

contact order, the property noted earlier that correlates with folding rate.[1483] Clearly a general description of the folding code is not in hand, but many small proteins can indeed be predicted with impressive accuracy. Even the most stringent critic of general rational protein prediction would have to concede that the folding problem has transformed from an apparently intractable quagmire into a challenge whose solution can be grasped, at least for proteins of moderate size.[1463] The success stories with modern energy minimization strategies themselves argue forcefully that the developers of these programs are "doing something right" in terms of reproducing the forces that mold proteins. And the other side of the coin to prediction is active protein design, which we should now take a look at.

Making Rational Proteins   Design can indeed be viewed as "inverse prediction," since it moves from a chosen specific protein structure to the corresponding sequence, rather than the sequence to structure pathway for prediction itself.[1484] But beyond the folding code, truly effective protein rational design must go up a step and also define "functional codes" for the task at hand. Defining a novel fold and then calculating the primary amino acid sequence required to generate such a structure is no mean feat, but this is not sufficient to produce a predefined function as well unless the necessary structure–function relationships are also defined. "Protein function space" encompassed by natural polypeptides has not yet been plumbed,[1485] and clearly if unnatural folds can be generated, the "functional ceiling" of proteins may yet be very high.

Comparably with protein structural predictions, rational protein design can be based on the remodeling of an existing polypeptide template, or fully *de novo* in the absence of a template guide. In the 1980s, pioneering *de novo* design efforts were based on the synthesis or expression of small proteins with defined simple secondary structural motifs. The "four-helix bundle" pattern is a natural motif incorporating four associated α-helices more or less arranged around a common axis. This was recapitulated in artificial structures using secondary structural algorithms and other knowledge of protein structural patterning, including placement of residues appropriate for ending ("capping") helical regions.[1486,1487] Small model β-sheet proteins were likewise designed.[1484] It was noted at an early stage of these endeavors that "negative design" (designing with an eye toward removing or destabilizing deleterious interactions) could be as important for ultimate success as "positive" design aspects.[1484]

More generalizable design efforts seek to calculate global energy minima for desired proteins or protein motifs. While computational protein design from first principles is definable as an NP-hard task, once again this is a "worst case scenario,"[1488] and suitable approximations can act as short cuts toward design ends. A useful means for computing the global free energy minima of proteins has been a deterministic process termed "dead-end elimination," which compares favorably with nondeterministic alternatives such as genetic algorithms.[1489] This method is based on the elimination of specific rotamers for each residue position which are incompatible with global energy minimization. If the contribution of a specific rotamer to the total energy summation can be reduced by replacement with an alternative rotamer, the initial rotamer can consequently be eliminated. Successive eliminations therefore converge toward computing the desired global energy minimum.[1489,1490] For reduction in the search space and maximizing the efficiency of this method, optimization of rotamer libraries is important,[1490,1491] and some versions of dead-end elimination compare clusters of rotamers rather than individual rotamers.[1492] This elimination approach was used in the first fully automated design of a small folded protein motif in 1997, based on the zinc finger motif.[1493] To use a more recent example that draws upon a favorite fold of Chapter 5, a putative "ideal" TIM-barrel fold has been generated by application of a dead-end elimination algorithm.[1494] It is notable that dead-end elimination and certain other successful protein design computational strategies do not explicitly address negative design.[1492]

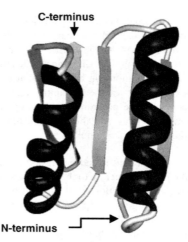

C-terminus

N-terminus

FIGURE 9.4 Structure of computationally designed protein Top7, with novel unnatural fold. α-helices, dark gray; β-sheets, light gray. *Source*: Protein Data Bank[2] (http://www.pdb.org); 1QYS[61]; Images generated with Protein Workshop.[5]

Fragment assembly processes used for protein structural predictions, including the Rosetta program noted above, can also be applied toward design challenges. Again, instead of a set target sequence whose structure must be deduced, design requires the appropriate sequence to be arrived at from a designated structure. Therefore, for design purposes, primary choice of fragments from known proteins in structural databases obviously cannot be based on sequence matching (since the desired sequence is initially unknown), but can be predicated on secondary structural criteria.[1471,1472] Searching for appropriate fragment combinations and systematic residue evaluations with the resulting model backbone can then proceed in an analogous manner as for predictive programs, until a sequence is generated. The utility of a fragment assembly based strategy (an adaptation of Rosetta) for *de novo* design was dramatically demonstrated with the first high-resolution design and production of a small protein with a fold not known in natural circumstances.[1495] This specific case ("Top7"; Fig. 9.4) was exceptionally stable, raising the interesting possibility that its absence from nature implies that many other "naturally unsampled" folds remain to be created, and these might have numerous useful applications.[1452,1495]

Artificially designed proteins with "unnatural" folds, of which Top7 is the first exemplar, are a powerful means to probe questions regarding how natural proteins are shaped by evolutionary pressures. Protein sequences that fold into functional forms are "special" with respect to the vastness of protein sequence space, but only functional polypeptides are subject to positive selection in any biosystems. Darwinian evolution thus readily accounts for the special structural features of biologically folded proteins (as noted in Chapter 2), but it is reasonable to wonder whether other aspects of protein behavior have also been accordingly selected. We have previously noted instances of folding modulation by steric foldases (Chapter 5), and pausing of translation on ribosomes (mediated by the presence of rare codons with corresponding low-abundance tRNAs) has recently been shown to be capable of modulating folding pathways with functional consequences.[31] Some aspects of protein folding are suggestive of natural "negative design" or selection against intermolecular aggregation events.[623] Evolution may also have disfavored the mixed "handedness" for amino acid stereochemistry in proteins in favor of homochirality (the exclusive use of L-amino acids), in order to facilitate compact helical folding states.[1496] Other evidence has suggested that natural selection does not act directly on folding rates *per se*.[1497]

So what has Top7 revealed with respect to the evolution of natural protein folding mechanisms? The folding pathways of natural small proteins reveal a smooth energy landscape with high cooperativity (Fig. 9.3), but a much more complex and less cooperative folding pattern was demonstrated with Top7.[1498] Moreover, fragments of Top7 that can be generated by mistranslation events in *E. coli* are also very stable.[1499] These results collectively suggest that the smooth and cooperative folding pathways of natural small proteins do not automatically arrive with stable conformations alone, but must be actively selected themselves through the functional benefits that cooperative folding evidently confers.* Also, it recalls the marginal stability of natural proteins (as noted in Chapter 5) as arising from their selection to be "just stable enough" in their working environments. (Excessive stability may discourage active turnover or have other deleterious effects.) Relatively stable folding intermediates in pathways with low cooperativity might also increase the likelihood of disadvantageous interactions in complex biosystems. A practical message for *de novo* design of therapeutic proteins then comes from these findings: it may not be sufficient to heed only the final folded state of the novel protein of interest (even if that is quite successful); *how* the protein folds is also likely to be important.

The strikingly high stability of the novel fold of Top7 may be due in part to its lack of functional constraints.[1495] In natural proteins, the demands of a functional role might force selection for specific protein segments that are suboptimal for stabilizing interactions, a dictate to which Top7 was never subjected. This consideration raises again the issue of *de novo* functional design, as noted at the beginning of this subsection. Truly *de novo* design of function can be distinguished from the redesign of functional binding or catalytic pockets for new activities, which bootstraps from the platforms provided by natural evolution. All aspects of functional design, but especially design aiming for "ground up" novelty, must use a broad knowledge base. The burgeoning genomic databases provide a resource in this regard by means of comparative multiple sequence alignment analyses, which can be exploited for information regarding amino acid residue combinations with specific functions.[1500] *De novo* design of molecular recognition necessitates the inclusion of protein–protein docking in the overall strategy[1471] or modeling of pockets for binding small molecules.[1501]

The biggest challenge for functional *de novo* design is evident when one considers the first-principles generation of novel enzymes. Despite the inherent difficulties, this arena has seen impressive recent advances.[1502–1504] The problem can be reduced to the design of a catalytic pocket appropriate for the envisaged task, and then modeling a suitable scaffold in which to locate the pocket itself. Quantum mechanical calculations enter into the design of residues in the active site that can act in transition-state stabilization, and the scaffold choice is based on its ability to stably support the idealized active site.[1503] Interestingly, modeling a scaffold for a specifically designed catalytic task (the "Kemp elimination") resulted in a convergence toward the TIM-barrel fold.[1503] As we noted in Chapter 5, this structure is the most widespread and diverse of catalytic scaffolds used in nature, suggesting that computational design may reproduce this "choice" as a common "best-fit" solution.

Although an impressive achievement, *de novo* design by creating a catalytic pocket and then finding a structural home for it does not take into account the full power of natural enzymes. Indeed, computationally designed enzymes have been considered as analogous with primordial biocatalysts more so than with modern enzymes, which have been sculpted and refined by

---

*Unlike many small natural proteins, Top7 has low contact order (as defined earlier), and thus a higher probability of significant stabilization of local motifs.[1498,1499]

megayears of natural evolution.[1502] By way of comparison, we have previously seen that catalytic antibodies are also limited in efficiency in comparison to their natural enzymatic counterparts. Unlike such artificially generated "abzymes," natural enzymes maximize their efficiencies through regions outside their direct substrate binding pockets; whole-enzyme dynamics are thus important for maximizing catalytic performance (Chapters 2 and 5). Accordingly, once a novel protein enzyme is computationally designed, directed evolution can step into the picture to obtain enhanced efficiency, as was achieved in the above "Kemp-elimination" catalytic example.[1503]

In fact, partnership between rational computational design and directed evolution of enzymes has acknowledged potential for future development.[576,1505] The basic argument is that design for a desired novel activity will permit the computational derivation of catalytic sites that leap beyond what is offered by nature. With a baseline of the new catalysis established, evolutionary approaches are then the most efficient means to reach toward its optimization. This two-tiered "best of both worlds" approach to design is similar to comparison made in Chapter 8 of small molecule development with evolutionary strategies. Even with the power of recombination included, from any given starting point directed evolution alone may not have the wherewithal to locate radically new design prototypes, at least in a time-frame reasonable for human activities. But human intervention, in the form of the highest levels of rational input, can in effect allow otherwise impossibly deep fitness valleys to be crossed. The future for protein design in general thus looks very bright, and the full impact of rational computational design is just beginning to be felt, with a huge range of potential applications.[1506,1507]

Before leaving this field, we should note as well that protein design is becoming increasingly assisted by advances in synthetic chemistry, which permit a greater range of novel structures than any natural proteins. An early synthetic avenue for avoiding the difficulty of designing increasingly complex motifs as single polypeptides has been the use of specific template molecules, to which small protein functional units can be selectively derivatized and thereby fixed in preset orientations. This "TASP" strategy (template-assembled synthetic proteins[1508]) has used a variety of template scaffolds and different secondary structural fragments, assembled in place as directed by the reactive groups on the scaffold itself. TASPs have been recently applied as artificial ion channels.[1509] In Chapter 6, we noted the utility of "mirror-image" aptamers, or spiegelmers, and chemical synthesis can allow D-amino acid versions of proteins with opposite chiralities to their natural counterparts.* Artificial syntheses can go far beyond even the capabilities of biosynthetically extended protein alphabets, but the synthetic products are usually limited in yield, and thus at present are mainly of use in laboratory studies (though immensely valuable for this contribution alone). Also, proteins of >200 residues are still formidable synthetic challenges.[981]

There is a continuum between the synthesis of purely natural proteins and increasingly modified versions of them, until a point is reached when the products might be more appropriately described as fully artificial foldamers (Chapter 7) rather than through a resemblance to proteins.† Previously, we have thought about foldamers as a general concept, and in the final chapter we will look at this topic once more. But an important point in the present context of rational design is that lessons learned through the application of increasingly sophisticated computational strategies for protein building will be applicable to many

---

*An example of this is the synthesis of the D-amino acid version of the HIV protease (a proteolytic enzyme essential for HIV maturation), which has activity only on peptide substrates that have the same inverted chirality.[981]

†Since foldamers are defined as nonbiological folded polymers, note the distinction between proteins synthesized on ribosomes and those made synthetically. A fully synthetic "normal" protein is nonetheless "biological" in the sense that it copies a biological sequence, but a synthetic protein modified to the extent that it can no longer be ribosomally synthesized even with sophisticated code extensions (Chapter 5) becomes nonbiological and qualifies as a foldamer.

peptidomimetic foldamers as well.[1471] Rational design is also relevant for the special area of functional nucleic acids, which we should now briefly inspect.

Designer Nucleic Acids

A folded functional nucleic acid differs fundamentally from a protein in that, given its sequence, it is a trivial matter to define its "antisense" strand by simple base-pairing rules. (In this context, we might recall the metaphorical Lucy sketch of Chapter 6 with respect to imaginary "antiproteins.") And of course this is the key to nucleic acid replication, realized from the time of the revelation of the structure of DNA in 1953. So, if we think of nucleic acid design in the broadest sense, it can be broken down into simple sequence-mediated design involving duplexes, or complex-folding design involving single strands. In Chapter 6, we also noted this distinction when considering molecular nucleic acid decoys, which can simply recapitulate binding sites in the form of short DNA duplexes or mimic the target duplex at the protein-binding site in the form of folded aptamers.

Sequence-based design of nucleic acids for functional purposes dates back to the use of hybridization probes, and then antisense RNAs or antisense DNA oligomers for targeting specific mRNAs to block gene expression. Although design of antisense sequences automatically follows from base-pairing rules against any nucleic acid target whose sequence is known, the specific choice of a target site *in vivo* in a long RNA molecule with complex secondary structure is not so trivial. Antisense inhibition proved to be notoriously case-dependent, sometimes giving good results but often unsatisfactory. For this reason, the development of small ribozymes (and later DNAzymes; Chapter 6) acting as site-specific ribonucleases was hailed as a breakthrough for artificial control of gene expression. Here, the technology is a combination of the products of natural or artificial evolution (to derive a catalytic core) and targeting of the ribozyme by simple target motifs and base pairing (as depicted in Fig. 9.5). At least, once a specific target RNA sequence is selected, the generation of flanking "arms" is in principle a trivial exercise, although once again the target site selection must be dictated by secondary structural considerations.

More recently RNAi (gene silencing by small RNAs) has gained most prominence as "the method of choice" for gene expression "knockdown," as introduced in Chapter 3. Both the strength and the potential weakness of RNAi are that it relies on specific processing pathways innate to eukaryotic cells. Its chief appeal lies in the fact that the "in-house" natural RNAi pathways are highly efficient and not exploited by earlier conventional antisense approaches. On the other hand, any innate defense system can itself be directly targeted for disruption by counter-strategies of invading viruses or other parasites, and viral systems for the

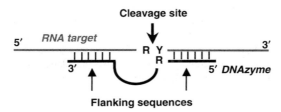

FIGURE 9.5   Schematic of RNA cleavage sequence requirements for the "10–23" DNAzyme.[62] R = purine; Y = pyrimidine. The flanking sequences are not essential for DNAzyme activity and thus can be determined by the target RNA. Other small nucleic acid enzymes have nonidentical but equally simple cleavage site preferences.[63]

suppression of host RNAi have been described.[1510,1511] Therefore, entirely artificial systems targeting invasive viral mRNAs still have appeal, and nucleic acid enzymes remain as candidates in this regard.[722,1512] But certainly for control of endogenous gene expression, including numerous genome-wide functional expression screens (ftp site), RNAi has emerged as the superior choice. Endogenous eukaryotic gene expression is now known to extensively rely on short or "micro"-RNAs (miRNAs) as well as protein-based mechanisms, and manipulation of miRNAs is also under intensive evaluation for its therapeutic potential.[*,1514,1515]

To activate the endogenous RNAi-processing system, artificial RNAs (whether directly provided or via cellular expression) must be presented as duplexes. Nevertheless, the recognition of the target mRNA by the input RNAi species still relies on base-pair complementarity, rendering RNAi design essentially a rational process once important additional design criteria for site selection in targets are defined.[1516] Yet even if ribozymes and DNAzymes were entirely overtaken by RNAi and/or miRNAs for the purpose of gene expression control, opportunities in biotechnology for therapeutic nucleic acid enzymes are by no means completely superseded. There is major interest in the use of ribozymes derived from Group I introns for making specific repairs or alterations to mRNAs via trans-splicing reactions, with potential applications in many areas of gene therapy.[1517] Here too an alternative exists that makes use of the endogenous RNA splicing machinery, by delivery of artificial RNAs with appropriate splice signals and desired exonic sequences, which direct desired trans-splicing events *in situ* at the spliceosome.[1518]

As we have seen, the secondary structures of single-stranded RNA molecules (usually an mRNA species) must very often be taken into account when they serve as targets for downregulation of gene expression. As with the folding of aptamers (Chapter 6), conventional Watson–Crick base pairing is important in determining these structures, but other types of interactions are also very significant. A variety of programs for predicting RNA secondary structures exist that are based on optimization of possible base interactions. More fundamental thermodynamic approaches by calculation of energy minima (analogously with proteins) have made progress with the accuracy of structural predictions for short RNA oligomers.[1519]

When a functional nucleic acid can be generated by a predetermined set of sequence-based rules and other established criteria, then it can be considered "rational"; as opposed to the selection and evolution of functional ribozymes and aptamers. But just as the empirical and rational often proceed hand-in-hand toward a difficult design goal, applications of functional nucleic acids can be based on exploitation of both strategies.[1520] As an example, a single RNA can be rendered as a fusion between a preselected aptamer binding a cell surface target, and a designed RNAi duplex against a gene expressed within the cell type of interest (Fig. 9.6). The aptamer segment then mediates targeting of desired cells, and the RNAi sequence targets the desired gene.[†,1522] Aptamer segments can also be used as switchable modulators of fused RNAi sequences through small-molecule allostery.[1523] Further input from rational design can enter at the level of chemical modifications of functional nucleic acids, as we saw also in Chapter 6.

Aptamers and RNAi in combination are useful owing to the need to consider different biosystem levels in a therapeutic context. No target exists in isolation, and the "system" aspect of cells and organisms is a compellingly important frontier for rational design. We can then

---

[*]Although miRNAs block gene expression by interfering with translation, multiple mechanisms may be involved.[1513] miRNAs themselves become the targets of therapeutic intervention when their aberrant expression is associated with pathological states such as tumors.

[†]Antibodies are also under investigation for analogous RNAi delivery purposes,[1521] but the ability to produce a single bifunctional RNA molecule (as in Fig. 9.6) transcriptionally or synthetically is clearly very convenient.

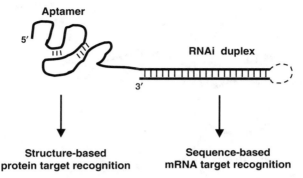

FIGURE 9.6   Aptamer–RNAi bifunctional molecule. The aptamer segment folds into a specific conformation enabling target molecular recognition (such as a cell-type specific surface structure enabling cell targeting and delivery); the RNAi exists as 21–23 bp duplex RNA recognizable by the interfering RNA processing pathway.[64] The enzyme Dicer cleaves longer duplex RNAs into the short forms then used in the RNAi pathway and can process short hairpin RNAs[65] (hairpin shown as dotted line).

finish our tour of this vast area by raising our eyes from specific targets and looking more at the "big picture" of biosystems in all their complexity.

Systems, Pathways, and Networks

A systems approach based upon computational analyses is needed whenever the complexity of a system is too large to be directly apprehended by human inspection. And this applies to biological systems much more so than for engineering tasks, which have long used a systems-based approach for simulating design before practical execution.[1524] This is even more emphatically the case if one wants to go beyond understanding the organization of a biosystem and make specific predictions about its perturbation by foreign entities, such as novel candidate drugs. System-level studies can be undertaken at multiple levels of biological organization, ranging from single cells to the development and regulation of whole organisms. Attempts to gain a global understanding of the intricacies of the mammalian immune system, for example, call for an embrace of the systems biological approach.[1525]

The complexity of biosystems is an emergent property of the unfolding of genomic information during cellular and organismal development. "Genomics" as a term for the study of the organization of the genome has led to the coining of many additional "omics" in reference to the global repertoires and activities of higher level biosystems. In particular, RNAs are encompassed by "transcriptome," proteins by the "proteome," and small molecules by the "metabolome" (or sometimes "metabonome"). It has been noted[1526] that "omics" carries within itself the syllable "Om," which through the Sanskrit word *Om* has connotations of universal inclusivity,* and indeed more and more specialized "omics" can be spun off as we go into deeper levels of detail. It seem clear, though, that there are a special set of primary "omes" that enable the replication of genomes, and these are depicted in Fig. 9.7a as a cycle that enables DNA molecules to replicate themselves. In this entangled web there are many and varied complexities, but one feature within Fig. 9.7a that should be noted is that "G" (Genome) only appears once, which is another way of highlighting genomic primacy. The genomic information

---

*This in turn might remind us of the imaginary OMspace raised in Chapter 1.

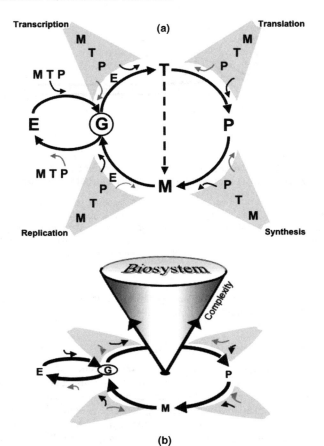

FIGURE 9.7   (a) "Omic" relationships as a cycle. Code: G = Genome, T = Transcriptome, P = Proteome, M = Metabolome, E = Epigenome. In this cycle, the genome gives rise to the transcriptome, and a portion of the latter in turn gives rise to the proteome. (For simplicity, in the case of eukaryotes, mitochondrial and other organelle genomes are subsumed into the general "Genome" category, though of course they are replicatively distinct.) A portion of the proteome is concerned with synthesizing and processing small molecules (the metabolome), which are required for a variety of purposes at various points of this cycle. The metabolome (M) is shown as directly cycling back to the genome because specific parts of the metabolome (dNTPs; synthesized via the proteome) are required for DNA synthesis and physically become incorporated into replicated genomes. Each of these steps requires the feeding back and participation of specific members of the proteome, the transcriptome, and the metabolome for completion, which proceeds in a forward positive direction (small black arrows) or is controlled and regulated (negative direction; small gray arrows). The epigenome is derived from the genome by chemical modification (especially methylation), and serves as an additional regulator of transcription and genomic replication itself. The dotted line from the transcriptome to the metabolome indicates the former pathway during the RNA World. Note that in a specific organism, not all small molecules required for life processes are synthesized, since many can be acquired from the environment, but all organisms produce a fraction of their required small molecules through their own agency. Further details and justification of the relationships are provided in ftp site. (b) The same diagram used to represent the great three-dimensional and temporal complexity of biosystems arising from the one-dimensional genomic information string.

repository thus encodes directly or indirectly the information allowing the "subomics" to unfold, which ultimately attend to its replication. All the other "omics" participate in multiple entangled ways throughout this cycle, but the genome is at the heart of all, and the ultimate evolutionary *raison d'être*. The "subsidiary omics" serve toward the perpetuation and replication of the "selfish" master genome, which itself can be viewed as a collaborative enterprise between multiple separate replicators.[9]

The complex summation of all functional molecular interactions occurring in a biosystem is often considered an as a special "ome" itself, the "interactome," and it is this area that is essentially the domain of systems biology. Of course, molecular interactions of significance in an operational biosystem are highly ordered, and can be classified as falling into specific pathways, which in turn intersect as networks. Biological synthetic pathways have long been recognized, but complex networks involved with signal transduction and gene regulation have been defined in more recent times. Relating the organization of pathways and networks to the constituent biomolecules that mediate them can then be considered the more specific objective of systems biology.[1527] Systems biology can be also considered as an antidote of sorts to excessive reductionism in the search for new drug targets.

Soon after its inception, DNA shuffling was shown to be successful in the evolutionary improvement of an *E. coli* operon,[1528] and directed evolution of cellular metabolic pathways is an ongoing and successful enterprise.[1529] *In vitro* compartmentalization (Chapter 4) is well suited for directed evolution of pathways and systems, given its maintenance of genotype–phenotype in artificial compartments without requiring physical linkages between encoding nucleic acids and participating proteins. By their nature, studies aimed at pathway improvements through directed evolution do not provide direct information concerning regulatory networks, but the "no-assumptions" approach itself has been seen as potentially finding novel solutions to complex interactive design questions that might escape rational intervention.[1528] On the other hand, there are potent reasons for pursuing the study, categorization and understanding of biosystems, whose aims in themselves are thus highly rational and seek a high level of rational input into biosystem manipulation. As we see again shortly, this is also good news for improvement of evolutionary methods as well. While systems biology is multidisciplinary, its knowledge base can be broadly divided into the two areas of experimental data acquisition and analytical modeling. This statement might seem obvious enough, but the latter area includes recent developments in network theory and analysis, whose impact is very broad indeed and extends beyond biology itself.

Conventional definition of metabolic pathways requires painstaking purifications of successive enzymatic components and analysis of their substrate preferences and catalytic properties. Although this extensive body of concerted biochemical effort laid the bedrock for further progress, high-throughput technologies have been vital in gathering information on the genome-wide scales needed for comprehensive biosystem analyses. Included among these are a variety of microarrays for global screens at the genomic DNA, mRNA, or protein levels. In metabolic studies, the rate of production of a metabolite product through a specific pathway (the "flux") is also of key relevance for gaining system-level insights. In order to understand biological pathways and networks, a detailed knowledge of the protein–protein interactome is also required, and great strides have been made toward this end in recent times. Here both genetic methods (principally the yeast two-hybrid system) and physical approaches (sensitive mass spectrometry) have been applied toward defining the global gamut of protein–protein interactions in model organisms (detailed in ftp site). The interactome, though, also includes protein–nucleic acid interactions, particularly important in gene regulation (as we noted earlier with respect to regulation based on various RNA interference mechanisms).

The interpretation of these and related data has greatly aided the general understanding of how the system components operate as complex networks, although there is still much to be learned. Networks in the most general sense can be classified through the linkage architectures of their connecting elements, or nodes. In a random network, the majority of nodes have the same number of linkages, but this pattern is not seen in real world networks, and particularly those observed within biosystems.[1530] The latter are generally "scale-free," where there is no "typical" network node but rather a series of highly interconnected "hubs" combined with many other nodes of low connectivity.* We have previously seen the importance of modularity in biological circumstances (Chapter 2), and modularity is an essential element of cellular network theory. Network modules have been termed "motifs," and defined as specific network node interconnection patterns that are overrepresented compared to frequencies expected in a corresponding random network.[1531] In turn, motifs themselves show clustering properties that may be another generalizable feature of biosystem networks.[1532] Another property of networks is path length, or the number of nodes and links that interconnect any two given nodes. Path lengths of cellular scale-free networks have been found to be particularly short, in both prokaryotes and eukaryotes.[1530]

The properties of protein interactive networks are directly correlatable with many of the factors noted in Chapter 2 as evolutionarily relevant for proteins. For example, connectivity as an evolutionary restraint was noted as significant only for the most heavily interconnected hubs in yeast networks. The evolutionary origin of cellular scale-free networks themselves has been attributed to an accretion process driven by gene duplications introducing new network members that retain old connections and gain new ones. Also introduced at the protein level in Chapter 2 was the concept of robustness, which has clear significance in a broader network sense as well. Nonrandom networks with connected hub structures are robust to random perturbations (random removal of any single node and its links), since there is a higher probability that a nonessential node will be hit than a crucial hub. On the other hand, important hubs can act as the Achilles' heel of such networks if they are selectively targeted.

Some features of natural biosystems are challenges for system-level analyses. Genomic "unfolding" (Fig. 9.7b) is inherently modular and parsimonious, such that many systems' components are used in multiple ways, both spatially and temporally. "Moonlighting" proteins (Chapter 5) can serve as one set of examples in this regard. Partial functional redundancy is also often observed,[†] and interpretable as a biological "safety net" for minimizing catastrophic mutational losses. In the context of gene regulation, these kinds of effects rule that without additional functional evidence, gene expression data alone cannot be used to construct accurate regulatory maps.[1533] The complexity of biological networks in higher multicellular organisms is in itself a challenge for assimilating all their scope and diversity.

Yet these are not unassailable problems, and there have already been numerous practical benefits from systems biology, as the perception of the importance of system-based approaches increases. Much of modern drug discovery has revolved around the premise that a specific target at the molecular level is the appropriate level of focus, and indeed attention can be directed to still higher levels of structural resolution when such information becomes available. Screening for drug-induced phenotypes then becomes a means to an end for precise target identification, which in turn serves as the springboard for obtaining superior drugs, either through rational design or through empirical screening with highly focused sets of candidate

---

*Scale-free networks conform to a simple power–law relationship, and where biological networks deviate from scale-free status, it is associated with an even greater disproportionality of linkage at key hubs. Also, unlike other known real-world networks, cellular networks tend to be "disassortative," where hubs are indirectly connected via low-connectivity intermediate nodes, rather than by direct linkages.[1530]

[†]As previously considered in Chapter 2.

compounds. Yet it has been rightly pointed out that the track record of final approvals of target-based drugs has been poor in the decade following widespread introduction of this avenue toward novel drugs.[1534,1535] As well as connectivity, important biological metabolic pathways have high degrees of robustness and redundancy.[1536,1537] A biological target can accordingly possess a well-defined function and show high sensitivity and selectivity to a specific drug *in vitro*, and yet the same promising drug candidate can still have unexpectedly weak activity *in vivo* owing to complex pathway redundancy and back-up systems. Much evidence has accrued showing often surprising robustness and versatility in metabolic pathways even when important enzymes are removed. An example of this is the persistence of functional glucose metabolism in *E. coli* despite lesions in genes formerly thought to be essential in this activity.[1538]

The upshot of these observations and arguments is that a purely reductionist approach to drug targeting and development will become counter-productive. No matter how efficient at the single-protein target level, a drug's utility is ultimately defined by its performance in highly complex and genetically diverse organisms (meaning us, for the most part). Although way ahead of his time, perhaps the poet John Donne was really trying to get across the message, "No Molecule Arising Naturally (MAN) is an island, entire of itself." The alternative to exclusively target-based searches is not a soft-focus holism but rather increased attention to the complex network milieu in which drug targets operate, or "pathway-based" drug discovery.[1537] Specific functional pathways lie within a spectrum ranging from simple bimolecular relationships to complex disease phenotypes, and it is practically simpler to establish a connection between a gene product as a component of a metabolic pathway than making a direct link to the ultimate higher order pathological outcome.[85]

The functional connections between a large set of human diseases and their underlying genetic bases have been used to generate a network representation of the human "diseasome."[1539] This led to the interesting conclusion that proteins corresponding to network hubs did not contribute significantly to genetic disease phenotypes, but were important in pathologies involving somatic mutational change such as cancer. (Mutational impairment of genes encoding critical highly connected hub proteins is unlikely to be passed on in the germline, since such defects severely affect maturation and reproduction itself. On the other hand, somatic mutational changes are only applicable within the affected subset of cells in an organism, where hub proteins can indeed be altered.) Known interactions between approved drugs and other proteins (beyond the designated drug targets) have also been subjected to a network-based analysis.[1540] Since drug specificity is rarely precise, target nodes in such a network link with multiple drugs, although sometimes "polypharmacology" is a desirable end. In any event, analysis of functional activities in a network context should become an important adjunct to rational drug design.

As noted above, knowledge of biosystem networks can feed back into evolutionary strategies for their modification, in part through distinguishing hubs from peripheral nodes that may be more amenable to tinkering. This also overlaps with ongoing aims for artificial derivation of molecular switches and circuits using directed evolution,[1541] where it is necessary to deploy novel regulatory entities usefully in the context of a pre-existing biosystem. But the ultimate success of systems biology will rest on the ability to make successful predictions as to the functional consequences of system perturbations or to define means for correcting system-level errors arising from mutations or other deleterious effects. Knowledge pertaining to metabolic networks can already (at least in some circumstances) suggest predictions for functional restoration of suboptimal mutant organisms, through suppression of inefficient pathways.[1542] In the nematode worm *Caenorhabditis elegans*, a single global genetic network has been devised, upon which useful functional predictions can be based.[1543]

But the hardest task for systems predictions arises when one addresses the ramifications of perturbing a specific gene product upon multiple higher level systems. For example, consider the case of the human *FOXP2* gene. Mutational inactivation of one chromosomal copy of *FOXP2* results in a dominant phenotype manifested by a speech disorder, whose familial occurrence facilitated the original mapping and cloning of *FOXP2* itself.[1544] This gene encodes a transcription factor (of the "forkhead" family), and the pathological effect of insufficient FOXP2 protein functional activity is apparently at the level of the muscular coordination necessary for speech generation.[1545] Now, this information was obtained through "top–down" systematic analyses of linkage patterns of the pathological phenotype with genetic markers, eventually pinpointing the responsible gene before any knowledge or available functional data for *FOXP2* itself. So, the question with respect to advanced systems biology is from the "bottom–up": given the gene, predict the consequences of its haploinsufficiency (50% inactivation) in the entire (human) organism. Moving from the genomic transcriptional targets of *FOXP2* all the way up to the higher level phenotype,* where a muscular coordination problem impacts directly on speech generation, might seem a daunting task. Yet we are only living in the infancy of systems biology as a science, so assigning any such undertaking as unfeasible might also seem premature. And given the direct relevance of systems biology to higher level molecular rational design, this question impinges directly on the immediate and ultimate limits of rational design itself, a major preoccupation of the next and final chapter.

---

*With respect to its significance for human language generation, it is interesting to note that the *FOXP2* coding sequence has two residue changes unique to humans (absent from chimpanzees and other primates).[1545] The alterations in this regulatory gene may therefore serve as another example of the importance of relatively small genetic changes in determining phenotype.

10 Exploring the Limits

OVERVIEW

We have seen that rational design of both small molecules and proteins is going ahead in leaps and bounds, in tandem with the general acceleration of technological progress. Is it possible to predict where this trend might stop? In other words, can we foresee inherent limits to rational design, where empirical input will still be required? This kind of pursuit will take us into several different arenas, where some fundamental issues will arise. The question itself is inherently entwined with the future of molecular discovery in general.

THE LIMITS OF RATIONAL DESIGN

Are empirical approaches for any molecular design problem useful only because theory and technology have not advanced enough to allow a rational approach? This could be likened to "god of the gaps" theology, where levels of ignorance (holes in knowledge) regarding natural phenomena are plugged by citing divine activity, and the requirement for this kind of supernatural explanatory intervention accordingly contracts as knowledge increases. The underlying issues here concern the limitations that fall upon rational design of molecules large or small, which can be immediately subdivided into problems whose solutions are not yet fully developed but clearly on the horizon, and those that may be more intractable, or even stand as absolute limits. In the case of the former, of course, it is assumed that an on-going "gap-filling" exercise will solve any remaining hindrances. At the same time, it is worth examining further the question of whether the latter "hard" problems will be solved more slowly or not within any reasonable period of time.

In thinking about hurdles for present and future rational design, we are returned again to the nature of the design process itself and how it spans a spectrum of difficulty. Although the rational design of both small ligands and protein effectors in isolation is far from a mature science, for biological purposes this is just the beginning of the story. A central problem is rational design of molecules showing predictable functions within a highly complex biological setting, itself a subset of the general goal of predicting the global behavior of an introduced molecule in the context of any complex interactive system. (Of course, biosystems are by far the most complex systems that are likely to be encountered in the natural world.) The biological context may determine what the optimal design of a mediator should be, where a balance sheet of factors may influence optimal ligand binding and many other potential variables. The importance of weak (and generally transient) interactions in biological networks in general is becoming increasingly appreciated and has clear implications for drug design in general.[1546] A suboptimal application of rational design might then produce an effective ligand at a specific

Searching for Molecular Solutions: Empirical Discovery and Its Future. By Ian S. Dunn
Copyright © 2010 John Wiley & Sons, Inc.

target site, but not necessarily one that has the best result in a complex *in vivo* system. (An empirical screen might circumvent this issue directly or reveal the best target site for subsequent rational development.) Strong rational design will also need to predict and/or avoid immune responses that may be mounted *in vivo* against protein therapeutics (or small molecule conjugates with macromolecular carriers.[1547]

One can only aspire toward truly "strong" rational design if complete knowledge of the operation of the biosystem of interest is available, which is obviously a very different level of understanding than defining (or synthesizing) its corresponding genomic sequence. As a case in point, complete synthesis of a bacterial genome is possible,[1548] but this is not the same as possessing a complete understanding of it. If the latter information was in hand, it should be possible in principle to design a wholly unique artificial genome from the ground up. On this knowledge-related note, we could briefly consider that relevant information gathering needs to operate at different levels: at both the reductionist level of the fine details of protein and functional nucleic acid structures, and the higher levels of cellular and organismal biosystems. Advances in the direct experimental acquisition of such information are matched by ongoing progress in theoretical predictive modeling of folded macromolecules and the behavior of interactive networks, as noted in the previous chapter.

Rational Design and Determinism

Rational design is always knowledge based, but we must also keep in mind the distinction between fully deterministic rational design and computational rational design based on nondeterministic algorithms (Chapter 1). At the highest level of biosystem complexity, insisting on a deterministic approach to "strong" rational design would demand a level of information that brings to mind the notion of Laplace's Demon, coined from a concept of the French mathematician and astronomer Pierre Simon Laplace (1749–1827). He envisaged that with cosmically all-encompassing information, an intellect (latter referred to as a "demon"[*] by others) could completely determine the causal progress of the universe through time. But while quantum indeterminacy formally precludes this proposal in general in the subatomic domain, could an isolated system (even of extreme complexity) be completely describable, understandable, and predictable in principle at the molecular level? In this consideration we might exclude large-scale phenomena subject to chaotic fluctuations such as weather and focus on ordered biosystems. The latter include complex subsystems for maintaining homeostasis, but an all-encompassing description of a biosystem must account for "emergent" properties[†] that are not apparent at the level of its individual components in isolation.

Emergent effects can result from cooperativity and allostery between system components that operate in a nonlinear fashion.[1549] The structures of some protein components of biosystems are themselves determined (either kinetically or absolutely) by the other system components themselves, as we have seen with chaperones and foldases. The routing of folding pathways through translational pausing (as noted in Chapters 2 and 9) is another case in point. Signal transduction, an essential requirement of all cellular organisms, also exhibits emergent properties when examined at the system network level.[1549] In this context we should note intricacies of ligand-induced allostery in receptors,[1550] which can produce a variety of

---

[*]Demons are a popular and versatile metaphoric device, spanning the nanodomain (Maxwell's Demon noted in Chapter 1) to the cosmic arena (Laplace).

[†]This is definable through information theory and entropy (considered briefly in Chapter 4). If the entropy of a system is smaller than the summation of the entropies of its individual components, then the system globally has emergent properties.[150] This can be rephrased by noting that in such a system, more order emerges collectively than from its parts in isolation.

differential signaling effects. Predicting these and many more such higher level phenomena from purely genomic sequence information is far beyond current capabilities, even if it is not formally impossible. If every component of a biosystem of interest is structurally characterized, and the complete biosystem interactome delineated, the prospects for strong rational design become more tractable, but predicting with complete accuracy all of the perturbing effects of introduced nonbiological molecules within such coordinated networks is still a daunting task. This is true even at the level of single cells, let alone at the organismal level when whole additional systems must be factored into the analysis (as with adaptive immunity, noted above).

Note that "predictive power" for rational design means predicting the effects of novelty, not simply "predicting" phenotypic markers from a given genotype. Certainly, knowledge of specific genetic markers for an individual genome can allow the assignment of some corresponding phenotypes, an increasingly useful option in forensics. (Known pigmentation and other markers can in principle allow the identification of ethnic origins of genomic DNAs detected at crime scenes, for example.) But this is a very trivial exercise compared with predicting *ab initio* how the information in a given (and unknown) genome will unfold during an organism's development, or how the complete operation of a known genome will be perturbed by unnatural molecular intervention.

Limitations on knowledge in an absolute sense can also be raised by consideration of molecular spaces and combinatorics, and this too necessarily impinges upon rational design. We have repeatedly seen that both small molecule chemical space and protein sequence space are hyperastronomically vast, and a deterministic search based upon a specific scaffold may be successful but not necessarily a globally universal optimum. For small molecule searches, scaffold-hopping to new "islands of activity" is a challenge for *de novo* design,[1434] and one can generalize this as "translating" an activity between molecular forms, a theme continued further in a later section. How can one know, then, whether a rational molecular solution in hand is indeed the best of all possible in a universal sense? Obtaining a satisfactory proof of this would seem to evade formal possibility, although in practical terms for many simplistic problems this consideration would not arise. (For example, design for an *in vitro* assay by searching for ligands binding to a fixed target pocket with affinities beyond certain high levels might return no further pragmatic benefits even if formally possible.[*]) Yet as more complex molecular problems are tackled, including whole system design, the issue of which, the "best of all possible worlds," is likely to become more relevant. In this respect, even the alphabets used for the combinatorial construction of macromolecular effectors become a significant factor. Here, the conventional protein and nucleic acid alphabets will compete with corresponding extended alphabets and fully artificial foldamers.

As we have seen (Chapters 1 and 5), molecular design that is still classifiable as rational can computationally make use of nondeterministic algorithms. Rationally recapitulating an empirical molecular search *in silico* with virtual screening and successive evolutionary cycles of diversification (as noted in Chapter 1) will return an answer that is definable as the "fittest" according to the search strategy itself. This approach is not immune to the real-world dilemma of becoming marooned on a local fitness peak,[†] unless additional information is fed in. But the general issue of virtual screening might make us wonder how far computational modeling and simulations can be taken, with an eye toward maximizing "strong" rational design. Could the emergent properties of biosystems be completely predicted by an advanced simulation? An

---

[*]This is simplistic in that as a molecular solution, only the consequences of binding a single target are considered. Once again, "strong" rational design for such a ligand (if it is to be functional in a complex biosystem) would need to predict its global behavior and potential undesirable cross-reactivities. Excluding the possibility that some other high-affinity binder could be a superior solution in this system-level sense is then much more difficult.

[†]Even Particle Swarm Optimization (Chapter 4) is limited by the encoding of the Solution Search Space.

ultimate "silicon cell" would be capable of instituting a full unfolding of encoded genomic information, completely simulating all cellular operations and including structural modeling for all biocomponents of the system as a whole. To be fully useful for rational design, the simulation should accurately predict the global effects of any introduced chemical perturbations, and also the consequences of mutational changes to any normal system components.

To function at the highest possible level, such a simulation would faithfully emulate real biological phenomena and proceed far beyond merely simulating the effects of pathway modulation. The hypothetical simulation should be capable of modeling the introduction of a completely novel molecule into the virtual biosystem, and then predict all significant primary interactions of the novel molecule with all cell constituents at a structural level. From this, it should predict all higher level cascade effects resulting from the primary interactions and move on to predicting the causal flow-on effects in multicellular systems. If the simulation has a complete "understanding" of encoded biological information, it should predict the "unfolding" of the genome alone without any other input. In other words, establish simulation for the genomic sequence and the appropriate initial conditions, and stand back. The starting conditions here equate to a minimal number of master transcription factors, or any other essential proteins or RNAs needed in order to "boot up" the system into "life." (This too would need to be precisely defined and understood.) The initial simulation conditions would thus need to set up an unfolding cascade of expression and coordinate regulation that allows virtual growth, cell division, and differentiation.

And to enable the highest level of rational design by computation, from a silicon cell one would need to progress to the "silicon organism." If this was attainable, all drug testing in animal models (and cell culture) would become superfluous. Taken to its ultimate limit with a virtual human cell model, this supersimulation should recapitulate human development entirely, including neural differentiation. An alternative route to artificial intelligence? The point of this heady thought experiment is to take the concept of biosystem simulation to its logical limits, but of course we can come back to earth a very long way before ethics committees have to be involved with respect to the rights of developing sentient simulations. But even a much more modest "silicon cell" (based on successful kinetic modeling of integrated metabolic pathways) will be a boon to systems biology and rational design.[1551]

Increasingly, sophisticated rational simulations of complex biosystems might superficially seem limited only by the available (and steadily increasing) computing power, but things are not as simple as that. As noted in the previous chapter, clearly there are practical nondeterministic ways around theoretical computational hurdles of NP-completeness for modeling small to medium proteins, but large multidomain proteins are still out of reach. And modeling a *structurally complete* system to include all network interactions along with the effects of introduced perturbations would thus seem quite unfeasible. Earlier we saw that a fundamental rational ideal would reduce chemical calculations to the quantum level, but this runs into excessive complexity long before the whole system level is encountered. And while a "magic computer" solving NP-complete problems would be the ultimate rational design answer by scanning all possible solutions to a problem, not even quantum computing is likely to be up to this task.[1393]

At this point, we should back up a little and note again that while the ultimate end point of the "strong" rational design ideal (Chapter 9) to cover the most complex biosystems may be never be completely realizable, lesser but still very impressive rational goals are certainly attainable. Just as approximation algorithms can produce practically useful solutions to NP-complete problems, rational design can produce solutions that are effective in performing a designated task, even if they cannot be proven as optimal in an absolute sense. In the design spectrum, such practices fall short of the "strong" ideal, but remain rational as real-world fixes. Such approximations will steadily improve and approach the ideal without actually precisely coinciding with it. In this regard we could recall the practical use of approximations in quantum chemistry and once again, protein-folding calculations. In areas such as these, quibbling about whether an effective solution is precise or merely a formal approximation is of little pragmatic benefit.

And pragmatism lends itself to whatever approach is the most effective. On this note, it is pertinent to reconsider the role of empirical approaches as adjuncts to rational strategies, and also as a recourse in the face of the most difficult of design challenges.

BOOTSTRAPPING AND INTERPLAY

The systematic acquisition of knowledge is the underpinning of rational design, and knowledge is generally obtained and compounded by empirical experimentation. This generalizable bootstrapping of rational approaches is especially evident where modern evolutionary and/ or selective techniques pave the way for implementation of highly rational design strategies. One good example among many is the development of expanded genetic codes *in vivo*, where the only practical ways to obtain tRNA and aminoacyl-tRNA synthetases with the required altered specificities were sophisticated selection techniques (Chapter 5). We have frequently noted this kind of entwinement between empirical and knowledge-based approaches, but it need not always flow in one direction. Another theme visited often has been the propensity of slavish conformity to an evolutionary algorithm to take molecular optimization only to a local fitness peak. Rational human direction based on empirically obtained information can lead to "jumping" to a new prototype, in a manner that could not occur in a natural evolutionary progression. (This type of interplay in the case of small molecule design was depicted in Fig. 8.14.) But let us now look at the empirical → rational bootstrapping pathway from a slightly different point of view.

Reverse-Engineering Biocomplexity for Design Insights

The knowledge required to implement rational design of complex molecules, molecular interactions, and molecular systems ultimately comes from a variety of sources, but the information directly relevant to molecular biosystem complexity is provided by biology itself. Gaining an understanding of a complex mechanism by pulling it apart and analyzing its components is often termed "reverse engineering." The intent of this exercise is usually to translate the obtained information into the construction of a functionally equivalent device or process, and this principle can be applied toward biological systems as much as human-engineered products.[1552] The initial phase of biological reverse-engineering necessarily requires fine-detailed knowledge of the relevant basic biological processes, and then a scale-up of understanding to systems of increasing complexity, ultimately at the levels of cells and organisms. Rational design follows as an applied outgrowth of this process.

Of course, not all correspondences between human invention and natural evolutionary design can be attributed to conscious reverse engineering, a point made in the Introduction. As with a recent example of dinosaur fossils with biplane-like flying adaptations[1553] whose discovery postdates human biplanes by many decades, until recently most instances of "natural precedents" were more akin to convergence toward optimal design solutions than any form of copying. And at the level of molecular systems, it is clear that even the possibility of reverse engineering cannot exist until basic chemical knowledge and experimental tools have been developed. At present, though, systematic biological investigations can feed back rapidly into innovations in molecular design. By such advances, rational design is increasingly empowered.

Consider an analogy with a human-engineered device, a conventional automobile, presented to a hypothetical preindustrial society untouched by outside influences. It is doubtful that the citizens of this land could reverse engineer this car, since they would lack the necessary tools to nondestructively disassemble it. At a higher level of technological development (yet before such devices were produced by their own insights), such people might gain a good functional understanding of the car's operation by pulling it to pieces, analyzing the roles of each car system and its components, and putting them back together.* Even so, this alone would be unlikely to provide them with a complete understanding of the physical principles that underlie the car's operation, such as electrochemistry (the car battery) or hydrocarbon combustion control (antiknock compounds in the fuel). But their reverse engineering efforts could certainly set them off in the right directions in order to acquire such knowledge, and more immediately allow them to begin reconstructing the functional parts that enable the car to work. They could then undertake "rational design" of cars of their own. A skeptic in this land might dispute that this process really deserved this rational accolade, since it could be more cynically viewed as a copying process. But the intrepid engineers could counter this accusation by producing cars with significant innovations of their own, ultimately with fundamentally different design strategies for internal combustion engines. At this point even the most hard-nosed critic would have to concede the existence of rational design, and acknowledge that for the engineers, "rev-eng" was indeed sweet.

Now, we can translate this analogy into an ongoing biological design issue. Synthetic biology and genomic sequence analyses have reached a point where it is not misguided hubris to speak of "artificial life" in a physical sense,† although realization of this ancient dream has not yet been attained. (Note that this is intended to refer to a cell with a synthetic genome, which can act as an independent free-living replicator. The relatively simple task of synthesis of small viral genomes (only replicable within host cells) is thus excluded from this definition). A significant step in this direction has been the synthesis of a prokaryotic genome (*Mycoplasma genitalium*; of ~0.58 Mbp) and its cloning in yeast.[1548] As presented in that report, it had not yet functioned as an independent replicator, but even when this is achieved, in the end it equates with copying (with small modifications) from a natural evolutionary design. Nevertheless, as with the above car analogy, increasing levels of creativity will follow through manipulation of the metabolic and synthetic pathways of such synthetic genomes until a stage is reached where the label of "artificial organism" becomes more accurate. This then truly brings rational design into the genomic sphere. In order to achieve this, a detailed understanding of all relevant biological pathway networks (or interactome) is required, and this is the domain of systems biology, as briefly encountered in the last chapter. We also previously noted that the design of molecules and molecular systems encompass a spectrum leading to rational design, and the latter too covers a range of achievements. In the area of genomic systems, the highest level of rational design will be reserved for a genome that encodes a completely novel biosystem. Arguably, only at this stage could one truly refer to the "creation of life." Yet if we recall the above considerations regarding rationally implementing a wholly complete biosystem simulation,

---

*But more than one copy of the car would be required in practice. Without relevant pre-existing technological knowledge, extensive empirical experimentation would be necessary to correlate function with performance of each component. This could be done by randomly destroying parts in separate cars and testing for the effects, a "hammer approach"[1554] analogous to functional genomic screening by random gene inactivation.

†As opposed to computational artificial life, which we have previously encountered in Chapter 4; "synthetic life" is thus preferred over the former when referring to artificially created biosystems, to avoid confusion.

whether an artificial actual embodiment of this could be engendered without recourse to real-world evolutionary methods is a moot point.

An even more dramatic example of reverse engineering of a biological system (if ultimately successful) would be the reverse engineering of the human brain for the construction of "strong" artificial intelligence.[*],[297] But even much less ambitious kinds of bootstrapping by biological reverse engineering is in effect a conscious exploitation of billions of years of empirical evolutionary design by one result of such design (human intelligence). In this sense, most current and future rational design is the beneficiary of such "natural empiricism," which we highlighted in Chapters 2 and 3.

Just as principles of reverse engineering can be applied to the products of natural evolutionary processes, so too can they be implemented, if necessary, for the outcomes of artificial evolutionary experimentation. An objection to the application of genetic algorithms has been along the lines of "It may work, but you won't understand what you have when you get it," and this could be generalized to the results of any evolutionary process where the outcome is useful but its mechanism is obscure. Since real-world evolutionary change is incremental, large mutational changes leading to quite novel processes tend to be very rare, but even small changes can in principle introduce surprising functional novelty. But where this occurs, reverse engineering is called for in order to understand the basis of the unexpected (and beneficial) alterations. And in turn, this newly acquired knowledge provides more feedback into the rational arena for future design challenges.

Time, Knowledge, and Molecules

Pragmatism in molecular discovery must also introduce the factor of anticipated speed of a satisfactory outcome, as a part of a general cost-benefit analysis. Of course, time is never an irrelevant factor, but it is more pressing in some circumstances than others. An extreme instance of urgency might occur with the need to respond to a suddenly emerging novel infectious agent, or unprecedented bioterrorist threat. But anticipating the rate of molecular discovery itself is highly knowledge based. An empirical screen for a phenotypic target in the absence of any information at all cannot even guarantee that a molecular solution is obtainable in principle, at least with a specific type of library. On the other hand, a rapid empirical screening program might yield a useful lead in a much shorter time period than that required to obtain the necessary information for instituting rational design. This is especially so where rational approaches themselves are still developmental. In practice, one would use whatever information was available in conjunction with empirical screening as required, in a compromise semi-rational strategy.

As in so many human endeavors, there are accordingly trade-offs involved here. One way to think about this is to use a representation inspired by the "Spreng triangle,"[†] where for our purposes the three relevant variables are time, library size required to ensure a high probability (P; where P approaches 1.0) of finding a lead molecule, and structural information pertaining to the target (Fig. 10.1). The "library size" in this context reflects a theoretical progressive decrease in the numbers of molecules that must be screened as information access allows the library content to be focused in increasingly directed ways toward the eventual molecular

---

[*]As noted above in the Quantum Chemistry section, some opinion has held that human consciousness and cognition is based on quantum principles, which has been logically discounted, since biomolecular machines are large enough to be considered within the classical domain.[1555] But even if quantum effects are significant in this regard, a comprehensive reverse engineering project of the human brain would eventually unveil such phenomena and incorporate them into an artificial intelligence model.

[†]This was originated by the physicist D. T. Spreng in 1978 (cited in[1556]) for time, information, and energy.

solution. At a hypothetical "pure empirical" level where no preconditions whatsover are prescribed and library members are chosen strictly at random, for a specified function a vast library within the universe of molecules (OMspace) might need to be screened. With significant available information, this huge random selection can be culled enormously. As an example, the syntopic antagonism strategy for finding histamine receptor antagonists (noted in the previous chapter) has the effect of vastly reducing the numbers of compounds that must be evaluated, considered as members of the set of all small molecules. Moving to the opposite end of the scale to the "strong rational," the ideal converges to a "library" of one: a single molecule successfully predicted in all its functional requirements from the outset and perfect for the task in mind.[*]

Within the triangle of Fig. 10.1, each vertex represents a hypothetical limit reached when one of the variables is zero and the other two are maximal. At the left vertex, a completely empirical search is indicated by a low-information state, requiring screening of extremely large molecular libraries over an indefinitely long time period. At the opposite right vertex, rational design necessitates an information-rich state, and the required library size has contracted to "zero," interpreted as meaning that no library is required in lieu of a single predicted lead molecule with the same high probability of success. But by the same token, the process of identifying such a single lead is time intensive.

What then is the significance of the third vertex, time-poor but still information-rich, and also associated with indefinitely large libraries? This is an extreme hypothetical circumstance of a semi-rational strategy as needed when time is the most demanding factor. Even if maximal

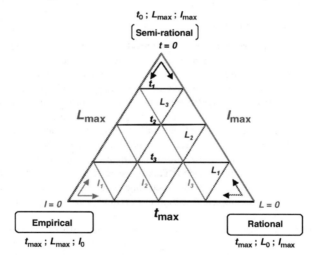

FIGURE 10.1 Empirical versus rational molecular design, modeled in the form of a Spreng triangle with respect to the roles of time (t), structural information (I), and library size required to ensure a constant high probability of finding a lead molecule (L). Here at each vertex one of the three variables is zero, and the opposite side of the triangle corresponds to the maximal value of the same variable, with graded values in between increasing from the vertex to the side (indicated by increasing numbers: L_1, L_2, L_3, etc.). Note that "structural information" refers to specific data applicable to the target molecule, and that background knowledge with respect to modern chemistry and molecular and cellular biology is available throughout.

---

[*]Note that these theoretical considerations do not necessarily mean that focused library sizes should be reduced in experimental practice. Increased library size may bring additional benefits, such as the identification of multiple alternative candidates.

structural information regarding the target is available, screening of large libraries may be appropriate in concert. This can also be interpreted as "making the most logical choice for a rapid outcome with information and materials on hand." In practice, of course, a more realistic semi-rational scenario would be located at an internal point within the triangle. Available information, even if incomplete, very often in principle allows the molecular library to be reduced in size while maintaining the same probability of finding the desired lead molecule.[*] (For example, structural definition and modeling of a target enzyme active site allows focusing on a vastly restricted set of potential inhibitory compounds in comparison to a completely random search, if the probability of finding a suitable molecular solution is held at the same level in each case.) So information available is used in the most productive manner to ensure the success of empirical strategies as needed, and under the dictates of time constraints.

If the set of all molecular problems are represented as triangles as in Fig. 10.1, will the future see a convergence toward the right rational vertex, even if not always attaining the ideal? As noted above, in part this question depends on the stringency of a definition of a rational molecular solution; whether it is intended to function in isolation or as part of a complex biosystem. So, for the "strong" definition of rational design, the time factor ("maximal" at the right vertex) extends to obtaining additional information relevant to the biosystem in which the molecular solution of interest must operate. The difficulties with this goal (referred to above) mean in effect that a compromise position between the empirical and rational vertices is the most realistic for the foreseeable future. But just as a conventional algorithm attempting to deliver a solution to an NP-complete problem may need more processing time than the age of the universe, so too a molecular discovery strategy requiring excessive time will be rejected as impractical. But if no molecular solution is deliverable even with a generous real-world definition of "maximal time" (along the base line of the triangle in Fig. 10.1) anywhere between the empirical and rational vertexes, then it could be considered unattainable with present resources and technology. Let us extend the latter notion a little in the next section, along with some additional empirical thoughts.

Outer Boundaries

One of the numbers of recurring themes in this book, and not least in this chapter, is the onrush of scientific and technological advances, which are expected to continually expand the domain of rational design. While this is not at all in dispute, one should not at the same time underestimate the impact of the same technological progress upon empirical molecular discovery. Certain aspects of empirical screening, such as high-throughput approaches, are sometimes seen as "brute-force" methods in comparison with the elegance of rational design. Ultimately, this is in the eye of the beholder, but it is instructive to consider how technological progress itself can raise a strategy to new thresholds of potential, brute-force or not. Consider an analogy with the time-honored biochemical technique of centrifugation. By spinning aqueous samples about a central axis, these devices increase the normal sedimentation rate due to gravity, measurable as the relative centrifugal force (RCF, or the g value). Centrifuges in early usage were manually driven, restricting them to a low RCF ranges suitable for such applications as the pelleting (concentration) of whole cells. But if far higher RCFs are attainable, whole new areas of centrifuge use become available. Ultracentrifuges, capable of rotary speeds generating up to

---

[*]Note that a "maximal" library here is considered as a completely random sampling of general molecular space, or OMspace. An alternative way of representing this would be with a constant library size in each case with a corresponding variable probability value for finding the appropriate molecule. As the library becomes more focused in the right direction based on available information, this probability steadily increases, until at the rational vertex it corresponds to a "library" of only one component and $P = 1.0$.

300,000 g, enable the fractionation of macromolecules such as proteins and can serve as very useful analytic tools for measuring molecular sedimentation properties. A very large increase in the basic task of centrifuges (spinning samples to increase their sedimentation rates) then enables whole new areas of application completely inaccessible to machines restricted to low-speed ranges. Another case in point is the development of DNA sequencing. In 1977, sequencing any gene was a major achievement, but 30 years later sequencing of whole genome sequences has proceeded to the point of ushering in the age of personalized human genomics.

The scientific and technical power of some experimental approaches is therefore directly linked to the physical limits to which they can be taken. In the field of empirical molecular discovery, analogies can be made with advances in high-throughput screening technologies that allow the practical evaluation of millions of compounds. Alternatively, we can consider the development of *in vitro* display techniques (ribosome and mRNA display) that enable library sizes to the order of 10^{15} and selection of novel proteins (Chapter 4). Sheer library size is accordingly very significant in empirical approaches where no assumptions as to the outcome are made, corresponding to the "low information" state of Fig. 10.1. And yet very quickly we return to sobering thoughts of the vastness of combinatorial protein sequence and small molecule chemical spaces, which confirm that the dream of pulling all molecular solutions directly from a sufficiently immense library is indeed but a dream. The most logical and powerful molecular discovery methods will then use the "cutting edge" of both rational design and empirical library-based approaches as they are available.

As we have seen, a solution to a molecular problem may not be ascertainable as an optimal result in a truly universal sense. Indeed, when can an alternative pathway to a solution be deemed truly impossible? Failure to find a desirable molecule from vast libraries may suggest (although never formally prove) unsuitability only for the types of scaffolds or foldamer designs evaluated for the task in question. Another structure "outside the box" might yet fulfill the desired performance criteria. This point can be levied at instances of apparent intractability of specific protein targets for the action of small molecules.[1557] We might compare this with the limits of natural evolution and the notion of "irreducible complexity," which was touched upon in Chapter 2. Identifying challenges that are truly irreducible for natural design is not simple, but it is clear that a great many examples will emerge from ongoing efforts toward the development of artificial nanostructures and machines based on biomolecules. In this regard, rational input enables facile crossing of fitness valleys insurmountable for natural evolution, and can "recombine" system components in ways out of reach for the natural world.

So if the blind designer of the tumultuous products of natural evolution could metaphorically express itself, it might conclude that many conceivable molecules and systems were impossible—but it would often be wrong. Is there an analogy to be made with a rational human molecular designer confronted with a seemingly impossible task? Placing something in the "impossible" basket is a flip side of the issue of universal optimization of a molecular solution, as referred to earlier. How can one demonstrate that a more efficient solution to the task at hand is impossible even in principle? We have previously seen that there is more to this than first meets the eye. For example, in Chapter 5 we noted that catalytic perfection was not the end of the story for enzyme efficiency, since "superefficient" enzymes have evolved. And could not these natural innovations themselves be further improved?

Obviously, rules of chemical bonding can be generalized, but it is worth recalling that chemical properties of specific residues in polymeric folded macromolecules can deviate from those of isolated monomers in ways unsuspected at the outset. (We have observed the importance of this in some instances of ribozyme catalysis.) These effects become also predictable and rationally designable in principle as the knowledge base grows, but the existence of other "special rules" in specific foldamer contexts is an unpredictable element. "Holes" in immunological repertoires and any others that are matched against potential targets are easy in principle to demonstrate, but filling holes in knowledge repertoires requires knowledge of where the holes exist, and whether they exist in the first place.

We can consider molecular syntheses in this regard. As previously noted, it can be reasonably argued that no chemically feasible small molecule can resist the synthetic attention of able modern organic chemists,[1558] but let us extend the definition of "small" beyond current boundaries* until synthesis becomes problematic, especially if the number of chiral (stereo-chemical) centers mushrooms. It might be imagined that some configurations were possible but synthetically hard to "get to," by analogy with the nuclear atomic domain and "islands of stability" of trans-uranic elements.[†] The synthesis of such molecules might be attempted for scientific purposes, but would there be compelling practical reasons for their creation? Scientific curiosity historically has a way of providing many useful spin-offs, but for generating useful functions beyond the conventional small molecule range, a more practical option is the use of combinatoric molecular alphabets. Indeed, if there was a need to synthesize very large nonpolymeric molecules, it is very likely that specific macromolecular catalytic tools would be needed, and these would be built from some kind of foldamer alphabet (if not the conventional protein alphabet). Let us examine this fundamentally important issue of combinatorics in more detail as a parting theme.

PATTERNS FOR THE FUTURE

We have seen that molecular solutions can be contemplated at the level of single molecules or complex systems, and with molecular entities ranging from the small to the exceedingly large. But repeatedly, we have noted that the most sophisticated functions in biological systems are performed by macromolecules, and these are routinely created from specific combinations of building blocks referred to as molecular alphabets. Living systems use not just one alphabet, but at least two (proteins as effectors and nucleic acids (primarily) as encoders and replicators). We can take the alphabet/building block analogy a step further to consider each biological alphabet as denoting a separate "language." This is a more restrictive definition than for real alphabets and languages, since obviously a single alphabet (such as the Roman script used for this writing) can serve for many other languages (French, Italian, etc). Also, some real languages (such as Japanese) use more than one writing system. But it is a useful device for some points raised in this and the following sections. The analogy of protein sequences with linguistics has indeed been well noted and has applications in the analysis of protein organization using similar approaches as with general computational linguistics.[1463,1560]

In order to think about the future of this process, let us return to some basic fundamentals, which takes us to a central issue for all living systems.

Primordial Alphabet Soup

Consider a hypothetical highly organized system from a distant alien world, which we would call "alive," and thus dignify with the term "organism." In order to qualify for such status, this organism would need to fulfill certain functional prerequisites, such as the ability to grow, replicate itself and interact with its environment. From knowledge of life on earth, we would in

---

*The range of "small" molecules was noted in Chapter 8. Nonpolymeric molecules of $>10,000$ Da would be larger than most current "small molecule" definitions.

[†]Artificially generated transuranic elements with atomic numbers >100 have very short half lives, but theory predicts the existence of superheavy nuclei as "islands of stability," which are relatively long-lived. This is supported by experimental observations, including the characterization of the superheavy element 112.[1559]

turn surmise that this organism possessed complex molecular systems for the fundamental operation of these essential living functions.[*] We would not, however, immediately assume that the biochemistry of such an organism would match anything found on earth (and studying these alien biosystems would be tremendously exciting). But could we make any generalizations at all? We could start by proposing that the principles of natural evolution defined on earth should logically be universal in their unfolding, though of course varying in the details of the evolutionary products that result.[†]

But what of the alien biology itself? As another thought experiment, we could first logically deduce that an organized living system must necessarily involve machines on a molecular scale to accomplish the necessary tasks for "living." In practice, most molecular recognition and catalytic tasks cannot be done with molecules of excessive simplicity, and growing molecular complexity correlates with molecular size.[††] Therefore, we can generalize the reference to "macromolecules" in an alien functional context. But then (for the present purposes) we could step out much further into the realm of speculation and imagine a scenario where each molecular machine is completely chemically distinct, whatever its physical size.

In such an imaginary world, each catalyst would be a large, complex molecule unrelated in its basic chemical nature to other catalysts, or other molecules composing the overall organized system. (Note here the intended meaning of "completely chemically distinct": by way of comparison, all biological proteins on earth are hugely diverse, but all share a common polypeptide chain chemistry, and are chemically related in this manner.) We could further clarify this imagined bio-world by comparing it to alphabetic written languages versus those using logographs. In alphabetic writing, words are composed of combinations of a limited number of letters in phonetically prescribed sequences, whereas true logographic writing uses a separate abstract character to represent each word. In fact, although written languages such as Chinese are often referred to as "logographic," strictly speaking this is not true, since a majority of Chinese writing has a semantic–phonetic ("morphosyllabic") basis.[1563,1564] Indeed, a "pure" logographic written language that totally ignored phonetics would logically have to invent novel characters for each unfamiliar foreign name or term it encountered and would become increasingly unwieldy. In any case, it is generally recognized that purely logographic writing cannot realistically represent a complete spoken natural language, although limited artificial logographic schemes have been devised.[**,1563,1564] Real language qualifications aside, writing of limited scope with true logographic representations is therefore feasible for humans, which serves to continue the molecular analogy.

The "chemically distinct" imaginary alien bio-world is then analogous to a logographic scheme, where each functional molecule is constructed in a distinct chemical manner. Could this kind of molecular arrangement be feasible for a real living system? In a word, no, at least for organisms of anything like the complexity of those found on earth. If every member of a complex organized biosystem was completely chemically distinct, thousands of distinct mediators would be required, just as a logographic writing system demands thousands of

---

[*]In fact, a completely satisfactory definition of "life" has been (perhaps surprisingly) contentious. An all-encompassing definition should include all known undisputed examples of life, exclude phenomena such as crystals and fire, and be broad enough to cover hypothetical nonterrestrial life. Most definitions include replication as a criterion, but then technically founder on the "mule" issue, or how to include sterile living organisms.[1561,1562]

[†]This is conceptually related to the thought experiment of "re-running the tape" of Earth evolution (Chapter 2), except of course for new biological and environmental variables in an alien context. Some definitions of life explicitly cite Darwinian evolution as a pre-eminent characteristic of living systems.[1562]

[††]Not all large molecules are complex (consider simple repetitive polymers), but very complex molecules cannot be very small, even if they are of a nonpolymeric nature.

[**]An example of such an artificial logographic scheme is "Blissymbolics," devised with supposedly "universal" applicability by C. K. Bliss (1897–1985).[1564]

separate characters. Each such molecular component would require a different replication/ synthetic system, each of which in turn would require unique molecules to assist the synthetic process, which accordingly would need their own replication-assisting molecules, *ad infinitum*. Life on earth does not work this way, and it is pretty safe bet that life nowhere else will, either. Accordingly, a fundamental molecular message from living systems is *the functional power of combinatorics*.

To enable a huge range of functions, you do not need to build a totally different tool in each case; it is possible to achieve this by different arrangements of a limited number of building blocks. It is not hard to deduce that the coupling of informational macromolecules (nucleic acids) and functional macromolecules (proteins) would be impossible without a combinatorial system. Our alien organism might use very different types of macromolecules, but we would comfortably predict that they would be polymers of a limited number of constituent monomers, as amino acids are to proteins and nucleotides are to DNA molecules the size of chromosomes. Here then is another definition of life: "Autonomous systems using complex (effector and informational) molecules combinatorially assembled from a limited set of simpler monomers in multiple diverse sequences, which can undergo adaptive evolution in response to environmental change."* Any life, in other words, *needs molecular alphabets*, and a limited number of them in turn. (Consider the 4-"letter" nucleic acid and 20-"letter" protein alphabets familiar to us.) While true logographic written communication systems are feasible (if cumbersome) for limited human applications short of real languages, complex "logographic life" cannot exist at all.

The analogy can in fact be extended precisely via this point. In some models for the emergence of protobiotic systems forming autocatalytic sets (as noted in Chapter 2), each participating catalytic molecule could be (in principle) a chemically distinct entity, if the system existed at a simple enough level. Such models are, however, seen only as precursors to the stage of intricate biotic development involving proteins and informational nucleic acids. Both complex human languages and any complex organized biosystem then cannot be based on "logographic" principles. Although molecular alphabets *per se* are a logical necessity, any specific alphabet is not necessarily written in stone, and we have already considered how the existing alphabets of life on this planet may be altered and extended (Chapters 6 and 7). The size of alphabets is another matter, though. In artificial life and digital genetic models, we find the simplest alphabets as binary bits encoding various functions, but we might recall from Chapter 6 that real-world functional ribozymes with only a 2-base alphabet have been devised. Nevertheless, for a range of complex functions, it is highly likely that much more extended alphabets are needed, as suggested by the "takeover" of the ancient RNA world by proteins. The question of what constitutes minimal informational and functional alphabets for a fully operational autonomous organism remains for future biologists to experimentally decide. In any event, emphasizing the centrality of alphabets is simply another way of framing the common notion of "life as information," where information equates with semantics.[1565]

Now, the above definition does not mention the slight imperfections in replication necessary to afford Darwinian evolution, which have been duly incorporated into some other definitions of life. You could, if you wished, attach such a clause, but it is likely that this is formally unnecessary, in the same way that it is unnecessary to specify that "molecules combinatorially assembled from a limited number of simpler monomeric molecules" should fold up into particular functional shapes. (If they are functional at all in a complex molecular system,

---

*Not that adaptive evolution implicitly requires diversification and replication, as we have noted often previously in both natural and artificial contexts. But what about the stubborn "mule" problem referred to earlier? Although natural or artificial selection is exerted at the level of individual replicators, evolution operates within populations of organisms, some of which will be reproductively defective (as with sterile hybrids such as mules). The latter still are autonomous "living" systems by virtue of the same molecular alphabets as their replication-competent confreres.

specific molecular shapes and configurations will "come with the territory.") Without very sophisticated means for error correction, mutations during replication are inevitable. And as we noted in Chapter 2, there is a trade-off between maintaining genomic integrity and investing increasing resources in energy-consuming proofreading and correction mechanisms. The logic for this too should be universal, such that any "living" replicative system will reach a high, but imperfect level of fidelity where Darwinian evolution will invariably occur.

The question then boils down to the choice of building blocks. In Chapter 2 we considered this from the point of view of protein capabilities, and noted that while pure sequences of polypeptides alone have certain limitations, systems of proteins can cooperate in expanding functional repertoires. Enzymes synthesizing cofactors thus enable other enzymes to function in diverse ways, and inorganic cofactors greatly extend proteins' functional abilities. But obtaining function from juggling sequence arrangements of molecular subunits is not limited to proteins, as we observed in Chapter 6: both DNA and RNA molecules can possess functional attributes, as well as being conduits of information.

What is the relevance of these ruminations to the future of molecular design? As the alphabets of proteins and nucleic acids are increasingly extended, an increasingly "alien" situation will be artificially generated in comparison to natural macromolecular effectors. Building upon what has already been achieved, nonbiological molecules will be generated with novel catalytic and other functions that mimic and extend those found naturally. It is a safe bet, though, that in the majority of cases these will not be unique "logographs" but rather foldamers of some description. And the latter will have their own alphabets, however artificial, just as "discovered" by natural biosystems on this planet billions of years ago.

On Molecular Translations

Replicating the functions of some small molecules with other structures has long been attempted, in order to marry the desired function with other useful properties such as stability. We have previously made note of this with respect to peptides (peptidomimetics), and peptides themselves can biologically mimic chemically unrelated molecules such as carbohydrates.[1566] Small molecule scaffold-hopping is another case in point mentioned in the previous chapter. But this "translation" effect can also be contemplated with macromolecules. Another reason for extending the notion of molecular alphabets to a "language" analogy is to consider how functions can be "translated" from one combinatorial alphabet into another, and to what extent some functions are "untranslatable," or at least suffer "lost in translation" compromise of functional properties. Effective relocation of a function from one foldamer alphabet to another has both theoretical and practical implications. As with mimetics in general, recapitulating a binding or catalytic function in the background of a different foldamer may be significant for stabilization purposes, but may also serve to improve the original function itself. The latter situation would arise if an alternative foldamer framework provided an improved recognition structure or catalytic site. At the same time, it should be recalled that we have noted that not all protein folds are equally adaptable to a given functional task, such that an improvement over a specific protein function with an artificial foldamer would not in itself prove that some other protein fold would be equally effective. By the same token, global optimization of a protein function would not disprove the existence of a completely distinct foldamer that could perform the task more effectively. (Again, recall the precedent of superefficient enzymes in this regard, and the danger of declaring that no further design improvements are possible.)

Yet it is clear that in many cases, equivalent functions can be reproduced with different combinations of the same biological alphabets. (In this respect, analogous enzymes noted in Chapter 5 come to mind.) In the "language" analogy, structurally distinct proteins with analogous functions would constitute synonymous words or phrases. The same function in one "biological language" (such as folded protein sequences) is then often "translatable" into another (such as folded RNA sequences). Here, we should pause for a moment to consider this terminology, which is potentially confusing owing to the universal use of "translation" in reference to the generation of proteins encoded by corresponding mRNA molecules. Accordingly, let us use "transcognate" as an alternative (and reasonably transparent) term to refer to distinct molecules with equivalent functional properties. Also, "translation" in this context may be misleading in another sense. In obtaining a similar function (such as ligand binding) between different foldamers, there is no simple correspondence between the required sequences for each structure, no simple "translation code." For example, we have seen (Chapter 6) that one cannot use a ribozyme or RNA aptamer to obtain a functionally equivalent DNA version by simply converting the RNA sequence to DNA, even given their close chemical similarities. Transcognate foldamers therefore converge functionally, but must be independently derived.[*] Finding transcognates for a known molecular function is then no different from searching for any desired function, but the end result is two or more biological macromolecules or artificial foldamers that can be usefully compared for relative efficiencies and stabilities.

As we have seen, the prediction of function where alphabetic combinatorics and folding is involved is a difficult rational design challenge, but solutions can be empirically obtained if the folded macromolecule is replicable (as for nucleic acids directly or proteins indirectly). Empirical generation of transcognates also has the advantage that no assumptions are made in advance. Recapitulation of functional criteria does not of course imply identity at the level of noncovalent binding contacts, and indeed empirically isolated transcognates might significantly differ in this aspect. Demonstration of this can be readily found in structurally characterized cases of molecular mimicry, such as anti-idiotypic "internal images" noted in Chapter 3. The instances of "multirecognition" of a single compound by multiple different natural receptors considered in Chapter 3 can be viewed as a set of transcognate recognition events, which could be extended further still artificially (as with aptamers, for example).

Searching for a transcognate molecule may be driven by perceived deficiencies in an existing alphabetic framework. Of course, a practical approach with proteins and nucleic acids is to extend the existing alphabet (as noted previously), rather than attempting to find a transcognate within an entirely novel artificial foldamer alphabet. The future, though, may approach the latter option more and more, as we will consider further in the next two sections. We should also recall a "natural solution" for shortcomings in the inherent protein chemical repertoire, in the use of organic and inorganic cofactors. Here, the "language analogy" can be pursued further as follows:

> Some languages may have a single word for a concept or action, which another language can only express with multiple words or phrases. For example, *schadenfreude* in German has no direct English word-equivalent, and its meaning in the English tongue must be rendered with a combination of several different words.[†] (The literal translation of "harm-joy" does not quite get the full flavor of it across; the meaning is generally given as "taking malicious pleasure in someone

---

[*]A loose parallel with real languages can be made by noting that top-quality linguistic translations must pay heed to all manners of subtle semantic information, as well as simple correspondences between nouns, verbs, and other parts of speech. Some items might appear untranslatable, and while the literal meaning of some wordplay, jokes, and the like may be truly language-specific, it will almost always be possible to devise a version in another language that is true to the spirit of the original.[1567]

[†]A neologism that encapsulates multiple utterances into one word serves a useful function, and borrowing single words that already exist in other languages is a frequently used alternative to this. A parallel could be made between this type of "lateral transfer" and evolutionary genetic lateral (horizontal) transfer (Chapter 2), such as the likely origins of the recombination-activating genes that mediate immunoglobulin and T cell receptor rearrangements (Chapter 3).

else's misfortune," or words to that effect.) Transporting this to a generalizable molecular analogy, a language (in one foldamer alphabet) that can provide a function efficiently in a single molecule has an advantage over another foldamer alphabet that requires cofactors (multiple components) to achieve the same function at an equivalent level of efficiency. Another German/English comparison can also serve to extend this analogy further. The early twentieth-century critic Friedrich Gundolf made the extraordinary claim that the works of Shakespeare, when translated into German, were "better" than the English of the original. Moreover, he contended that Shakespeare was "accidentally" forced to use English by dint of his birthplace.[1567,1568] These notions would strike most people (probably of any nationality) as absurd, and obviously unresolvably subjective in any case (Maybe Shakespeare's *ouevre* would have been even more outstanding had his mother tongue been Basque, for example?) In the movie *Star Trek VI*, a Klingon alien aficionado of Shakespeare alleges that the immortal bard is best served "In the original Klingon," perhaps inspired by the above Herr Gundolf. It could indeed be proposed that the "optimal" language for Shakespearean expression has yet to be discovered, but this is hardly a verifiable assertion.

But this strange idea becomes much more acceptable when we filter it through the molecular language analogy. Once again, a specific functional biological macromolecule (or an artificial foldamer) may be the best possible solution of its type, but not necessarily the best of all possible solutions in the universe of all folded molecules. This argument applies also to complex functional systems, as alternative transcognate molecules in different foldamer "languages" might offer objectively definable advantages. A complex natural cellular biosystem likened to a work of Shakespeare might then indeed be "written" in superior form in a different "language" from which it originated through megayears of evolution. Conversely, some functions may prove "untranslatable" between different foldamer frameworks, at least with comparable levels of efficiency. With the same linguistic analogy, the German *schadenfreude* has no direct single-word English cognate and is "untranslatable" by single-word correspondence alone. English, of course, uses the "cofactor" approach to describe the same concept in multiple words, but has also co-opted the more efficient German single-word solution.

Molecular translations can be conceived at higher levels as well, where sophisticated multicellular systems are functionally reproduced (and perhaps improved) using a different set of structural subunits. As an ultimate example, by this logic the human brain and mind could be rendered as a "transcognate" system using some alternative molecular framework to underlie its informational processing capabilities. The aim *per se* of reproducing human cognition corresponds of course with "strong" artificial intelligence (AI) projects. Although AI is not usually conceived of in molecular computational terms, we noted earlier that reverse engineering the operational principles of human cognition may prove to be a fast-track route toward realization of these goals. In any case, the essential issue for the present purposes is that, given that the human mind arises from a highly complex biosystem, it should logically be capable of recapitulation in some other complex system using different structural components and novel foldamers.

In projecting future developments, as previously noted, the term "foldamers" refers to nonbiological artificial folded polymeric compounds, and for complex folding and function alphabets of some description are required. But to date, foldamers are not amenable to *in vitro* evolutionary approaches in the same manner as for nucleic acid effector molecules. Let us briefly explore this issue in the next section.

Effectors and Replicators

Proteins are believed to have superseded the RNA world by virtue of their enhanced functional capabilities conferred though their extended alphabet. Yet proteins do not replicate themselves directly, but only through their "entanglement" with informational nucleic acids. While a single-stranded nucleic acid aptamer will fold into a specific shape important for its activity, it can be denatured and replicated (and indefinitely amplified) by a template–primer system. Proteins *in vitro* must be accompanied by some form of linkage with their encoding nucleic acids if repeated cycles of evolutionary selection are to be contemplated. Nucleic acids are amplifiable through cycles of template-directed polymerization and denaturation of mutually complementary duplexes, but this is not a practical option for proteins. (Recall Lucy's discoveries in the Virtual Molecular Museum of Chapter 6.) And by the same token, neither could an artificial foldamer be expected to be amplifiable in a template-directed manner, as noted in Chapter 7.

In Fig. 10.2, a foldamer with an alphabet of six "letters" is depicted, imagined as functionally selected from a random library of sequences of 15 residues using the same alphabet. In principle, such a library might correspond to nucleic acids with an extended alphabet (as in Fig. 7.13), but for the present purposes let us picture it as entirely nonbiological. Although the specific foldamer sequence of interest is nonamplifiable, it can be indirectly identified through an amplifiable (nucleic acid) tag, which encodes the vital foldamer sequence information in ways described in Chapter 8. While a direct means for amplifying artificial foldamers with defined alphabets would tremendously facilitate directed evolutionary studies within such a framework, this remains a formidable challenge. For template-directed polymerization on a linear (denatured) artificial foldamer template, not only specifically complementary monomeric "antisense" building blocks but also an exotic polymerase enzymatic machinery for

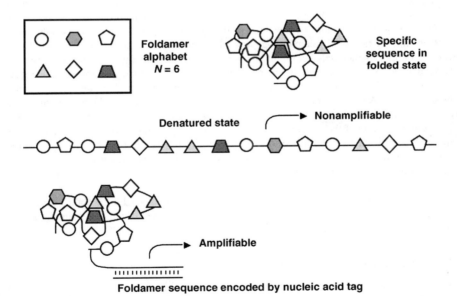

FIGURE 10.2   Depiction of a nonbiological foldamer with an alphabet of six polymerizable side chain groups. From a random chemically generated library of length 15 residues, a specific sequence (as shown) is functionally selected but cannot be amplified as such. If a set of complementary groups for each "letter" of the foldamer alphabet exists which is also polymerizable, mutual template-directed polymerization and amplification is conceivable if a polymerase capable of directing this exists. In lieu of this, other informational tags can be used to obtain the needed foldamer sequence, as detailed in Chapter 8.

performing the necessary sequential coupling steps would be required. In the short-term future, the best prospects for novel amplifiable foldamers come from modified nucleic acids with extended alphabets and (if necessary) engineered natural polymerases with acquired compatibility with the novel nucleobases (Chapter 6).

One feature of artificial foldamers in general is of great fascination, since an implicit question in this research and related work on alphabetic extensions is the evolutionary origin of the specific protein alphabet for making varied effector molecules. As we have seen (Chapter 5), extensions to the protein alphabet have been proven to be mechanistically valuable for functional applications, and perhaps certain completely novel foldamer frameworks could "go one better" even further. But these existing alphabetic expansions are of course still compatible with directed evolutionary approaches, by virtue of the expanded genetic codes that enable the enhanced alphabets to be deployed. But when considering leaping off into a wild blue yonder of completely novel foldamers, the evolutionary aspects of selecting or screening novel foldamer activities should give some pause.

In this context, we might note from Chapter 2 that functional proteins are highly nonrandomly distributed in sequence space, and evolutionarily speaking, this is not a trivial point.[172] Access to altered protein functions can be obtained by evolution through sequences closely clustered in sequence space, or by the shuffling of a limited number of protein domains. Natural selection would founder if every incremental functional improvement required protein sequences remote from each other in sequence space.

Is the clustering of functional protein sequences in sequence space another special property of the protein alphabet itself or is it inevitable with any foldamer? That is, is this "functional clustering" another property that one would have to demand from a totally different alternative foldamer (in order to fulfill the range of functions required for an effectively evolvable system), or would it come "automatically"? This question of course would apply with equal force to a totally alien biology as well as human molecular ventures on this planet. A nonprotein foldamer alphabet might be imagined where even related functions necessarily corresponded to sequences widely separated from each other in their own sequence space. If this were true, Darwinian evolution with such a foldamer type would be difficult if it could proceed at all. (And by definition, such a hypothetical "nonevolvable" foldamer could not be employed within an alien exobiology.) This issue would presumably hinge at least to some extent on the nature of the novel foldamer folding pathways and stabilization, whether it was indeed "protein-like" or conformed to a different set of rules entirely, subject to the specific chemistry of the foldamer alphabet in question. If the former condition was at least approximated, it might be presumed that many of these concerns as to foldamer evolvability would prove groundless.

To conclude, let us return to the general issue of combinatorics, which is at the heart of any work dealing with foldamers. This can take us seemingly far afield from molecular discovery, but will return us there nonetheless.

The Universal Power of Combinatorics

We have repeatedly observed combinatorics at work in the generation of random libraries of proteins and nucleic acids, and this of course operates at the level of their molecular alphabets. But this fundamental layer of combinatorial assortment underlies progressively higher stages of biological organization, which also have distinct combinatorial aspects (Fig. 10.3). The combinatorial application of both small molecule scaffolds and biological macromolecules in multiple different functional contexts clearly has advantages in efficiency over single-application "dedicated" components, and this universal feature of living organisms may also be

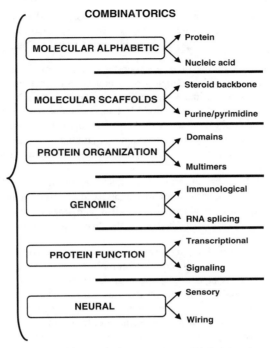

FIGURE 10.3   Biological combinatorics at different levels. All categories show two examples out of many possible cases (except for the *Molecular Alphabetics*, where the two natural effector (protein) and informational/effector/replicator (nucleic acid) alphabets are indicated. *Molecular Scaffold combinatorics*: Small molecule backbones that are used in multiple contexts, such as the steroid core structure (hormones, cholesterol), and purines and pyrimidines (nucleic acids, many cofactors and mediators). *Protein Organizational combinatorics*: This can result from the shuffling of single domains or domain sets ("supradomains"), or at the level of whole proteins in multimeric complexes. *Genomic combinatorics*: Mechanisms that combinatorially increase information directly encoded within the genome, such as immunological germline rearrangements (Chapter 3) and diverse RNA splicing processes. *Protein Functional combinatorics* refers to the combinatorial use of proteins in multiple functional situations, such as transcription (combinatorial use of transcription factors) and signal translation (combinatorial use of kinases). *Neural combinatorics* covers diverse processes ranging from chemosensory signaling (Chapter 3) to neural wiring processes.

an inevitable consequence of evolutionary pressures. Combinatorics goes hand in hand with modularity, a biological theme we have visited repeatedly, and which is becoming increasingly emphasized following the accumulation of massive amounts of genomic information in recent times.[1569]

All of the mechanisms for natural biological diversification and its laboratory counterpart that we inspected in Chapters 2–4 are founded upon combinatorics at the level of either nucleotides (mutation) or larger blocks (recombination in various forms). In combination with screening, combinatorics is applicable to diverse human needs, ranging from materials design[1570] to therapeutics.[1571] But an additional higher level of functional combinatorics of obvious relevance to humans is possible, if one considers cognition itself. The fundamental notion that the creative formation of novel thoughts and ideas has a combinatorial basis is very old, as exemplified by the elaborate combination-based "Great Art" of the fourteenth-century theologian Ramon Llull.[1572] Yet its appeal is still strong as a cognitive theory in our time.[1573] To briefly consider this type of concept, let us co-opt an old friend for one final visit. . .

A Kaleidoscopic View   At the end of her long quest in molecular discovery, Lucy consults with another Molecular Sage of indeterminate origin and waxes somewhat philosophical. She ventures the thought that the ultimate source of molecular discovery is Mind.

"MIND?" said the Sage, "You mean—Mechanisms Inducing Neural Diversity? Yes, although of course you could say that about the generation of any productive thoughts."

Picking up on this theme, Lucy suggests that the generation of diversity and combinatorics are equally important at the cognitive level of creative molecular discovery as with the practical implementation of it.

"Necessarily, GOD in general encompasses MIND," replied the Sage, "And it would indeed seem logical that cognitive combinatorics is involved in creativity. For example, the great scientist Linus Pauling is attributed with the aphorism: *The best way to have a good idea is to have lots of ideas.* I interpret this to mean that one should encourage a veritable kaleidoscope of ideas and pick the best."

Ignorant of kaleidoscopes, Lucy required further explanation from the Sage, who added, "Lucy, your fascination with diversification and combinatorics makes me think of you in that light yourself—you have, metaphorically speaking, kaleidoscope eyes."

But Lucy suddenly raised the issue of how the "best" ideas, or kaleidoscopic patterns, could be "selected." And indeed some kind of *selection* might seem to be required, since even "high-throughput screening" of millions of combinatorial "idea fragments" would presumably take an excessively long time for a brain to evaluate, unless some kind of massive parallel processing takes place. What then is the agent of selection, and how does it operate?

"Ah, *that* is the question," said the Sage, and she smiled patiently at Lucy, whose bright eyes flashed back.

And the "selection" of novel ideas from combinatorial "libraries" of thoughts is indeed a crucial limiting (and mysterious) factor. Combinatorics of ideas without a useful selection process was amusingly lampooned over 280 years ago by Jonathan Swift in one of Gulliver's travels (to the Grand Academy of Lagardo[1574]). And without an accounting for this kind of selectivity, the role of combinatorics in thought processes becomes "true but trivial."[1572] In both natural and artificial evolution, combinatorics is inherent in the necessary diversity generation, and the selection processes involved are not enigmatic. Nevertheless, the potential parallel between the normal evolutionary algorithm and the "evolution" of thought and ideas has long been an enticing prospect.[1575] Applying such an evolutionary model to concept generation in the context of artificial evolution creates a loop of sorts. In this regard, it is interesting that computational evolutionary algorithms for *de novo* design of small molecules can themselves be viewed as "idea generators" for subsequent design innovations.[1576]

Diversity generation in the world of human thought is necessary as a basic precondition to allow combinatorial associations to take place. And just as human diversity has enriched scientific and intellectual development in the past, so too it will for molecular design into the future. But on this note, one might wonder to what extent the generation of novel strategies for molecular design will continue to rely on human intelligence, as opposed to the artificial variety. In rational design, the levels of basic science logistics, design itself, and analyses will increasingly be sourced from advanced AI systems. "Robot scientists" are here, and here to stay.[1577,1578] Are jobs threatened? Traditionally, robotics progressively replaces repetitive tasks requiring low intellectual input, but as AI advances the theoretical "job replacement level" becomes higher and higher.* A superior robot scientist applied toward protein or drug design might seem the pinnacle of rationality in the laboratory, casting aside the annoying distractions and peccadilloes associated with humans. Moreover, eventually on-going developments should lead to robots capable of self-reproduction.[1579] A robot processor may be perfectly logical and systematic in its approaches to its assigned problems, but it could not by the same token ignore logical problems of design, especially with respect to proteins. It would encounter the same problems of exploring novel regions of protein and chemical sequence space as human designers and might conclude that certain problems were indeed best approached by evolutionary methods, if they themselves are applied in the most rational possible manner.

The future of molecular design itself will be heavily linked with the derivation of complex molecular switches, circuits, and pathways. We have only lightly touched upon some aspects of this big field, which overlaps further with the general area of nanotechnology and nanomaterials design. Yet at least as much as anywhere else, the development of artificial biological circuitry demonstrates the power of combining empirical evolutionary design and rational principles in

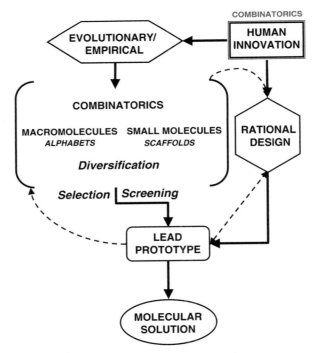

FIGURE 10.4 Rational and empirical interplay in molecular discovery. The "rational design" box includes input from artificial intelligence, as an indirect outgrowth from human innovation. "Solutions" here are also intended to include the artificial development of molecular switches, circuits, and pathways.

---

*One day, a book such as this could be written entirely by strong AI (such as a system based on Intelligent Artificial Neural Design Using Nested Networks). According to some, the "one day" will arrive much sooner than many of us realize.[297] An AI version of this book might also be significantly improved, such that the AI product is a superior copy within the metaphorical Library of Babel referred to in Chapter 1. If "strong" AI is achievable, by definition all human activities could be replaced. It is not at all certain, though, that this ideal can be attained, at least with digital computational systems.

action.[1541] It is this ongoing interplay that, we can be confident, will continue into the foreseeable future, depicted in Fig. 10.4.

Biology instructs us that evolution can deliver whole edifices whose functions have an evolutionary underpinning themselves, as we have seen with immune systems. Evolution has led to the parsimonious use of combinatorics and modularity at a variety of different operational levels. But the highest level of this exists in human neural systems, from which minds that conceive of molecular solutions can arise.

GLOSSARY

Abzyme   See Catalytic Antibody.

Adaptive Immune Systems   Immune systems of vertebrates that can mount a response to a foreign antigen through an adaptive process, where new combining specificities of T cells and antibodies are selected and amplified.

Affinity/Avidity   Affinity of an antibody measures the strength of one of its combining sites toward a specific antigen epitope; avidity is a measure of the net binding strength of an entire antibody, which is usually at least bivalent. A structure with polymeric epitopes may thus be bound with high avidity, but not necessarily high affinity.

Allele   Alternative versions of genes (encoding the same protein or RNA but with genetic polymorphisms [q.v.]) found within populations of the same species.

Allogeneic   Pertaining to an organism or its component cells in their status as genetically distinct from another member of their own species. Mammalian organisms normally show extensive allelic differences with other members of their species, except for identical twins or highly inbred animals, which become genetically identical, or syngeneic.

Alloimmunity   See Allorecognition.

Allorecognition   Specific recognition of allogeneic molecules as nonself by immune systems.

Allostery   A conformational change in a protein or nucleic acid that is induced by the binding of another molecule, and which has significant functional consequences.

Alphabets   See Molecular Alphabets.

Amino Acid Code, Single-Letter   Alanine, A; Arginine, R; Aspartic Acid, D; Asparagine, N; Cysteine, C; Glycine, G; Glutamic Acid, E; Glutamine, Q; Histidine, H; Isoleucine, I; Leucine, L; Lysine, K; Methionine, M; Phenylalanine, F; Proline, P; Serine, S; Threonine, T; Tryptophan, W; Tyrosine, Y; Valine, V.

Aminoacyl tRNA Synthetases   Enzymes crucial for protein synthesis, which act by "charging" specific tRNA molecules (q.v.) with their cognate amino acids.

Antibody   Proteins of the "humoral" arm of adaptive immune systems that recognize and bind antigens. Antigen binding occurs through their variable regions, and their constant regions group them into several different classes (such as IgM, IgG, IgA, and IgE). Antibodies can be actively secreted or serve as surface receptors on B cells (q.v.)

Antigen, Immunogen   An antigen is any molecular structure recognizable by the immune system, whether or not such a structure can generate an immune response itself. An immunogen is both recognized by an immune system and capable of stimulating an immune response against itself. Thus, the aphorism, "All immunogens are antigens, but not all antigens are immunogens."

Searching for Molecular Solutions: Empirical Discovery and Its Future. By Ian S. Dunn
Copyright © 2010 John Wiley & Sons, Inc.

Anticodon   Under normal circumstances, a triplet sequence in a specific region of tRNA molecules (q.v.; in their anticodon loops), which is complementary to a specific codon (or small set of codons) in mRNA molecules (q.v.)

Aptamer   A nucleic acid molecule that can fold into a shape that confers a useful binding function toward another molecule. Aptamers are usually obtained through SELEX (q.v.) or related procedures.

Archaea   Single-celled prokaryotic (q.v.) organisms of ancient evolutionary origins, which constitute one of the three major domains of life. (The other two Domains cover eukaryotes (q.v.) and bacteria (see Eubacteria; Prokaryotes).)

B cells   Cells of the adaptive immune system (q.v.) that express surface immunoglobulin (q.v.), which enables recognition of antigens. Following activation, B cells differentiate to plasma cells that secrete large amounts of antibody.

Bimolecular   Involving the participation of two specific molecules, especially with respect to reaction mechanisms and kinetics. Not to be confused with Biomolecular (q.v.).

Biomolecular   Pertaining to any molecular entity (large or small) of biological origin.

Catalytic Promiscuity   Enzyme activity that is not perfectly specific either at the level of substrate (q.v.) or at that the level of product generation.

Catalytic Antibody   An antibody raised against a low molecular weight hapten (q.v.) designed as a mimic of a transition-state intermediate for an enzymatic reaction, such that the resulting antibody itself has the desired catalytic activity.

CDR   See Complementarity-Determining Region.

Chaperone   A protein that assists the folding of another protein or an RNA molecule. Chaperones accelerate protein folding and prevent misfolding under cellular conditions but are not strictly essential for folding *per se*. (See also Foldase).

Chaperonin   Chaperonins are chaperones that do not directly interact with newly formed ribosomal polypeptides, and which form large ring structures where protein folding guiding takes place. An example is the *GroEL* chaperonin of *E. coli*.

Chiral, Chirality   A chiral molecule has stereoisomers, which are isomers (q.v.) of the molecule that differ at one or more carbon atoms whose three-dimensional arrangement of bonded groups is not superimposable on the other.

Codon   A base triplet in a specific reading frame on an mRNA molecule (q.v.), which encodes an amino acid.

Complementarity-Determining Region (CDR)   Specific regions of antibody heavy and light chains that make contact with antigen; or specific regions of T cell receptors (usually the α and β chains) that make contact with peptides in the context of MHC (q.v.) In each case, three CDRs are defined (CDR1; CDR2; CDR3).

Directed Evolution   Any process that uses artificial diversification and screening or selection (q.v.) in an iterated manner in order to obtain a desired molecular or molecular system property. Often termed *in vitro* evolution, although strictly speaking this should not include procedures that use bacterial vectors for screening or selection.

Designability   The number of variant protein sequences whose folding converges toward a single state with the lowest free energy (q.v.). Compare with robustness (q.v.).

DNA Shuffling   A technique for *in vitro* diversification using PCR (q.v.) to shuffle sequences; equivalent to a process of homologous recombination (q.v.) or sexual exchange between multiple sets of genes in one experiment.

DNAzyme   See Ribozyme.

Domain, Protein   The smallest functional segment of a protein as an evolutionary module, and that often acts as an independent folding unit.

Druggable   Susceptible to the action of conventional low molecular weight drugs.

E. coli   *Escherichia coli*, a bacterial organism of the human intestine. Specific strains of *E. coli* are used ubiquitously in biochemistry, molecular biology and biotechnology.

Enantiomer   An enantiomer is a molecule that has a stereoisomer (see Chiral), which is a mirror image of it.

Epigenetics   An epigenetic change is the modification of a genome (q.v.) that does not involve an alteration of the nucleotide base sequence. This is often via methylation of cytosine residues, but can also involve changes in the genome's protein packaging (chromatin). Epigenetic changes can modulate gene expression. Related term: epigenome.

Episome   A nucleic acid sequence, often a circular or linear plasmid, which can replicate independently of the genome of its host cell.

Epitope   The specific region of a macromolecule (usually a protein) that is recognized by antibody or T cell variable regions.

Eubacteria   The domain of life that encompasses the bacterial world, especially as opposed to the Archaea (q.v.), which are also prokaryotes but nonetheless very distinct.

Eukaryote   One of the three major domains of life (see Archaea), characterized by increased complexity and size, nuclear membranes, and usually (but not always) organelles such as mitochondria (q.v.) and chloroplasts.

Fab   Monovalent antigen-binding fragment of an immunoglobulin (q.v.) molecule.

(Fab)$_2$ Divalent antigen-binding fragment of an immunoglobulin (q.v.) molecule.

Flagellin   A protein component of bacterial flagella (a bacterial motor structure), which is recognized by the innate immune system (q.v.).

Fold   A stable tertiary structure that a macromolecule assumes after its synthesis. Such molecules (including proteins and nucleic acids) are composed of smaller monomeric units of different types.

Foldamer   Any nonbiological molecule composed of a limited number of subunits (see Molecular Alphabets) that can take up a stable fold, which may have functional properties.

Foldase   Enzymes for which some proteins have an obligate requirement for folding assistance, in order to attain a folded state that is not at the lowest free energy (q.v.) level. Compare Chaperones (q.v.).

Free Energy   A thermodynamic property (G) whose measurement is useful for studying macromolecular folding. The change in free energy over a folding pathway is expressed as ΔG, and the great majority of proteins reach their lowest free energy states when they attain their normal folds.

Gene   Originally defined as the unit of heredity and then as a specific nucleic acid sequence encoding a protein or a functional RNA. The original simple picture has become much more complex from recognition of phenomena such as differential mRNA splicing.

Gene Targeting   A method based on homologous recombination (q.v.) whereby gene sequences in pluripotent embryonal stem cells can be replaced by input homologous sequences with any specific alterations. Once the modified stem cells are generated, they can be used to produce mice whose genomes are altered in the corresponding manner. A great many genetic modifications in mice and other species have been produced in this manner.

General Acid–Base Catalysis   An enzymatic mechanism where a specific functional group can act as a general acid and base through donating or accepting protons. In order to function in a general acid–base role, a functional group must have a pK_a value (q.v.) near physiological pH (in the range of 6–7).

Genome; Genomics   The genome is the entire complement of informational nucleic acid for an organism. This is usually DNA, but RNA viral genomes are well known. Genomics is the science that studies the sequences, organization, and properties of genomes.

Genotype   The sequence patterns of genomic nucleic acids (usually DNA) for specific alleles, or sets of alleles, for an individual organism. This can be extended to include its complete genome, to show complete patterns of allelic and other polymorphisms as defined by all conserved regions of the genome for its species.

Germline   The genetic information of a multicellular organism that is specifically transmitted to the succeeding generation, usually associated with sexual reproduction following fusion of haploid (q.v.) sperm and egg cells resulting from meiosis (q.v.). Contrast with somatic (q.v.).

GOD   Generation Of Diversity, in the context of natural evolution, or making molecular libraries for screening or selection, especially where the latter itself is an artificial evolutionary process. GOD can result from a number of different processes, including mutation and recombination.

GPCR/G-Protein-Coupled Receptor   A large family of protein cell-surface receptors with seven transmembrane segments. Their signaling mechanism is coupled with heterotrimeric proteins that switch between active and inactive forms based on differential binding of guanine nucleotides (hence "G-proteins").

Guest   A molecule (usually a small entity) that is bound or interacts with a specific site or pocket within a large chemical structure (its "host"; q.v.).

Haploid   Having half the normal complement of chromosomes.

Hapten   A small molecule recognized by mediators of the immune system, especially antibodies. Haptens are antigens, but are not immunogenic themselves unless in polymeric form or (more usually) when conjugated with a protein carrier molecule.

Heteroclitic   An immune response that generates antibody or T cell responses that have higher affinity for a different molecule to that actually used for immunization itself.

High-throughput screening   Screening systems for molecular libraries based on automated high-volume processing, typically processing tens of thousands of library members per day at minimum.

Host   See Guest.

Hydrolysis   An enzymatic reaction that involves the breakdown of a water molecule as an inherent component of the catalytic process.

Hydrophobic/Hydrophilic   Amino acid side chains that repel or attract water molecules, respectively. Residues with hydrophobic side chains pack into a protein's hydrophobic core.

Idiotype   A structure within an immunoglobulin variable region that is unique to that molecule. As such, it is not self (q.v.) and can provoke an immune response.

Immunogen   See Antigen.

Immunoglobulin   An antibody (q.v.), especially when in reference to its nature as a protein.

Innate Immune System   The front-line defense of vertebrates against pathogens, involving receptors that are fixed in the germline (q.v.) rather than derived through adaptive processes, as with the adaptive immune system (q.v.). Among the receptors of the innate immune system, a family termed toll-like receptors (TLRs) is of special importance.

In Vitro **Evolution**   See Directed Evolution.

Isomers   Molecules with the same constituent atoms but different structures.

Isoform   An alternate form of the same protein, which can be produced by various means, including alternate mRNA (q.v.) splicing.

Isostere   A molecule that has similar molecular size and shape to another molecule.

k_{cat} The rate constant for an enzyme, equivalent to turnover number.

K_d The dissociation constant for a binding reaction in equilibrium, expressed in molarity (moles/liter; M). The lower the K_d value, the higher the affinity.

kDa   Kilodaltons, a measure of molecular weight. One Dalton is approximately the mass of a hydrogen atom. Molecular weights are also expressed as grams per mole (6.023×10^{23}) of molecules. Since the weight of hydrogen in one mole of hydrogen atoms equals one gram, a molecular weight of N Daltons is equivalent to N grams/mole for that molecule. A typical medium-sized protein has a molecular weight of about 50,000 Da or 50 kDa.

K_m In enzyme kinetics, the Michaelis constant, equal to the substrate (q.v.) concentration at $V_{max}/2$, where V_{max} is the maximal reaction velocity obtainable in the limit as substrate concentration increases indefinitely. K_m gives an indication of the tightness of substrate binding by an enzyme.

Library   A collection of molecules used for screening or selecting a desired functional property. A library can range from large proteins and nucleic acids to small chemical compounds and can be based on chemically similar members or deliberately designed to maximize diversity, depending on the application. A library can be completely artificial, or a collection of natural molecules deriving from a common source, as with complementary DNA (cDNA) libraries for all mRNAs (q.v.) expressed by specific cell types.

Ligand   Any molecule (but usually a relatively small one) that specifically binds to a protein or functional nucleic acid.

Major Histocompatibility Complex; MHC   Cell-surface proteins of great importance in vertebrate immune responses. Foreign protein antigens are processed into short peptides and bound in clefts of MHC proteins (of which there are two major types, Class I and II), which allows recognition by T cell receptors (q.v.).

Mass Spectometry   A highly sensitive technique for molecular analysis, where samples are ionized and molecular fragments distinguished on the basis of their mass to charge ratios.

Meiosis   A special form of cell division in sex cells, characterized by division of the normal number of paired chromosomes into half the number of unpaired chromosomes, or the haploid state (q.v.).

Microarray   A high-density arrayed set of nucleic acids or proteins designed to be tested with a molecular probe in order to gain information.

MIP   See Molecularly Imprinted Polymer.

Mitochondrion (pl. mitochondria)   An organelle present in almost all eukaryotic cells (absent in certain protozoans), which is chiefly involved with energy metabolism and the synthesis of adenosine triphosphate, and which has its own DNA genome.

Molecular Alphabets   The range of monomeric subunits used to construct heteropolymeric folded macromolecules.

Molecularly Imprinted Polymer (MIP)   A polymeric structure formed around a target molecule or higher order structure (which can be of widely varying size) such that an imprint (or "cast") of the target shape is formed. When the original target is removed, the MIP can recognize and bind the original molecule or structure.

Moonlighting   The ability of a protein to fulfill two (or more) entirely separate functional roles in a biosystem.

mRNA   Messenger RNA, or RNA that is produced as an RNA copy of a DNA protein-coding sequence strand, and that serves as a program for protein synthesis on ribosomes (q.v.).

microRNAs (miRNAs)   Small biological RNA molecules involved with gene regulation, usually through blocking of translation.

Neofunctionalization   The acquisition of novel gene function in one copy following a duplication event.

NP-complete   A class of computational problems for which algorithms for solving them in a realistic time frame have not been derived and may be impossible to obtain. These are then "NP-hard" (q.v.) problems, where "NP" denotes "nondeterministic polynomial time."

NP-hard   See NP-complete.

Nucleic acid   See Nucleoside/nucleotide.

Nucleobase   See Nucleoside/nucleotide.

Nucleoside/nucleotide   Nucleosides are composed of a sugar (ribose) joined to a nucleobase (q.v.) (compounds with heterocyclic ring structures, which include adenine, cytosine, thymine, guanine, and uracil); nucleotides are nucleosides with attached phosphate groups. Nucleotides joined through phosphodiester bonds form polymeric nucleic acids, and both nucleosides and nucleotides participate in a variety of metabolic and signaling transactions in biosystems.

Oligonucleotide   A short nucleic acid sequence, usually single-stranded and synthetic.

"Omics"   A suffix coined for genome-wide ("global") aspects of biosystems, all of which arise from expression of the genome or through the activities of its expression products. The major "omics" are the transcriptome (the transcribed genome), the proteome (the expressed genome at the protein level), and the metabolome (or metabonome, all small molecule products of enzymes encoded by the genome). Many other "omic" terms have been coined.

OMspace   An imaginary universal molecular space introduced in Chapter 1.

Operon   An organized gene regulation system in prokaryotes. The first defined was the lactose ("lac") operon in *E. coli* (q.v.), but numerous others exist in this and other prokaryotic organisms.

ORF   Open Reading Frame; a DNA sequence read in a specific triplet-codon reading frame that is not "closed" by the appearance of a stop codon. An ORF of significant length therefore has a good chance of corresponding to a genetic coding sequence for a protein.

Orthogonal   A molecular reaction or a molecular system that is self-contained and operates independently and without interfering with another reaction or system operating concurrently within the same physical space.

Orthologs   Homologous genes in different species that have evolved from a common ancestor.

PAMPs   Pathogen-Associated Molecular Patterns; conserved structures on pathogenic organisms that are recognized by the Innate Immune System (q.v.).

Paralogs   Genes of the same organism with divergent functions, which derive from a common ancestor and a gene duplication event.

PCR   See Polymerase Chain Reaction.

Peptide   A short string of amino acids joined by peptide bonds (q.v.)

Peptide Bond   The chemical linkage between amino acids as found in proteins and peptides (q.v.), characterized by the formation of an amide bond between the carboxyl group of one amino acid and the amino group of another.

Peptidomimetic   A molecule that acts as a structural mimic of a natural peptide, usually created with the aim of improving its properties in terms of stability, cell penetration, or other factors.

Phenotype   The functional outcome of the expression of specific genes or genomes (q.v.).

pK_a value   The pK_a is equivalent to the value of the pH when a weak acid is at a state of 50% ionization (a point of half dissociation into negatively charged ions and hydrogen ions [or strictly speaking, H_3O^+ hydronium ions]). It is useful as an index of the acidity of a compound or functional group through its propensity for acid dissociation. Acid strength is inversely proportional to pK_a value.

PNA   Peptide Nucleic Acid, an artificial nucleic acid with a variant backbone characterized by peptide bond linkages rather than phosphodiester bonds.

Polymerase Chain Reaction   An important molecular biological technique for amplifying DNA sequences through repeated cycles of duplex denaturation followed by extension from specific primers by thermostable DNA polymerases.

Polymorphisms   Literally "many forms"; having multiple alternative alleles (q.v.) within a population. Polymorphisms can occur through many types of sequence differences, but are common at the single-nucleotide level. The major histocompatibility complex genes (q.v.) are the most polymorphic genes in humans.

Primer   An oligonucleotide (q.v.) that allows the initiation of nucleic acid replication from a template strand mediated by a polymerase enzyme; especially significant in the polymerase chain reaction (q.v.).

Probe   A molecule used to "interrogate" a library in order to find molecules that bind to it. Usually "probe" is synonymous with Target (q.v.). "Probe" is also used in other molecular biological contexts, such as with microarrays (q.v.).

Prokaryote   Single-celled organisms that lack a defined nuclear membrane and other features characteristic of eukaryotes (q.v.). Prokaryotes include bacteria and Archaea (q.v.).

QSAR   Quantitative structure–activity relationships; the correlation of molecular structures with measurements of the effects of structural change on a molecule's functional properties.

Recombination   The exchange of sequences between nucleic acids. This can occur between related sequences as homologous recombination or by joining of unrelated sequences as nonhomologous (or "illegitimate") recombination.

Restriction enzyme/restriction endonuclease   Bacterial enzymes that cleave at specific DNA sequences and act as protective agents for bacteria against invasion by foreign nucleic acids. These enzymes are heavily used by molecular biology and biotechnology.

Retrotransposon   A transposable genetic element whose mechanism of transfer between genomic sites involves a reverse transcription step (conversion of RNA to DNA through a reverse transcriptase polymerase).

Ribosome   The protein synthesis factory of all cells, consisting of large and small ribonu-cleoprotein subunits. The key step in protein synthesis within ribosomes is mediated by a peptidyl transferase ribozyme (q.v.).

Ribozyme   An enzyme composed solely of ribonucleic acid, of which both natural and artificial precedents exist. Deoxyribonucleic acid enzymes (DNAzymes) have also been artificially selected but have not been observed in nature.

RNA Interference/RNAi   A normal biological process for blocking invading nucleic acids (usually from viruses or foreign transposable genetic elements) involving the processing of RNAs into short segments ("siRNAs" or short interfering RNAs). It has been widely exploited in molecular biology and biotechnology.

RNA World   A hypothetical developmental phase between nonliving systems and protein/nucleic acid-based biosystems when both informational and effector molecules were composed of RNA.

Robustness   The ability of proteins to withstand mutation without functional loss. Translational robustness refers to the ability of certain proteins to tolerate translational mistakes that result in the misincorporation of amino acids.

Rotamer   A rotational isomer of a compound. In proteins, it refers to alternative rotational configurations of each amino acid side chain.

Scaffold   For small molecules, a defined structure (often based on a ring configuration) that serves as the starting point for the synthesis of many compounds; for proteins, a defined folding pattern into which peptide segments (such as loops) can be transferred without disruption of the entire structure.

Secondary Structure   A local region of a macromolecule (usually protein or nucleic acid) that forms a well-defined structural motif. Examples in proteins are α-helices and β-sheets.

SELEX   Systematic Evolution of Ligands by Exponential enrichment; an *in vitro* technique for iterated cycles of functional selection from molecular libraries.

Self   Molecular structures encoded by an individual organism toward which its developing adaptive immune system specifically learns to avoid activating responses, or from which its immune system is functionally shielded. Molecules outside this definition are accordingly nonself.

Shuffling   See DNA Shuffling.

Somatic   Pertaining to cells of the body of an organism, or events occurring during the lifetime of an organism that are not passed on through the germline.

Spiegelmer   An aptamer whose ribose sugar moieties are in the unnatural L-stereoisomeric configurations, conferring high resistance to nucleases. From German *spiegel*, mirror.

Spliceosome   A large protein/RNA complex involved in the splicing of intervening RNA sequences (introns) from transcribed mRNAs in eukaryotic cells.

Substrate   The molecular entity that is the catalytic target of a specific enzyme.

Subfunctionalization   The splitting of functions formerly performed by one protein into two, following a gene duplication event.

Target   A specific objective molecule or system toward which one seeks another molecule that will modify the properties of the objective molecule or system in a desired manner. Libraries (q.v.) are used to find such suitable molecules for a given target. While libraries are by definition variable and diverse, a target is a constant structure or system.

Tautomer   An isomer (q.v.) of a compound that can rapidly interchange with another isomer under normal conditions.

T Cell receptor   The recognition structures on the surface of T cells, which mediate many aspects of immunity and assist B cells in the response to antigens. Unlike antibodies (q.v.), T cell receptors only recognize short peptide antigens bound in the context of major histocompatibility complex (MHC) proteins (q.v.).

TNA   Threose Nucleic Acid, an artificial nucleic acid with a variant backbone where the sugar moeity is threose rather than ribose.

Toll-like receptor/TLR   See Innate Immune Systems.

Transcognate   A folded functional macromolecule composed of a specific molecular alphabet, for which an equivalent function can be found either with a different fold in the same alphabet, or with an entirely different alphabet (Chapter 10).

tRNA   Small folded RNA molecules that function as adaptors for amino acids and mRNA (q.v.) coding sequences during protein synthesis. Specific tRNAs are covalently joined with their cognate amino acids by aminoacyl-tRNA synthetase enzymes (q.v).

Two-Hybrid System   A widely used *in vivo* method for detecting protein–protein interactions, originally developed for yeast cells. Many variants of it exist for detecting other types of interactions.

Vector   In a molecular biological context, a vector is a replicable nucleic acid element, often a plasmid episome (q.v.), which can allow the propagation of other nucleic acid sequences linked with it.

Xenobiotic   Literally "stranger to life," but here denotes molecules that are of foreign or nonbiological origin. The so-called "xenobiotic immune system" is the system of vertebrate organisms for recognizing and detoxifying such potentially dangerous chemicals.

BIBLIOGRAPHY

1. Selfridge, O., in *Symposium on the Mechanization of Thought Processes* (HM Stationary Office, London, 1959).

2. Fontana, W. & Schuster, P. Continuity in evolution: on the nature of transitions. *Science* **280**, 1451–1455 (1998).

3. Savchuk, N. P., Balakin, K. V., & Tkachenko, S. E. Exploring the chemogenomic knowledge space with annotated chemical libraries. *Curr Opin Chem Biol* **8**, 412–417 (2004).

4. Beroza, P., Damodaran, K., & Lum, R. T. Target-related affinity profiling: Telik's lead discovery technology. *Curr Top Med Chem* **5**, 371–381 (2005).

5. Perelson, A. S. & Oster, G. F. Theoretical studies of clonal selection: minimal antibody repertoire size and reliability of self–non-self discrimination. *J Theor Biol* **81**, 645–670 (1979).

6. Terwilliger, T. C. & Berendzen, J. Exploring structure space. A protein structure initiative. *Genetica* **106**, 141–147 (1999).

7. Yahyanejad, M., Kardar, M., & Chang, C. Structure space of model proteins: a principle component analysis. *J Chem Phys* **118**, 4277–4284 (2003).

8. Xia, Y. & Levitt, M. Simulating protein evolution in sequence and structure space. *Curr Opin Struct Biol* **14**, 202–207 (2004).

9. Dawkins, R. *The Selfish Gene* (Oxford University Press, Oxford, 1976).

10. Dawkins, R. *The Blind Watchmaker* (Longman, London, 1986).

11. Borges, J. L. in *Labyrinths: Selected Stories and Other Writings of Jorge Luis Borges* (eds. Yates, D. A. & Irby, J. E.) (New Directions, New York, 1962).

12. Lederberg, J. Instructive selection and immunological theory. *Immunol Rev* **185**, 50–53 (2002).

13. Holland, J. H. *Adaptation in Natural and Artificial Systems: An Introductory Analysis with Applications to Biology, Control, and Artificial Intelligence* (MIT Press, Cambridge, MA, 1992).

14. Poli, R., Langdon, W. B., & McPhee, N. F. *A Field Guide to Genetic Programming* (Lulu.com, 2008).

15. Weber, L. Applications of genetic algorithms in molecular diversity. *Curr Opin Chem Biol* **2**, 381–385 (1998).

16. Dennett, D. *Darwin's Dangerous Idea. Evolution and the Meanings of Life* (Simon and Schuster, 1995).

17. Dawkins, R. *The God Delusion* (Bantam Press, 2006).

18. Kauffman, S. *At Home in the Universe. The Search for the Laws of Self-Organization and Complexity* (Oxford University Press, 1995).

19. Maynard Smith, J. The units of selection. *Novartis Found Symp* **213**, 203–211; discussion 211–217 (1998).

20. Nowak, M. A. Five rules for the evolution of cooperation. *Science* **314**, 1560–1563 (2006).

Searching for Molecular Solutions: Empirical Discovery and Its Future. By Ian S. Dunn
Copyright © 2010 John Wiley & Sons, Inc.

21. Lehmann, L., Keller, L., West, S., & Roze, D. Group selection and kin selection: two concepts but one process. *Proc Natl Acad Sci USA* **104**, 6736–6739 (2007).

22. Lynch, M. The frailty of adaptive hypotheses for the origins of organismal complexity. *Proc Natl Acad Sci USA* **104** (Suppl 1), 8597–8604 (2007).

23. Voigt, C. A., Kauffman, S., & Wang, Z. G. Rational evolutionary design: the theory of *in vitro* protein evolution. *Adv Protein Chem* **55**, 79–160 (2000).

24. Jackel, C., Kast, P., & Hilvert, D. Protein design by directed evolution. *Annu Rev Biophys* **37**, 153–173 (2008).

25. Ayala, F. J. Darwin's greatest discovery: design without designer. *Proc Natl Acad Sci USA* **104** (Suppl 1), 8567–8573 (2007).

26. Sniegowski, P. D., Gerrish, P. J., Johnson, T., & Shaver, A. The evolution of mutation rates: separating causes from consequences. *Bioessays* **22**, 1057–1066 (2000).

27. Ellegren, H., Smith, N. G., & Webster, M. T. Mutation rate variation in the mammalian genome. *Curr Opin Genet Dev* **13**, 562–568 (2003).

28. Ellegren, H. Characteristics, causes and evolutionary consequences of male-biased mutation. *Proc Biol Sci* **274**, 1–10 (2007).

29. Kimura, M. Recent development of the neutral theory viewed from the Wrightian tradition of theoretical population genetics. *Proc Natl Acad Sci USA* **88**, 5969–5973 (1991).

30. Chamary, J. V., Parmley, J. L., & Hurst, L. D. Hearing silence: non-neutral evolution at synonymous sites in mammals. *Nat Rev Genet* **7**, 98–108 (2006).

31. Tsai, C. J., et al. Synonymous mutations and ribosome stalling can lead to altered folding pathways and distinct minima. *J Mol Biol* **383**, 281–291 (2008).

32. Kijak, G. H., et al. Lost in translation: implications of HIV-1 codon usage for immune escape and drug resistance. *AIDS Rev* **6**, 54–60 (2004).

33. Rutherford, S. L. & Lindquist, S. Hsp90 as a capacitor for morphological evolution. *Nature* **396**, 336–342 (1998).

34. Carey, C. C., Gorman, K. F., & Rutherford, S. Modularity and intrinsic evolvability of Hsp90-buffered change. *PLoS ONE* **1**, e76 (2006).

35. Lehman, N. A case for the extreme antiquity of recombination. *J Mol Evol* **56**, 770–777 (2003).

36. Hurles, M. How homologous recombination generates a mutable genome. *Hum Genomics* **2**, 179–186 (2005).

37. Kauffman, S. *Investigations* (Oxford University Press, 2000).

38. Szathmary, E. Why are there four letters in the genetic alphabet? *Nat Rev Genet* **4**, 995–1001 (2003).

39. Diamond, J. The curse of QWERTY. *Discover* **18**, 34–42 (1997).

40. Liebowitz, S. & Margolis, S. The fable of the keys. *J Law Econ* **33**, 1–25 (1990).

41. Hiesinger, P. R. The evolution of the millennium bug. *Riv Biol* **93**, 169–174 (2000).

42. McGrew, D. A. & Knight, K. L. Molecular design and functional organization of the RecA protein. *Crit Rev Biochem Mol Biol* **38**, 385–432 (2003).

43. Ikeda, H., Shiraishi, K., & Ogata, Y. Illegitimate recombination mediated by double-strand break and end-joining in *Escherichia coli*. *Adv Biophys* **38**, 3–20 (2004).

44. Long, M. Evolution of novel genes. *Curr Opin Genet Dev* **11**, 673–680 (2001).

45. Marcon, E. & Moens, P. B. The evolution of meiosis: recruitment and modification of somatic DNA-repair proteins. *Bioessays* **27**, 795–808 (2005).

46. Agrawal, A. F. Evolution of sex: why do organisms shuffle their genotypes? *Curr Biol* **16**, R696–R704 (2006).

47. Felsenstein, J. The evolutionary advantage of recombination. *Genetics* **78**, 737–756 (1974).

48. Chao, L. Fitness of RNA virus decreased by Muller's ratchet. *Nature* **348**, 454–455 (1990).

49. Duarte, E., Clarke, D., Moya, A., Domingo, E., & Holland, J. Rapid fitness losses in mammalian RNA virus clones due to Muller's ratchet. *Proc Natl Acad Sci USA* **89**, 6015–6019 (1992).

50. Paland, S. & Lynch, M. Transitions to asexuality result in excess amino acid substitutions. *Science* **311**, 990–992 (2006).

51. de Visser, J. A. & Rozen, D. E. Limits to adaptation in asexual populations. *J Evol Biol* **18**, 779–788 (2005).

52. Keightley, P. D. & Otto, S. P. Interference among deleterious mutations favours sex and recombination in finite populations. *Nature* **443**, 89–92 (2006).

53. Ridley, M. *The Red Queen. Sex and the Evolution of Human Nature* (Penguin Books, 1993).

54. Birky, C. W., Jr. Sex: is giardia doing it in the dark? *Curr Biol* **15**, R56–R58 (2005).

55. Smith, R. J., Kamiya, T., & Horne, D. J. Living males of the 'ancient asexual' Darwinulidae (Ostracoda: Crustacea). *Proc Biol Sci* **273**, 1569–1578 (2006).

56. Butlin, R. Evolution of sex: the costs and benefits of sex: new insights from old asexual lineages. *Nat Rev Genet* **3**, 311–317 (2002).

57. Welch, D. M. & Meselson, M. Evidence for the evolution of bdelloid rotifers without sexual reproduction or genetic exchange. *Science* **288**, 1211–1215 (2000).

58. Arkhipova, I. & Meselson, M. Transposable elements in sexual and ancient asexual taxa. *Proc Natl Acad Sci USA* **97**, 14473–14477 (2000).

59. Dolgin, E. S. & Charlesworth, B. The fate of transposable elements in asexual populations. *Genetics* **174**, 817–827 (2006).

60. Schon, I. & Martens, K. No slave to sex. *Proc Biol Sci* **270**, 827–833 (2003).

61. Gandolfi, A., Sanders, I. R., Rossi, V., & Menozzi, P. Evidence of recombination in putative ancient asexuals. *Mol Biol Evol* **20**, 754–761 (2003).

62. Page, S. L. & Hawley, R. S. Chromosome choreography: the meiotic ballet. *Science* **301**, 785–789 (2003).

63. Cavalier-Smith, T. Origins of the machinery of recombination and sex. *Heredity* **88**, 125–141 (2002).

64. Nouvel, P. The mammalian genome shaping activity of reverse transcriptase. *Genetica* **93**, 191–201 (1994).

65. Hughes, A. L. The evolution of functionally novel proteins after gene duplication. *Proc Biol Sci* **256**, 119–124 (1994).

66. Cheng, Z., et al. A genome-wide comparison of recent chimpanzee and human segmental duplications. *Nature* **437**, 88–93 (2005).

67. Simillion, C., Vandepoele, K., Van Montagu, M. C., Zabeau, M., & Van de Peer, Y. The hidden duplication past of *Arabidopsis thaliana*. *Proc Natl Acad Sci USA* **99**, 13627–13632 (2002).

68. Semon, M. & Wolfe, K. H. Rearrangement rate following the whole-genome duplication in teleosts. *Mol Biol Evol* **24**, 860–867 (2007).

69. Cavalier-Smith, T. Economy, speed and size matter: evolutionary forces driving nuclear genome miniaturization and expansion. *Ann Bot (Lond)* **95**, 147–175 (2005).

70. Hurles, M. Gene duplication: the genomic trade in spare parts. *PLoS Biol* **2**, E206 (2004).

71. Shakhnovich, B. E. & Koonin, E. V. Origins and impact of constraints in evolution of gene families. *Genome Res* **16**, 1529–1536 (2006).

72. Sugino, R. P. & Innan, H. Selection for more of the same product as a force to enhance concerted evolution of duplicated genes. *Trends Genet* **22**, 642–644 (2006).

73. Papp, B., Pal, C., & Hurst, L. D. Dosage sensitivity and the evolution of gene families in yeast. *Nature* **424**, 194–197 (2003).

74. Lynch, M. & Conery, J. S. The evolutionary fate and consequences of duplicate genes. *Science* **290**, 1151–1155 (2000).

75. Cusack, B. P. & Wolfe, K. H. When gene marriages don't work out: divorce by subfunctionalization. *Trends Genet* **23**, 270–272 (2007).

76. Force, A., et al. Preservation of duplicate genes by complementary, degenerative mutations. *Genetics* **151**, 1531–1545 (1999).

77. Chothia, C., Gough, J., Vogel, C., & Teichmann, S. A. Evolution of the protein repertoire. *Science* **300**, 1701–1703 (2003).

78. Andreeva, A. & Murzin, A. G. Evolution of protein fold in the presence of functional constraints. *Curr Opin Struct Biol* **16**, 399–408 (2006).

79. Vogel, C. & Morea, V. Duplication, divergence and formation of novel protein topologies. *Bioessays* **28**, 973–978 (2006).

80. Martin, A. C., et al. Protein folds and functions. *Structure* **6**, 875–884 (1998).

81. Orengo, C. A. & Thornton, J. M. Protein families and their evolution—a structural perspective. *Annu Rev Biochem* **74**, 867–900 (2005).

82. Kolodny, R., Petrey, D., & Honig, B. Protein structure comparison: implications for the nature of 'fold space', and structure and function prediction. *Curr Opin Struct Biol* **16**, 393–398 (2006).

83. Peisajovich, S. G., Rockah, L., & Tawfik, D. S. Evolution of new protein topologies through multistep gene rearrangements. *Nat Genet* **38**, 168–174 (2006).

84. He, X. & Zhang, J. Transcriptional reprogramming and backup between duplicate genes: is it a genomewide phenomenon? *Genetics* **172**, 1363–1367 (2006).

85. Lander, E. S. & Schork, N. J. Genetic dissection of complex traits. *Science* **265**, 2037–2048 (1994).

86. Yadav, D. & Sarvetnick, N. Cytokines and autoimmunity: redundancy defines their complex nature. *Curr Opin Immunol* **15**, 697–703 (2003).

87. Copp, A. J. Death before birth: clues from gene knockouts and mutations. *Trends Genet* **11**, 87–93 (1995).

88. Gu, Z., et al. Role of duplicate genes in genetic robustness against null mutations. *Nature* **421**, 63–66 (2003).

89. Wagner, A. & Wright, J. Alternative routes and mutational robustness in complex regulatory networks. *Biosystems* **88**, 163–172 (2007).

90. Harrison, R., Papp, B., Pal, C., Oliver, S. G., & Delneri, D. Plasticity of genetic interactions in metabolic networks of yeast. *Proc Natl Acad Sci USA* **104**, 2307–2312 (2007).

91. Gould, S. J. & Eldredge, N. Punctuated equilibrium comes of age. *Nature* **366**, 223–227 (1993).

92. Elena, S. F., Cooper, V. S., & Lenski, R. E. Punctuated evolution caused by selection of rare beneficial mutations. *Science* **272**, 1802–1804 (1996).

93. Gehring, W. J. Homeo boxes in the study of development. *Science* **236**, 1245–1252 (1987).

94. Dietrich, M. R. Richard Goldschmidt: hopeful monsters and other 'heresies'. *Nat Rev Genet* **4**, 68–74 (2003).

95. Geant, E., Mouchel-Vielh, E., Coutanceau, J. P., Ozouf-Costaz, C. & Deutsch, J. S. Are Cirripedia hopeful monsters? Cytogenetic approach and evidence for a Hox gene cluster in the cirripede crustacean *Sacculina carcini*. *Dev Genes Evol* **216**, 443–449 (2006).

96. Kazazian, H. H., Jr. Mobile elements: drivers of genome evolution. *Science* **303**, 1626–1632 (2004).

97. Erwin, D. H. & Valentine, J. W. "Hopeful monsters," transposons, and metazoan radiation. *Proc Natl Acad Sci USA* **81**, 5482–5483 (1984).

98. Hughes, A. L. Genomic catastrophism and the origin of vertebrate immunity. *Arch Immunol Ther Exp (Warsz)* **47**, 347–353 (1999).

99. Ochman, H., Lerat, E., & Daubin, V. Examining bacterial species under the specter of gene transfer and exchange. *Proc Natl Acad Sci USA* **102** (Suppl 1), 6595–6599 (2005).

100. Butlin, R. K. Recombination and speciation. *Mol Ecol* **14**, 2621–2635 (2005).

101. Woese, C. The universal ancestor. *Proc Natl Acad Sci USA* **95**, 6854–6859 (1998).

102. Matic, I., Rayssiguier, C., & Radman, M. Interspecies gene exchange in bacteria: the role of SOS and mismatch repair systems in evolution of species. *Cell* **80**, 507–515 (1995).

103. Margulis, L. & Bermudes, D. Symbiosis as a mechanism of evolution: status of cell symbiosis theory. *Symbiosis* **1**, 101–124 (1985).

104. Ponting, C. P. Plagiarized bacterial genes in the human book of life. *Trends Genet* **17**, 235–237 (2001).

105. Skaar, E. P. & Seifert, H. S. The misidentification of bacterial genes as human cDNAs: was the human D-1 tumor antigen gene acquired from bacteria? *Genomics* **79**, 625–627 (2002).

106. Stanhope, M. J., et al. Phylogenetic analyses do not support horizontal gene transfers from bacteria to vertebrates. *Nature* **411**, 940–944 (2001).

107. Ge, F., Wang, L. S., & Kim, J. The cobweb of life revealed by genome-scale estimates of horizontal gene transfer. *PLoS Biol* **3**, e316 (2005).

108. Keller, E. F. *A Feeling for the Organism. The Life and Work of Barbara McClintock* (W. H. Freeman, New York, 1983).

109. Schatz, D. G. Antigen receptor genes and the evolution of a recombinase. *Semin Immunol* **16**, 245–256 (2004).

110. Lirrmiyarri, G. M. & Lofts, P. *How the Kangaroos Got Their Tails* (Scholastic, 1992).

111. Klose, R. J. & Bird, A. P. Genomic DNA methylation: the mark and its mediators. *Trends Biochem Sci* **31**, 89–97 (2006).

112. Morgan, H. D., Santos, F., Green, K., Dean, W., & Reik, W. Epigenetic reprogramming in mammals. *Hum Mol Genet* **14** (Spec No 1), R47–R58 (2005).

113. Zilberman, D. & Henikoff, S. Epigenetic inheritance in *Arabidopsis*: selective silence. *Curr Opin Genet Dev* **15**, 557–562 (2005).

114. Molinier, J., Ries, G., Zipfel, C., & Hohn, B. Transgeneration memory of stress in plants. *Nature* **442**, 1046–1049 (2006).

115. Chong, S. & Whitelaw, E. Epigenetic germline inheritance. *Curr Opin Genet Dev* **14**, 692–696 (2004).

116. Kaneda, M., et al. Essential role for *de novo* DNA methyltransferase Dnmt3a in paternal and maternal imprinting. *Nature* **429**, 900–903 (2004).

117. Arima, T., et al. Loss of the maternal imprint in Dnmt3Lmat-/- mice leads to a differentiation defect in the extraembryonic tissue. *Dev Biol* **297**, 361–373 (2006).

118. Wrzeska, M. & Rejduch, B. Genomic imprinting in mammals. *J Appl Genet* **45**, 427–433 (2004).

119. Lander, E. S., et al. Initial sequencing and analysis of the human genome. *Nature* **409**, 860–921 (2001).

120. Venter, J. C., et al. The sequence of the human genome. *Science* **291**, 1304–1351 (2001).

121. Chimpanzee Sequencing and Analysis Consortium. Initial sequence of the chimpanzee genome and comparison with the human genome. *Nature* **437**, 69–87 (2005).

122. Khaitovich, P., Enard, W., Lachmann, M., & Paabo, S. Evolution of primate gene expression. *Nat Rev Genet* **7**, 693–702 (2006).

123. Sakate, R., et al. Mapping of chimpanzee full-length cDNAs onto the human genome unveils large potential divergence of the transcriptome. *Gene* (2007).

124. Kauffman, S. *The Origins of Order. Self-Organization and Selection in Evolution* (Oxford University Press, 1993).

125. Hodges, A. *Alan Turing: The Enigma* (Simon and Schuster, New York, 1983).

126. Dulos, E., Boissonade, J., Perraud, J. J., Rudovics, B., & De Kepper, P. Chemical morphogenesis: turing patterns in an experimental chemical system. *Acta Biotheor* **44**, 249–261 (1996).

127. Sick, S., Reinker, S., Timmer, J., & Schlake, T. WNT and DKK determine hair follicle spacing through a reaction–diffusion mechanism. *Science* **314**, 1447–1450 (2006).

128. Wang, Y., Badea, T., & Nathans, J. Order from disorder: self-organization in mammalian hair patterning. *Proc Natl Acad Sci USA* **103**, 19800–19805 (2006).

129. Wolfram, S. Cellular automata as models of complexity. *Nature* **341**, 419–424 (1984).

130. Gordon, R. Evolution escapes rugged fitness landscapes by gene or genome doubling: the blessing of higher dimensionality. *Comput Chem* **18**, 325–331 (1994).

131. Sievers, D. & von Kiedrowski, G. Self-replication of complementary nucleotide-based oligomers. *Nature* **369**, 221–224 (1994).

132. Lee, D. H., Severin, K., & Ghadiri, M. R. Autocatalytic networks: the transition from molecular self-replication to molecular ecosystems. *Curr Opin Chem Biol* **1**, 491–496 (1997).

133. Gilbert, W. The RNA world. *Nature* **319**, 618 (1986).

134. Goldstein, R. A. Emergent robustness in competition between autocatalytic chemical networks. *Orig Life Evol Biosph* **36**, 381–389 (2006).

135. Shapiro, R. Small molecule interactions were central to the origin of life. *Q Rev Biol* **81**, 105–125 (2006).

136. Chen, I. A., Roberts, R. W., & Szostak, J. W. The emergence of competition between model protocells. *Science* **305**, 1474–1476 (2004).

137. Anet, F. A. The place of metabolism in the origin of life. *Curr Opin Chem Biol* **8**, 654–659 (2004).

138. McFadden, J. *Quantum Evolution. Life in the Multiverse* (HarperCollins, 2000).

139. Davies, P. C. Does quantum mechanics play a non-trivial role in life? *Biosystems* **78**, 69–79 (2004).

140. Gould, S. J. *Wonderful Life. The Burgess Shale and the Nature of History* (Penguin Books, 1991).

141. Fontana, W. & Buss, L. W. What would be conserved if "the tape were played twice"? *Proc Natl Acad Sci USA* **91**, 757–761 (1994).

142. Travisano, M., Mongold, J. A., Bennett, A. F., & Lenski, R. E. Experimental tests of the roles of adaptation, chance, and history in evolution. *Science* **267**, 87–90 (1995).

143. Woods, R., Schneider, D., Winkworth, C. L., Riley, M. A., & Lenski, R. E. Tests of parallel molecular evolution in a long-term experiment with *Escherichia coli. Proc Natl Acad Sci USA* **103**, 9107–9112 (2006).

144. Riley, M. S., Cooper, V. S., Lenski, R. E., Forney, L. J., & Marsh, T. L. Rapid phenotypic change and diversification of a soil bacterium during 1000 generations of experimental evolution. *Microbiology* **147**, 995–1006 (2001).

145. Adami, C. What is complexity? *Bioessays* **24**, 1085–1094 (2002).

146. Szathmary, E. & Smith, J. M. The major evolutionary transitions. *Nature* **374**, 227–232 (1995).

147. Gregory, T. R. Coincidence, coevolution, or causation? DNA content, cell size, and the *C*-value enigma. *Biol Rev Camb Philos Soc* **76**, 65–101 (2001).

148. Shannon, C. A mathematical theory of communication. *Bell Syst Tech J* **27**, 379–423, 623–656 (1948).

149. Adami, C., Ofria, C., & Collier, T. C. Evolution of biological complexity. *Proc Natl Acad Sci USA* **97**, 4463–4468 (2000).

150. Ricard, J. What do we mean by biological complexity? *C R Biol* **326**, 133–140 (2003).

151. He, X. & Zhang, J. Gene complexity and gene duplicability. *Curr Biol* **15**, 1016–1021 (2005).

152. Soyer, O. S. & Bonhoeffer, S. Evolution of complexity in signaling pathways. *Proc Natl Acad Sci USA* **103**, 16337–16342 (2006).

153. Zuckerkandl, E. Natural restoration can generate biological complexity. *Complexity* **11**, 14–27 (2005).

154. Kirschner, M. & Gerhart, J. Evolvability. *Proc Natl Acad Sci USA* **95**, 8420–8427 (1998).

155. Earl, D. J. & Deem, M. W. Evolvability is a selectable trait. *Proc Natl Acad Sci USA* **101**, 11531–11536 (2004).

156. Kelley, B. P., et al. Conserved pathways within bacteria and yeast as revealed by global protein network alignment. *Proc Natl Acad Sci USA* **100**, 11394–11399 (2003).

157. Conant, G. C. & Wagner, A. Convergent evolution of gene circuits. *Nat Genet* **34**, 264–266 (2003).

158. Bull, J. J., et al. Exceptional convergent evolution in a virus. *Genetics* **147**, 1497–1507 (1997).

159. Wichman, H. A., Millstein, J., & Bull, J. J. Adaptive molecular evolution for 13,000 phage generations: a possible arms race. *Genetics* **170**, 19–31 (2005).

160. Arnold, F. H. When blind is better: protein design by evolution. *Nat Biotechnol* **16**, 617–618 (1998).

161. Arnold, F. H., Giver, L., Gershenson, A., Zhao, H., & Miyazaki, K. Directed evolution of mesophilic enzymes into their thermophilic counterparts. *Ann NY Acad Sci* **870**, 400–403 (1999).

162. Trifonov, E. N. The triplet code from first principles. *J Biomol Struct Dyn* **22**, 1–11 (2004).

163. Jordan, I. K., et al. A universal trend of amino acid gain and loss in protein evolution. *Nature* **433**, 633–638 (2005).

164. Hurst, L. D., Feil, E. J., & Rocha, E. P. Protein evolution. Causes of trends in amino-acid gain and loss. *Nature* **442**, E11–E12 (2005).

165. McDonald, J. H. Apparent trends of amino acid gain and loss in protein evolution due to nearly neutral variation. *Mol Biol Evol* **23**, 240–244 (2006).

166. Labeit, S. & Kolmerer, B. Titins: giant proteins in charge of muscle ultrastructure and elasticity. *Science* **270**, 293–296 (1995).

167. Vogel, C. & Chothia, C. Protein family expansions and biological complexity. *PLoS Comput Biol* **2**, e48 (2006).

168. Vogel, C., Berzuini, C., Bashton, M., Gough, J., & Teichmann, S. A. Supra-domains: evolutionary units larger than single protein domains. *J Mol Biol* **336**, 809–823 (2004).

169. Andreeva, A., et al. Data growth and its impact on the SCOP database: new developments. *Nucleic Acids Res* **36**, D419–D425 (2008).

170. Yan, Y. & Moult, J. Protein family clustering for structural genomics. *J Mol Biol* **353**, 744–759 (2005).

171. Yooseph, S., et al. The Sorcerer II Global Ocean Sampling Expedition: expanding the universe of protein families. *PLoS Biol* **5**, e16 (2007).

172. Arnold, F. H. Fancy footwork in the sequence space shuffle. *Nat Biotechnol* **24**, 328–330 (2006).

173. Herbeck, J. T. & Wall, D. P. Converging on a general model of protein evolution. *Trends Biotechnol* **23**, 485–487 (2005).

174. Bloom, J. D., Labthavikul, S. T., Otey, C. R., & Arnold, F. H. Protein stability promotes evolvability. *Proc Natl Acad Sci USA* **103**, 5869–5874 (2006).

175. Li, H., Helling, R., Tang, C., & Wingreen, N. Emergence of preferred structures in a simple model of protein folding. *Science* **273**, 666–669 (1996).

176. Hirsh, A. E. & Fraser, H. B. Protein dispensability and rate of evolution. *Nature* **411**, 1046–1049 (2001).

177. Fang, G., Rocha, E., & Danchin, A. How essential are nonessential genes? *Mol Biol Evol* **22**, 2147–2156 (2005).

178. Jordan, I. K., Rogozin, I. B., Wolf, Y. I., & Koonin, E. V. Essential genes are more evolutionarily conserved than are nonessential genes in bacteria. *Genome Res* **12**, 962–968 (2002).

179. Wall, D. P., et al. Functional genomic analysis of the rates of protein evolution. *Proc Natl Acad Sci USA* **102**, 5483–5488 (2005).

180. Zhang, J. & He, X. Significant impact of protein dispensability on the instantaneous rate of protein evolution. *Mol Biol Evol* **22**, 1147–1155 (2005).

181. Yang, J., Gu, Z., & Li, W. H. Rate of protein evolution versus fitness effect of gene deletion. *Mol Biol Evol* **20**, 772–774 (2003).

182. Zuckerkandl, E. Evolutionary processes and evolutionary noise at the molecular level. I. Functional density in proteins. *J Mol Evol* **7**, 167–183 (1976).

183. Lin, Y. S., Hsu, W. L., Hwang, J. K., & Li, W. H. Proportion of solvent-exposed amino acids in a protein and rate of protein evolution. *Mol Biol Evol* **24**, 1005–1011 (2007).

184. Drummond, D. A., Bloom, J. D., Adami, C., Wilke, C. O., & Arnold, F. H. Why highly expressed proteins evolve slowly. *Proc Natl Acad Sci USA* **102**, 14338–14343 (2005).

185. Hahn, M. W., Conant, G. C., & Wagner, A. Molecular evolution in large genetic networks: does connectivity equal constraint? *J Mol Evol* **58**, 203–211 (2004).

186. Jordan, I. K., Wolf, Y. I., & Koonin, E. V. No simple dependence between protein evolution rate and the number of protein–protein interactions: only the most prolific interactors tend to evolve slowly. *BMC Evol Biol* **3**, 1 (2003).

187. Shi, T., Bibby, T. S., Jiang, L., Irwin, A. J., & Falkowski, P. G. Protein interactions limit the rate of evolution of photosynthetic genes in cyanobacteria. *Mol Biol Evol* **22**, 2179–2189 (2005).

188. Mintseris, J. & Weng, Z. Structure, function, and evolution of transient and obligate protein–protein interactions. *Proc Natl Acad Sci USA* **102**, 10930–10935 (2005).

189. Price, N. P. & Stevens, L. *Fundamentals of Enzymology: The Cell and Molecular Biology of Catalytic Proteins*, 3rd edition (Oxford University Press, 1999).

190. Glasner, M. E., Gerlt, J. A., & Babbitt, P. C. Evolution of enzyme superfamilies. *Curr Opin Chem Biol* **10**, 492–497 (2006).

191. Howell, E. E. Searching sequence space: two different approaches to dihydrofolate reductase catalysis. *Chembiochem* **6**, 590–600 (2005).

192. Anantharaman, V., Aravind, L., & Koonin, E. V. Emergence of diverse biochemical activities in evolutionarily conserved structural scaffolds of proteins. *Curr Opin Chem Biol* **7**, 12–20 (2003).

193. Jadhav, V. R. & Yarus, M. Coenzymes as coribozymes. *Biochimie* **84**, 877–888 (2002).

194. Doudna, J. A. & Lorsch, J. R. Ribozyme catalysis: not different, just worse. *Nat Struct Mol Biol* **12**, 395–402 (2005).

195. Gutteridge, A. & Thornton, J. M. Understanding nature's catalytic toolkit. *Trends Biochem Sci* **30**, 622–629 (2005).

196. Lee, S. C. & Holm, R. H. Speculative synthetic chemistry and the nitrogenase problem. *Proc Natl Acad Sci USA* **100**, 3595–3600 (2003).

197. Rees, D. C. & Howard, J. B. The interface between the biological and inorganic worlds: iron–sulfur metalloclusters. *Science* **300**, 929–931 (2003).

198. Fetzner, S. Oxygenases without requirement for cofactors or metal ions. *Appl Microbiol Biotechnol* **60**, 243–257 (2002).

199. Petrounia, I. P. & Arnold, F. H. Designed evolution of enzymatic properties. *Curr Opin Biotechnol* **11**, 325–330 (2000).

200. Joo, H., Lin, Z., & Arnold, F. H. Laboratory evolution of peroxide-mediated cytochrome P450 hydroxylation. *Nature* **399**, 670–673 (1999).

201. Gonzalez, J. C., Banerjee, R. V., Huang, S., Sumner, J. S., & Matthews, R. G. Comparison of cobalamin-independent and cobalamin-dependent methionine synthases from *Escherichia coli*: two solutions to the same chemical problem. *Biochemistry* **31**, 6045–6056 (1992).

202. Ma, B. G., et al. Characters of very ancient proteins. *Biochem Biophys Res Commun* **366**, 607–611 (2008).

203. Bottoms, C. A., Smith, P. E., & Tanner, J. J. A structurally conserved water molecule in Rossmann dinucleotide-binding domains. *Protein Sci* **11**, 2125–2137 (2002).

204. Tang, M., et al. Antiplatelet agents aspirin and clopidogrel are hydrolyzed by distinct carboxylesterases, and clopidogrel is transesterificated in the presence of ethyl alcohol. *J Pharmacol Exp Ther* **319**, 1467–1476 (2006).

205. Imai, T., Taketani, M., Shii, M., Hosokawa, M., & Chiba, K. Substrate specificity of carboxylesterase isozymes and their contribution to hydrolase activity in human liver and small intestine. *Drug Metab Dispos* **34**, 1734–1741 (2006).

206. Klibanov, A. M. Enzymatic catalysis in anhydrous organic solvents. *Trends Biochem Sci* **14**, 141–144 (1989).

207. Schindelin, H., Kisker, C., & Rajagopalan, K. V. Molybdopterin from molybdenum and tungsten enzymes. *Adv Protein Chem* **58**, 47–94 (2001).

208. Corbett, M. C., et al. Structural insights into a protein-bound iron–molybdenum cofactor precursor. *Proc Natl Acad Sci USA* **103**, 1238–1243 (2006).

209. Schuster, P. Evolution and design. The Darwinian view of evolution is a scientific fact and not an ideology. *Complexity* **11**, 12–15 (2006).

210. Sober, E. What is wrong with intelligent design? *Q Rev Biol* **82**, 3–8 (2007).

211. Gophna, U., Ron, E. Z., & Graur, D. Bacterial type III secretion systems are ancient and evolved by multiple horizontal-transfer events. *Gene* **312**, 151–163 (2003).

212. Lenski, R. E., Ofria, C., Pennock, R. T., & Adami, C. The evolutionary origin of complex features. *Nature* **423**, 139–144 (2003).

213. Adami, C. Digital genetics: unravelling the genetic basis of evolution. *Nat Rev Genet* **7**, 109–118 (2006).

214. Wilson, E. O. *The Diversity of Life* (Penguin Books, 1993).

215. Asimov, I. *The Roving Mind* (Prometheus Books, Buffalo, NY, 1983).

216. Boehm, T. Quality control in self/nonself discrimination. *Cell* **125**, 845–858 (2006).

217. Sonoda, J. & Evans, R. M. Biological function and mode of action of nuclear xenobiotic receptors. *Pure Appl Chem* **75**, 1733–1742 (2003).

218. Tenbaum, S. & Baniahmad, A. Nuclear receptors: structure, function and involvement in disease. *Int J Biochem Cell Biol* **29**, 1325–1341 (1997).

219. Hager, G. L., Lim, C. S., Elbi, C., & Baumann, C. T. Trafficking of nuclear receptors in living cells. *J Steroid Biochem Mol Biol* **74**, 249–254 (2000).

220. Francis, G. A., Fayard, E., Picard, F., & Auwerx, J. Nuclear receptors and the control of metabolism. *Annu Rev Physiol* **65**, 261–311 (2003).

221. Berkenstam, A. & Gustafsson, J. A. Nuclear receptors and their relevance to diseases related to lipid metabolism. *Curr Opin Pharmacol* **5**, 171–176 (2005).

222. Kliewer, S. A., Lehmann, J. M., & Willson, T. M. Orphan nuclear receptors: shifting endocrinology into reverse. *Science* **284**, 757–760 (1999).

223. Danielson, P. B. The cytochrome P450 superfamily: biochemistry, evolution and drug metabolism in humans. *Curr Drug Metab* **3**, 561–597 (2002).

224. Wang, H. & LeCluyse, E. L. Role of orphan nuclear receptors in the regulation of drug-metabolising enzymes. *Clin Pharmacokinet* **42**, 1331–1357 (2003).

225. Mohan, R. & Heyman, R. A. Orphan nuclear receptor modulators. *Curr Top Med Chem* **3**, 1637–1647 (2003).

226. Reschly, E. J. & Krasowski, M. D. Evolution and function of the NR1I nuclear hormone receptor subfamily (VDR, PXR, and CAR) with respect to metabolism of xenobiotics and endogenous compounds. *Curr Drug Metab* **7**, 349–365 (2006).

227. Kliewer, S. A., Goodwin, B., & Willson, T. M. The nuclear pregnane X receptor: a key regulator of xenobiotic metabolism. *Endocr Rev* **23**, 687–702 (2002).

228. Orans, J., Teotico, D. G., & Redinbo, M. R. The nuclear xenobiotic receptor pregnane X receptor: recent insights and new challenges. *Mol Endocrinol* **19**, 2891–2900 (2005).

229. Timsit, Y. E. & Negishi, M. CAR and PXR: the xenobiotic-sensing receptors. *Steroids* **72**, 231–246 (2007).

230. Willson, T. M. & Moore, J. T. Genomics versus orphan nuclear receptors: a half-time report. *Mol Endocrinol* **16**, 1135–1144 (2002).

231. Carnahan, V. E. & Redinbo, M. R. Structure and function of the human nuclear xenobiotic receptor PXR. *Curr Drug Metab* **6**, 357–367 (2005).

232. Watkins, R. E., Noble, S. M., & Redinbo, M. R. Structural insights into the promiscuity and function of the human pregnane X receptor. *Curr Opin Drug Discov Devel* **5**, 150–158 (2002).

233. Zhang, Z., et al. Genomic analysis of the nuclear receptor family: new insights into structure, regulation, and evolution from the rat genome. *Genome Res* **14**, 580–590 (2004).

234. Moore, J. T. & Kliewer, S. A. Use of the nuclear receptor PXR to predict drug interactions. *Toxicology* **153**, 1–10 (2000).

235. Lamba, J., Lamba, V., & Schuetz, E. Genetic variants of PXR (NR1I2) and CAR (NR1I3) and their implications in drug metabolism and pharmacogenetics. *Curr Drug Metab* **6**, 369–383 (2005).

236. Borges-Walmsley, M. I., McKeegan, K. S. & Walmsley, A. R. Structure and function of efflux pumps that confer resistance to drugs. *Biochem J* **376**, 313–338 (2003).

237. Donadio, S., Staver, M. J., McAlpine, J. B., Swanson, S. J., & Katz, L. Modular organization of genes required for complex polyketide biosynthesis. *Science* **252**, 675–679 (1991).

238. Cerutti, H. & Casas-Mollano, J. A. On the origin and functions of RNA-mediated silencing: from protists to man. *Curr Genet* **50**, 81–99 (2006).

239. Turelli, P. & Trono, D. Editing at the crossroad of innate and adaptive immunity. *Science* **307**, 1061–1065 (2005).

240. Bass, B. L. RNA editing by adenosine deaminases that act on RNA. *Annu Rev Biochem* **71**, 817–846 (2002).

241. Kawahara, Y., et al. Redirection of silencing targets by adenosine-to-inosine editing of miRNAs. *Science* **315**, 1137–1140 (2007).

242. Malim, M. H. Natural resistance to HIV infection: the Vif–APOBEC interaction. *C R Biol* **329**, 871–875 (2006).

243. Janeway, C. A., Jr., & Medzhitov, R. Innate immune recognition. *Annu Rev Immunol* **20**, 197–216 (2002).

244. Zipfel, C., et al. Bacterial disease resistance in *Arabidopsis* through flagellin perception. *Nature* **428**, 764–767 (2004).

245. Li, X., et al. Flagellin induces innate immunity in nonhost interactions that is suppressed by *Pseudomonas syringae* effectors. *Proc Natl Acad Sci USA* **102**, 12990–12995 (2005).

246. Jones, J. D. & Dangl, J. L. The plant immune system. *Nature* **444**, 323–329 (2006).

247. Paul, W. E. *Fundamental Immunology* (Lippincott Williams & Wilkins, Philadelphia, PA, 2003).

248. McFall-Ngai, M. Adaptive immunity: care for the community. *Nature* **445**, 153 (2007).

249. Loker, E. S., Adema, C. M., Zhang, S. M., & Kepler, T. B. Invertebrate immune systems: not homogeneous, not simple, not well understood. *Immunol Rev* **198**, 10–24 (2004).

250. Cannon, J. P., Haire, R. N., & Litman, G. W. Identification of diversified genes that contain immunoglobulin-like variable regions in a protochordate. *Nat Immunol* **3**, 1200–1207 (2002).

251. Sadd, B. M. & Siva-Jothy, M. T. Self-harm caused by an insect's innate immunity. *Proc Biol Sci* **273**, 2571–2574 (2006).

252. Medzhitov, R. & Janeway, C. A., Jr. Decoding the patterns of self and nonself by the innate immune system. *Science* **296**, 298–300 (2002).

253. Jackson, K. J., Gaeta, B., Sewell, W., & Collins, A. M. Exonuclease activity and P nucleotide addition in the generation of the expressed immunoglobulin repertoire. *BMC Immunol* **5**, 19 (2004).

254. Benedict, C. L., Gilfillan, S., Thai, T. H. & Kearney, J. F. Terminal deoxynucleotidyl transferase and repertoire development. *Immunol Rev* **175**, 150–157 (2000).

255. Gerdes, T. & Wabl, M. Autoreactivity and allelic inclusion in a B cell nuclear transfer mouse. *Nat Immunol* **5**, 1282–1287 (2004).

256. Padovan, E., et al. Normal T lymphocytes can express two different T cell receptor beta chains: implications for the mechanism of allelic exclusion. *J Exp Med* **181**, 1587–1591 (1995).

257. Di Noia, J. M. & Neuberger, M. S. Molecular mechanisms of antibody somatic hypermutation. *Annu Rev Biochem* (2007).

258. Janeway, C. A., Jr., Travers, P., Walport, M., & Shlomchik, M. *Immunobiology. The Immune System in Health and Disease* (Garland Science Publishers, 2004).

259. Longo, N. S. & Lipsky, P. E. Why do B cells mutate their immunoglobulin receptors? *Trends Immunol* **27**, 374–380 (2006).

260. Shapiro, G. S. & Wysocki, L. J. DNA target motifs of somatic mutagenesis in antibody genes. *Crit Rev Immunol* **22**, 183–200 (2002).

261. Clark, L. A., Ganesan, S., Papp, S., & van Vlijmen, H. W. Trends in antibody sequence changes during the somatic hypermutation process. *J Immunol* **177**, 333–340 (2006).

262. Wedemayer, G. J., Patten, P. A., Wang, L. H., Schultz, P. G., & Stevens, R. C. Structural insights into the evolution of an antibody combining site. *Science* **276**, 1665–1669 (1997).

263. Schultz, P. G., Yin, J., & Lerner, R. A. The chemistry of the antibody molecule. *Angew Chem Int Ed Engl* **41**, 4427–4437 (2002).

264. Sundberg, E. J. & Mariuzza, R. A. Molecular recognition in antibody–antigen complexes. *Adv Protein Chem* **61**, 119–160 (2002).

265. Zimmermann, J., et al. Antibody evolution constrains conformational heterogeneity by tailoring protein dynamics. *Proc Natl Acad Sci USA* **103**, 13722–13727 (2006).

266. Nikolich-Zugich, J., Slifka, M. K. & Messaoudi, I. The many important facets of T-cell repertoire diversity. *Nat Rev Immunol* **4**, 123–132 (2004).

267. Davis, M. M. & Bjorkman, P. J. T-cell antigen receptor genes and T-cell recognition. *Nature* **334**, 395–402 (1988).

268. Huseby, E. S., et al. How the T cell repertoire becomes peptide and MHC specific. *Cell* **122**, 247–260 (2005).

269. Xu, J. L. & Davis, M. M. Diversity in the CDR3 region of V(H) is sufficient for most antibody specificities. *Immunity* **13**, 37–45 (2000).

270. Bankovich, A. J., Girvin, A. T., Moesta, A. K., & Garcia, K. C. Peptide register shifting within the MHC groove: theory becomes reality. *Mol Immunol* **40**, 1033–1039 (2004).

271. Falk, K., et al. Identification of naturally processed viral nonapeptides allows their quantification in infected cells and suggests an allele-specific T cell epitope forecast. *J Exp Med* **174**, 425–434 (1991).

272. Rudolph, M. G., Stanfield, R. L., & Wilson, I. A. How TCRs bind MHCs, peptides, and coreceptors. *Annu Rev Immunol* **24**, 419–466 (2006).

273. Cooper, M. D. & Alder, M. N. The evolution of adaptive immune systems. *Cell* **124**, 815–822 (2006).

274. Matsuda, F., et al. The complete nucleotide sequence of the human immunoglobulin heavy chain variable region locus. *J Exp Med* **188**, 2151–2162 (1998).

275. McMahon, S. A., et al. The C-type lectin fold as an evolutionary solution for massive sequence variation. *Nat Struct Mol Biol* **12**, 886–892 (2005).

276. Segel, L. A. & Perelson, A. S. Shape space: an approach to the evaluation of cross-reactivity effects, stability and controllability in the immune system. *Immunol Lett* **22**, 91–99 (1989).

277. James, L. C., Roversi, P., & Tawfik, D. S. Antibody multispecificity mediated by conformational diversity. *Science* **299**, 1362–1367 (2003).

278. Carneiro, J. & Stewart, J. Rethinking "shape space": evidence from simulated docking suggests that steric shape complementarity is not limiting for antibody–antigen recognition and idiotypic interactions. *J Theor Biol* **169**, 391–402 (1994).

279. Zhang, C. & Kim, S. H. A comprehensive analysis of the Greek key motifs in protein beta-barrels and beta-sandwiches. *Proteins* **40**, 409–419 (2000).

280. Batista, F. D. & Neuberger, M. S. Affinity dependence of the B cell response to antigen: a threshold, a ceiling, and the importance of off-rate. *Immunity* **8**, 751–759 (1998).

281. Boder, E. T., Midelfort, K. S., & Wittrup, K. D. Directed evolution of antibody fragments with monovalent femtomolar antigen-binding affinity. *Proc Natl Acad Sci USA* **97**, 10701–10705 (2000).

282. Lawrence, M. C. & Colman, P. M. Shape complementarity at protein/protein interfaces. *J Mol Biol* **234**, 946–950 (1993).

283. Difilippantonio, M. J., McMahan, C. J., Eastman, Q. M., Spanopoulou, E., & Schatz, D. G. RAG1 mediates signal sequence recognition and recruitment of RAG2 in V(D)J recombination. *Cell* **87**, 253–262 (1996).

284. Stoddard, B. L. Homing endonuclease structure and function. *Q Rev Biophys* **38**, 49–95 (2005).

285. Garvie, C. W. & Wolberger, C. Recognition of specific DNA sequences. *Mol Cell* **8**, 937–946 (2001).

286. Di Pietro, S. M., et al. Specific antibody–DNA interaction: a novel strategy for tight DNA recognition. *Biochemistry* **42**, 6218–6227 (2003).

287. Liu, C. C., et al. Protein evolution with an expanded genetic code. *Proc Natl Acad Sci USA* **105**, 17688–17693 (2008).

288. Bogen, B. & Weiss, S. Processing and presentation of idiotypes to MHC-restricted T cells. *Int Rev Immunol* **10**, 337–355 (1993).

289. Jerne, N. K. The generative grammar of the immune system. *Science* **229**, 1057–1059 (1985).

290. Perelson, A. S. Immune network theory. *Immunol Rev* **110**, 5–36 (1989).

291. Fields, B. A., Goldbaum, F. A., Ysern, X., Poljak, R. J., & Mariuzza, R. A. Molecular basis of antigen mimicry by an anti-idiotope. *Nature* **374**, 739–742 (1995).

292. Bentley, G. A., Boulot, G., Riottot, M. M., & Poljak, R. J. Three-dimensional structure of an idiotope–anti-idiotope complex. *Nature* **348**, 254–257 (1990).

293. Lundkvist, I., Coutinho, A., Varela, F., & Holmberg, D. Evidence for a functional idiotypic network among natural antibodies in normal mice. *Proc Natl Acad Sci USA* **86**, 5074–5078 (1989).

294. Prinz, D. M., Smithson, S. L., Kieber-Emmons, T. & Westerink, M. A. Induction of a protective capsular polysaccharide antibody response to a multiepitope DNA vaccine encoding a peptide mimic of meningococcal serogroup C capsular polysaccharide. *Immunology* **110**, 242–249 (2003).

295. Goldbaum, F. A., et al. Characterization of anti-anti-idiotypic antibodies that bind antigen and an anti-idiotype. *Proc Natl Acad Sci USA* **94**, 8697–8701 (1997).

296. Maruyama, H., et al. Cancer vaccines: single-epitope anti-idiotype vaccine versus multiple-epitope antigen vaccine. *Cancer Immunol Immunother* **49**, 123–132 (2000).

297. Kurzweil, R. *The Singularity is Near* (Penguin Books, New York, 2004).

298. Blalock, J. E. & Smith, E. M. Conceptual development of the immune system as a sixth sense. *Brain Behav Immun* **21**, 23–33 (2007).

299. Boehm, T. & Zufall, F. MHC peptides and the sensory evaluation of genotype. *Trends Neurosci* **29**, 100–107 (2006).

300. Thomas, L. *The Lives of a Cell* (Viking Press, 1974).

301. Li, X., et al. Pseudogenization of a sweet-receptor gene accounts for cats' indifference toward sugar. *PLoS Genet* **1**, 27–35 (2005).

302. Huang, A. L., et al. The cells and logic for mammalian sour taste detection. *Nature* **442**, 934–938 (2006).

303. Gabius, H. J. Biological information transfer beyond the genetic code: the sugar code. *Naturwissenschaften* **87**, 108–121 (2000).

304. Holt, C. E. & Dickson, B. J. Sugar codes for axons? *Neuron* **46**, 169–172 (2005).

305. Chandrashekar, J., Hoon, M. A., Ryba, N. J., & Zuker, C. S. The receptors and cells for mammalian taste. *Nature* **444**, 288–294 (2006).

306. Mueller, K. L., et al. The receptors and coding logic for bitter taste. *Nature* **434**, 225–229 (2005).

307. Inoue, M., Miyazaki, K., Uehara, H., Maruyama, M., & Hirama, M. First- and second-generation total synthesis of ciguatoxin CTX3C. *Proc Natl Acad Sci USA* **101**, 12013–12018 (2004).

308. Buck, L. B. Olfactory receptors and odor coding in mammals. *Nutr Rev* **62**, S184–S188; discussion S224–S241 (2004).

309. Malnic, B., Godfrey, P. A., & Buck, L. B. The human olfactory receptor gene family. *Proc Natl Acad Sci USA* **101**, 2584–2589 (2004).

310. Eggan, K., et al. Mice cloned from olfactory sensory neurons. *Nature* **428**, 44–49 (2004).

311. Hochedlinger, K. & Jaenisch, R. Monoclonal mice generated by nuclear transfer from mature B and T donor cells. *Nature* **415**, 1035–1038 (2002).

312. Malnic, B., Hirono, J., Sato, T., & Buck, L. B. Combinatorial receptor codes for odors. *Cell* **96**, 713–723 (1999).

313. Oka, Y., Omura, M., Kataoka, H., & Touhara, K. Olfactory receptor antagonism between odorants. *Embo J* **23**, 120–126 (2004).

314. Haddad, R., Lapid, H., Harel, D., & Sobel, N. Measuring smells. *Curr Opin Neurobiol* **18**, 438–444 (2008).

315. Elenkov, I. J., Wilder, R. L., Chrousos, G. P., & Vizi, E. S. The sympathetic nerve: an integrative interface between two supersystems: the brain and the immune system. *Pharmacol Rev* **52**, 595–638 (2000).

316. Pelkonen, O., Rautio, A., Raunio, H., & Pasanen, M. CYP2A6: a human coumarin 7-hydroxylase. *Toxicology* **144**, 139–147 (2000).

317. Killard, A. J., Keating, G. J., & O'Kennedy, R. Production and characterization of anti-coumarin scFv antibodies. *Biochem Soc Trans* **26**, S33 (1998).

318. Ui, M., et al. How protein recognizes ladder-like polycyclic ethers. Interactions between ciguatoxin (CTX3C) fragments and its specific antibody 10C9. *J Biol Chem* **283**, 19440–19447 (2008).

319. Trut, L. N. Early canid domestication: the farm fox experiment. *Am Sci* **87**, 160–169 (1999).

320. Lal, R., et al. Engineering antibiotic producers to overcome the limitations of classical strain improvement programs. *Crit Rev Microbiol* **22**, 201–255 (1996).

321. Brakmann, S. & Schweinhorst, A. (eds.) *Evolutionary Methods in Biotechnology. Clever Tricks for Directed Evolution* (Wiley-VCH, 2004).

322. Kunkel, T. A. & Bebenek, K. DNA replication fidelity. *Annu Rev Biochem* **69**, 497–529 (2000).

323. Horst, J. P., Wu, T. H., & Marinus, M. G. *Escherichia coli* mutator genes. *Trends Microbiol* **7**, 29–36 (1999).

324. Nguyen, A. W. & Daugherty, P. S. Production of randomly mutated plasmid libraries using mutator strains. *Methods Mol Biol* **231**, 39–44 (2003).

325. Fabret, C., et al. Efficient gene targeted random mutagenesis in genetically stable *Escherichia coli* strains. *Nucleic Acids Res* **28**, E95 (2000).

326. Neylon, C. Chemical and biochemical strategies for the randomization of protein encoding DNA sequences: library construction methods for directed evolution. *Nucleic Acids Res* **32**, 1448–1459 (2004).

327. Biles, B. D. & Connolly, B. A. Low-fidelity *Pyrococcus furiosus* DNA polymerase mutants useful in error-prone PCR. *Nucleic Acids Res* **32**, e176 (2004).

328. Murakami, H., Hohsaka, T., & Sisido, M. Random insertion and deletion of arbitrary number of bases for codon-based random mutation of DNAs. *Nat Biotechnol* **20**, 76–81 (2002).

329. Kashiwagi, K., Isogai, Y., Nishiguchi, K., & Shiba, K. Frame shuffling: a novel method for *in vitro* protein evolution. *Protein Eng Des Sel* **19**, 135–140 (2006).

330. Wong, T. S., Tee, K. L., Hauer, B., & Schwaneberg, U. Sequence saturation mutagenesis with tunable mutation frequencies. *Anal Biochem* **341**, 187–189 (2005).

331. Debrauwere, H., Gendrel, C. G., Lechat, S., & Dutreix, M. Differences and similarities between various tandem repeat sequences: minisatellites and microsatellites. *Biochimie* **79**, 577–586 (1997).

332. Dunn, I. S., Cowan, R., & Jennings, P. A. Improved peptide function from random mutagenesis over short 'windows'. *Protein Eng* **2**, 283–291 (1988).

333. Wells, J. A., Vasser, M., & Powers, D. B. Cassette mutagenesis: an efficient method for generation of multiple mutations at defined sites. *Gene* **34**, 315–323 (1985).

334. Reetz, M. T. & Carballeira, J. D. Iterative saturation mutagenesis (ISM) for rapid directed evolution of functional enzymes. *Nat Protoc* **2**, 891–903 (2007).

335. Virnekas, B., et al. Trinucleotide phosphoramidites: ideal reagents for the synthesis of mixed oligonucleotides for random mutagenesis. *Nucleic Acids Res* **22**, 5600–5607 (1994).

336. Wong, T. S., Roccatano, D., Zacharias, M., & Schwaneberg, U. A statistical analysis of random mutagenesis methods used for directed protein evolution. *J Mol Biol* **355**, 858–871 (2006).

337. Daugherty, P. S., Chen, G., Iverson, B. L., & Georgiou, G. Quantitative analysis of the effect of the mutation frequency on the affinity maturation of single chain Fv antibodies. *Proc Natl Acad Sci USA* **97**, 2029–2034 (2000).

338. Drummond, D. A., Iverson, B. L., Georgiou, G., & Arnold, F. H. Why high-error-rate random mutagenesis libraries are enriched in functional and improved proteins. *J Mol Biol* **350**, 806–816 (2005).

339. Zaccolo, M. & Gherardi, E. The effect of high-frequency random mutagenesis on *in vitro* protein evolution: a study on TEM-1 beta-lactamase. *J Mol Biol* **285**, 775–783 (1999).

340. Friedberg, E. C., Wagner, R., & Radman, M. Specialized DNA polymerases, cellular survival, and the genesis of mutations. *Science* **296**, 1627–1630 (2002).

341. Tippin, B., Pham, P., & Goodman, M. F. Error-prone replication for better or worse. *Trends Microbiol* **12**, 288–295 (2004).

342. Livneh, Z. Keeping mammalian mutation load in check: regulation of the activity of error-prone DNA polymerases by p53 and p21. *Cell Cycle* **5**, 1918–1922 (2006).

343. Smith, G. P. Applied evolution. The progeny of sexual PCR. *Nature* **370**, 324–325 (1994).

344. Stemmer, W. P. Rapid evolution of a protein *in vitro* by DNA shuffling. *Nature* **370**, 389–391 (1994).

345. Stemmer, W. P. DNA shuffling by random fragmentation and reassembly: *in vitro* recombination for molecular evolution. *Proc Natl Acad Sci USA* **91**, 10747–10751 (1994).

346. Ness, J. E., et al. DNA shuffling of subgenomic sequences of subtilisin. *Nat Biotechnol* **17**, 893–896 (1999).

347. Eijsink, V. G., Gaseidnes, S., Borchert, T. V., & van den Burg, B. Directed evolution of enzyme stability. *Biomol Eng* **22**, 21–30 (2005).

348. Crameri, A., Raillard, S. A., Bermudez, E., & Stemmer, W. P. DNA shuffling of a family of genes from diverse species accelerates directed evolution. *Nature* **391**, 288–291 (1998).

349. Ness, J. E., Del Cardayre, S. B., Minshull, J., & Stemmer, W. P. Molecular breeding: the natural approach to protein design. *Adv Protein Chem* **55**, 261–292 (2000).

350. Ness, J. E., et al. Synthetic shuffling expands functional protein diversity by allowing amino acids to recombine independently. *Nat Biotechnol* **20**, 1251–1255 (2002).

351. Zhao, H. & Arnold, F. H. Optimization of DNA shuffling for high fidelity recombination. *Nucleic Acids Res* **25**, 1307–1308 (1997).

352. Eggert, T., et al. Multiplex-PCR-based recombination as a novel high-fidelity method for directed evolution. *Chembiochem* **6**, 1062–1067 (2005).

353. Bogarad, L. D. & Deem, M. W. A hierarchical approach to protein molecular evolution. *Proc Natl Acad Sci USA* **96**, 2591–2595 (1999).

354. Gilbert, W. Why genes in pieces? *Nature* **271**, 501 (1978).

355. de Roos, A. D. Conserved intron positions in ancient protein modules. *Biol Direct* **2**, 7 (2007).

356. Kolkman, J. A. & Stemmer, W. P. Directed evolution of proteins by exon shuffling. *Nat Biotechnol* **19**, 423–428 (2001).

357. Bittker, J. A., Le, B. V., Liu, J. M., & Liu, D. R. Directed evolution of protein enzymes using nonhomologous random recombination. *Proc Natl Acad Sci USA* **101**, 7011–7016 (2004).

358. O'Maille, P. E., Bakhtina, M. & Tsai, M. D. Structure-based combinatorial protein engineering (SCOPE). *J Mol Biol* **321**, 677–691 (2002).

359. Silberg, J. J., Endelman, J. B., & Arnold, F. H. SCHEMA-guided protein recombination. *Methods Enzymol* **388**, 35–42 (2004).

360. Saraf, M. C., Horswill, A. R., Benkovic, S. J., & Maranas, C. D. FamClash: a method for ranking the activity of engineered enzymes. *Proc Natl Acad Sci USA* **101**, 4142–4147 (2004).

361. Moore, G. L. & Maranas, C. D. Identifying residue–residue clashes in protein hybrids by using a second-order mean-field approach. *Proc Natl Acad Sci USA* **100**, 5091–5096 (2003).

362. Binkowski, B. F., Richmond, K. E., Kaysen, J., Sussman, M. R., & Belshaw, P. J. Correcting errors in synthetic DNA through consensus shuffling. *Nucleic Acids Res* **33**, e55 (2005).

363. Smith, H. O., Hutchison, C. A., 3rd, Pfannkoch, C., & Venter, J. C. Generating a synthetic genome by whole genome assembly: phiX174 bacteriophage from synthetic oligonucleotides. *Proc Natl Acad Sci USA* **100**, 15440–15445 (2003).

364. Reyes Santos, J. & Haimes, Y. Y. Applying the partitioned multiobjective risk method (PMRM) to portfolio selection. *Risk Anal* **24**, 697–713 (2004).

365. Tabuchi, I., Soramoto, S., Ueno, S., & Husimi, Y. Multi-line split DNA synthesis: a novel combinatorial method to make high quality peptide libraries. *BMC Biotechnol* **4**, 19 (2004).

366. Kennedy, J., Eberhart, R. C., & Shi, Y. *Swarm Intelligence* (Morgan Kaufmann, San Francisco and London, 2001).

367. Schneider, G., et al. Voyages to the (un)known: adaptive design of bioactive compounds. *Trends Biotechnol* **27**, 18–26 (2009).

368. Waldrop, M. M. *Complexity: The Emerging Science at the Edge of Order and Chaos* (Simon & Schuster, 1992).

369. Santoro, S. W. & Schultz, P. G. Directed evolution of the substrate specificities of a site-specific recombinase and an aminoacyl-tRNA synthetase using fluorescence-activated cell sorting (FACS). *Methods Mol Biol* **230**, 291–312 (2003).

370. Schmidt-Dannert, C. & Arnold, F. H. Directed evolution of industrial enzymes. *Trends Biotechnol* **17**, 135–136 (1999).

371. Zhao, H. & Arnold, F. H. Directed evolution converts subtilisin E into a functional equivalent of thermitase. *Protein Eng* **12**, 47–53 (1999).

372. Van den Burg, B., Vriend, G., Veltman, O. R., Venema, G., & Eijsink, V. G. Engineering an enzyme to resist boiling. *Proc Natl Acad Sci USA* **95**, 2056–2060 (1998).

373. Van den Burg, B., de Kreij, A., Van der Veek, P., Mansfeld, J., & Venema, G. Characterization of a novel stable biocatalyst obtained by protein engineering. *Biotechnol Appl Biochem* **30** (Pt 1), 35–40 (1999).

374. Arnott, M. A., Michael, R. A., Thompson, C. R., Hough, D. W., & Danson, M. J. Thermostability and thermoactivity of citrate synthases from the thermophilic and hyperthermophilic archaea, *Thermoplasma acidophilum* and *Pyrococcus furiosus*. *J Mol Biol* **304**, 657–668 (2000).

375. Bjork, A., Dalhus, B., Mantzilas, D., Eijsink, V. G., & Sirevag, R. Stabilization of a tetrameric malate dehydrogenase by introduction of a disulfide bridge at the dimer–dimer interface. *J Mol Biol* **334**, 811–821 (2003).

376. Pedersen, J. S., Otzen, D. E., & Kristensen, P. Directed evolution of barnase stability using proteolytic selection. *J Mol Biol* **323**, 115–123 (2002).

377. Andrews, S. R., et al. The use of forced protein evolution to investigate and improve stability of family 10 xylanases. The production of Ca^{2+}-independent stable xylanases. *J Biol Chem* **279**, 54369–54379 (2004).

378. Zhang, J. H., Dawes, G., & Stemmer, W. P. Directed evolution of a fucosidase from a galactosidase by DNA shuffling and screening. *Proc Natl Acad Sci USA* **94**, 4504–4509 (1997).

379. Christians, F. C., Scapozza, L., Crameri, A., Folkers, G., & Stemmer, W. P. Directed evolution of thymidine kinase for AZT phosphorylation using DNA family shuffling. *Nat Biotechnol* **17**, 259–264 (1999).

380. Lutz, S. & Patrick, W. M. Novel methods for directed evolution of enzymes: quality, not quantity. *Curr Opin Biotechnol* **15**, 291–297 (2004).

381. McCarthy, J. AI as sport. *Science* 1518–1519 (1997).

382. McClain, D. L. Once Again, Machine Beats Human Champion at Chess. *The New York Times* (New York, December 5, 2006).

383. Cui, Y., Wong, W. H., Bornberg-Bauer, E. & Chan, H. S. Recombinatoric exploration of novel folded structures: a heteropolymer-based model of protein evolutionary landscapes. *Proc Natl Acad Sci USA* **99**, 809–814 (2002).

384. Drummond, D. A., Silberg, J. J., Meyer, M. M., Wilke, C. O., & Arnold, F. H. On the conservative nature of intragenic recombination. *Proc Natl Acad Sci USA* **102**, 5380–5385 (2005).

385. Joern, J. M., Meinhold, P., & Arnold, F. H. Analysis of shuffled gene libraries. *J Mol Biol* **316**, 643–656 (2002).

386. Moore, G. L. & Maranas, C. D. eCodonOpt: a systematic computational framework for optimizing codon usage in directed evolution experiments. *Nucleic Acids Res* **30**, 2407–2416 (2002).

387. Moore, G. L., Maranas, C. D., Lutz, S., & Benkovic, S. J. Predicting crossover generation in DNA shuffling. *Proc Natl Acad Sci USA* **98**, 3226–3231 (2001).

388. Maheshri, N. & Schaffer, D. V. Computational and experimental analysis of DNA shuffling. *Proc Natl Acad Sci USA* **100**, 3071–3076 (2003).

389. Moore, G. L. & Maranas, C. D. Predicting out-of-sequence reassembly in DNA shuffling. *J Theor Biol* **219**, 9–17 (2002).

390. Moore, J. C., Jin, H. M., Kuchner, O., & Arnold, F. H. Strategies for the *in vitro* evolution of protein function: enzyme evolution by random recombination of improved sequences. *J Mol Biol* **272**, 336–347 (1997).

391. Gregoret, L. M. & Sauer, R. T. Additivity of mutant effects assessed by binomial mutagenesis. *Proc Natl Acad Sci USA* **90**, 4246–4250 (1993).

392. Sandberg, W. S. & Terwilliger, T. C. Engineering multiple properties of a protein by combinatorial mutagenesis. *Proc Natl Acad Sci USA* **90**, 8367–8371 (1993).

393. Delagrave, S., Goldman, E. R., & Youvan, D. C. Recursive ensemble mutagenesis. *Protein Eng* **6**, 327–331 (1993).

394. Fox, R., et al. Optimizing the search algorithm for protein engineering by directed evolution. *Protein Eng* **16**, 589–597 (2003).

395. Fox, R. J., et al. Improving catalytic function by ProSAR-driven enzyme evolution. *Nat Biotechnol* **25**, 338–344 (2007).

396. Voigt, C. A., Mayo, S. L., Arnold, F. H., & Wang, Z. G. Computational method to reduce the search space for directed protein evolution. *Proc Natl Acad Sci USA* **98**, 3778–3783 (2001).

397. Kono, H. & Saven, J. G. Statistical theory for protein combinatorial libraries. Packing interactions, backbone flexibility, and the sequence variability of a main-chain structure. *J Mol Biol* **306**, 607–628 (2001).

398. Patrick, W. M., Firth, A. E., & Blackburn, J. M. User-friendly algorithms for estimating completeness and diversity in randomized protein-encoding libraries. *Protein Eng* **16**, 451–457 (2003).

399. Volles, M. J. & Lansbury, P. T., Jr. A computer program for the estimation of protein and nucleic acid sequence diversity in random point mutagenesis libraries. *Nucleic Acids Res* **33**, 3667–3677 (2005).

400. Lipovsek, D. & Pluckthun, A. *In-vitro* protein evolution by ribosome display and mRNA display. *J Immunol Methods* **290**, 51–67 (2004).

401. Chen, K. & Arnold, F. H. Tuning the activity of an enzyme for unusual environments: sequential random mutagenesis of subtilisin E for catalysis in dimethylformamide. *Proc Natl Acad Sci USA* **90**, 5618–5622 (1993).

402. Matsuura, T., et al. Nonadditivity of mutational effects on the properties of catalase I and its application to efficient directed evolution. *Protein Eng* **11**, 789–795 (1998).

403. Hayashi, Y., et al. Experimental rugged fitness landscape in protein sequence space. *PLoS ONE* **1**, e96 (2006).

404. Treynor, T. P., Vizcarra, C. L., Nedelcu, D., & Mayo, S. L. Computationally designed libraries of fluorescent proteins evaluated by preservation and diversity of function. *Proc Natl Acad Sci USA* **104**, 48–53 (2007).

405. Saraf, M. C., Gupta, A., & Maranas, C. D. Design of combinatorial protein libraries of optimal size. *Proteins* **60**, 769–777 (2005).

406. Weinreich, D. M., Delaney, N. F., Depristo, M. A., & Hartl, D. L. Darwinian evolution can follow only very few mutational paths to fitter proteins. *Science* **312**, 111–114 (2006).

407. Matsumura, I. & Ellington, A. D. *In vitro* evolution of beta-glucuronidase into a beta-galactosidase proceeds through non-specific intermediates. *J Mol Biol* **305**, 331–339 (2001).

408. Wagner, A. Robustness, evolvability, and neutrality. *FEBS Lett* **579**, 1772–1778 (2005).

409. Gupta, R. D. & Tawfik, D. S. Directed enzyme evolution via small and effective neutral drift libraries. *Nat Methods* **5**, 939–942 (2008).

410. Nelson, H. C. & Sauer, R. T. Lambda repressor mutations that increase the affinity and specificity of operator binding. *Cell* **42**, 549–558 (1985).

411. Oxender, D. L. & Gibson, A. L. Second-site reversion as means of enhancing DNA-binding affinity. *Methods Enzymol* **208**, 641–651 (1991).

412. Sideraki, V., Huang, W., Palzkill, T., & Gilbert, H. F. A secondary drug resistance mutation of TEM-1 beta-lactamase that suppresses misfolding and aggregation. *Proc Natl Acad Sci USA* **98**, 283–288 (2001).

413. Hecky, J., Mason, J. M., Arndt, K. M., & Muller, K. M. A general method of terminal truncation, evolution, and re-elongation to generate enzymes of enhanced stability. *Methods Mol Biol* **352**, 275–304 (2007).

414. Smith, G. P. Filamentous fusion phage: novel expression vectors that display cloned antigens on the virion surface. *Science* **228**, 1315–1317 (1985).

415. Marvin, D. A. Filamentous phage structure, infection and assembly. *Curr Opin Struct Biol* **8**, 150–158 (1998).

416. Lubkowski, J., Hennecke, F., Pluckthun, A., & Wlodawer, A. The structural basis of phage display elucidated by the crystal structure of the N-terminal domains of g3p. *Nat Struct Biol* **5**, 140–147 (1998).

417. Kwasnikowski, P., Kristensen, P., & Markiewicz, W. T. Multivalent display system on filamentous bacteriophage pVII minor coat protein. *J Immunol Methods* **307**, 135–143 (2005).

418. Deng, L. W. & Perham, R. N. Delineating the site of interaction on the pIII protein of filamentous bacteriophage fd with the F-pilus of *Escherichia coli*. *J Mol Biol* **319**, 603–614 (2002).

419. Malik, P., et al. Role of capsid structure and membrane protein processing in determining the size and copy number of peptides displayed on the major coat protein of filamentous bacteriophage. *J Mol Biol* **260**, 9–21 (1996).

420. Hufton, S. E., et al. Phage display of cDNA repertoires: the pVI display system and its applications for the selection of immunogenic ligands. *J Immunol Methods* **231**, 39–51 (1999).

421. Gao, C., et al. Making artificial antibodies: a format for phage display of combinatorial heterodimeric arrays. *Proc Natl Acad Sci USA* **96**, 6025–6030 (1999).

422. Sidhu, S. S., Weiss, G. A., & Wells, J. A. High copy display of large proteins on phage for functional selections. *J Mol Biol* **296**, 487–495 (2000).

423. Bass, S., Greene, R., & Wells, J. A. Hormone phage: an enrichment method for variant proteins with altered binding properties. *Proteins* **8**, 309–314 (1990).

424. Cwirla, S. E., Peters, E. A., Barrett, R. W., & Dower, W. J. Peptides on phage: a vast library of peptides for identifying ligands. *Proc Natl Acad Sci USA* **87**, 6378–6382 (1990).

425. Devlin, J. J., Panganiban, L. C., & Devlin, P. E. Random peptide libraries: a source of specific protein binding molecules. *Science* **249**, 404–406 (1990).

426. Scott, J. K. & Smith, G. P. Searching for peptide ligands with an epitope library. *Science* **249**, 386–390 (1990).

427. Jestin, J. L., Volioti, G., & Winter, G. Improving the display of proteins on filamentous phage. *Res Microbiol* **152**, 187–191 (2001).

428. Thammawong, P., Kasinrerk, W., Turner, R. J., & Tayapiwatana, C. Twin-arginine signal peptide attributes effective display of CD147 to filamentous phage. *Appl Microbiol Biotechnol* **69**, 697–703 (2006).

429. Steiner, D., Forrer, P., Stumpp, M. T., & Pluckthun, A. Signal sequences directing cotranslational translocation expand the range of proteins amenable to phage display. *Nat Biotechnol* **24**, 823–831 (2006).

430. Peters, E. A., Schatz, P. J., Johnson, S. S., & Dower, W. J. Membrane insertion defects caused by positive charges in the early mature region of protein pIII of filamentous phage fd can be corrected by prlA suppressors. *J Bacteriol* **176**, 4296–4305 (1994).

431. Loset, G. A., Lunde, E., Bogen, B., Brekke, O. H., & Sandlie, I. Functional phage display of two murine {alpha}/{beta} T-cell receptors is strongly dependent on fusion format, mode and periplasmic folding assistance. *Protein Eng Des Sel* **20**, 461–472 (2007).

432. Maruyama, I. N., Maruyama, H. I., & Brenner, S. Lambda foo: a lambda phage vector for the expression of foreign proteins. *Proc Natl Acad Sci USA* **91**, 8273–8277 (1994).

433. Dunn, I. S. Assembly of functional bacteriophage lambda virions incorporating C-terminal peptide or protein fusions with the major tail protein. *J Mol Biol* **248**, 497–506 (1995).

434. Sternberg, N. & Hoess, R. H. Display of peptides and proteins on the surface of bacteriophage lambda. *Proc Natl Acad Sci USA* **92**, 1609–1613 (1995).

435. Zanghi, C. N., Sapinoro, R., Bradel-Tretheway, B. & Dewhurst, S. A tractable method for simultaneous modifications to the head and tail of bacteriophage lambda and its application to enhancing phage-mediated gene delivery. *Nucleic Acids Res* **35**, e59 (2007).

436. Krumpe, L. R., et al. T7 lytic phage-displayed peptide libraries exhibit less sequence bias than M13 filamentous phage-displayed peptide libraries. *Proteomics* **6**, 4210–4222 (2006).

437. Smith, G. P. & Yu, J. In search of dark horses: affinity maturation of phage-displayed ligands. *Mol Divers* **2**, 2–4 (1996).

438. Wang, C. L., Yang, D. C., & Wabl, M. Directed molecular evolution by somatic hypermutation. *Protein Eng Des Sel* **17**, 659–664 (2004).

439. Tuerk, C. & Gold, L. Systematic evolution of ligands by exponential enrichment: RNA ligands to bacteriophage T4 DNA polymerase. *Science* **249**, 505–510 (1990).

440. Mattheakis, L. C., Bhatt, R. R., & Dower, W. J. An *in vitro* polysome display system for identifying ligands from very large peptide libraries. *Proc Natl Acad Sci USA* **91**, 9022–9026 (1994).

441. Hanes, J. & Pluckthun, A. *In vitro* selection and evolution of functional proteins by using ribosome display. *Proc Natl Acad Sci USA* **94**, 4937–4942 (1997).

442. Keiler, K. C., Waller, P. R., & Sauer, R. T. Role of a peptide tagging system in degradation of proteins synthesized from damaged messenger RNA. *Science* **271**, 990–993 (1996).

443. Schimmele, B., Grafe, N., & Pluckthun, A. Ribosome display of mammalian receptor domains. *Protein Eng Des Sel* **18**, 285–294 (2005).

444. Yan, X. & Xu, Z. Ribosome-display technology: applications for directed evolution of functional proteins. *Drug Discov Today* **11**, 911–916 (2006).

445. Lamla, T. & Erdmann, V. A. Searching sequence space for high-affinity binding peptides using ribosome display. *J Mol Biol* **329**, 381–388 (2003).

446. Gold, L. mRNA display: diversity matters during *in vitro* selection. *Proc Natl Acad Sci USA* **98**, 4825–4826 (2001).

447. Roberts, R. W. & Szostak, J. W. RNA-peptide fusions for the *in vitro* selection of peptides and proteins. *Proc Natl Acad Sci USA* **94**, 12297–12302 (1997).

448. Nemoto, N., Miyamoto-Sato, E., Husimi, Y. & Yanagawa, H. *In vitro* virus: bonding of mRNA bearing puromycin at the 3'-terminal end to the C-terminal end of its encoded protein on the ribosome *in vitro*. *FEBS Lett* **414**, 405–408 (1997).

449. Miyamoto-Sato, E., Nemoto, N., Kobayashi, K., & Yanagawa, H. Specific bonding of puromycin to full-length protein at the C-terminus. *Nucleic Acids Res* **28**, 1176–1182 (2000).

450. Wilson, D. S., Keefe, A. D., & Szostak, J. W. The use of mRNA display to select high-affinity protein-binding peptides. *Proc Natl Acad Sci USA* **98**, 3750–3755 (2001).

451. McPherson, M., Yang, Y., Hammond, P. W., & Kreider, B. L. Drug receptor identification from multiple tissues using cellular-derived mRNA display libraries. *Chem Biol* **9**, 691–698 (2002).

452. Shen, X., Valencia, C. A., Szostak, J. W., Dong, B., & Liu, R. Scanning the human proteome for calmodulin-binding proteins. *Proc Natl Acad Sci USA* **102**, 5969–5974 (2005).

453. Horisawa, K., et al. *In vitro* selection of Jun-associated proteins using mRNA display. *Nucleic Acids Res* **32**, e169 (2004).

454. Tateyama, S., et al. Affinity selection of DNA-binding protein complexes using mRNA display. *Nucleic Acids Res* **34**, e27 (2006).

455. Sieber, V., Pluckthun, A., & Schmid, F. X. Selecting proteins with improved stability by a phage-based method. *Nat Biotechnol* **16**, 955–960 (1998).

456. Tawfik, D. S. & Griffiths, A. D. Man-made cell-like compartments for molecular evolution. *Nat Biotechnol* **16**, 652–656 (1998).

457. Griffiths, A. D. & Tawfik, D. S. Miniaturising the laboratory in emulsion droplets. *Trends Biotechnol* **24**, 395–402 (2006).

458. Mastrobattista, E., et al. High-throughput screening of enzyme libraries: *in vitro* evolution of a beta-galactosidase by fluorescence-activated sorting of double emulsions. *Chem Biol* **12**, 1291–1300 (2005).

459. Leemhuis, H., Stein, V., Griffiths, A. D., & Hollfelder, F. New genotype–phenotype linkages for directed evolution of functional proteins. *Curr Opin Struct Biol* **15**, 472–478 (2005).

460. Rothe, A., Surjadi, R. N., & Power, B. E. Novel proteins in emulsions using *in vitro* compartmentalization. *Trends Biotechnol* **24**, 587–592 (2006).

461. England, J. L., Shakhnovich, B. E., & Shakhnovich, E. I. Natural selection of more designable folds: a mechanism for thermophilic adaptation. *Proc Natl Acad Sci USA* **100**, 8727–8731 (2003).

462. Bloom, J. D., et al. Evolution favors protein mutational robustness in sufficiently large populations. *BMC Biology* **5**, 29 (2007).

463. Bloom, J. D., et al. Thermodynamic prediction of protein neutrality. *Proc Natl Acad Sci USA* **102**, 606–611 (2005).

464. Besenmatter, W., Kast, P., & Hilvert, D. Relative tolerance of mesostable and thermostable protein homologs to extensive mutation. *Proteins* **66**, 500–506 (2007).

465. Bloom, J. D., Wilke, C. O., Arnold, F. H., & Adami, C. Stability and the evolvability of function in a model protein. *Biophys J* **86**, 2758–2764 (2004).

466. O'Loughlin, T. L., Patrick, W. M. & Matsumura, I. Natural history as a predictor of protein evolvability. *Protein Eng Des Sel* **19**, 439–442 (2006).

467. Arnold, F. H., Wintrode, P. L., Miyazaki, K., & Gershenson, A. How enzymes adapt: lessons from directed evolution. *Trends Biochem Sci* **26**, 100–106 (2001).

468. Rothman, S. C. & Kirsch, J. F. How does an enzyme evolved *in vitro* compare to naturally occurring homologs possessing the targeted function? Tyrosine aminotransferase from aspartate aminotransferase. *J Mol Biol* **327**, 593–608 (2003).

469. Jaenicke, R. Stability and stabilization of globular proteins in solution. *J Biotechnol* **79**, 193–203 (2000).

470. Taverna, D. M. & Goldstein, R. A. Why are proteins marginally stable? *Proteins* **46**, 105–109 (2002).

471. Vieille, C. & Zeikus, G. J. Hyperthermophilic enzymes: sources, uses, and molecular mechanisms for thermostability. *Microbiol Mol Biol Rev* **65**, 1–43 (2001).

472. Somero, G. N. Proteins and temperature. *Annu Rev Physiol* **57**, 43–68 (1995).

473. Hiller, R., Zhou, Z. H., Adams, M. W., & Englander, S. W. Stability and dynamics in a hyperthermophilic protein with melting temperature close to 200°C. *Proc Natl Acad Sci USA* **94**, 11329–11332 (1997).

474. Karshikoff, A. & Ladenstein, R. Proteins from thermophilic and mesophilic organisms essentially do not differ in packing. *Protein Eng* **11**, 867–872 (1998).

475. Arnold, F. H. Enzyme engineering reaches the boiling point. *Proc Natl Acad Sci USA* **95**, 2035–2036 (1998).

476. Jaenicke, R. What ultrastable globular proteins teach us about protein stabilization. *Biochemistry (Mosc)* **63**, 312–321 (1998).

477. Vetriani, C., et al. Protein thermostability above 100°C: a key role for ionic interactions. *Proc Natl Acad Sci USA* **95**, 12300–12305 (1998).

478. Dominy, B. N., Minoux, H., & Brooks, C. L., 3rd. An electrostatic basis for the stability of thermophilic proteins. *Proteins* **57**, 128–141 (2004).

479. Kim, Y. W., et al. Directed evolution of *Thermus* maltogenic amylase toward enhanced thermal resistance. *Appl Environ Microbiol* **69**, 4866–4874 (2003).

480. Miyazaki, K., et al. Thermal stabilization of *Bacillus subtilis* family-11 xylanase by directed evolution. *J Biol Chem* **281**, 10236–10242 (2006).

481. Minagawa, H., et al. Improving the thermal stability of lactate oxidase by directed evolution. *Cell Mol Life Sci* **64**, 77–81 (2007).

482. Palackal, N., et al. An evolutionary route to xylanase process fitness. *Protein Sci* **13**, 494–503 (2004).

483. Chica, R. A., Doucet, N., & Pelletier, J. N. Semi-rational approaches to engineering enzyme activity: combining the benefits of directed evolution and rational design. *Curr Opin Biotechnol* **16**, 378–384 (2005).

484. Acharya, P., Rajakumara, E., Sankaranarayanan, R., & Rao, N. M. Structural basis of selection and thermostability of laboratory evolved *Bacillus subtilis* lipase. *J Mol Biol* **341**, 1271–1281 (2004).

485. Kellis, J. T., Jr., Todd, R. J., & Arnold, F. H. Protein stabilization by engineered metal chelation. *Biotechnology (NY)* **9**, 994–995 (1991).

486. Regan, L. Protein design: novel metal-binding sites. *Trends Biochem Sci* **20**, 280–285 (1995).

487. Baik, S. H., Ide, T., Yoshida, H., Kagami, O., & Harayama, S. Significantly enhanced stability of glucose dehydrogenase by directed evolution. *Appl Microbiol Biotechnol* **61**, 329–335 (2003).

488. Buchner, J. Supervising the fold: functional principles of molecular chaperones. *Faseb J* **10**, 10–19 (1996).

489. Yura, T. & Nakahigashi, K. Regulation of the heat-shock response. *Curr Opin Microbiol* **2**, 153–158 (1999).

490. Ohtsuka, K. & Hata, M. Molecular chaperone function of mammalian Hsp70 and Hsp40: a review. *Int J Hyperthermia* **16**, 231–245 (2000).

491. Kohda, J., Kawanishi, H., Suehara, K., Nakano, Y., & Yano, T. Stabilization of free and immobilized enzymes using hyperthermophilic chaperonin. *J Biosci Bioeng* **101**, 131–136 (2006).

492. Laksanalamai, P., Pavlov, A. R., Slesarev, A. I., & Robb, F. T. Stabilization of Taq DNA polymerase at high temperature by protein folding pathways from a hyperthermophilic archaeon, *Pyrococcus furiosus*. *Biotechnol Bioeng* **93**, 1–5 (2006).

493. Cannio, R., Contursi, P., Rossi, M., & Bartolucci, S. Thermoadaptation of a mesophilic hygromycin B phosphotransferase by directed evolution in hyperthermophilic archaea: selection of a stable genetic marker for DNA transfer into *Sulfolobus solfataricus*. *Extremophiles* **5**, 153–159 (2001).

494. Clarke, A. R. Cytosolic chaperonins: a question of promiscuity. *Mol Cell* **24**, 165–167 (2006).

495. Makino, Y., Amada, K., Taguchi, H., & Yoshida, M. Chaperonin-mediated folding of green fluorescent protein. *J Biol Chem* **272**, 12468–12474 (1997).

496. Braig, K. Chaperonins. *Curr Opin Struct Biol* **8**, 159–165 (1998).

497. Wang, J. D., Herman, C., Tipton, K. A., Gross, C. A., & Weissman, J. S. Directed evolution of substrate-optimized GroEL/S chaperonins. *Cell* **111**, 1027–1039 (2002).

498. Kirk, O., Borchert, T. V., & Fuglsang, C. C. Industrial enzyme applications. *Curr Opin Biotechnol* **13**, 345–351 (2002).

499. Brouns, S. J., et al. Engineering a selectable marker for hyperthermophiles. *J Biol Chem* **280**, 11422–11431 (2005).

500. Rothschild, L. J. & Mancinelli, R. L. Life in extreme environments. *Nature* **409**, 1092–1101 (2001).

501. Baker-Austin, C. & Dopson, M. Life in acid: pH homeostasis in acidophiles. *Trends Microbiol* **15**, 165–171 (2007).

502. Demirjian, D. C., Moris-Varas, F. & Cassidy, C. S. Enzymes from extremophiles. *Curr Opin Chem Biol* **5**, 144–151 (2001).

503. Bessler, C., Schmitt, J., Maurer, K. H., & Schmid, R. D. Directed evolution of a bacterial alpha-amylase: toward enhanced pH-performance and higher specific activity. *Protein Sci* **12**, 2141–2149 (2003).

504. Danielsen, S., et al. *In vitro* selection of enzymatically active lipase variants from phage libraries using a mechanism-based inhibitor. *Gene* **272**, 267–274 (2001).

505. Cherry, J. R. Directed evolution of microbial oxidative enzymes. *Curr Opin Biotechnol* **11**, 250–254 (2000).

506. Boer, H. & Koivula, A. The relationship between thermal stability and pH optimum studied with wild-type and mutant *Trichoderma reesei* cellobiohydrolase Cel7A. *Eur J Biochem* **270**, 841–848 (2003).

507. Kumar, S., Sun, L., Liu, H., Muralidhara, B. K., & Halpert, J. R. Engineering mammalian cytochrome P450 2B1 by directed evolution for enhanced catalytic tolerance to temperature and dimethyl sulfoxide. *Protein Eng Des Sel* **19**, 547–554 (2006).

508. Bugg, T. *Introduction to Enzyme and Coenzyme Chemistry*, 2nd edition (Blackwell, 2004).

509. Stroppolo, M. E., Falconi, M., Caccuri, A. M., & Desideri, A. Superefficient enzymes. *Cell Mol Life Sci* **58**, 1451–1460 (2001).

510. Berg, O. G. Orientation constraints in diffusion-limited macromolecular association. The role of surface diffusion as a rate-enhancing mechanism. *Biophys J* **47**, 1–14 (1985).

511. Getzoff, E. D., et al. Faster superoxide dismutase mutants designed by enhancing electrostatic guidance. *Nature* **358**, 347–351 (1992).

512. Miller, B. G. & Wolfenden, R. Catalytic proficiency: the unusual case of OMP decarboxylase. *Annu Rev Biochem* **71**, 847–885 (2002).

513. Radzicka, A. & Wolfenden, R. A proficient enzyme. *Science* **267**, 90–93 (1995).

514. Griffiths, A. D. & Tawfik, D. S. Directed evolution of an extremely fast phosphotriesterase by *in vitro* compartmentalization. *Embo J* **22**, 24–35 (2003).

515. Shimotohno, A., Oue, S., Yano, T., Kuramitsu, S., & Kagamiyama, H. Demonstration of the importance and usefulness of manipulating non-active-site residues in protein design. *J Biochem (Tokyo)* **129**, 943–948 (2001).

516. Tomatis, P. E., Rasia, R. M., Segovia, L., & Vila, A. J. Mimicking natural evolution in metallo-beta-lactamases through second-shell ligand mutations. *Proc Natl Acad Sci USA* **102**, 13761–13766 (2005).

517. Benkovic, S. J. & Hammes-Schiffer, S. A perspective on enzyme catalysis. *Science* **301**, 1196–1202 (2003).

518. Eisenmesser, E. Z., et al. Intrinsic dynamics of an enzyme underlies catalysis. *Nature* **438**, 117–121 (2005).

519. Miller, B. G. The mutability of enzyme active-site shape determinants. *Protein Sci* **16**, 1965–1968 (2007).

520. Wolf-Watz, M., et al. Linkage between dynamics and catalysis in a thermophilic–mesophilic enzyme pair. *Nat Struct Mol Biol* **11**, 945–949 (2004).

521. Antosiewicz, J., Gilson, M. K., Lee, I. H., & McCammon, J. A. Acetylcholinesterase: diffusional encounter rate constants for dumbbell models of ligand. *Biophys J* **68**, 62–68 (1995).

522. Weinbroum, A. A. Pathophysiological and clinical aspects of combat anticholinesterase poisoning. *Br Med Bull* **72**, 119–133 (2004).

523. Siddiqui, K. S. & Cavicchioli, R. Cold-adapted enzymes. *Annu Rev Biochem* **75**, 403–433 (2006).

524. Lebbink, J. H., Kaper, T., Bron, P., van der Oost, J., & de Vos, W. M. Improving low-temperature catalysis in the hyperthermostable *Pyrococcus furiosus* beta-glucosidase CelB by directed evolution. *Biochemistry* **39**, 3656–3665 (2000).

525. Gershenson, A., Schauerte, J. A., Giver, L., & Arnold, F. H. Tryptophan phosphorescence study of enzyme flexibility and unfolding in laboratory-evolved thermostable esterases. *Biochemistry* **39**, 4658–4665 (2000).

526. Georlette, D., et al. Structural and functional adaptations to extreme temperatures in psychrophilic, mesophilic, and thermophilic DNA ligases. *J Biol Chem* **278**, 37015–37023 (2003).

527. Miyazaki, K., Wintrode, P. L., Grayling, R. A., Rubingh, D. N., & Arnold, F. H. Directed evolution study of temperature adaptation in a psychrophilic enzyme. *J Mol Biol* **297**, 1015–1026 (2000).

528. Taguchi, S., Ozaki, A., Nonaka, T., Mitsui, Y., & Momose, H. A cold-adapted protease engineered by experimental evolution system. *J Biochem (Tokyo)* **126**, 689–693 (1999).

529. Arrizubieta, M. J. & Polaina, J. Increased thermal resistance and modification of the catalytic properties of a beta-glucosidase by random mutagenesis and *in vitro* recombination. *J Biol Chem* **275**, 28843–28848 (2000).

530. Suen, W. C., Zhang, N., Xiao, L., Madison, V., & Zaks, A. Improved activity and thermostability of *Candida antarctica* lipase B by DNA family shuffling. *Protein Eng Des Sel* **17**, 133–140 (2004).

531. Gonzalez-Blasco, G., Sanz-Aparicio, J., Gonzalez, B., Hermoso, J. A. & Polaina, J. Directed evolution of beta-glucosidase A from *Paenibacillus polymyxa* to thermal resistance. *J Biol Chem* **275**, 13708–13712 (2000).

532. Sakaue, R. & Kajiyama, N. Thermostabilization of bacterial fructosyl-amino acid oxidase by directed evolution. *Appl Environ Microbiol* **69**, 139–145 (2003).

533. Hamamatsu, N., et al. Directed evolution by accumulating tailored mutations: thermostabilization of lactate oxidase with less trade-off with catalytic activity. *Protein Eng Des Sel* **19**, 483–489 (2006).

534. Giver, L., Gershenson, A., Freskgard, P. O., & Arnold, F. H. Directed evolution of a thermostable esterase. *Proc Natl Acad Sci USA* **95**, 12809–12813 (1998).

535. Cherry, J. R. & Fidantsef, A. L. Directed evolution of industrial enzymes: an update. *Curr Opin Biotechnol* **14**, 438–443 (2003).

536. Turner, N. J. Controlling chirality. *Curr Opin Biotechnol* **14**, 401–406 (2003).

537. Jaeger, K. E. & Eggert, T. Enantioselective biocatalysis optimized by directed evolution. *Curr Opin Biotechnol* **15**, 305–313 (2004).

538. Babbitt, P. C. & Gerlt, J. A. New functions from old scaffolds: how nature reengineers enzymes for new functions. *Adv Protein Chem* **55**, 1–28 (2000).

539. Bartlett, G. J., Borkakoti, N., & Thornton, J. M. Catalysing new reactions during evolution: economy of residues and mechanism. *J Mol Biol* **331**, 829–860 (2003).

540. Wierenga, R. K. The TIM-barrel fold: a versatile framework for efficient enzymes. *FEBS Lett* **492**, 193–198 (2001).

541. Gerlt, J. A. & Raushel, F. M. Evolution of function in (beta/alpha)8-barrel enzymes. *Curr Opin Chem Biol* **7**, 252–264 (2003).

542. Schmidt, D. M., et al. Evolutionary potential of (beta/alpha)8-barrels: functional promiscuity produced by single substitutions in the enolase superfamily. *Biochemistry* **42**, 8387–8393 (2003).

543. Hocker, B., Beismann-Driemeyer, S., Hettwer, S., Lustig, A., & Sterner, R. Dissection of a (betaalpha)8-barrel enzyme into two folded halves. *Nat Struct Biol* **8**, 32–36 (2001).

544. Seitz, T., Bocola, M., Claren, J., & Sterner, R. Stabilisation of a (betaalpha)8-barrel protein designed from identical half barrels. *J Mol Biol* **372**, 114–129 (2007).

545. Aharoni, A., et al. The 'evolvability' of promiscuous protein functions. *Nat Genet* **37**, 73–76 (2005).

546. Bloom, J. D., Romero, P. A., Lu, Z., & Arnold, F. H. Neutral genetic drift can alter promiscuous protein functions, potentially aiding functional evolution. *Biol Direct* **2**, 17 (2007).

547. Khersonsky, O., Roodveldt, C., & Tawfik, D. S. Enzyme promiscuity: evolutionary and mechanistic aspects. *Curr Opin Chem Biol* **10**, 498–508 (2006).

548. Yasutake, Y., Yao, M., Sakai, N., Kirita, T., & Tanaka, I. Crystal structure of the *Pyrococcus horikoshii* isopropylmalate isomerase small subunit provides insight into the dual substrate specificity of the enzyme. *J Mol Biol* **344**, 325–333 (2004).

549. Kazlauskas, R. J. Enhancing catalytic promiscuity for biocatalysis. *Curr Opin Chem Biol* **9**, 195–201 (2005).

550. Glasner, M. E., Gerlt, J. A., & Babbitt, P. C. Mechanisms of protein evolution and their application to protein engineering. *Adv Enzymol Relat Areas Mol Biol* **75**, 193–239, xii–xiii (2007).

551. Gould, S. M. & Tawfik, D. S. Directed evolution of the promiscuous esterase activity of carbonic anhydrase II. *Biochemistry* **44**, 5444–5452 (2005).

552. Williams, G. J., Zhang, C., & Thorson, J. S. Expanding the promiscuity of a natural-product glycosyltransferase by directed evolution. *Nat Chem Biol* **3**, 657–662 (2007).

553. Toscano, M. D., Woycechowsky, K. J., & Hilvert, D. Minimalist active-site redesign: teaching old enzymes new tricks. *Angew Chem Int Ed Engl* **46**, 3212–3236 (2007).

554. Jurgens, C., et al. Directed evolution of a (beta alpha)8-barrel enzyme to catalyze related reactions in two different metabolic pathways. *Proc Natl Acad Sci USA* **97**, 9925–9930 (2000).

555. Vick, J. E., Schmidt, D. M., & Gerlt, J. A. Evolutionary potential of (beta/alpha)8-barrels: *in vitro* enhancement of a "new" reaction in the enolase superfamily. *Biochemistry* **44**, 11722–11729 (2005).

556. Dumas, D. P., Caldwell, S. R., Wild, J. R., & Raushel, F. M. Purification and properties of the phosphotriesterase from *Pseudomonas diminuta*. *J Biol Chem* **264**, 19659–19665 (1989).

557. Raushel, F. M. & Holden, H. M. Phosphotriesterase: an enzyme in search of its natural substrate. *Adv Enzymol Relat Areas Mol Biol* **74**, 51–93 (2000).

558. Hocker, B. Directed evolution of (betaalpha)(8)-barrel enzymes. *Biomol Eng* **22**, 31–38 (2005).

559. Roodveldt, C. & Tawfik, D. S. Shared promiscuous activities and evolutionary features in various members of the amidohydrolase superfamily. *Biochemistry* **44**, 12728–12736 (2005).

560. Hill, C. M., Li, W. S., Thoden, J. B., Holden, H. M., & Raushel, F. M. Enhanced degradation of chemical warfare agents through molecular engineering of the phosphotriesterase active site. *J Am Chem Soc* **125**, 8990–8991 (2003).

561. Reetz, M. T., et al. Learning from directed evolution: further lessons from theoretical investigations into cooperative mutations in lipase enantioselectivity. *Chembiochem* **8**, 106–112 (2007).

562. Zha, D., Wilensek, S., Hermes, M., Jaeger, K.-E. & Reetz, M. T. Complete reversal of enantio-selectivity of an enzyme-catalysed reaction by directed evolution. *Chem Commun* 2664–2665 (2001).

563. Wada, M., et al. Directed evolution of *N*-acetylneuraminic acid aldolase to catalyze enantiomeric aldol reactions. *Bioorg Med Chem* **11**, 2091–2098 (2003).

564. Chen-Goodspeed, M., Sogorb, M. A., Wu, F., & Raushel, F. M. Enhancement, relaxation, and reversal of the stereoselectivity for phosphotriesterase by rational evolution of active site residues. *Biochemistry* **40**, 1332–1339 (2001).

565. Reetz, M. T. Controlling the enantioselectivity of enzymes by directed evolution: practical and theoretical ramifications. *Proc Natl Acad Sci USA* **101**, 5716–5722 (2004).

566. Hashmi, A. S. Catalysis: raising the gold standard. *Nature* **449**, 292–293 (2007).

567. Cheng, H., Kim, B. H., & Grishin, N. V. MALISAM: a database of structurally analogous motifs in proteins. *Nucleic Acids Res* **36**, D211–D217 (2008).

568. Galperin, M. Y., Walker, D. R., & Koonin, E. V. Analogous enzymes: independent inventions in enzyme evolution. *Genome Res* **8**, 779–790 (1998).

569. Copley, R. R. & Bork, P. Homology among (betaalpha)(8) barrels: implications for the evolution of metabolic pathways. *J Mol Biol* **303**, 627–641 (2000).

570. Nagano, N., Orengo, C. A., & Thornton, J. M. One fold with many functions: the evolutionary relationships between TIM barrel families based on their sequences, structures and functions. *J Mol Biol* **321**, 741–765 (2002).

571. Nair, P. A., et al. Structural basis for nick recognition by a minimal pluripotent DNA ligase. *Nat Struct Mol Biol* **14**, 770–778 (2007).

572. Hilvert, D. Critical analysis of antibody catalysis. *Annu Rev Biochem* **69**, 751–793 (2000).

573. Charbonnier, J. B., et al. Structural convergence in the active sites of a family of catalytic antibodies. *Science* **275**, 1140–1142 (1997).

574. Park, H. S., et al. Design and evolution of new catalytic activity with an existing protein scaffold. *Science* **311**, 535–538 (2006).

575. Tawfik, D. S. Biochemistry. Loop grafting and the origins of enzyme species. *Science* **311**, 475–476 (2006).

576. Woycechowsky, K. J., Vamvaca, K., & Hilvert, D. Novel enzymes through design and evolution. *Adv Enzymol Relat Areas Mol Biol* **75**, 241–294 xiii (2007).

577. Hocker, B., Claren, J., & Sterner, R. Mimicking enzyme evolution by generating new (betaalpha) 8-barrels from (betaalpha)4-half-barrels. *Proc Natl Acad Sci USA* **101**, 16448–16453 (2004).

578. Smiley, J. A. & Benkovic, S. J. Selection of catalytic antibodies for a biosynthetic reaction from a combinatorial cDNA library by complementation of an auxotrophic *Escherichia coli*: antibodies for orotate decarboxylation. *Proc Natl Acad Sci USA* **91**, 8319–8323 (1994).

579. Anfinsen, C. B. Principles that govern the folding of protein chains. *Science* **181**, 223–230 (1973).

580. Rose, G. D., Fleming, P. J., Banavar, J. R., & Maritan, A. A backbone-based theory of protein folding. *Proc Natl Acad Sci USA* **103**, 16623–16633 (2006).

581. Ellis, R. J., Dobson, C., & Hartl, U. Sequence does specify protein conformation. *Trends Biochem Sci* **23**, 468 (1998).

582. Netzer, W. J. & Hartl, F. U. Protein folding in the cytosol: chaperonin-dependent and -independent mechanisms. *Trends Biochem Sci* **23**, 68–73 (1998).

583. Netzer, W. J. & Hartl, F. U. Recombination of protein domains facilitated by co-translational folding in eukaryotes. *Nature* **388**, 343–349 (1997).

584. Hirano, N., Sawasaki, T., Tozawa, Y., Endo, Y., & Takai, K. Tolerance for random recombination of domains in prokaryotic and eukaryotic translation systems: limited interdomain misfolding in a eukaryotic translation system. *Proteins* **64**, 343–354 (2006).

585. Wegrzyn, R. D. & Deuerling, E. Molecular guardians for newborn proteins: ribosome-associated chaperones and their role in protein folding. *Cell Mol Life Sci* **62**, 2727–2738 (2005).

586. Wulf, G., Finn, G., Suizu, F., & Lu, K. P. Phosphorylation-specific prolyl isomerization: is there an underlying theme? *Nat Cell Biol* **7**, 435–441 (2005).

587. Nagradova, N. Enzymes catalyzing protein folding and their cellular functions. *Curr Protein Pept Sci* **8**, 273–282 (2007).

588. Pauwels, K., Van Molle, I., Tommassen, J., & Van Gelder, P. Chaperoning Anfinsen: the steric foldases. *Mol Microbiol* **64**, 917–922 (2007).

589. Ellis, R. J. Steric chaperones. *Trends Biochem Sci* **23**, 43–45 (1998).

590. Shinde, U., Fu, X., & Inouye, M. A pathway for conformational diversity in proteins mediated by intramolecular chaperones. *J Biol Chem* **274**, 15615–15621 (1999).

591. Dalal, S., Balasubramanian, S., & Regan, L. Protein alchemy: changing beta-sheet into alpha-helix. *Nat Struct Biol* **4**, 548–552 (1997).

592. Alexander, P. A., Rozak, D. A., Orban, J., & Bryan, P. N. Directed evolution of highly homologous proteins with different folds by phage display: implications for the protein folding code. *Biochemistry* **44**, 14045–14054 (2005).

593. Pagel, K., et al. Random coils, beta-sheet ribbons, and alpha-helical fibers: one peptide adopting three different secondary structures at will. *J Am Chem Soc* **128**, 2196–2197 (2006).

594. Kabsch, W. & Sander, C. On the use of sequence homologies to predict protein structure: identical pentapeptides can have completely different conformations. *Proc Natl Acad Sci USA* **81**, 1075–1078 (1984).

595. Minor, D. L., Jr., & Kim, P. S. Context-dependent secondary structure formation of a designed protein sequence. *Nature* **380**, 730–734 (1996).

596. Mapelli, M., et al. Determinants of conformational dimerization of Mad2 and its inhibition by p31comet. *Embo J* **25**, 1273–1284 (2006).

597. Tompa, P., Szasz, C., & Buday, L. Structural disorder throws new light on moonlighting. *Trends Biochem Sci* **30**, 484–489 (2005).

598. Caughey, B. & Baron, G. S. Prions and their partners in crime. *Nature* **443**, 803–810 (2006).

599. Gerber, R., Tahiri-Alaoui, A., Hore, P. J. & James, W. Oligomerization of the human prion protein proceeds via a molten globule intermediate. *J Biol Chem* **282**, 6300–6307 (2007).

600. Wickner, R. B., et al. Prion genetics: new rules for a new kind of gene. *Annu Rev Genet* **38**, 681–707 (2004).

601. Shorter, J. & Lindquist, S. Prions as adaptive conduits of memory and inheritance. *Nat Rev Genet* **6**, 435–450 (2005).

602. Wong, P. & Frishman, D. Fold designability, distribution, and disease. *PLoS Comput Biol* **2**, e40 (2006).

603. Li, H., Tang, C., & Wingreen, N. S. Are protein folds atypical? *Proc Natl Acad Sci USA* **95**, 4987–4990 (1998).

604. Chevalier, B., Turmel, M., Lemieux, C., Monnat, R. J., Jr., & Stoddard, B. L. Flexible DNA target site recognition by divergent homing endonuclease isoschizomers I-CreI and I-MsoI. *J Mol Biol* **329**, 253–269 (2003).

605. Ofran, Y. & Margalit, H. Proteins of the same fold and unrelated sequences have similar amino acid composition. *Proteins* **64**, 275–279 (2006).

606. Davidson, A. R. & Sauer, R. T. Folded proteins occur frequently in libraries of random amino acid sequences. *Proc Natl Acad Sci USA* **91**, 2146–2150 (1994).

607. Davidson, A. R., Lumb, K. J., & Sauer, R. T. Cooperatively folded proteins in random sequence libraries. *Nat Struct Biol* **2**, 856–864 (1995).

608. Babajide, A., Hofacker, I. L., Sippl, M. J., & Stadler, P. F. Neutral networks in protein space: a computational study based on knowledge-based potentials of mean force. *Fold Des* **2**, 261–269 (1997).

609. Brooks, D. J., Fresco, J. R., Lesk, A. M., & Singh, M. Evolution of amino acid frequencies in proteins over deep time: inferred order of introduction of amino acids into the genetic code. *Mol Biol Evol* **19**, 1645–1655 (2002).

610. Doi, N., Kakukawa, K., Oishi, Y., & Yanagawa, H. High solubility of random-sequence proteins consisting of five kinds of primitive amino acids. *Protein Eng Des Sel* **18**, 279–284 (2005).

611. Riddle, D. S., et al. Functional rapidly folding proteins from simplified amino acid sequences. *Nat Struct Biol* **4**, 805–809 (1997).

612. Taylor, S. V., Walter, K. U., Kast, P., & Hilvert, D. Searching sequence space for protein catalysts. *Proc Natl Acad Sci USA* **98**, 10596–10601 (2001).

613. Walter, K. U., Vamvaca, K., & Hilvert, D. An active enzyme constructed from a 9-amino acid alphabet. *J Biol Chem* **280**, 37742–37746 (2005).

614. Axe, D. D., Foster, N. W., & Fersht, A. R. Active barnase variants with completely random hydrophobic cores. *Proc Natl Acad Sci USA* **93**, 5590–5594 (1996).

615. Munson, M., et al. What makes a protein a protein? Hydrophobic core designs that specify stability and structural properties. *Protein Sci* **5**, 1584–1593 (1996).

616. Gassner, N. C., Baase, W. A., & Matthews, B. W. A test of the "jigsaw puzzle" model for protein folding by multiple methionine substitutions within the core of T4 lysozyme. *Proc Natl Acad Sci USA* **93**, 12155–12158 (1996).

617. Xu, J., Baase, W. A., Baldwin, E., & Matthews, B. W. The response of T4 lysozyme to large-to-small substitutions within the core and its relation to the hydrophobic effect. *Protein Sci* **7**, 158–177 (1998).

618. Kamtekar, S., Schiffer, J. M., Xiong, H., Babik, J. M., & Hecht, M. H. Protein design by binary patterning of polar and nonpolar amino acids. *Science* **262**, 1680–1685 (1993).

619. Hecht, M. H., Das, A., Go, A., Bradley, L. H., & Wei, Y. De novo proteins from designed combinatorial libraries. *Protein Sci* **13**, 1711–1723 (2004).

620. West, M. W., et al. De novo amyloid proteins from designed combinatorial libraries. *Proc Natl Acad Sci USA* **96**, 11211–11216 (1999).

621. Beasley, J. R. & Hecht, M. H. Protein design: the choice of de novo sequences. *J Biol Chem* **272**, 2031–2034 (1997).

622. Hecht, M. H. De novo design of beta-sheet proteins. *Proc Natl Acad Sci USA* **91**, 8729–8730 (1994).

623. Richardson, J. S. & Richardson, D. C. Natural beta-sheet proteins use negative design to avoid edge-to-edge aggregation. *Proc Natl Acad Sci USA* **99**, 2754–2759 (2002).

624. Cho, G., Keefe, A. D., Liu, R., Wilson, D. S., & Szostak, J. W. Constructing high complexity synthetic libraries of long ORFs using in vitro selection. *J Mol Biol* **297**, 309–319 (2000).

625. Matsuura, T. & Pluckthun, A. Strategies for selection from protein libraries composed of de novo designed secondary structure modules. *Origins Life Evol Biosphere* **34**, 151–157 (2004).

626. Graziano, J. J., et al. Selecting folded proteins from a library of secondary structural elements. *J Am Chem Soc* **130**, 176–185 (2008).

627. Matsuura, T. & Pluckthun, A. Selection based on the folding properties of proteins with ribosome display. *FEBS Lett* **539**, 24–28 (2003).

628. Keefe, A. D. & Szostak, J. W. Functional proteins from a random-sequence library. *Nature* **410**, 715–718 (2001).

629. Ohno, S. Birth of a unique enzyme from an alternative reading frame of the preexisted, internally repetitious coding sequence. *Proc Natl Acad Sci USA* **81**, 2421–2425 (1984).

630. Shiba, K., Takahashi, Y., & Noda, T. On the role of periodism in the origin of proteins. *J Mol Biol* **320**, 833–840 (2002).

631. Forrer, P., Binz, H. K., Stumpp, M. T., & Pluckthun, A. Consensus design of repeat proteins. *Chembiochem* **5**, 183–189 (2004).

632. Kristensen, P. & Winter, G. Proteolytic selection for protein folding using filamentous bacteriophages. *Fold Des* **3**, 321–328 (1998).

633. Minard, P., Scalley-Kim, M., Watters, A. & Baker, D. A "loop entropy reduction" phage-display selection for folded amino acid sequences. *Protein Sci* **10**, 129–134 (2001).

634. Watters, A. L. & Baker, D. Searching for folded proteins in vitro and in silico. *Eur J Biochem* **271**, 1615–1622 (2004).

635. Cho, G. S. & Szostak, J. W. Directed evolution of ATP binding proteins from a zinc finger domain by using mRNA display. *Chem Biol* **13**, 139–147 (2006).

636. Chaput, J. C. & Szostak, J. W. Evolutionary optimization of a nonbiological ATP binding protein for improved folding stability. *Chem Biol* **11**, 865–874 (2004).

637. Mansy, S. S., et al. Structure and evolutionary analysis of a non-biological ATP-binding protein. *J Mol Biol* **371**, 501–513 (2007).

638. Smith, M. D., et al. Structural insights into the evolution of a non-biological protein: importance of surface residues in protein fold optimization. *PLoS ONE* **2**, e467 (2007).

639. Lo Surdo, P., Walsh, M. A., & Sollazzo, M. A novel ADP- and zinc-binding fold from function-directed *in vitro* evolution. *Nat Struct Mol Biol* **11**, 382–383 (2004).

640. Krishna, S. S. & Grishin, N. V. Structurally analogous proteins do exist! *Structure* **12**, 1125–1127 (2004).

641. Wei, Y., Kim, S., Fela, D., Baum, J., & Hecht, M. H. Solution structure of a *de novo* protein from a designed combinatorial library. *Proc Natl Acad Sci USA* **100**, 13270–13273 (2003).

642. Klepeis, J. L., Wei, Y., Hecht, M. H., & Floudas, C. A. *Ab initio* prediction of the three-dimensional structure of a *de novo* designed protein: a double-blind case study. *Proteins* **58**, 560–570 (2005).

643. Yamauchi, A., et al. Evolvability of random polypeptides through functional selection within a small library. *Protein Eng* **15**, 619–626 (2002).

644. Axe, D. D. Estimating the prevalence of protein sequences adopting functional enzyme folds. *J Mol Biol* **341**, 1295–1315 (2004).

645. Wei, Y. & Hecht, M. H. Enzyme-like proteins from an unselected library of designed amino acid sequences. *Protein Eng Des Sel* **17**, 67–75 (2004).

646. Rozinov, M. N. & Nolan, G. P. Evolution of peptides that modulate the spectral qualities of bound, small-molecule fluorophores. *Chem Biol* **5**, 713–728 (1998).

647. Corey, M. J. & Corey, E. On the failure of *de novo*-designed peptides as biocatalysts. *Proc Natl Acad Sci USA* **93**, 11428–11434 (1996).

648. Johnsson, K., Allemann, R. K., Widmer, H., & Benner, S. A. Synthesis, structure and activity of artificial, rationally designed catalytic polypeptides. *Nature* **365**, 530–532 (1993).

649. Weston, C. J., Cureton, C. H., Calvert, M. J., Smart, O. S., & Allemann, R. K. A stable miniature protein with oxaloacetate decarboxylase activity. *Chembiochem* **5**, 1075–1080 (2004).

650. Tanaka, F., Fuller, R., & Barbas, C. F., 3rd. Development of small designer aldolase enzymes: catalytic activity, folding, and substrate specificity. *Biochemistry* **44**, 7583–7592 (2005).

651. Brack, A. From interstellar amino acids to prebiotic catalytic peptides: a review. *Chem Biodivers* **4**, 665–679 (2007).

652. Kirby, A. J., Hollfelder, F., & Tawfik, D. S. Nonspecific catalysis by protein surfaces. *Appl Biochem Biotechnol* **83**, 173–180; discussion 180–181, 297–313 (2000).

653. Menger, F. M. & Ladika, M. Origin of rate accelerations in an enzyme model: the *p*-nitrophenyl ester syndrome. *J Am Chem Soc* **109**, 3145–3146 (1987).

654. MacBeath, G., Kast, P., & Hilvert, D. Redesigning enzyme topology by directed evolution. *Science* **279**, 1958–1961 (1998).

655. Seelig, B. & Szostak, J. W. Selection and evolution of enzymes from a partially randomized non-catalytic scaffold. *Nature* **448**, 828–831 (2007).

656. Kaelin, W. G. Proline hydroxylation and gene expression. *Annu Rev Biochem* **74**, 115–128 (2005).

657. Myllyharju, J. & Kivirikko, K. I. Collagens, modifying enzymes and their mutations in humans, flies and worms. *Trends Genet* **20**, 33–43 (2004).

658. Kryukov, G. V., et al. Characterization of mammalian selenoproteomes. *Science* **300**, 1439–1443 (2003).

659. Blattner, F. R., et al. The complete genome sequence of *Escherichia coli* K-12. *Science* **277**, 1453–1474 (1997).

660. Karp, P. D., et al. Multidimensional annotation of the *Escherichia coli* K-12 genome. *Nucleic Acids Res* (2007).

661. Zhouravleva, G., et al. Termination of translation in eukaryotes is governed by two interacting polypeptide chain release factors, eRF1 and eRF3. *Embo J* **14**, 4065–4072 (1995).

662. Inge-Vechtomov, S., Zhouravleva, G. & Philippe, M. Eukaryotic release factors (eRFs) history. *Biol Cell* **95**, 195–209 (2003).

663. Leinfelder, W., Zehelein, E., Mandrand-Berthelot, M. A. & Bock, A. Gene for a novel tRNA species that accepts L-serine and cotranslationally inserts selenocysteine. *Nature* **331**, 723–725 (1988).

664. Low, S. C. & Berry, M. J. Knowing when not to stop: selenocysteine incorporation in eukaryotes. *Trends Biochem Sci* **21**, 203–208 (1996).

665. Copeland, P. R. Regulation of gene expression by stop codon recoding: selenocysteine. *Gene* **312**, 17–25 (2003).

666. Srinivasan, G., James, C. M., & Krzycki, J. A. Pyrrolysine encoded by UAG in archaea: charging of a UAG-decoding specialized tRNA. *Science* **296**, 1459–1462 (2002).

667. Blight, S. K., et al. Direct charging of tRNA(CUA) with pyrrolysine *in vitro* and *in vivo*. *Nature* **431**, 333–335 (2004).

668. Zhang, Y., Baranov, P. V., Atkins, J. F., & Gladyshev, V. N. Pyrrolysine and selenocysteine use dissimilar decoding strategies. *J Biol Chem* **280**, 20740–20751 (2005).

669. Longstaff, D. G., et al. A natural genetic code expansion cassette enables transmissible biosynthesis and genetic encoding of pyrrolysine. *Proc Natl Acad Sci USA* **104**, 1021–1026 (2007).

670. Lobanov, A. V., Kryukov, G. V., Hatfield, D. L., & Gladyshev, V. N. Is there a twenty third amino acid in the genetic code? *Trends Genet* **22**, 357–360 (2006).

671. Kimmerlin, T. & Seebach, D. '100 years of peptide synthesis': ligation methods for peptide and protein synthesis with applications to beta-peptide assemblies. *J Pept Res* **65**, 229–260 (2005).

672. Link, A. J. & Tirrell, D. A. Reassignment of sense codons *in vivo*. *Methods* **36**, 291–298 (2005).

673. Wang, L., Xie, J., & Schultz, P. G. Expanding the genetic code. *Annu Rev Biophys Biomol Struct* **35**, 225–249 (2006).

674. Chen, G. T. & Inouye, M. Role of the AGA/AGG codons, the rarest codons in global gene expression in *Escherichia coli. Genes Dev* **8**, 2641–2652 (1994).

675. Zahn, K. & Landy, A. Modulation of lambda integrase synthesis by rare arginine tRNA. *Mol Microbiol* **21**, 69–76 (1996).

676. Saier, M. H., Jr. Differential codon usage: a safeguard against inappropriate expression of specialized genes? *FEBS Lett* **362**, 1–4 (1995).

677. Heckler, T. G., et al. T4 RNA ligase mediated preparation of novel "chemically misacylated" tRNAPheS. *Biochemistry* **23**, 1468–1473 (1984).

678. Baldini, G., Martoglio, B., Schachenmann, A., Zugliani, C., & Brunner, J. Mischarging *Escherichia coli* tRNAPhe with L-4′-[3-(trifluoromethyl)-3H-diazirin-3-yl]phenylalanine, a photoactivatable analogue of phenylalanine. *Biochemistry* **27**, 7951–7959 (1988).

679. Noren, C. J., Anthony-Cahill, S. J., Griffith, M. C., & Schultz, P. G. A general method for site-specific incorporation of unnatural amino acids into proteins. *Science* **244**, 182–188 (1989).

680. Ibba, M. & Soll, D. Aminoacyl-tRNA synthesis. *Annu Rev Biochem* **69**, 617–650 (2000).

681. Southern, E. M. Detection of specific sequences among DNA fragments separated by gel electrophoresis. *J Mol Biol* **98**, 503–517 (1975).

682. Anderson, J. C., et al. An expanded genetic code with a functional quadruplet codon. *Proc Natl Acad Sci USA* **101**, 7566–7571 (2004).

683. Anderson, J. C., Magliery, T. J., & Schultz, P. G. Exploring the limits of codon and anticodon size. *Chem Biol* **9**, 237–244 (2002).

684. Bain, J. D., et al. Site-specific incorporation of nonnatural residues during *in vitro* protein biosynthesis with semisynthetic aminoacyl-tRNAs. *Biochemistry* **30**, 5411–5421 (1991).

685. Hartman, M. C., Josephson, K., Lin, C. W., & Szostak, J. W. An expanded set of amino acid analogs for the ribosomal translation of unnatural peptides. *PLoS ONE* **2**, e972 (2007).

686. Lee, N., Bessho, Y., Wei, K., Szostak, J. W., & Suga, H. Ribozyme-catalyzed tRNA aminoacylation. *Nat Struct Biol* **7**, 28–33 (2000).

687. Murakami, H., Ohta, A., Ashigai, H., & Suga, H. A highly flexible tRNA acylation method for non-natural polypeptide synthesis. *Nat Methods* **3**, 357–359 (2006).

688. Ohuchi, M., Murakami, H., & Suga, H. The flexizyme system: a highly flexible tRNA aminoacylation tool for the translation apparatus. *Curr Opin Chem Biol* **11**, 537–542 (2007).

689. Frankel, A., Li, S., Starck, S. R., & Roberts, R. W. Unnatural RNA display libraries. *Curr Opin Struct Biol* **13**, 506–512 (2003).

690. Muranaka, N., Hohsaka, T., & Sisido, M. Four-base codon mediated mRNA display to construct peptide libraries that contain multiple nonnatural amino acids. *Nucleic Acids Res* **34**, e7 (2006).

691. Tan, Z., Blacklow, S. C., Cornish, V. W., & Forster, A. C. *De novo* genetic codes and pure translation display. *Methods* **36**, 279–290 (2005).

692. Xie, J. & Schultz, P. G. A chemical toolkit for proteins: an expanded genetic code. *Nat Rev Mol Cell Biol* **7**, 775–782 (2006).

693. Tan, Z., Forster, A. C., Blacklow, S. C., & Cornish, V. W. Amino acid backbone specificity of the *Escherichia coli* translation machinery. *J Am Chem Soc* **126**, 12752–12753 (2004).

694. Kohrer, C., Xie, L., Kellerer, S., Varshney, U., & RajBhandary, U. L. Import of amber and ochre suppressor tRNAs into mammalian cells: a general approach to site-specific insertion of amino acid analogues into proteins. *Proc Natl Acad Sci USA* **98**, 14310–14315 (2001).

695. Steer, B. A. & Schimmel, P. Major anticodon-binding region missing from an archaebacterial tRNA synthetase. *J Biol Chem* **274**, 35601–35606 (1999).

696. Mehl, R. A., et al. Generation of a bacterium with a 21 amino acid genetic code. *J Am Chem Soc* **125**, 935–939 (2003).

697. Paule, M. R. & White, R. J. Survey and summary: transcription by RNA polymerases I and III. *Nucleic Acids Res* **28**, 1283–1298 (2000).

698. Dieci, G., Fiorino, G., Castelnuovo, M., Teichmann, M., & Pagano, A. The expanding RNA polymerase III transcriptome. *Trends Genet* **23**, 614–622 (2007).

699. Rodriguez, M. S., Dargemont, C., & Stutz, F. Nuclear export of RNA. *Biol Cell* **96**, 639–655 (2004).

700. Kiga, D., et al. An engineered *Escherichia coli* tyrosyl-tRNA synthetase for site-specific incorporation of an unnatural amino acid into proteins in eukaryotic translation and its application in a wheat germ cell-free system. *Proc Natl Acad Sci USA* **99**, 9715–9720 (2002).

701. Sakamoto, K., et al. Site-specific incorporation of an unnatural amino acid into proteins in mammalian cells. *Nucleic Acids Res* **30**, 4692–4699 (2002).

702. Cropp, T. A., Anderson, J. C., & Chin, J. W. Reprogramming the amino-acid substrate specificity of orthogonal aminoacyl-tRNA synthetases to expand the genetic code of eukaryotic cells. *Nat Protoc* **2**, 2590–2600 (2007).

703. Kwon, I., Kirshenbaum, K., & Tirrell, D. A. Breaking the degeneracy of the genetic code. *J Am Chem Soc* **125**, 7512–7513 (2003).

704. Kwon, I. & Tirrell, D. A. Site-specific incorporation of tryptophan analogues into recombinant proteins in bacterial cells. *J Am Chem Soc* **129**, 10431–10437 (2007).

705. Tian, F., Tsao, M. L., & Schultz, P. G. A phage display system with unnatural amino acids. *J Am Chem Soc* **126**, 15962–15963 (2004).

706. Jackson, J. C., Duffy, S. P., Hess, K. R., & Mehl, R. A. Improving nature's enzyme active site with genetically encoded unnatural amino acids. *J Am Chem Soc* **128**, 11124–11127 (2006).

707. Cavalcanti, A. R. & Landweber, L. F. Genetic code: what nature missed. *Curr Biol* **13**, R884–R885 (2003).

708. Budisa, N., et al. Toward the experimental codon reassignment *in vivo*: protein building with an expanded amino acid repertoire. *Faseb J* **13**, 41–51 (1999).

709. Wu, N., Deiters, A., Cropp, T. A., King, D., & Schultz, P. G. A genetically encoded photocaged amino acid. *J Am Chem Soc* **126**, 14306–14307 (2004).

710. Dobkin, C., Mills, D. R., Kramer, F. R., & Spiegelman, S. RNA replication: required intermediates and the dissociation of template, product, and Q beta replicase. *Biochemistry* **18**, 2038–2044 (1979).

711. Kacian, D. L., Mills, D. R., Kramer, F. R., & Spiegelman, S. A replicating RNA molecule suitable for a detailed analysis of extracellular evolution and replication. *Proc Natl Acad Sci USA* **69**, 3038–3042 (1972).

712. Joyce, G. F. Forty years of *in vitro* evolution. *Angew Chem Int Ed Engl* **46**, 6420–6436 (2007).

713. Kruger, K., et al. Self-splicing RNA: autoexcision and autocyclization of the ribosomal RNA intervening sequence of *Tetrahymena*. *Cell* **31**, 147–157 (1982).

714. Guerrier-Takada, C., Gardiner, K., Marsh, T., Pace, N., & Altman, S. The RNA moiety of ribonuclease P is the catalytic subunit of the enzyme. *Cell* **35**, 849–857 (1983).

715. Guerrier-Takada, C. & Altman, S. Catalytic activity of an RNA molecule prepared by transcription *in vitro*. *Science* **223**, 285–286 (1984).

716. Zaug, A. J. & Cech, T. R. The intervening sequence RNA of *Tetrahymena* is an enzyme. *Science* **231**, 470–475 (1986).

717. Cech, T. R. The efficiency and versatility of catalytic RNA: implications for an RNA world. *Gene* **135**, 33–36 (1993).

718. Valadkhan, S. The spliceosome: a ribozyme at heart? *Biol Chem* **388**, 693–697 (2007).

719. Teixeira, A., et al. Autocatalytic RNA cleavage in the human beta-globin pre-mRNA promotes transcription termination. *Nature* **432**, 526–530 (2004).

720. Salehi-Ashtiani, K., Luptak, A., Litovchick, A., & Szostak, J. W. A genomewide search for ribozymes reveals an HDV-like sequence in the human CPEB3 gene. *Science* **313**, 1788–1792 (2006).

721. Altman, S. A view of RNase P. *Mol Biosyst* **3**, 604–607 (2007).

722. Bagheri, S. & Kashani-Sabet, M. Ribozymes in the age of molecular therapeutics. *Curr Mol Med* **4**, 489–506 (2004).

723. Weeks, K. M. & Cech, T. R. Efficient protein-facilitated splicing of the yeast mitochondrial bI5 intron. *Biochemistry* **34**, 7728–7738 (1995).

724. Weeks, K. M. & Cech, T. R. Protein facilitation of group I intron splicing by assembly of the catalytic core and the 5′ splice site domain. *Cell* **82**, 221–230 (1995).

725. Maguire, B. A., Beniaminov, A. D., Ramu, H., Mankin, A. S., & Zimmermann, R. A. A protein component at the heart of an RNA machine: the importance of protein l27 for the function of the bacterial ribosome. *Mol Cell* **20**, 427–435 (2005).

726. Anderson, R. M., Kwon, M., & Strobel, S. A. Toward ribosomal RNA catalytic activity in the absence of protein. *J Mol Evol* **64**, 472–483 (2007).

727. Doudna, J. A. & Cech, T. R. The chemical repertoire of natural ribozymes. *Nature* **418**, 222–228 (2002).

728. Sun, L., Cui, Z., Gottlieb, R. L., & Zhang, B. A selected ribozyme catalyzing diverse dipeptide synthesis. *Chem Biol* **9**, 619–628 (2002).

729. Challis, G. L. & Naismith, J. H. Structural aspects of non-ribosomal peptide biosynthesis. *Curr Opin Struct Biol* **14**, 748–756 (2004).

730. McCall, M. J., Hendry, P., & Jennings, P. A. Minimal sequence requirements for ribozyme activity. *Proc Natl Acad Sci USA* **89**, 5710–5714 (1992).

731. Haseloff, J. & Gerlach, W. L. Simple RNA enzymes with new and highly specific endoribonuclease activities. *Nature* **334**, 585–591 (1988).

732. Shimayama, T., Nishikawa, S., & Taira, K. Generality of the NUX rule: kinetic analysis of the results of systematic mutations in the trinucleotide at the cleavage site of hammerhead ribozymes. *Biochemistry* **34**, 3649–3654 (1995).

733. Penedo, J. C., Wilson, T. J., Jayasena, S. D., Khvorova, A., & Lilley, D. M. Folding of the natural hammerhead ribozyme is enhanced by interaction of auxiliary elements. *RNA* **10**, 880–888 (2004).

734. Robertson, M. P. & Scott, W. G. The structural basis of ribozyme-catalyzed RNA assembly. *Science* **315**, 1549–1553 (2007).

735. Phylactou, L. A., Kilpatrick, M. W., & Wood, M. J. Ribozymes as therapeutic tools for genetic disease. *Hum Mol Genet* **7**, 1649–1653 (1998).

736. Citti, L. & Rainaldi, G. Synthetic hammerhead ribozymes as therapeutic tools to control disease genes. *Curr Gene Ther* **5**, 11–24 (2005).

737. Ellington, A. D. & Szostak, J. W. *In vitro* selection of RNA molecules that bind specific ligands. *Nature* **346**, 818–822 (1990).

738. Berezovski, M., Musheev, M., Drabovich, A., & Krylov, S. N. Non-SELEX selection of aptamers. *J Am Chem Soc* **128**, 1410–1411 (2006).

739. Joyce, G. F. Directed evolution of nucleic acid enzymes. *Annu Rev Biochem* **73**, 791–836 (2004).

740. Hirao, I. & Ellington, A. D. Re-creating the RNA world. *Curr Biol* **5**, 1017–1022 (1995).

741. Muller, U. F. Re-creating an RNA world. *Cell Mol Life Sci* **63**, 1278–1293 (2006).

742. Breaker, R. R. Are engineered proteins getting competition from RNA? *Curr Opin Biotechnol* **7**, 442–448 (1996).

743. Szostak, J. W., Bartel, D. P., & Luisi, P. L. Synthesizing life. *Nature* **409**, 387–390 (2001).

744. Joyce, G. F. The antiquity of RNA-based evolution. *Nature* **418**, 214–221 (2002).

745. Been, M. D. & Cech, T. R. RNA as an RNA polymerase: net elongation of an RNA primer catalyzed by the *Tetrahymena* ribozyme. *Science* **239**, 1412–1416 (1988).

746. Bartel, D. P., Doudna, J. A., Usman, N., & Szostak, J. W. Template-directed primer extension catalyzed by the *Tetrahymena* ribozyme. *Mol Cell Biol* **11**, 3390–3394 (1991).

747. Doudna, J. A. & Szostak, J. W. RNA-catalysed synthesis of complementary-strand RNA. *Nature* **339**, 519–522 (1989).

748. McGinness, K. E. & Joyce, G. F. RNA-catalyzed RNA ligation on an external RNA template. *Chem Biol* **9**, 297–307 (2002).

749. Zaug, A. J., Grabowski, P. J., & Cech, T. R. Autocatalytic cyclization of an excised intervening sequence RNA is a cleavage–ligation reaction. *Nature* **301**, 578–583 (1983).

750. Saldanha, R., Mohr, G., Belfort, M., & Lambowitz, A. M. Group I and group II introns. *Faseb J* **7**, 15–24 (1993).

751. Ekland, E. H., Szostak, J. W., & Bartel, D. P. Structurally complex and highly active RNA ligases derived from random RNA sequences. *Science* **269**, 364–370 (1995).

752. Bartel, D. P. & Szostak, J. W. Isolation of new ribozymes from a large pool of random sequences [see comment]. *Science* **261**, 1411–1418 (1993).

753. Levy, M., Griswold, K. E., & Ellington, A. D. Direct selection of trans-acting ligase ribozymes by *in vitro* compartmentalization. *RNA* **11**, 1555–1562 (2005).

754. Paul, N. & Joyce, G. F. Inaugural article: a self-replicating ligase ribozyme. *Proc Natl Acad Sci USA* **99**, 12733–12740 (2002).

755. Kim, D. E. & Joyce, G. F. Cross-catalytic replication of an RNA ligase ribozyme. *Chem Biol* **11**, 1505–1512 (2004).

756. Lawrence, M. S. & Bartel, D. P. New ligase-derived RNA polymerase ribozymes. *RNA* **11**, 1173–1180 (2005).

757. Zaher, H. S. & Unrau, P. J. Selection of an improved RNA polymerase ribozyme with superior extension and fidelity. *RNA* **13**, 1017–1026 (2007).

758. McGinness, K. E. & Joyce, G. F. In search of an RNA replicase ribozyme. *Chem Biol* **10**, 5–14 (2003).

759. Salehi-Ashtiani, K. & Szostak, J. W. *In vitro* evolution suggests multiple origins for the hammerhead ribozyme. *Nature* **414**, 82–84 (2001).

760. Fedor, M. J. & Williamson, J. R. The catalytic diversity of RNAs. *Nat Rev Mol Cell Biol* **6**, 399–412 (2005).

761. Lazarev, D., Puskarz, I., & Breaker, R. R. Substrate specificity and reaction kinetics of an X-motif ribozyme. *RNA* **9**, 688–697 (2003).

762. Lorsch, J. R. & Szostak, J. W. *In vitro* evolution of new ribozymes with polynucleotide kinase activity. *Nature* **371**, 31–36 (1994).

763. Tarasow, T. M., Tarasow, S. L., & Eaton, B. E. RNA-catalysed carbon–carbon bond formation. *Nature* **389**, 54–57 (1997).

764. Agresti, J. J., Kelly, B. T., Jaschke, A., & Griffiths, A. D. Selection of ribozymes that catalyse multiple-turnover Diels–Alder cycloadditions by using *in vitro* compartmentalization. *Proc Natl Acad Sci USA* **102**, 16170–16175 (2005).

765. Fusz, S., Eisenfuhr, A., Srivatsan, S. G., Heckel, A., & Famulok, M. A ribozyme for the aldol reaction. *Chem Biol* **12**, 941–950 (2005).

766. Moulton, V., et al. RNA folding argues against a hot-start origin of life. *J Mol Evol* **51**, 416–421 (2000).

767. Schultes, E. A., Hraber, P. T., & LaBean, T. H. Estimating the contributions of selection and self-organization in RNA secondary structure. *J Mol Evol* **49**, 76–83 (1999).

768. Le, S. Y., Zhang, K., & Maizel, J. V., Jr. RNA molecules with structure dependent functions are uniquely folded. *Nucleic Acids Res* **30**, 3574–3582 (2002).

769. Schultes, E. A., Spasic, A., Mohanty, U., & Bartel, D. P. Compact and ordered collapse of randomly generated RNA sequences. *Nat Struct Mol Biol* **12**, 1130–1136 (2005).

770. Uhlenbeck, O. C. Keeping RNA happy. *RNA* **1**, 4–6 (1995).

771. Takagi, Y. & Taira, K. Temperature-dependent change in the rate-determining step in a reaction catalyzed by a hammerhead ribozyme. *FEBS Lett* **361**, 273–276 (1995).

772. Peracchi, A. Origins of the temperature dependence of hammerhead ribozyme catalysis. *Nucleic Acids Res* **27**, 2875–2882 (1999).

773. Vazquez-Tello, A., et al. Efficient trans-cleavage by the *Schistosoma mansoni* SMalpha1 hammerhead ribozyme in the extreme thermophile *Thermus thermophilus*. *Nucleic Acids Res* **30**, 1606–1612 (2002).

774. Li, Y. L., et al. Self-association of adenine-dependent hairpin ribozymes. *Eur Biophys J* **37**, 173–182 (2008).

775. Meyers, L. A., Lee, J. F., Cowperthwaite, M., & Ellington, A. D. The robustness of naturally and artificially selected nucleic acid secondary structures. *J Mol Evol* **58**, 681–691 (2004).

776. Jabri, E. & Cech, T. R. *In vitro* selection of the *Naegleria* GIR1 ribozyme identifies three base changes that dramatically improve activity. *RNA* **4**, 1481–1492 (1998).

777. Guo, F. & Cech, T. R. Evolution of *Tetrahymena* ribozyme mutants with increased structural stability. *Nat Struct Biol* **9**, 855–861 (2002).

778. Khvorova, A., Lescoute, A., Westhof, E., & Jayasena, S. D. Sequence elements outside the hammerhead ribozyme catalytic core enable intracellular activity. *Nat Struct Biol* **10**, 708–712 (2003).

779. Saksmerprome, V., Roychowdhury-Saha, M., Jayasena, S., Khvorova, A., & Burke, D. H. Artificial tertiary motifs stabilize trans-cleaving hammerhead ribozymes under conditions of submillimolar divalent ions and high temperatures. *RNA* **10**, 1916–1924 (2004).

780. Burke, D. H. & Greathouse, S. T. Low-magnesium, trans-cleavage activity by type III, tertiary stabilized hammerhead ribozymes with stem 1 discontinuities. *BMC Biochem* **6**, 14 (2005).

781. Cho, B. & Burke, D. H. Topological rearrangement yields structural stabilization and interhelical distance constraints in the Kin.46 self-phosphorylating ribozyme. *RNA* **12**, 2118–2125 (2006).

782. Stage-Zimmermann, T. K. & Uhlenbeck, O. C. A covalent crosslink converts the hammerhead ribozyme from a ribonuclease to an RNA ligase. *Nat Struct Biol* **8**, 863–867 (2001).

783. Torres-Larios, A., Swinger, K. K., Pan, T., & Mondragon, A. Structure of ribonuclease P: a universal ribozyme. *Curr Opin Struct Biol* **16**, 327–335 (2006).

784. Pannucci, J. A., Haas, E. S., Hall, T. A., Harris, J. K., & Brown, J. W. RNase P RNAs from some archaea are catalytically active. *Proc Natl Acad Sci USA* **96**, 7803–7808 (1999).

785. Haas, E. S., Armbruster, D. W., Vucson, B. M., Daniels, C. J., & Brown, J. W. Comparative analysis of ribonuclease P RNA structure in archaea. *Nucleic Acids Res* **24**, 1252–1259 (1996).

786. Kikovska, E., Svard, S. G., & Kirsebom, L. A. Eukaryotic RNase P RNA mediates cleavage in the absence of protein. *Proc Natl Acad Sci USA* **104**, 2062–2067 (2007).

787. Willkomm, D. K. & Hartmann, R. K. An important piece of the RNase P jigsaw solved. *Trends Biochem Sci* **32**, 247–250 (2007).

788. Darr, S. C., Pace, B., & Pace, N. R. Characterization of ribonuclease P from the archaebacterium *Sulfolobus solfataricus*. *J Biol Chem* **265**, 12927–12932 (1990).

789. LaGrandeur, T. E., Darr, S. C., Haas, E. S., & Pace, N. R. Characterization of the RNase P RNA of *Sulfolobus acidocaldarius*. *J Bacteriol* **175**, 5043–5048 (1993).

790. Paul, R., Lazarev, D., & Altman, S. Characterization of RNase P from *Thermotoga maritima*. *Nucleic Acids Res* **29**, 880–885 (2001).

791. Marszalkowski, M., Willkomm, D. K., & Hartmann, R. K. 5′-End maturation of tRNA in *Aquifex aeolicus*. *Biol Chem* **389**, 395–403 (2008).

792. Gold, T. The deep, hot biosphere. *Proc Natl Acad Sci USA* **89**, 6045–6049 (1992).

793. Schwartzman, D. W. & Lineweaver, C. H. The hyperthermophilic origin of life revisited. *Biochem Soc Trans* **32**, 168–171 (2004).

794. Forterre, P. Looking for the most "primitive" organism(s) on earth today: the state of the art. *Planet Space Sci* **43**, 167–177 (1995).

795. Galtier, N., Tourasse, N., & Gouy, M. A nonhyperthermophilic common ancestor to extant life forms. *Science* **283**, 220–221 (1999).

796. Islas, S., Velasco, A. M., Becerra, A., Delaye, L., & Lazcano, A. Hyperthermophily and the origin and earliest evolution of life. *Int Microbiol* **6**, 87–94 (2003).

797. Vlassov, A. V., Kazakov, S. A., Johnston, B. H., & Landweber, L. F. The RNA world on ice: a new scenario for the emergence of RNA information. *J Mol Evol* **61**, 264–273 (2005).

798. Price, P. B. Microbial life in glacial ice and implications for a cold origin of life. *FEMS Microbiol Ecol* **59**, 217–231 (2007).

799. Schmitt, T. & Lehman, N. Non-unity molecular heritability demonstrated by continuous evolution *in vitro*. *Chem Biol* **6**, 857–869 (1999).

800. Lehman, N. & Joyce, G. F. Evolution *in vitro* of an RNA enzyme with altered metal dependence. *Nature* **361**, 182–185 (1993).

801. Frank, D. N. & Pace, N. R. *In vitro* selection for altered divalent metal specificity in the RNase P RNA. *Proc Natl Acad Sci USA* **94**, 14355–14360 (1997).

802. Perrotta, A. T. & Been, M. D. A single nucleotide linked to a switch in metal ion reactivity preference in the HDV ribozymes. *Biochemistry* **46**, 5124–5130 (2007).

803. Miyamoto, Y., Teramoto, N., Imanishi, Y., & Ito, Y. *In vitro* adaptation of a ligase ribozyme for activity under a low-pH condition. *Biotechnol Bioeng* **75**, 590–596 (2001).

804. Miyamoto, Y., Teramoto, N., Imanishi, Y., & Ito, Y. *In vitro* evolution and characterization of a ligase ribozyme adapted to acidic conditions: effect of further rounds of evolution. *Biotechnol Bioeng* **90**, 36–45 (2005).

805. Striggles, J. C., Martin, M. B., & Schmidt, F. J. Frequency of RNA–RNA interaction in a model of the RNA World. *RNA* **12**, 353–359 (2006).

806. Lai, M. M. RNA recombination in animal and plant viruses. *Microbiol Rev* **56**, 61–79 (1992).

807. Chetverin, A. B. The puzzle of RNA recombination. *FEBS Lett* **460**, 1–5 (1999).

808. Riley, C. A. & Lehman, N. Generalized RNA-directed recombination of RNA. *Chem Biol* **10**, 1233–1243 (2003).

809. Hayden, E. J., Riley, C. A., Burton, A. S., & Lehman, N. RNA-directed construction of structurally complex and active ligase ribozymes through recombination. *RNA* **11**, 1678–1687 (2005).

810. Burke, D. H. & Willis, J. H. Recombination, RNA evolution, and bifunctional RNA molecules isolated through chimeric SELEX. *RNA* **4**, 1165–1175 (1998).

811. Wang, Q. S. & Unrau, P. J. Ribozyme motif structure mapped using random recombination and selection. *RNA* **11**, 404–411 (2005).

812. Bittker, J. A., Le, B. V., & Liu, D. R. Nucleic acid evolution and minimization by nonhomologous random recombination. *Nat Biotechnol* **20**, 1024–1029 (2002).

813. Chapple, K. E., Bartel, D. P., & Unrau, P. J. Combinatorial minimization and secondary structure determination of a nucleotide synthase ribozyme. *RNA* **9**, 1208–1220 (2003).

814. Curtis, E. A. & Bartel, D. P. New catalytic structures from an existing ribozyme. *Nat Struct Mol Biol* **12**, 994–1000 (2005).

815. Le, S. Y., Chen, J. H., Konings, D., & Maizel, J. V., Jr. Discovering well-ordered folding patterns in nucleotide sequences. *Bioinformatics* **19**, 354–361 (2003).

816. Uzilov, A. V., Keegan, J. M., & Mathews, D. H. Detection of non-coding RNAs on the basis of predicted secondary structure formation free energy change. *BMC Bioinformatics* **7**, 173 (2006).

817. Schultes, E. A. & Bartel, D. P. One sequence, two ribozymes: implications for the emergence of new ribozyme folds. *Science* **289**, 448–452 (2000).

818. Reidys, C., Forst, C. V., & Schuster, P. Replication and mutation on neutral networks. *Bull Math Biol* **63**, 57–94 (2001).

819. Lisacek, F., Diaz, Y., & Michel, F. Automatic identification of group I intron cores in genomic DNA sequences. *J Mol Biol* **235**, 1206–1217 (1994).

820. Haugen, P., Simon, D. M., & Bhattacharya, D. The natural history of group I introns. *Trends Genet* **21**, 111–119 (2005).

821. Herschlag, D. RNA chaperones and the RNA folding problem. *J Biol Chem* **270**, 20871–20874 (1995).

822. Mohr, S., Stryker, J. M., & Lambowitz, A. M. A DEAD-box protein functions as an ATP-dependent RNA chaperone in group I intron splicing. *Cell* **109**, 769–779 (2002).

823. Mohr, S., Matsuura, M., Perlman, P. S., & Lambowitz, A. M. A DEAD-box protein alone promotes group II intron splicing and reverse splicing by acting as an RNA chaperone. *Proc Natl Acad Sci USA* **103**, 3569–3574 (2006).

824. Halls, C., et al. Involvement of DEAD-box proteins in group I and group II intron splicing. Biochemical characterization of Mss116p, ATP hydrolysis-dependent and -independent mechanisms, and general RNA chaperone activity. *J Mol Biol* **365**, 835–855 (2007).

825. Johnston, W. K., Unrau, P. J., Lawrence, M. S., Glasner, M. E., & Bartel, D. P. RNA-catalyzed RNA polymerization: accurate and general RNA-templated primer extension. *Science* **292**, 1319–1325 (2001).

826. Fedor, M. J. The role of metal ions in RNA catalysis. *Curr Opin Struct Biol* **12**, 289–295 (2002).

827. Scott, W. G. Ribozymes. *Curr Opin Struct Biol* **17**, 280–286 (2007).

828. Perrotta, A. T. & Been, M. D. HDV ribozyme activity in monovalent cations. *Biochemistry* **45**, 11357–11365 (2006).

829. Curtis, E. A. & Bartel, D. P. The hammerhead cleavage reaction in monovalent cations. *RNA* **7**, 546–552 (2001).

830. Murray, J. B., Seyhan, A. A., Walter, N. G., Burke, J. M., & Scott, W. G. The hammerhead, hairpin and VS ribozymes are catalytically proficient in monovalent cations alone. *Chem Biol* **5**, 587–595 (1998).

831. Bevilacqua, P. C. Mechanistic considerations for general acid–base catalysis by RNA: revisiting the mechanism of the hairpin ribozyme. *Biochemistry* **42**, 2259–2265 (2003).

832. Das, S. R. & Piccirilli, J. A. General acid catalysis by the hepatitis delta virus ribozyme. *Nat Chem Biol* **1**, 45–52 (2005).

833. Bevilacqua, P. C. & Yajima, R. Nucleobase catalysis in ribozyme mechanism. *Curr Opin Chem Biol* **10**, 455–464 (2006).

834. Gong, B., et al. Direct measurement of a $pK(a)$ near neutrality for the catalytic cytosine in the genomic HDV ribozyme using Raman crystallography. *J Am Chem Soc* **129**, 13335–13342 (2007).

835. Walter, N. G. Ribozyme catalysis revisited: is water involved? *Mol Cell* **28**, 923–929 (2007).

836. Pan, T. & Uhlenbeck, O. C. A small metalloribozyme with a two-step mechanism. *Nature* **358**, 560–563 (1992).

837. Pan, T. & Uhlenbeck, O. C. *In vitro* selection of RNAs that undergo autolytic cleavage with Pb^{2+}. *Biochemistry* **31**, 3887–3895 (1992).

838. Komatsu, Y. & Ohtsuka, E. Regulation of ribozyme cleavage activity by oligonucleotides. *Methods Mol Biol* **252**, 165–177 (2004).

839. Penchovsky, R. & Breaker, R. R. Computational design and experimental validation of oligonucleotide-sensing allosteric ribozymes. *Nat Biotechnol* **23**, 1424–1433 (2005).

840. Zhao, Z. Y., et al. Nucleobase participation in ribozyme catalysis. *J Am Chem Soc* **127**, 5026–5027 (2005).

841. Wilson, T. J., et al. Nucleobase catalysis in the hairpin ribozyme. *RNA* **12**, 980–987 (2006).

842. Perrotta, A. T., Wadkins, T. S., & Been, M. D. Chemical rescue, multiple ionizable groups, and general acid–base catalysis in the HDV genomic ribozyme. *RNA* **12**, 1282–1291 (2006).

843. Tsukiji, S., Pattnaik, S. B., & Suga, H. An alcohol dehydrogenase ribozyme. *Nat Struct Biol* **10**, 713–717 (2003).

844. Rogers, J. & Joyce, G. F. A ribozyme that lacks cytidine. *Nature* **402**, 323–325 (1999).

845. Rogers, J. & Joyce, G. F. The effect of cytidine on the structure and function of an RNA ligase ribozyme. *RNA* **7**, 395–404 (2001).

846. Reader, J. S. & Joyce, G. F. A ribozyme composed of only two different nucleotides. *Nature* **420**, 841–844 (2002).

847. Baines, I. C. & Colas, P. Peptide aptamers as guides for small-molecule drug discovery. *Drug Discov Today* **11**, 334–341 (2006).

848. Wang, A. J., et al. Left-handed double helical DNA: variations in the backbone conformation. *Science* **211**, 171–176 (1981).

849. Drew, H. R., et al. Structure of a B-DNA dodecamer: conformation and dynamics. *Proc Natl Acad Sci USA* **78**, 2179–2183 (1981).

850. Hall, K., Cruz, P., Tinoco, I., Jr., Jovin, T. M., & van de Sande, J. H. 'Z-RNA': a left-handed RNA double helix. *Nature* **311**, 584–586 (1984).

851. Davis, P. W., Adamiak, R. W., & Tinoco, I., Jr. Z-RNA: the solution NMR structure of r(CGCGCG). *Biopolymers* **29**, 109–122 (1990).

852. Zahn, K. & Blattner, F. R. Sequence-induced DNA curvature at the bacteriophage lambda origin of replication. *Nature* **317**, 451–453 (1985).

853. Maher, L. J., 3rd. Mechanisms of DNA bending. *Curr Opin Chem Biol* **2**, 688–694 (1998).

854. Mirkin, S. M. Discovery of alternative DNA structures: a heroic decade (1979–1989). *Front Biosci* **13**, 1064–1071 (2008).

855. Volk, D. E., et al. Solution structure and design of dithiophosphate backbone aptamers targeting transcription factor NF-kappaB. *Bioorg Chem* **30**, 396–419 (2002).

856. Burge, S., Parkinson, G. N., Hazel, P., Todd, A. K., & Neidle, S. Quadruplex DNA: sequence, topology and structure. *Nucleic Acids Res* **34**, 5402–5415 (2006).

857. Nishikawa, F., Murakami, K., Noda, K., Yokoyama, T., & Nishikawa, S. Detection of structural changes of RNA aptamer containing GGA repeats under the ionic condition using the microchip electrophoresis. *Nucleic Acids Symp Ser (Oxf)* 397–398 (2007).

858. Phan, A. T., et al. An interlocked dimeric parallel-stranded DNA quadruplex: a potent inhibitor of HIV-1 integrase. *Proc Natl Acad Sci USA* **102**, 634–639 (2005).

859. Chou, S. H., Chin, K. H., & Wang, A. H. DNA aptamers as potential anti-HIV agents. *Trends Biochem Sci* **30**, 231–234 (2005).

860. Lato, S. M., Boles, A. R., & Ellington, A. D. *In vitro* selection of RNA lectins: using combinatorial chemistry to interpret ribozyme evolution. *Chem Biol* **2**, 291–303 (1995).

861. Xu, W. & Ellington, A. D. Anti-peptide aptamers recognize amino acid sequence and bind a protein epitope. *Proc Natl Acad Sci USA* **93**, 7475–7480 (1996).

862. Eaton, B. E. The joys of *in vitro* selection: chemically dressing oligonucleotides to satiate protein targets. *Curr Opin Chem Biol* **1**, 10–16 (1997).

863. Weill, L., Louis, D., & Sargueil, B. Selection and evolution of NTP-specific aptamers. *Nucleic Acids Res* **32**, 5045–5058 (2004).

864. Jay, D. G. Selective destruction of protein function by chromophore-assisted laser inactivation. *Proc Natl Acad Sci USA* **85**, 5454–5458 (1988).

865. Liao, J. C., Roider, J., & Jay, D. G. Chromophore-assisted laser inactivation of proteins is mediated by the photogeneration of free radicals. *Proc Natl Acad Sci USA* **91**, 2659–2663 (1994).

866. Hoffman-Kim, D., Diefenbach, T. J., Eustace, B. K., & Jay, D. G. Chromophore-assisted laser inactivation. *Methods Cell Biol* **82**, 335–354 (2007).

867. Grate, D. & Wilson, C. Laser-mediated, site-specific inactivation of RNA transcripts. *Proc Natl Acad Sci USA* **96**, 6131–6136 (1999).

868. Flinders, J., et al. Recognition of planar and nonplanar ligands in the malachite green–RNA aptamer complex. *Chembiochem* **5**, 62–72 (2004).

869. Brackett, D. M. & Dieckmann, T. Aptamer to ribozyme: the intrinsic catalytic potential of a small RNA. *Chembiochem* **7**, 839–843 (2006).

870. Patel, D. J. Structural analysis of nucleic acid aptamers. *Curr Opin Chem Biol* **1**, 32–46 (1997).

871. Held, D. M., Greathouse, S. T., Agrawal, A., & Burke, D. H. Evolutionary landscapes for the acquisition of new ligand recognition by RNA aptamers. *J Mol Evol* **57**, 299–308 (2003).

872. Staple, D. W. & Butcher, S. E. Pseudoknots: RNA structures with diverse functions. *PLoS Biol* **3**, e213 (2005).

873. Hermann, T. & Patel, D. J. Adaptive recognition by nucleic acid aptamers. *Science* **287**, 820–825 (2000).

874. Marshall, K. A., Robertson, M. P., & Ellington, A. D. A biopolymer by any other name would bind as well: a comparison of the ligand-binding pockets of nucleic acids and proteins. *Structure* **5**, 729–734 (1997).

875. Noeske, J., et al. Interplay of 'induced fit' and preorganization in the ligand induced folding of the aptamer domain of the guanine binding riboswitch. *Nucleic Acids Res* **35**, 572–583 (2007).

876. Jenison, R. D., Gill, S. C., Pardi, A., & Polisky, B. High-resolution molecular discrimination by RNA. *Science* **263**, 1425–1429 (1994).

877. Wallace, S. T. & Schroeder, R. *In vitro* selection and characterization of streptomycin-binding RNAs: recognition discrimination between antibiotics. *RNA* **4**, 112–123 (1998).

878. Nguyen, D. H., DeFina, S. C., Fink, W. H., & Dieckmann, T. Binding to an RNA aptamer changes the charge distribution and conformation of malachite green. *J Am Chem Soc* **124**, 15081–15084 (2002).

879. Baugh, C., Grate, D., & Wilson, C. 2.8 A crystal structure of the malachite green aptamer. *J Mol Biol* **301**, 117–128 (2000).

880. Knight, R. & Yarus, M. Finding specific RNA motifs: function in a zeptomole world? *RNA* **9**, 218–230 (2003).

881. Knight, R., et al. Abundance of correctly folded RNA motifs in sequence space, calculated on computational grids. *Nucleic Acids Res* **33**, 5924–5935 (2005).

882. Yang, Y., Kochoyan, M., Burgstaller, P., Westhof, E., & Famulok, M. Structural basis of ligand discrimination by two related RNA aptamers resolved by NMR spectroscopy. *Science* **272**, 1343–1347 (1996).

883. Pavski, V. & Le, X. C. Detection of human immunodeficiency virus type 1 reverse transcriptase using aptamers as probes in affinity capillary electrophoresis. *Anal Chem* **73**, 6070–6076 (2001).

884. Gopinath, S. C., Balasundaresan, D., Akitomi, J., & Mizuno, H. An RNA aptamer that discriminates bovine factor IX from human factor IX. *J Biochem* **140**, 667–676 (2006).

885. Eaton, B. E., Gold, L., & Zichi, D. A. Let's get specific: the relationship between specificity and affinity. *Chem Biol* **2**, 633–638 (1995).

886. Carothers, J. M., Davis, J. H., Chou, J. J., & Szostak, J. W. Solution structure of an informationally complex high-affinity RNA aptamer to GTP. *RNA* **12**, 567–579 (2006).

887. Carothers, J. M., Oestreich, S. C., & Szostak, J. W. Aptamers selected for higher-affinity binding are not more specific for the target ligand. *J Am Chem Soc* **128**, 7929–7937 (2006).

888. Hirao, I., Spingola, M., Peabody, D., & Ellington, A. D. The limits of specificity: an experimental analysis with RNA aptamers to MS2 coat protein variants. *Mol Divers* **4**, 75–89 (1999).

889. Bock, L. C., Griffin, L. C., Latham, J. A., Vermaas, E. H., & Toole, J. J. Selection of single-stranded DNA molecules that bind and inhibit human thrombin. *Nature* **355**, 564–566 (1992).

890. Ellington, A. D. & Szostak, J. W. Selection *in vitro* of single-stranded DNA molecules that fold into specific ligand-binding structures. *Nature* **355**, 850–852 (1992).

891. Huizenga, D. E. & Szostak, J. W. A DNA aptamer that binds adenosine and ATP. *Biochemistry* **34**, 656–665 (1995).

892. Jiang, F., Kumar, R. A., Jones, R. A., & Patel, D. J. Structural basis of RNA folding and recognition in an AMP–RNA aptamer complex. *Nature* **382**, 183–186 (1996).

893. Lin, C. H. & Patel, D. J. Structural basis of DNA folding and recognition in an AMP-DNA aptamer complex: distinct architectures but common recognition motifs for DNA and RNA aptamers complexed to AMP. *Chem Biol* **4**, 817–832 (1997).

894. Michaud, M., et al. A DNA aptamer as a new target-specific chiral selector for HPLC. *J Am Chem Soc* **125**, 8672–8679 (2003).

895. Michaud, M., et al. Immobilized DNA aptamers as target-specific chiral stationary phases for resolution of nucleoside and amino acid derivative enantiomers. *Anal Chem* **76**, 1015–1020 (2004).

896. Shoji, A., Kuwahara, M., Ozaki, H., & Sawai, H. Modified DNA aptamer that binds the (R)-isomer of a thalidomide derivative with high enantioselectivity. *J Am Chem Soc* **129**, 1456–1464 (2007).

897. Cech, T. R. The chemistry of self-splicing RNA and RNA enzymes. *Science* **236**, 1532–1539 (1987).

898. Paul, N., Springsteen, G., & Joyce, G. F. Conversion of a ribozyme to a deoxyribozyme through *in vitro* evolution. *Chem Biol* **13**, 329–338 (2006).

899. Gold, L., et al. From oligonucleotide shapes to genomic SELEX: novel biological regulatory loops. *Proc Natl Acad Sci USA* **94**, 59–64 (1997).

900. Gold, L., Brody, E., Heilig, J., & Singer, B. One, two, infinity: genomes filled with aptamers. *Chem Biol* **9**, 1259–1264 (2002).

901. Guttman, M., et al. Chromatin signature reveals over a thousand highly conserved large non-coding RNAs in mammals. *Nature* **458**, 223–227 (2009).

902. Valegard, K., Murray, J. B., Stockley, P. G., Stonehouse, N. J., & Liljas, L. Crystal structure of an RNA bacteriophage coat protein–operator complex. *Nature* **371**, 623–626 (1994).

903. Convery, M. A., et al. Crystal structure of an RNA aptamer–protein complex at 2.8 Å resolution. *Nat Struct Biol* **5**, 133–139 (1998).

904. Lato, S. M. & Ellington, A. D. Screening chemical libraries for nucleic-acid-binding drugs by *in vitro* selection: a test case with lividomycin. *Mol Divers* **2**, 103–110 (1996).

905. Bourdeau, V., Ferbeyre, G., Pageau, M., Paquin, B., & Cedergren, R. The distribution of RNA motifs in natural sequences. *Nucleic Acids Res* **27**, 4457–4467 (1999).

906. Patte, J. C., Akrim, M., & Mejean, V. The leader sequence of the *Escherichia coli* lysC gene is involved in the regulation of LysC synthesis. *FEMS Microbiol Lett* **169**, 165–170 (1998).

907. Winkler, W., Nahvi, A., & Breaker, R. R. Thiamine derivatives bind messenger RNAs directly to regulate bacterial gene expression. *Nature* **419**, 952–956 (2002).

908. Sudarsan, N., Barrick, J. E., & Breaker, R. R. Metabolite-binding RNA domains are present in the genes of eukaryotes. *RNA* **9**, 644–647 (2003).

909. Bocobza, S., et al. Riboswitch-dependent gene regulation and its evolution in the plant kingdom. *Genes Dev* **21**, 2874–2879 (2007).

910. Winkler, W. C. Riboswitches and the role of noncoding RNAs in bacterial metabolic control. *Curr Opin Chem Biol* **9**, 594–602 (2005).

911. Wakeman, C. A., Winkler, W. C., & Dann, C. E., 3rd. Structural features of metabolite-sensing riboswitches. *Trends Biochem Sci* **32**, 415–424 (2007).

912. Edwards, T. E., Klein, D. J., & Ferre-D' Amare, A. R. Riboswitches: small-molecule recognition by gene regulatory RNAs. *Curr Opin Struct Biol* **17**, 273–279 (2007).

913. Mandal, M. & Breaker, R. R. Adenine riboswitches and gene activation by disruption of a transcription terminator. *Nat Struct Mol Biol* **11**, 29–35 (2004).

914. Mandal, M., et al. A glycine-dependent riboswitch that uses cooperative binding to control gene expression. *Science* **306**, 275–279 (2004).

915. Cochrane, J. C., Lipchock, S. V., & Strobel, S. A. Structural investigation of the GlmS ribozyme bound to its catalytic cofactor. *Chem Biol* **14**, 97–105 (2007).

916. Tinsley, R. A., Furchak, J. R., & Walter, N. G. Trans-acting glmS catalytic riboswitch: locked and loaded. *RNA* **13**, 468–477 (2007).

917. Kazanov, M. D., Vitreschak, A. G., & Gelfand, M. S. Abundance and functional diversity of riboswitches in microbial communities. *BMC Genomics* **8**, 347 (2007).

918. Urban, J. H., Papenfort, K., Thomsen, J., Schmitz, R. A., & Vogel, J. A conserved small RNA promotes discoordinate expression of the glmUS operon mRNA to activate GlmS synthesis. *J Mol Biol* **373**, 521–528 (2007).

919. Soukup, G. A. Aptamers meet allostery. *Chem Biol* **11**, 1031–1032 (2004).

920. Soukup, G. A. & Breaker, R. R. Engineering precision RNA molecular switches. *Proc Natl Acad Sci USA* **96**, 3584–3589 (1999).

921. Soukup, G. A., Emilsson, G. A., & Breaker, R. R. Altering molecular recognition of RNA aptamers by allosteric selection. *J Mol Biol* **298**, 623–632 (2000).

922. Soukup, G. A., DeRose, E. C., Koizumi, M., & Breaker, R. R. Generating new ligand-binding RNAs by affinity maturation and disintegration of allosteric ribozymes. *RNA* **7**, 524–536 (2001).

923. Scarabino, D., Crisari, A., Lorenzini, S., Williams, K., & Tocchini-Valentini, G. P. tRNA prefers to kiss. *Embo J* **18**, 4571–4578 (1999).

924. Toulme, J. J., Darfeuille, F., Kolb, G., Chabas, S., & Staedel, C. Modulating viral gene expression by aptamers to RNA structures. *Biol Cell* **95**, 229–238 (2003).

925. Van Melckebeke, H., et al. Liquid–crystal NMR structure of HIV TAR RNA bound to its SELEX RNA aptamer reveals the origins of the high stability of the complex. *Proc Natl Acad Sci USA* **105**, 9210–9215 (2008).

926. Eckstein, F. Small non-coding RNAs as magic bullets. *Trends Biochem Sci* **30**, 445–452 (2005).

927. Ulrich, H. DNA and RNA aptamers as modulators of protein function. *Med Chem* **1**, 199–208 (2005).

928. Blank, M. & Blind, M. Aptamers as tools for target validation. *Curr Opin Chem Biol* **9**, 336–342 (2005).

929. Tsai, D. E., Kenan, D. J., & Keene, J. D. *In vitro* selection of an RNA epitope immunologically cross-reactive with a peptide. *Proc Natl Acad Sci USA* **89**, 8864–8868 (1992).

930. Stocks, M. Intrabodies as drug discovery tools and therapeutics. *Curr Opin Chem Biol* **9**, 359–365 (2005).

931. Famulok, M., Blind, M., & Mayer, G. Intramers as promising new tools in functional proteomics. *Chem Biol* **8**, 931–939 (2001).

932. Famulok, M. & Mayer, G. Intramers and aptamers: applications in protein–function analyses and potential for drug screening. *Chembiochem* **6**, 19–26 (2005).

933. Shi, H., Hoffman, B. E., & Lis, J. T. RNA aptamers as effective protein antagonists in a multicellular organism. *Proc Natl Acad Sci USA* **96**, 10033–10038 (1999).

934. Yang, C., et al. RNA aptamers targeting the cell death inhibitor CED-9 induce cell killing in *Caenorhabditis elegans*. *J Biol Chem* **281**, 9137–9144 (2006).

935. Bunka, D. H. & Stockley, P. G. Aptamers come of age: at last. *Nat Rev Microbiol* **4**, 588–596 (2006).

936. Gilmore, T. D. Introduction to NF-kappaB: players, pathways, perspectives. *Oncogene* **25**, 6680–6684 (2006).

937. Isomura, I. & Morita, A. Regulation of NF-kappaB signaling by decoy oligodeoxynucleotides. *Microbiol Immunol* **50**, 559–563 (2006).

938. Lebruska, L. L. & Maher, L. J., 3rd. Selection and characterization of an RNA decoy for transcription factor NF-kappa B. *Biochemistry* **38**, 3168–3174 (1999).

939. Ghosh, G., Huang, D. B., & Huxford, T. Molecular mimicry of the NF-kappaB DNA target site by a selected RNA aptamer. *Curr Opin Struct Biol* **14**, 21–27 (2004).

940. Reiter, N. J., Maher, L. J., 3rd & Butcher, S. E. DNA mimicry by a high-affinity anti-NF-kappaB RNA aptamer. *Nucleic Acids Res* **36**, 1227–1236 (2008).

941. Lee, J. F., Stovall, G. M., & Ellington, A. D. Aptamer therapeutics advance. *Curr Opin Chem Biol* **10**, 282–289 (2006).

942. Ferreira, C. S., Matthews, C. S., & Missailidis, S. DNA aptamers that bind to MUC1 tumour marker: design and characterization of MUC1-binding single-stranded DNA aptamers. *Tumour Biol* **27**, 289–301 (2006).

943. Ireson, C. R. & Kelland, L. R. Discovery and development of anticancer aptamers. *Mol Cancer Ther* **5**, 2957–2962 (2006).

944. Lee, J. H., et al. A therapeutic aptamer inhibits angiogenesis by specifically targeting the heparin binding domain of VEGF165. *Proc Natl Acad Sci USA* **102**, 18902–18907 (2005).

945. Waisbourd, M., Loewenstein, A., Goldstein, M., & Leibovitch, I. Targeting vascular endothelial growth factor: a promising strategy for treating age-related macular degeneration. *Drugs Aging* **24**, 643–662 (2007).

946. Sudarsan, N., Cohen-Chalamish, S., Nakamura, S., Emilsson, G. M. & Breaker, R. R. Thiamine pyrophosphate riboswitches are targets for the antimicrobial compound pyrithiamine. *Chem Biol* **12**, 1325–1335 (2005).

947. Blount, K. F. & Breaker, R. R. Riboswitches as antibacterial drug targets. *Nat Biotechnol* **24**, 1558–1564 (2006).

948. Beaudry, A., DeFoe, J., Zinnen, S., Burgin, A., & Beigelman, L. *In vitro* selection of a novel nuclease-resistant RNA phosphodiesterase. *Chem Biol* **7**, 323–334 (2000).

949. Piccirilli, J. A., Krauch, T., Moroney, S. E., & Benner, S. A. Enzymatic incorporation of a new base pair into DNA and RNA extends the genetic alphabet. *Nature* **343**, 33–37 (1990).

950. Henry, A. A. & Romesberg, F. E. Beyond A, C, G and T: augmenting nature's alphabet. *Curr Opin Chem Biol* **7**, 727–733 (2003).

951. Hirao, I. Unnatural base pair systems for DNA/RNA-based biotechnology. *Curr Opin Chem Biol* **10**, 622–627 (2006).

952. Endo, M., et al. Unnatural base pairs mediate the site-specific incorporation of an unnatural hydrophobic component into RNA transcripts. *Bioorg Med Chem Lett* **14**, 2593–2596 (2004).

953. Hirao, I., et al. An unnatural hydrophobic base pair system: site-specific incorporation of nucleotide analogs into DNA and RNA. *Nat Methods* **3**, 729–735 (2006).

954. Matsuda, S., et al. Efforts toward expansion of the genetic alphabet: structure and replication of unnatural base pairs. *J Am Chem Soc* **129**, 10466–10473 (2007).

955. Liu, H., Lynch, S. R., & Kool, E. T. Solution structure of xDNA: a paired genetic helix with increased diameter. *J Am Chem Soc* **126**, 6900–6905 (2004).

956. Fa, M., Radeghieri, A., Henry, A. A., & Romesberg, F. E. Expanding the substrate repertoire of a DNA polymerase by directed evolution. *J Am Chem Soc* **126**, 1748–1754 (2004).

957. Henry, A. A. & Romesberg, F. E. The evolution of DNA polymerases with novel activities. *Curr Opin Biotechnol* **16**, 370–377 (2005).

958. Leconte, A. M., Chen, L., & Romesberg, F. E. Polymerase evolution: efforts toward expansion of the genetic code. *J Am Chem Soc* **127**, 12470–12471 (2005).

959. Szathmary, E. What is the optimum size for the genetic alphabet? *Proc Natl Acad Sci USA* **89**, 2614–2618 (1992).

960. Gardner, P. P., Holland, B. R., Moulton, V., Hendy, M., & Penny, D. Optimal alphabets for an RNA world. *Proc Biol Sci* **270**, 1177–1182 (2003).

961. Kawai, R., et al. Site-specific fluorescent labeling of RNA molecules by specific transcription using unnatural base pairs. *J Am Chem Soc* **127**, 17286–17295 (2005).

962. Moriyama, K., Kimoto, M., Mitsui, T., Yokoyama, S., & Hirao, I. Site-specific biotinylation of RNA molecules by transcription using unnatural base pairs. *Nucleic Acids Res* **33**, e129 (2005).

963. Schoning, K., et al. Chemical etiology of nucleic acid structure: the alpha-threofuranosyl-(3′ → 2′) oligonucleotide system. *Science* **290**, 1347–1351 (2000).

964. Joyce, G. F., Schwartz, A. W., Miller, S. L., & Orgel, L. E. The case for an ancestral genetic system involving simple analogues of the nucleotides. *Proc Natl Acad Sci USA* **84**, 4398–4402 (1987).

965. Schneider, K. C. & Benner, S. A. Oligonucleotides containing flexible nucleoside analogues. *J Am Chem Soc* **112**, 453–455 (1990).

966. Nielsen, P. E. PNA technology. *Mol Biotechnol* **26**, 233–248 (2004).

967. Wittung, P., Nielsen, P. E., Buchardt, O., Egholm, M., & Norden, B. DNA-like double helix formed by peptide nucleic acid. *Nature* **368**, 561–563 (1994).

968. Egholm, M., et al. PNA hybridizes to complementary oligonucleotides obeying the Watson–Crick hydrogen-bonding rules. *Nature* **365**, 566–568 (1993).

969. Nielsen, P. E. Peptide nucleic acids and the origin of life. *Chem Biodivers* **4**, 1996–2002 (2007).

970. Bohler, C., Nielsen, P. E., & Orgel, L. E. Template switching between PNA and RNA oligonucleotides. *Nature* **376**, 578–581 (1995).

971. Schmidt, J. G., Christensen, L., Nielsen, P. E., & Orgel, L. E. Information transfer from DNA to peptide nucleic acids by template-directed syntheses. *Nucleic Acids Res* **25**, 4792–4796 (1997).

972. Almarsson, O. & Bruice, T. C. Peptide nucleic acid (PNA) conformation and polymorphism in PNA–DNA and PNA–RNA hybrids. *Proc Natl Acad Sci USA* **90**, 9542–9546 (1993).

973. Nielsen, P. E. Targeting double stranded DNA with peptide nucleic acid (PNA). *Curr Med Chem* **8**, 545–550 (2001).

974. Wang, G. & Xu, X. S. Peptide nucleic acid (PNA) binding-mediated gene regulation. *Cell Res* **14**, 111–116 (2004).

975. Koppelhus, U. & Nielsen, P. E. Cellular delivery of peptide nucleic acid (PNA). *Adv Drug Deliv Rev* **55**, 267–280 (2003).

976. Hauser, N. C., et al. Utilising the left-helical conformation of L-DNA for analysing different marker types on a single universal microarray platform. *Nucleic Acids Res* **34**, 5101–5111 (2006).

977. Nolte, A., Klussmann, S., Bald, R., Erdmann, V. A., & Furste, J. P. Mirror-design of L-oligonucleotide ligands binding to L-arginine. *Nat Biotechnol* **14**, 1116–1119 (1996).

978. Klussmann, S., Nolte, A., Bald, R., Erdmann, V. A., & Furste, J. P. Mirror-image RNA that binds D-adenosine. *Nat Biotechnol* **14**, 1112–1115 (1996).

979. Bolik, S., et al. First experimental evidence for the preferential stabilization of the natural D- over the nonnatural L-configuration in nucleic acids. *RNA* **13**, 1877–1880 (2007).

980. Williams, K. P., et al. Bioactive and nuclease-resistant L-DNA ligand of vasopressin. *Proc Natl Acad Sci USA* **94**, 11285–11290 (1997).

981. Kent, S. Total chemical synthesis of enzymes. *J Pept Sci* **9**, 574–593 (2003).

982. Schumacher, T. N., et al. Identification of D-peptide ligands through mirror-image phage display. *Science* **271**, 1854–1857 (1996).

983. Schlatterer, J. C., Stuhlmann, F., & Jaschke, A. Stereoselective synthesis using immobilized Diels–Alderase ribozymes. *Chembiochem* **4**, 1089–1092 (2003).

984. Kempeneers, V., Vastmans, K., Rozenski, J., & Herdewijn, P. Recognition of threosyl nucleotides by DNA and RNA polymerases. *Nucleic Acids Res* **31**, 6221–6226 (2003).

985. Horhota, A., et al. Kinetic analysis of an efficient DNA-dependent TNA polymerase. *J Am Chem Soc* **127**, 7427–7434 (2005).

986. Wojciechowski, F. & Hudson, R. H. Nucleobase modifications in peptide nucleic acids. *Curr Top Med Chem* **7**, 667–679 (2007).

987. Wilson, C. & Keefe, A. D. Building oligonucleotide therapeutics using non-natural chemistries. *Curr Opin Chem Biol* **10**, 607–614 (2006).

988. Smith, J. K., Hsieh, J., & Fierke, C. A. Importance of RNA-protein interactions in bacterial ribonuclease P structure and catalysis. *Biopolymers* **87**, 329–338 (2007).

989. Tsuchihashi, Z., Khosla, M., & Herschlag, D. Protein enhancement of hammerhead ribozyme catalysis. *Science* **262**, 99–102 (1993).

990. Herschlag, D., Khosla, M., Tsuchihashi, Z., & Karpel, R. L. An RNA chaperone activity of non-specific RNA binding proteins in hammerhead ribozyme catalysis. *Embo J* **13**, 2913–2924 (1994).

991. Liu, F. & Altman, S. Differential evolution of substrates for an RNA enzyme in the presence and absence of its protein cofactor. *Cell* **77**, 1093–1100 (1994).

992. Kim, J. J., Kilani, A. F., Zhan, X., Altman, S., & Liu, F. The protein cofactor allows the sequence of an RNase P ribozyme to diversify by maintaining the catalytically active structure of the enzyme. *RNA* **3**, 613–623 (1997).

993. Cole, K. B. & Dorit, R. L. Protein cofactor-dependent acquisition of novel catalytic activity by the RNase P ribonucleoprotein of *E. coli*. *J Mol Biol* **307**, 1181–1212 (2001).

994. Sargueil, B., Pecchia, D. B., & Burke, J. M. An improved version of the hairpin ribozyme functions as a ribonucleoprotein complex. *Biochemistry* **34**, 7739–7748 (1995).

995. Atsumi, S., Ikawa, Y., Shiraishi, H., & Inoue, T. Design and development of a catalytic ribonucleoprotein. *Embo J* **20**, 5453–5460 (2001).

996. Ikawa, Y., Tsuda, K., Matsumura, S., Atsumi, S., & Inoue, T. Putative intermediary stages for the molecular evolution from a ribozyme to a catalytic RNP. *Nucleic Acids Res* **31**, 1488–1496 (2003).

997. Saito, H. & Inoue, T. RNA and RNP as new molecular parts in synthetic biology. *J Biotechnol* **132**, 1–7 (2007).

998. Zappulla, D. C. & Cech, T. R. RNA as a flexible scaffold for proteins: yeast telomerase and beyond. *Cold Spring Harb Symp Quant Biol* **71**, 217–224 (2006).

999. Zappulla, D. C. & Cech, T. R. Yeast telomerase RNA: a flexible scaffold for protein subunits. *Proc Natl Acad Sci USA* **101**, 10024–10029 (2004).

1000. Lerner, R. A. Manufacturing immunity to disease in a test tube: the magic bullet realized. *Angew Chem Int Ed Engl* **45**, 8106–8125 (2006).

1001. Kohler, G. & Milstein, C. Continuous cultures of fused cells secreting antibody of predefined specificity. *Nature* **256**, 495–497 (1975).

1002. Milstein, C. Monoclonal antibodies. *Sci Am* **243**, 66–74 (1980).

1003. Hwang, W. Y. & Foote, J. Immunogenicity of engineered antibodies. *Methods* **36**, 3–10 (2005).

1004. Nowinski, R., et al. Human monoclonal antibody against Forssman antigen. *Science* **210**, 537–539 (1980).

1005. Croce, C. M., Linnenbach, A., Hall, W., Steplewski, Z., & Koprowski, H. Production of human hybridomas secreting antibodies to measles virus. *Nature* **288**, 488–489 (1980).

1006. Olsson, L. & Kaplan, H. S. Human–human hybridomas producing monoclonal antibodies of predefined antigenic specificity. *Proc Natl Acad Sci USA* **77**, 5429–5431 (1980).

1007. Riechmann, L., Clark, M., Waldmann, H., & Winter, G. Reshaping human antibodies for therapy. *Nature* **332**, 323–327 (1988).

1008. Hansson, L., Rabbani, H., Fagerberg, J., Osterborg, A., & Mellstedt, H. T-cell epitopes within the complementarity-determining and framework regions of the tumor-derived immunoglobulin heavy chain in multiple myeloma. *Blood* **101**, 4930–4936 (2003).

1009. Presta, L. G. Engineering of therapeutic antibodies to minimize immunogenicity and optimize function. *Adv Drug Deliv Rev* **58**, 640–656 (2006).

1010. Kashmiri, S. V., De Pascalis, R., Gonzales, N. R. & Schlom, J. SDR grafting: a new approach to antibody humanization. *Methods* **36**, 25–34 (2005).

1011. Zhu, L., et al. Production of human monoclonal antibody in eggs of chimeric chickens. *Nat Biotechnol* **23**, 1159–1169 (2005).

1012. Lonberg, N. Human monoclonal antibodies from transgenic mice. *Handb Exp Pharmacol* (181), 69–97 (2008).

1013. Robl, J. M., et al. Artificial chromosome vectors and expression of complex proteins in transgenic animals. *Theriogenology* **59**, 107–113 (2003).

1014. Jakobovits, A., Amado, R. G., Yang, X., Roskos, L., & Schwab, G. From XenoMouse technology to panitumumab, the first fully human antibody product from transgenic mice. *Nat Biotechnol* **25**, 1134–1143 (2007).

1015. Haurum, J. S. Recombinant polyclonal antibodies: the next generation of antibody therapeutics? *Drug Discov Today* **11**, 655–660 (2006).

1016. Kuroiwa, Y., et al. Cloned transchromosomic calves producing human immunoglobulin. *Nat Biotechnol* **20**, 889–894 (2002).

1017. Meijer, P. J., et al. Isolation of human antibody repertoires with preservation of the natural heavy and light chain pairing. *J Mol Biol* **358**, 764–772 (2006).

1018. Midelfort, K. S., et al. Substantial energetic improvement with minimal structural perturbation in a high affinity mutant antibody. *J Mol Biol* **343**, 685–701 (2004).

1019. Clark, L. A., et al. Affinity enhancement of an *in vivo* matured therapeutic antibody using structure-based computational design. *Protein Sci* **15**, 949–960 (2006).

1020. Huse, W. D., et al. Generation of a large combinatorial library of the immunoglobulin repertoire in phage lambda. *Science* **246**, 1275–1281 (1989).

1021. McCafferty, J., Griffiths, A. D., Winter, G., & Chiswell, D. J. Phage antibodies: filamentous phage displaying antibody variable domains. *Nature* **348**, 552–554 (1990).

1022. Kang, A. S., Barbas, C. F., Janda, K. D., Benkovic, S. J., & Lerner, R. A. Linkage of recognition and replication functions by assembling combinatorial antibody Fab libraries along phage surfaces. *Proc Natl Acad Sci USA* **88**, 4363–4366 (1991).

1023. Hoogenboom, H. R. Selecting and screening recombinant antibody libraries. *Nat Biotechnol* **23**, 1105–1116 (2005).

1024. Ward, E. S., Gussow, D., Griffiths, A. D., Jones, P. T., & Winter, G. Binding activities of a repertoire of single immunoglobulin variable domains secreted from *Escherichia coli*. *Nature* **341**, 544–546 (1989).

1025. Griffiths, A. D., et al. Human anti-self antibodies with high specificity from phage display libraries. *Embo J* **12**, 725–734 (1993).

1026. Rauchenberger, R., et al. Human combinatorial Fab library yielding specific and functional antibodies against the human fibroblast growth factor receptor 3. *J Biol Chem* **278**, 38194–38205 (2003).

1027. Hust, M., et al. Single chain Fab (scFab) fragment. *BMC Biotechnol* **7**, 14 (2007).

1028. de Haard, H. J., et al. A large non-immunized human Fab fragment phage library that permits rapid isolation and kinetic analysis of high affinity antibodies. *J Biol Chem* **274**, 18218–18230 (1999).

1029. Klein, U., Rajewsky, K., & Kuppers, R. Human immunoglobulin (Ig)M$^+$IgD$^+$ peripheral blood B cells expressing the CD27 cell surface antigen carry somatically mutated variable region genes: CD27 as a general marker for somatically mutated (memory) B cells. *J Exp Med* **188**, 1679–1689 (1998).

1030. Winter, G. & Milstein, C. Man-made antibodies. *Nature* **349**, 293–299 (1991).

1031. Vaughan, T. J., et al. Human antibodies with sub-nanomolar affinities isolated from a large non-immunized phage display library. *Nat Biotechnol* **14**, 309–314 (1996).

1032. Kramer, R. A., et al. The human antibody repertoire specific for rabies virus glycoprotein as selected from immune libraries. *Eur J Immunol* **35**, 2131–2145 (2005).

1033. Moulard, M., et al. Broadly cross-reactive HIV-1-neutralizing human monoclonal Fab selected for binding to gp120–CD4–CCR5 complexes. *Proc Natl Acad Sci USA* **99**, 6913–6918 (2002).

1034. Marzari, R., et al. Molecular dissection of the tissue transglutaminase autoantibody response in celiac disease. *J Immunol* **166**, 4170–4176 (2001).

1035. Chowdhury, P. S. & Pastan, I. Improving antibody affinity by mimicking somatic hypermutation *in vitro*. *Nat Biotechnol* **17**, 568–572 (1999).

1036. Lamminmaki, U., et al. Expanding the conformational diversity by random insertions to CDRH2 results in improved anti-estradiol antibodies. *J Mol Biol* **291**, 589–602 (1999).

1037. Lee, C. V., et al. High-affinity human antibodies from phage-displayed synthetic Fab libraries with a single framework scaffold. *J Mol Biol* **340**, 1073–1093 (2004).

1038. Barbas, C. F., 3rd, Bain, J. D., Hoekstra, D. M., & Lerner, R. A. Semisynthetic combinatorial antibody libraries: a chemical solution to the diversity problem. *Proc Natl Acad Sci USA* **89**, 4457–4461 (1992).

1039. Hoet, R. M., et al. Generation of high-affinity human antibodies by combining donor-derived and synthetic complementarity-determining-region diversity. *Nat Biotechnol* **23**, 344–348 (2005).

1040. Azriel-Rosenfeld, R., Valensi, M. & Benhar, I. A human synthetic combinatorial library of arrayable single-chain antibodies based on shuffling *in vivo* formed CDRs into general framework regions. *J Mol Biol* **335**, 177–192 (2004).

1041. Crameri, A., Cwirla, S., & Stemmer, W. P. Construction and evolution of antibody–phage libraries by DNA shuffling. *Nat Med* **2**, 100–102 (1996).

1042. Wang, X. B., Zhou, B., Yin, C. C., Lin, Q., & Huang, H. L. A new approach for rapidly reshaping single-chain antibody *in vitro* by combining DNA shuffling with ribosome display. *J Biochem* **136**, 19–28 (2004).

1043. Kang, A. S., Jones, T. M., & Burton, D. R. Antibody redesign by chain shuffling from random combinatorial immunoglobulin libraries. *Proc Natl Acad Sci USA* **88**, 11120–11123 (1991).

1044. Kodaira, M., et al. Organization and evolution of variable region genes of the human immunoglobulin heavy chain. *J Mol Biol* **190**, 529–541 (1986).

1045. Chothia, C., et al. Structural repertoire of the human VH segments. *J Mol Biol* **227**, 799–817 (1992).

1046. Ewert, S., Honegger, A., & Pluckthun, A. Structure-based improvement of the biophysical properties of immunoglobulin VH domains with a generalizable approach. *Biochemistry* **42**, 1517–1528 (2003).

1047. Knappik, A., et al. Fully synthetic human combinatorial antibody libraries (HuCAL) based on modular consensus frameworks and CDRs randomized with trinucleotides. *J Mol Biol* **296**, 57–86 (2000).

1048. Dall' Acqua, W. F., et al. Antibody humanization by framework shuffling. *Methods* **36**, 43–60 (2005).

1049. Damschroder, M. M., et al. Framework shuffling of antibodies to reduce immunogenicity and manipulate functional and biophysical properties. *Mol Immunol* **44**, 3049–3060 (2007).

1050. Desiderio, A., et al. A semi-synthetic repertoire of intrinsically stable antibody fragments derived from a single-framework scaffold. *J Mol Biol* **310**, 603–615 (2001).

1051. Weiner, L. M. & Carter, P. Tunable antibodies. *Nat Biotechnol* **23**, 556–557 (2005).

1052. Liu, X. Y., Pop, L. M., & Vitetta, E. S. Engineering therapeutic monoclonal antibodies. *Immunol Rev* **222**, 9–27 (2008).

1053. Braden, B. C., et al. Three-dimensional structures of the free and the antigen-complexed Fab from monoclonal anti-lysozyme antibody D44.1. *J Mol Biol* **243**, 767–781 (1994).

1054. Wucherpfennig, K. W., et al. Polyspecificity of T cell and B cell receptor recognition. *Semin Immunol* **19**, 216–224 (2007).

1055. Soltes, G., et al. On the influence of vector design on antibody phage display. *J Biotechnol* **127**, 626–637 (2007).

1056. Wark, K. L. & Hudson, P. J. Latest technologies for the enhancement of antibody affinity. *Adv Drug Deliv Rev* **58**, 657–670 (2006).

1057. Bond, C. J., Wiesmann, C., Marsters, J. C., Jr., & Sidhu, S. S. A structure-based database of antibody variable domain diversity. *J Mol Biol* **348**, 699–709 (2005).

1058. Thom, G., et al. Probing a protein–protein interaction by *in vitro* evolution. *Proc Natl Acad Sci USA* **103**, 7619–7624 (2006).

1059. Groves, M., et al. Affinity maturation of phage display antibody populations using ribosome display. *J Immunol Methods* **313**, 129–139 (2006).

1060. Fukuda, I., et al. *In vitro* evolution of single-chain antibodies using mRNA display. *Nucleic Acids Res* **34**, e127 (2006).

1061. Wu, H., et al. Development of motavizumab, an ultra-potent antibody for the prevention of respiratory syncytial virus infection in the upper and lower respiratory tract. *J Mol Biol* **368**, 652–665 (2007).

1062. Harmsen, M. M. & De Haard, H. J. Properties, production, and applications of camelid single-domain antibody fragments. *Appl Microbiol Biotechnol* **77**, 13–22 (2007).

1063. Stanfield, R. L., Dooley, H., Flajnik, M. F., & Wilson, I. A. Crystal structure of a shark single-domain antibody V region in complex with lysozyme. *Science* **305**, 1770–1773 (2004).

1064. Holliger, P. & Hudson, P. J. Engineered antibody fragments and the rise of single domains. *Nat Biotechnol* **23**, 1126–1136 (2005).

1065. Rumfelt, L. L., Lohr, R. L., Dooley, H., & Flajnik, M. F. Diversity and repertoire of IgW and IgM VH families in the newborn nurse shark. *BMC Immunol* **5**, 8 (2004).

1066. Greenberg, A. S., et al. A new antigen receptor gene family that undergoes rearrangement and extensive somatic diversification in sharks. *Nature* **374**, 168–173 (1995).

1067. Streltsov, V. A., et al. Structural evidence for evolution of shark Ig new antigen receptor variable domain antibodies from a cell-surface receptor. *Proc Natl Acad Sci USA* **101**, 12444–12449 (2004).

1068. Streltsov, V. A., Carmichael, J. A., & Nuttall, S. D. Structure of a shark IgNAR antibody variable domain and modeling of an early-developmental isotype. *Protein Sci* **14**, 2901–2909 (2005).

1069. Hamers-Casterman, C., et al. Naturally occurring antibodies devoid of light chains. *Nature* **363**, 446–448 (1993).

1070. Roux, K. H., et al. Structural analysis of the nurse shark (new) antigen receptor (NAR): molecular convergence of NAR and unusual mammalian immunoglobulins. *Proc Natl Acad Sci USA* **95**, 11804–11809 (1998).

1071. Lauwereys, M., et al. Potent enzyme inhibitors derived from dromedary heavy-chain antibodies. *Embo J* **17**, 3512–3520 (1998).

1072. De Genst, E., et al. Molecular basis for the preferential cleft recognition by dromedary heavy-chain antibodies. *Proc Natl Acad Sci USA* **103**, 4586–4591 (2006).

1073. Raso, V. & Stollar, B. D. The antibody-enzyme analogy. Comparison of enzymes and antibodies specific for phosphopyridoxyltyrosine. *Biochemistry* **14**, 591–599 (1975).

1074. Slobin, L. I. Preparation and some properties of antibodies with specificity toward rho-nitrophenyl esters. *Biochemistry* **5**, 2836–2844 (1966).

1075. Gallacher, G., et al. A polyclonal antibody preparation with Michaelian catalytic properties. *Biochem J* **279** (Pt 3), 871–881 (1991).

1076. Stephens, D. B. & Iverson, B. L. Catalytic polyclonal antibodies. *Biochem Biophys Res Commun* **192**, 1439–1444 (1993).

1077. Kohen, F., Kim, J. B., Lindner, H. R., Eshhar, Z., & Green, B. Monoclonal immunoglobulin G augments hydrolysis of an ester of the homologous hapten: an esterase-like activity of the antibody-containing site? *FEBS Lett* **111**, 427–431 (1980).

1078. Pollack, S. J., Jacobs, J. W., & Schultz, P. G. Selective chemical catalysis by an antibody. *Science* **234**, 1570–1573 (1986).

1079. Tramontano, A., Janda, K. D., & Lerner, R. A. Catalytic antibodies. *Science* **234**, 1566–1570 (1986).

1080. Jacobsen, J. R. & Schultz, P. G. The scope of antibody catalysis. *Curr Opin Struct Biol* **5**, 818–824 (1995).

1081. Ose, T., et al. Insight into a natural Diels–Alder reaction from the structure of macrophomate synthase. *Nature* **422**, 185–189 (2003).

1082. Romesberg, F. E., Spiller, B., Schultz, P. G., & Stevens, R. C. Immunological origins of binding and catalysis in a Diels–Alderase antibody. *Science* **279**, 1929–1933 (1998).

1083. Xu, J., et al. Evolution of shape complementarity and catalytic efficiency from a primordial antibody template. *Science* **286**, 2345–2348 (1999).

1084. Golinelli-Pimpaneau, B., et al. Structural evidence for a programmed general base in the active site of a catalytic antibody. *Proc Natl Acad Sci USA* **97**, 9892–9895 (2000).

1085. Shokat, K. M., Leumann, C. J., Sugasawara, R., & Schultz, P. G. A new strategy for the generation of catalytic antibodies. *Nature* **338**, 269–271 (1989).

1086. Thorn, S. N., Daniels, R. G., Auditor, M. T., & Hilvert, D. Large rate accelerations in antibody catalysis by strategic use of haptenic charge. *Nature* **373**, 228–230 (1995).

1087. Wentworth, P., Jr., et al. A bait and switch hapten strategy generates catalytic antibodies for phosphodiester hydrolysis. *Proc Natl Acad Sci USA* **95**, 5971–5975 (1998).

1088. Izadyar, L., Friboulet, A., Remy, M. H., Roseto, A., & Thomas, D. Monoclonal anti-idiotypic antibodies as functional internal images of enzyme active sites: production of a catalytic antibody with a cholinesterase activity. *Proc Natl Acad Sci USA* **90**, 8876–8880 (1993).

1089. Kolesnikov, A. V., et al. Enzyme mimicry by the antiidiotypic antibody approach. *Proc Natl Acad Sci USA* **97**, 13526–13531 (2000).

1090. Ponomarenko, N. A., et al. Anti-idiotypic antibody mimics proteolytic function of parent antigen. *Biochemistry* **46**, 14598–14609 (2007).

1091. Seebeck, F. P. & Hilvert, D. Positional ordering of reacting groups contributes significantly to the efficiency of proton transfer at an antibody active site. *J Am Chem Soc* **127**, 1307–1312 (2005).

1092. Baca, M., Scanlan, T. S., Stephenson, R. C., & Wells, J. A. Phage display of a catalytic antibody to optimize affinity for transition-state analog binding. *Proc Natl Acad Sci USA* **94**, 10063–10068 (1997).

1093. Takahashi, N., Kakinuma, H., Liu, L., Nishi, Y., & Fujii, I. *In vitro* abzyme evolution to optimize antibody recognition for catalysis. *Nat Biotechnol* **19**, 563–567 (2001).

1094. Tanaka, F., Fuller, R., Shim, H., Lerner, R. A., & Barbas, C. F., 3rd. Evolution of aldolase antibodies *in vitro*: correlation of catalytic activity and reaction-based selection. *J Mol Biol* **335**, 1007–1018 (2004).

1095. Janda, K. D., et al. Direct selection for a catalytic mechanism from combinatorial antibody libraries. *Proc Natl Acad Sci USA* **91**, 2532–2536 (1994).

1096. Janda, K. D., et al. Chemical selection for catalysis in combinatorial antibody libraries. *Science* **275**, 945–948 (1997).

1097. Iverson, B. L. & Lerner, R. A. Sequence-specific peptide cleavage catalyzed by an antibody. *Science* **243**, 1184–1188 (1989).

1098. Gramatikova, S., Mouratou, B., Stetefeld, J., Mehta, P. K., & Christen, P. Pyridoxal-5′-phosphate-dependent catalytic antibodies. *J Immunol Methods* **269**, 99–110 (2002).

1099. Iverson, B. L., et al. Metalloantibodies. *Science* **249**, 659–662 (1990).

1100. Barbas, C. F., 3rd, Rosenblum, J. S. & Lerner, R. A. Direct selection of antibodies that coordinate metals from semisynthetic combinatorial libraries. *Proc Natl Acad Sci USA* **90**, 6385–6389 (1993).

1101. Wentworth, P., et al. Toward antibody-directed "abzyme" prodrug therapy, ADAPT: carbamate prodrug activation by a catalytic antibody and its *in vitro* application to human tumor cell killing. *Proc Natl Acad Sci USA* **93**, 799–803 (1996).

1102. Shabat, D., et al. *In vivo* activity in a catalytic antibody–prodrug system: antibody catalyzed etoposide prodrug activation for selective chemotherapy. *Proc Natl Acad Sci USA* **98**, 7528–7533 (2001).

1103. Melton, R. G. & Sherwood, R. F. Antibody–enzyme conjugates for cancer therapy. *J Natl Cancer Inst* **88**, 153–165 (1996).

1104. Schrama, D., Reisfeld, R. A., & Becker, J. C. Antibody targeted drugs as cancer therapeutics. *Nat Rev Drug Discov* **5**, 147–159 (2006).

1105. Hanson, C. V., Nishiyama, Y., & Paul, S. Catalytic antibodies and their applications. *Curr Opin Biotechnol* **16**, 631–636 (2005).

1106. Gill, D. S. & Damle, N. K. Biopharmaceutical drug discovery using novel protein scaffolds. *Curr Opin Biotechnol* **17**, 653–658 (2006).

1107. Wu, Y., Li, Q., & Chen, X. Z. Detecting protein–protein interactions by far western blotting. *Nat Protoc* **2**, 3278–3284 (2007).

1108. Schett, G. Review: immune cells and mediators of inflammatory arthritis. *Autoimmunity* **41**, 224–229 (2008).

1109. Reichert, J. M., Rosensweig, C. J., Faden, L. B., & Dewitz, M. C. Monoclonal antibody successes in the clinic. *Nat Biotechnol* **23**, 1073–1078 (2005).

1110. Mease, P. J. Adalimumab in the treatment of arthritis. *Ther Clin Risk Manag* **3**, 133–148 (2007).

1111. Rigby, W. F. Drug insight: different mechanisms of action of tumor necrosis factor antagonists–passive–aggressive behavior? *Nat Clin Pract Rheumatol* **3**, 227–233 (2007).

1112. Callen, J. P. Complications and adverse reactions in the use of newer biologic agents. *Semin Cutan Med Surg* **26**, 6–14 (2007).

1113. Arend, W. P., Malyak, M., Guthridge, C. J., & Gabay, C. Interleukin-1 receptor antagonist: role in biology. *Annu Rev Immunol* **16**, 27–55 (1998).

1114. Calabrese, L. H. Anakinra treatment of patients with rheumatoid arthritis. *Ann Pharmacother* **36**, 1204–1209 (2002).

1115. O'Neil, K. T., et al. Identification of novel peptide antagonists for GPIIb/IIIa from a conformationally constrained phage peptide library. *Proteins* **14**, 509–515 (1992).

1116. Houston, M. E., Jr., Wallace, A., Bianchi, E., Pessi, A., & Hodges, R. S. Use of a conformationally restricted secondary structural element to display peptide libraries: a two-stranded alpha-helical coiled-coil stabilized by lactam bridges. *J Mol Biol* **262**, 270–282 (1996).

1117. Cochran, A. G., et al. A minimal peptide scaffold for beta-turn display: optimizing a strand position in disulfide-cyclized beta-hairpins. *J Am Chem Soc* **123**, 625–632 (2001).

1118. Iannolo, G., Minenkova, O., Gonfloni, S., Castagnoli, L., & Cesareni, G. Construction, exploitation and evolution of a new peptide library displayed at high density by fusion to the major coat protein of filamentous phage. *Biol Chem* **378**, 517–521 (1997).

1119. Terry, T. D., Malik, P., & Perham, R. N. Accessibility of peptides displayed on filamentous bacteriophage virions: susceptibility to proteinases. *Biol Chem* **378**, 523–530 (1997).

1120. Dunn, I. S. Total modification of the bacteriophage lambda tail tube major subunit protein with foreign peptides. *Gene* **183**, 15–21 (1996).

1121. Petrenko, V. A. & Smith, G. P. Phages from landscape libraries as substitute antibodies. *Protein Eng* **13**, 589–592 (2000).

1122. Hosse, R. J., Rothe, A., & Power, B. E. A new generation of protein display scaffolds for molecular recognition. *Protein Sci* **15**, 14–27 (2006).

1123. Skerra, A. Alternative non-antibody scaffolds for molecular recognition. *Curr Opin Biotechnol* **18**, 295–304 (2007).

1124. Nord, K., et al. Binding proteins selected from combinatorial libraries of an alpha-helical bacterial receptor domain. *Nat Biotechnol* **15**, 772–777 (1997).

1125. Nygren, P. A. Alternative binding proteins: affibody binding proteins developed from a small three-helix bundle scaffold. *Febs J* **275**, 2668–2676 (2008).

1126. Kronqvist, N., Lofblom, J., Jonsson, A., Wernerus, H., & Stahl, S. A novel affinity protein selection system based on staphylococcal cell surface display and flow cytometry. *Protein Eng Des Sel* **21**, 247–255 (2008).

1127. Ku, J. & Schultz, P. G. Alternate protein frameworks for molecular recognition. *Proc Natl Acad Sci USA* **92**, 6552–6556 (1995).

1128. Binz, H. K., et al. High-affinity binders selected from designed ankyrin repeat protein libraries. *Nat Biotechnol* **22**, 575–582 (2004).

1129. Binz, H. K., Kohl, A., Pluckthun, A., & Grutter, M. G. Crystal structure of a consensus-designed ankyrin repeat protein: implications for stability. *Proteins* **65**, 280–284 (2006).

1130. Billings, K. S., Best, R. B., Rutherford, T. J., & Clarke, J. Crosstalk between the protein surface and hydrophobic core in a core-swapped fibronectin type III domain. *J Mol Biol* **375**, 560–571 (2008).

1131. Danen, E. H., Sonneveld, P., Brakebusch, C., Fassler, R., & Sonnenberg, A. The fibronectin-binding integrins alpha5beta1 and alphavbeta3 differentially modulate RhoA–GTP loading, organization of cell matrix adhesions, and fibronectin fibrillogenesis. *J Cell Biol* **159**, 1071–1086 (2002).

1132. Plaxco, K. W., Spitzfaden, C., Campbell, I. D., & Dobson, C. M. Rapid refolding of a proline-rich all-beta-sheet fibronectin type III module. *Proc Natl Acad Sci USA* **93**, 10703–10706 (1996).

1133. Koide, A., Bailey, C. W., Huang, X., & Koide, S. The fibronectin type III domain as a scaffold for novel binding proteins. *J Mol Biol* **284**, 1141–1151 (1998).

1134. Koide, A. & Koide, S. Monobodies: antibody mimics based on the scaffold of the fibronectin type III domain. *Methods Mol Biol* **352**, 95–109 (2007).

1135. Batori, V., Koide, A., & Koide, S. Exploring the potential of the monobody scaffold: effects of loop elongation on the stability of a fibronectin type III domain. *Protein Eng* **15**, 1015–1020 (2002).

1136. Xu, L., et al. Directed evolution of high-affinity antibody mimics using mRNA display. *Chem Biol* **9**, 933–942 (2002).

1137. Beste, G., Schmidt, F. S., Stibora, T., & Skerra, A. Small antibody-like proteins with prescribed ligand specificities derived from the lipocalin fold. *Proc Natl Acad Sci USA* **96**, 1898–1903 (1999).

1138. Skerra, A. Anticalins as alternative binding proteins for therapeutic use. *Curr Opin Mol Ther* **9**, 336–344 (2007).

1139. Silverman, J., et al. Multivalent avimer proteins evolved by exon shuffling of a family of human receptor domains. *Nat Biotechnol* **23**, 1556–1561 (2005).

1140. Sheridan, C. Pharma consolidates its grip on post-antibody landscape. *Nat Biotechnol* **25**, 365–366 (2007).

1141. Thogersen, H. C. & Holldack, J. A tetranectin-based platform for protein engineering. *Innovations Pharmaceutical Technol* 27–30 (2006).

1142. Kastrup, J. S., et al. Structure of the C-type lectin carbohydrate recognition domain of human tetranectin. *Acta Crystallogr D Biol Crystallogr* **54**, 757–766 (1998).

1143. Rothe, A., Hosse, R. J., & Power, B. E. *In vitro* display technologies reveal novel biopharmaceutics. *Faseb J* **20**, 1599–1610 (2006).

1144. Sheridan, C. Small molecule challenges dominance of TNF-alpha inhibitors. *Nat Biotechnol* **26**, 143–144 (2008).

1145. Nielsen, B. B., et al. Crystal structure of tetranectin, a trimeric plasminogen-binding protein with an alpha-helical coiled coil. *FEBS Lett* **412**, 388–396 (1997).

1146. Ameloot, P., Declercq, W., Fiers, W., Vandenabeele, P., & Brouckaert, P. Heterotrimers formed by tumor necrosis factors of different species or muteins. *J Biol Chem* **276**, 27098–27103 (2001).

1147. Zhao, Y., Russ, M., Morgan, C., Muller, S., & Kohler, H. Therapeutic applications of superantibodies. *Drug Discov Today* **10**, 1231–1236 (2005).

1148. Dickey, F. H. The preparation of specific adsorbents. *Proc Natl Acad Sci USA* **35**, 227–229 (1949).

1149. Alexander, C., et al. Molecular imprinting science and technology: a survey of the literature for the years up to and including 2003. *J Mol Recognit* **19**, 106–180 (2006).

1150. Pauling, L. & Campbell, D. H. The production of antibodies *in vitro*. *Science* **95**, 440–441 (1942).

1151. Braco, L., Dabulis, K., & Klibanov, A. M. Production of abiotic receptors by molecular imprinting of proteins. *Proc Natl Acad Sci USA* **87**, 274–277 (1990).

1152. Dabulis, K. & Klibanov, A. M. Molecular imprinting of proteins and other macromolecules resulting in new adsorbents. *Biotech Bioeng* **39**, 176–185 (1992).

1153. Stahl, M., Mansson, M., & Mosbach, K. The synthesis of a D-amino acid ester in an organic media with alpha-chymotrypsin modified by a bio-imprinting procedure. *Biotech Letters* **12**, 161–166 (1990).

1154. Stahl, M., Jeppsson-Wistrand, U., Mansson, M. & Mosbach, K. Induced stereoselectivity and substrate selectivity of bio-imprinted alpha-chymotrypsin in anhydrous organic media. *J Am Chem Soc* **113**, 9366–9368 (1991).

1155. Friggeri, A., Kobayashi, H., Shinkai, S., & Reinhoudt, D. N. From solutions to surfaces: a novel molecular imprinting method based on the Conformational changes of boronic-acid-appended poly (L-lysine). *Angew Chem Int Ed Engl* **40**, 4729–4731 (2001).

1156. Riesewijk, A. M., et al. Absence of an obvious molecular imprinting mechanism in a human fetus with monoallelic IGF2R expression. *Biochem Biophys Res Commun* **245**, 272–277 (1998).

1157. Kawabe, A., Fujimoto, R., & Charlesworth, D. High diversity due to balancing selection in the promoter region of the Medea gene in *Arabidopsis lyrata*. *Curr Biol* **17**, 1885–1889 (2007).

1158. Wulff, G. The role of binding-site interactions in the molecular imprinting of polymers. *Trends Biotechnol* **11**, 85–87 (1993).

1159. Haupt, K. & Mosbach, K. Plastic antibodies: developments and applications. *Trends Biotechnol* **16**, 468–475 (1998).

1160. Zhang, H., Ye, L., & Mosbach, K. Non-covalent molecular imprinting with emphasis on its application in separation and drug development. *J Mol Recognit* **19**, 248–259 (2006).

1161. Ansell, R. J., Kriz, D., & Mosbach, K. Molecularly imprinted polymers for bioanalysis: chromatography, binding assays and biomimetic sensors. *Curr Opin Biotechnol* **7**, 89–94 (1996).

1162. Zimmerman, S. C. & Lemcoff, N. G. Synthetic hosts via molecular imprinting: are universal synthetic antibodies realistically possible? *Chem Commun (Camb)* 5–14 (2004).

1163. Hansen, D. E. Recent developments in the molecular imprinting of proteins. *Biomaterials* **28**, 4178–4191 (2007).

1164. Batra, D. & Shea, K. J. Combinatorial methods in molecular imprinting. *Curr Opin Chem Biol* **7**, 434–442 (2003).

1165. Karim, K., et al. How to find effective functional monomers for effective molecularly imprinted polymers? *Adv Drug Deliv Rev* **57**, 1795–1808 (2005).

1166. Umpleby, R. J., 2nd, Baxter, S. C., Chen, Y., Shah, R. N., & Shimizu, K. D. Characterization of molecularly imprinted polymers with the Langmuir–Freundlich isotherm. *Anal Chem* **73**, 4584–4591 (2001).

1167. Umpleby, R. J., 2nd, et al. Characterization of the heterogeneous binding site affinity distributions in molecularly imprinted polymers. *J Chromatogr B Analyt Technol Biomed Life Sci* **804**, 141–149 (2004).

1168. Piletska, E. V., et al. Design of molecular imprinted polymers compatible with aqueous environment. *Anal Chim Acta* **607**, 54–60 (2008).

1169. Andersson, L. I., Muller, R., Vlatakis, G., & Mosbach, K. Mimics of the binding sites of opioid receptors obtained by molecular imprinting of enkephalin and morphine. *Proc Natl Acad Sci USA* **92**, 4788–4792 (1995).

1170. Ansell, R. J., Ramstrom, O., & Mosbach, K. Towards artificial antibodies prepared by molecular imprinting. *Clin Chem* **42**, 1506–1512 (1996).

1171. Bruggemann, O. Molecularly imprinted materials: receptors more durable than nature can provide. *Adv Biochem Eng Biotechnol* **76**, 127–163 (2002).

1172. Vlatakis, G., Andersson, L. I., Muller, R., & Mosbach, K. Drug assay using antibody mimics made by molecular imprinting. *Nature* **361**, 645–647 (1993).

1173. Ansell, R. J. Molecularly imprinted polymers in pseudoimmunoassay. *J Chromatogr B Analyt Technol Biomed Life Sci* **804**, 151–165 (2004).

1174. Ye, L. & Haupt, K. Molecularly imprinted polymers as antibody and receptor mimics for assays, sensors and drug discovery. *Anal Bioanal Chem* **378**, 1887–1897 (2004).

1175. Rathbone, D. L. Molecularly imprinted polymers in the drug discovery process. *Adv Drug Deliv Rev* **57**, 1854–1874 (2005).

1176. Allender, C. J., Richardson, C., Woodhouse, B., Heard, C. M., & Brain, K. R. Pharmaceutical applications for molecularly imprinted polymers. *Int J Pharm* **195**, 39–43 (2000).

1177. Sellergren, B. & Allender, C. J. Molecularly imprinted polymers: a bridge to advanced drug delivery. *Adv Drug Deliv Rev* **57**, 1733–1741 (2005).

1178. Liu, J. Q. & Wulff, G. Molecularly imprinted polymers with strong carboxypeptidase A-like activity: combination of an amidinium function with a zinc–ion binding site in transition-state imprinted cavities. *Angew Chem Int Ed Engl* **43**, 1287–1290 (2004).

1179. Liu, J. Q. & Wulff, G. Functional mimicry of the active site of carboxypeptidase a by a molecular imprinting strategy: cooperativity of an amidinium and a copper ion in a transition-state imprinted cavity giving rise to high catalytic activity. *J Am Chem Soc* **126**, 7452–7453 (2004).

1180. Strikovsky, A. G., et al. Catalytic molecularly imprinted polymers using conventional bulk polymerization or suspension polymerization: selective hydrolysis of diphenyl carbonate and diphenyl carbamate. *J Am Chem Soc* **122**, 6295–6296 (2000).

1181. Yu, Y., Ye, L., Haupt, K., & Mosbach, K. Formation of a class of enzyme inhibitors (drugs), including a chiral compound, by using imprinted polymers or biomolecules as molecular-scale reaction vessels. *Angew Chem Int Ed Engl* **41**, 4459–4463 (2002).

1182. Boldt, G. E., Dickerson, T. J., & Janda, K. D. Emerging chemical and biological approaches for the preparation of discovery libraries. *Drug Discov Today* **11**, 143–148 (2006).

1183. Mosbach, K., Yu, Y., Andersch, J., & Ye, L. Generation of new enzyme inhibitors using imprinted binding sites: the anti-idiotypic approach, a step toward the next generation of molecular imprinting. *J Am Chem Soc* **123**, 12420–12421 (2001).

1184. Paterson, Y., Englander, S. W., & Roder, H. An antibody binding site on cytochrome c defined by hydrogen exchange and two-dimensional NMR. *Science* **249**, 755–759 (1990).

1185. Che, Z., et al. Antibody-mediated neutralization of human rhinovirus 14 explored by means of cryoelectron microscopy and X-ray crystallography of virus–Fab complexes. *J Virol* **72**, 4610–4622 (1998).

1186. Lewin, B. *Genes IX* (Jones and Bartlett, 2007).

1187. Dickert, F. L., Hayden, O., & Halikias, K. P. Synthetic receptors as sensor coatings for molecules and living cells. *Analyst* **126**, 766–771 (2001).

1188. Takatsy, A., Sedzik, J., Kilar, F., & Hjerten, S. Universal method for synthesis of artificial gel antibodies by the imprinting approach combined with a unique electrophoresis technique for detection of minute structural differences of proteins, viruses, and cells (bacteria): II. Gel antibodies against virus (Semliki Forest virus). *J Sep Sci* **29**, 2810–2815 (2006).

1189. Bacskay, I., et al. Universal method for synthesis of artificial gel antibodies by the imprinting approach combined with a unique electrophoresis technique for detection of minute structural differences of proteins, viruses, and cells (bacteria). III: Gel antibodies against cells (bacteria). *Electrophoresis* **27**, 4682–4687 (2006).

1190. Ge, Y. & Turner, A. P. Too large to fit? Recent developments in macromolecular imprinting. *Trends Biotechnol* **26**, 218–224 (2008).

1191. Rachkov, A. & Minoura, N. Towards molecularly imprinted polymers selective to peptides and proteins. The epitope approach. *Biochim Biophys Acta* **1544**, 255–266 (2001).

1192. Titirici, M. M. & Sellergren, B. Peptide recognition via hierarchical imprinting. *Anal Bioanal Chem* **378**, 1913–1921 (2004).

1193. Shinkawa, T., et al. The absence of fucose but not the presence of galactose or bisecting *N*-acetylglucosamine of human IgG1 complex-type oligosaccharides shows the critical role of enhancing antibody-dependent cellular cytotoxicity. *J Biol Chem* **278**, 3466–3473 (2003).

1194. Satoh, M., Iida, S., & Shitara, K. Non-fucosylated therapeutic antibodies as next-generation therapeutic antibodies. *Expert Opin Biol Ther* **6**, 1161–1173 (2006).

1195. Jefferis, R. Antibody therapeutics: isotype and glycoform selection. *Expert Opin Biol Ther* **7**, 1401–1413 (2007).

1196. Cox, J. C., et al. Automated selection of aptamers against protein targets translated *in vitro*: from gene to aptamer. *Nucleic Acids Res* **30**, e108 (2002).

1197. Breaker, R. R. Natural and engineered nucleic acids as tools to explore biology. *Nature* **432**, 838–845 (2004).

1198. Cheng, R. P. Beyond *de novo* protein design: *de novo* design of non-natural folded oligomers. *Curr Opin Struct Biol* **14**, 512–520 (2004).

1199. Goodman, C. M., Choi, S., Shandler, S., & DeGrado, W. F. Foldamers as versatile frameworks for the design and evolution of function. *Nat Chem Biol* **3**, 252–262 (2007).

1200. Hecht, S. & Huc, I. (eds.) *Foldamers: Structure, Properties, and Applications* (Wiley-VCH, 2007).

1201. Kirshenbaum, K. Book review: foldamers: structure, properties, and applications. *Chembiochem* **9**, 157–158 (2008).

1202. Schreiber, S. L. Small molecules: the missing link in the central dogma. *Nat Chem Biol* **1**, 64–66 (2005).

1203. Hager, G. L. Understanding nuclear receptor function: from DNA to chromatin to the interphase nucleus. *Prog Nucleic Acid Res Mol Biol* **66**, 279–305 (2001).

1204. Waring, M. Binding of antibiotics to DNA. *Ciba Found Symp* **158**, 128–142. discussion 142–146, 204–212 (1991).

1205. Gillet, V. J. New directions in library design and analysis. *Curr Opin Chem Biol* **12**, 372–378 (2008).

1206. Gorse, A. D. Diversity in medicinal chemistry space. *Curr Top Med Chem* **6**, 3–18 (2006).

1207. Murata, M. & Yasumoto, T. The structure elucidation and biological activities of high molecular weight algal toxins: maitotoxin, prymnesins and zooxanthellatoxins. *Nat Prod Rep* **17**, 293–314 (2000).

1208. Nicolaou, K. C., Frederick, M. O., & Aversa, R. J. The continuing saga of the marine polyether biotoxins. *Angew Chem Int Ed Engl* **47**, 7182–7225 (2008).

1209. Raffin-Sanson, M. L., de Keyzer, Y. & Bertagna, X. Proopiomelanocortin, a polypeptide precursor with multiple functions: from physiology to pathological conditions. *Eur J Endocrinol* **149**, 79–90 (2003).

1210. Autelitano, D. J., et al. The cryptome: a subset of the proteome, comprising cryptic peptides with distinct bioactivities. *Drug Discov Today* **11**, 306–314 (2006).

1211. Bolitho, C., et al. The anti-apoptotic activity of albumin for endothelium is mediated by a partially cryptic protein domain and reduced by inhibitors of G-coupled protein and PI-3 kinase, but is independent of radical scavenging or bound lipid. *J Vasc Res* **44**, 313–324 (2007).

1212. Hammes, A., et al. Role of endocytosis in cellular uptake of sex steroids. *Cell* **122**, 751–762 (2005).

1213. Chen, I. A., Salehi-Ashtiani, K. & Szostak, J. W. RNA catalysis in model protocell vesicles. *J Am Chem Soc* **127**, 13213–13219 (2005).

1214. Anand, N., Bindra, J. S., & Ranganathan, S. *Art in Organic Synthesis*, 2nd edition (John Wiley & Sons, 1988).

1215. Mislow, K. & Siegel, J. Stereoisomerism and local chirality. *J Am Chem Soc* **106**, 3319–3328 (1984).

1216. Kishi, Y. Complete structure of maitotoxin. *Pure Appl Chem* **70**, 339–344 (1998).

1217. Martens, J. H., Barg, H., Warren, M. J., & Jahn, D. Microbial production of vitamin B_{12}. *Appl Microbiol Biotechnol* **58**, 275–285 (2002).

1218. Yamanishi, M., Vlasie, M., & Banerjee, R. Adenosyltransferase: an enzyme and an escort for coenzyme B_{12}? *Trends Biochem Sci* **30**, 304–308 (2005).

1219. Warren, M. J., Raux, E., Schubert, H. L., & Escalante-Semerena, J. C. The biosynthesis of adenosylcobalamin (vitamin B_{12}). *Nat Prod Rep* **19**, 390–412 (2002).

1220. Pearson, H. The bitterest pill. *Nature* **444**, 532–533 (2006).

1221. Sun, D., et al. *In vitro* testing of drug absorption for drug 'developability' assessment: forming an interface between *in vitro* preclinical data and clinical outcome. *Curr Opin Drug Discov Devel* **7**, 75–85 (2004).

1222. Lin, J., et al. The role of absorption, distribution, metabolism, excretion and toxicity in drug discovery. *Curr Top Med Chem* **3**, 1125–1154 (2003).

1223. Wishart, D. S. Improving early drug discovery through ADME modelling: an overview. *Drugs R D* **8**, 349–362 (2007).

1224. Walters, W. P., Ajay, & Murcko, M. A. Recognizing molecules with drug-like properties. *Curr Opin Chem Biol* **3**, 384–387 (1999).

1225. Lipinski, C. A., Lombardo, F., Dominy, B. W., & Feeney, P. J. Experimental and computational approaches to estimate solubility and permeability in drug discovery and development settings. *Adv Drug Deliv Rev* **23**, 3–25 (1997).

1226. Lipinski, C. A. Drug-like properties and the causes of poor solubility and poor permeability. *J Pharmacol Toxicol Methods* **44**, 235–249 (2000).

1227. Vistoli, G., Pedretti, A., & Testa, B. Assessing drug-likeness: what are we missing? *Drug Discov Today* **13**, 285–294 (2008).

1228. Zhang, M. Q. & Wilkinson, B. Drug discovery beyond the 'rule-of-five'. *Curr Opin Biotechnol* **18**, 478–488 (2007).

1229. Lajiness, M. S., Vieth, M., & Erickson, J. Molecular properties that influence oral drug-like behavior. *Curr Opin Drug Discov Devel* **7**, 470–477 (2004).

1230. Overington, J. P., Al-Lazikani, B. & Hopkins, A. L. How many drug targets are there? *Nat Rev Drug Discov* **5**, 993–996 (2006).

1231. Guerin, G. A., Pratuangdejkul, J., Alemany, M., Launay, J. M., & Manivet, P. Rational and efficient geometric definition of pharmacophores is essential for the patent process. *Drug Discov Today* **11**, 991–998 (2006).

1232. Sheridan, R. P., Rusinko, A., 3rd, Nilakantan, R. & Venkataraghavan, R. Searching for pharmacophores in large coordinate data bases and its use in drug design. *Proc Natl Acad Sci USA* **86**, 8165–8169 (1989).

1233. Langer, T. & Hoffmann, R. D. (eds.) *Pharmacophores and Pharmacophore Searches* (Wiley-VCH, 2006).

1234. Stoll, F., et al. Pharmacophore definition and three-dimensional quantitative structure–activity relationship study on structurally diverse prostacyclin receptor agonists. *Mol Pharmacol* **62**, 1103–1111 (2002).

1235. Sakharkar, M. K. & Sakharkar, K. R. Targetability of human disease genes. *Curr Drug Discov Technol* **4**, 48–58 (2007).

1236. Keller, T. H., Pichota, A., & Yin, Z. A practical view of 'druggability'. *Curr Opin Chem Biol* **10**, 357–361 (2006).

1237. Hajduk, P. J., Huth, J. R., & Tse, C. Predicting protein druggability. *Drug Discov Today* **10**, 1675–1682 (2005).

1238. Russ, A. P. & Lampel, S. The druggable genome: an update. *Drug Discov Today* **10**, 1607–1610 (2005).

1239. Imming, P., Sinning, C., & Meyer, A. Drugs, their targets and the nature and number of drug targets. *Nat Rev Drug Discov* **5**, 821–834 (2006).

1240. Visintin, M., Melchionna, T., Cannistraci, I., & Cattaneo, A. *In vivo* selection of intrabodies specifically targeting protein–protein interactions: a general platform for an "undruggable" class of disease targets. *J Biotechnol* **135**, 1–15 (2008).

1241. Dev, K. K. Making protein interactions druggable: targeting PDZ domains. *Nat Rev Drug Discov* **3**, 1047–1056 (2004).

1242. Domling, A. Small molecular weight protein–protein interaction antagonists: an insurmountable challenge? *Curr Opin Chem Biol* **12**, 281–291 (2008).

1243. Erlanson, D. A., Wells, J. A., & Braisted, A. C. Tethering: fragment-based drug discovery. *Annu Rev Biophys Biomol Struct* **33**, 199–223 (2004).

1244. Hwang, S., et al. Inhibition of gene expression in human cells through small molecule–RNA interactions. *Proc Natl Acad Sci USA* **96**, 12997–13002 (1999).

1245. Kaufmann, S. H. Paul Ehrlich: founder of chemotherapy. *Nat Rev Drug Discov* **7**, 373 (2008).

1246. Diller, D. J. The synergy between combinatorial chemistry and high-throughput screening. *Curr Opin Drug Discov Devel* **11**, 346–355 (2008).

1247. Koesdjojo, M. T., Rasmussen, H. T., Fermier, A. M., Patel, P., & Remcho, V. T. The development of a semiautomated procedure for the synthesis and screening of a large group of molecularly imprinted polymers. *J Comb Chem* **9**, 929–934 (2007).

1248. Webb, T. R. Current directions in the evolution of compound libraries. *Curr Opin Drug Discov Devel* **8**, 303–308 (2005).

1249. Hann, M. M. & Oprea, T. I. Pursuing the leadlikeness concept in pharmaceutical research. *Curr Opin Chem Biol* **8**, 255–263 (2004).

1250. Verheij, H. J. Leadlikeness and structural diversity of synthetic screening libraries. *Mol Divers* **10**, 377–388 (2006).

1251. Ganesan, A. The impact of natural products upon modern drug discovery. *Curr Opin Chem Biol* **12**, 306–317 (2008).

1252. Feher, M. & Schmidt, J. M. Property distributions: differences between drugs, natural products, and molecules from combinatorial chemistry. *J Chem Inf Comput Sci* **43**, 218–227 (2003).

1253. Messer, R., Fuhrer, C. A., & Haner, R. Natural product-like libraries based on non-aromatic, polycyclic motifs. *Curr Opin Chem Biol* **9**, 259–265 (2005).

1254. Tan, D. S. Diversity-oriented synthesis: exploring the intersections between chemistry and biology. *Nat Chem Biol* **1**, 74–84 (2005).

1255. Fergus, S., Bender, A., & Spring, D. R. Assessment of structural diversity in combinatorial synthesis. *Curr Opin Chem Biol* **9**, 304–309 (2005).

1256. Hann, M. M., Leach, A. R., & Harper, G. Molecular complexity and its impact on the probability of finding leads for drug discovery. *J Chem Inf Comput Sci* **41**, 856–864 (2001).

1257. Selzer, P., Roth, H. J., Ertl, P., & Schuffenhauer, A. Complex molecules: do they add value? *Curr Opin Chem Biol* **9**, 310–316 (2005).

1258. Schuffenhauer, A., Brown, N., Selzer, P., Ertl, P., & Jacoby, E. Relationships between molecular complexity, biological activity, and structural diversity. *J Chem Inf Model* **46**, 525–535 (2006).

1259. Abel, U., Koch, C., Speitling, M., & Hansske, F. G. Modern methods to produce natural-product libraries. *Curr Opin Chem Biol* **6**, 453–458 (2002).

1260. Bugni, T. S., et al. Marine natural product libraries for high-throughput screening and rapid drug discovery. *J Nat Prod* **71**, 1095–1098 (2008).

1261. Baker, D. D., Chu, M., Oza, U., & Rajgarhia, V. The value of natural products to future pharmaceutical discovery. *Nat Prod Rep* **24**, 1225–1244 (2007).

1262. Lam, K. S. New aspects of natural products in drug discovery. *Trends Microbiol* **15**, 279–289 (2007).

1263. Ortholand, J. Y. & Ganesan, A. Natural products and combinatorial chemistry: back to the future. *Curr Opin Chem Biol* **8**, 271–280 (2004).

1264. Jacoby, E., et al. Key aspects of the Novartis compound collection enhancement project for the compilation of a comprehensive chemogenomics drug discovery screening collection. *Curr Top Med Chem* **5**, 397–411 (2005).

1265. Schopfer, U., et al. The Novartis compound archive: from concept to reality. *Comb Chem High Throughput Screen* **8**, 513–519 (2005).

1266. Baurin, N., et al. Drug-like annotation and duplicate analysis of a 23-supplier chemical database totalling 2.7 million compounds. *J Chem Inf Comput Sci* **44**, 643–651 (2004).

1267. Hergenrother, P. J. Obtaining and screening compound collections: a user's guide and a call to chemists. *Curr Opin Chem Biol* **10**, 213–218 (2006).

1268. Monge, A., Arrault, A., Marot, C., & Morin-Allory, L. Managing, profiling and analyzing a library of 2.6 million compounds gathered from 32 chemical providers. *Mol Divers* **10**, 389–403 (2006).

1269. Lajiness, M. & Watson, I. Dissimilarity-based approaches to compound acquisition. *Curr Opin Chem Biol* **12**, 366–371 (2008).

1270. Merrifield, R. B. Solid-phase peptide synthesis. *Adv Enzymol Relat Areas Mol Biol* **32**, 221–296 (1969).

1271. Geysen, H. M., Meloen, R. H., & Barteling, S. J. Use of peptide synthesis to probe viral antigens for epitopes to a resolution of a single amino acid. *Proc Natl Acad Sci USA* **81**, 3998–4002 (1984).

1272. de Koster, H. S., et al. The use of dedicated peptide libraries permits the discovery of high affinity binding peptides. *J Immunol Methods* **187**, 179–188 (1995).

1273. Affleck, R. L. Solutions for library encoding to create collections of discrete compounds. *Curr Opin Chem Biol* **5**, 257–263 (2001).

1274. Dobson, C. M. Chemical space and biology. *Nature* **432**, 824–828 (2004).

1275. Gribbon, P. & Sewing, A. High-throughput drug discovery: what can we expect from HTS? *Drug Discov Today* **10**, 17–22 (2005).

1276. Fink, T., Bruggesser, H., & Reymond, J. L. Virtual exploration of the small-molecule chemical universe below 160 Daltons. *Angew Chem Int Ed Engl* **44**, 1504–1508 (2005).

1277. Merritt, A. Rumors of the demise of combinatorial chemistry are greatly exaggerated. *Curr Opin Chem Biol* **7**, 305–307 (2003).

1278. Schnur, D. M. Recent trends in library design: 'rational design' revisited. *Curr Opin Drug Discov Devel* **11**, 375–380 (2008).

1279. Huwe, C. M. Synthetic library design. *Drug Discov Today* **11**, 763–767 (2006).

1280. Zhou, J. Z. Structure-directed combinatorial library design. *Curr Opin Chem Biol* **12**, 379–385 (2008).

1281. Schreiber, S. L. Target-oriented and diversity-oriented organic synthesis in drug discovery. *Science* **287**, 1964–1969 (2000).

1282. Martin, Y. C., Kofron, J. L., & Traphagen, L. M. Do structurally similar molecules have similar biological activity? *J Med Chem* **45**, 4350–4358 (2002).

1283. Koehn, F. E. & Carter, G. T. The evolving role of natural products in drug discovery. *Nat Rev Drug Discov* **4**, 206–220 (2005).

1284. Koch, M. A., et al. Charting biologically relevant chemical space: a structural classification of natural products (SCONP). *Proc Natl Acad Sci USA* **102**, 17272–17277 (2005).

1285. Boldi, A. M. Libraries from natural product-like scaffolds. *Curr Opin Chem Biol* **8**, 281–286 (2004).

1286. Shang, S. & Tan, D. S. Advancing chemistry and biology through diversity-oriented synthesis of natural product-like libraries. *Curr Opin Chem Biol* **9**, 248–258 (2005).

1287. Ertl, P. & Schuffenhauer, A. Cheminformatics analysis of natural products: lessons from nature inspiring the design of new drugs. *Prog Drug Res* **66** 217, 219–235 (2008).

1288. Kingston, D. G. & Newman, D. J. Natural products as drug leads: an old process or the new hope for drug discovery? *IDrugs* **8**, 990–992 (2005).

1289. Reayi, A. & Arya, P. Natural product-like chemical space: search for chemical dissectors of macromolecular interactions. *Curr Opin Chem Biol* **9**, 240–247 (2005).

1290. DeSimone, R. W., Currie, K. S., Mitchell, S. A., Darrow, J. W., & Pippin, D. A. Privileged structures: applications in drug discovery. *Comb Chem High Throughput Screen* **7**, 473–494 (2004).

1291. Costantino, L. & Barlocco, D. Privileged structures as leads in medicinal chemistry. *Curr Med Chem* **13**, 65–85 (2006).

1292. Duarte, C. D., Barreiro, E. J., & Fraga, C. A. Privileged structures: a useful concept for the rational design of new lead drug candidates. *Mini Rev Med Chem* **7**, 1108–1119 (2007).

1293. Bondensgaard, K., et al. Recognition of privileged structures by G-protein coupled receptors. *J Med Chem* **47**, 888–899 (2004).

1294. Bywater, R. P. Privileged structures in GPCRs. *Ernst Schering Found Symp Proc* 75–91 (2006).

1295. Guo, T. & Hobbs, D. W. Privileged structure-based combinatorial libraries targeting G protein-coupled receptors. *Assay Drug Dev Technol* **1**, 579–592 (2003).

1296. Muller, G. Medicinal chemistry of target family-directed masterkeys. *Drug Discov Today* **8**, 681–691 (2003).

1297. Schnur, D. M., Hermsmeier, M. A., & Tebben, A. J. Are target-family-privileged substructures truly privileged? *J Med Chem* **49**, 2000–2009 (2006).

1298. Burke, M. D. & Lalic, G. Teaching target-oriented and diversity-oriented organic synthesis at Harvard University. *Chem Biol* **9**, 535–541 (2002).

1299. Ulaczyk-Lesanko, A. & Hall, D. G. Wanted: new multicomponent reactions for generating libraries of polycyclic natural products. *Curr Opin Chem Biol* **9**, 266–276 (2005).

1300. Marcaurelle, L. A. & Johannes, C. W. Application of natural product-inspired diversity-oriented synthesis to drug discovery. *Prog Drug Res* **66**, 187, 189–216 (2008).

1301. Hulme, C. & Nixey, T. Rapid assembly of molecular diversity via exploitation of isocyanide-based multi-component reactions. *Curr Opin Drug Discov Devel* **6**, 921–929 (2003).

1302. Bender, A., et al. Diversity oriented synthesis: a challenge for synthetic chemists. *Ernst Schering Res Found Workshop* 47–60 (2006).

1303. Weber, L. The application of multi-component reactions in drug discovery. *Curr Med Chem* **9**, 2085–2093 (2002).

1304. Hesterkamp, T. & Whittaker, M. Fragment-based activity space: smaller is better. *Curr Opin Chem Biol* **12**, 260–268 (2008).

1305. Siegel, M. G. & Vieth, M. Drugs in other drugs: a new look at drugs as fragments. *Drug Discov Today* **12**, 71–79 (2007).

1306. Congreve, M., Carr, R., Murray, C., & Jhoti, H. A 'rule of three' for fragment-based lead discovery? *Drug Discov Today* **8**, 876–877 (2003).

1307. Hajduk, P. J. Fragment-based drug design: how big is too big? *J Med Chem* **49**, 6972–6976 (2006).

1308. Hajduk, P. J. & Greer, J. A decade of fragment-based drug design: strategic advances and lessons learned. *Nat Rev Drug Discov* **6**, 211–219 (2007).

1309. Fattori, D., Squarcia, A., & Bartoli, S. Fragment-based approach to drug lead discovery: overview and advances in various techniques. *Drugs R D* **9**, 217–227 (2008).

1310. Schuffenhauer, A., et al. Library design for fragment based screening. *Curr Top Med Chem* **5**, 751–762 (2005).

1311. Irwin, J. J. How good is your screening library? *Curr Opin Chem Biol* **10**, 352–356 (2006).

1312. Shuker, S. B., Hajduk, P. J., Meadows, R. P., & Fesik, S. W. Discovering high-affinity ligands for proteins: SAR by NMR. *Science* **274**, 1531–1534 (1996).

1313. Erlanson, D. A. Fragment-based lead discovery: a chemical update. *Curr Opin Biotechnol* **17**, 643–652 (2006).

1314. Zartler, E. R. & Shapiro, M. J. Fragonomics: fragment-based drug discovery. *Curr Opin Chem Biol* **9**, 366–370 (2005).

1315. Keseru, G. M. & Makara, G. M. Hit discovery and hit-to-lead approaches. *Drug Discov Today* **11**, 741–748 (2006).

1316. Nguyen, R. & Huc, I. Using an enzyme's active site to template inhibitors. *Angew Chem Int Ed Engl* **40**, 1774–1776 (2001).

1317. Ramstrom, O. & Lehn, J. M. Drug discovery by dynamic combinatorial libraries. *Nat Rev Drug Discov* **1**, 26–36 (2002).

1318. Cheeseman, J. D., Corbett, A. D., Gleason, J. L., & Kazlauskas, R. J. Receptor-assisted combinatorial chemistry: thermodynamics and kinetics in drug discovery. *Chemistry* **11**, 1708–1716 (2005).

1319. Rowan, S. J., Cantrill, S. J., Cousins, G. R., Sanders, J. K., & Stoddart, J. F. Dynamic covalent chemistry. *Angew Chem Int Ed Engl* **41**, 898–952 (2002).

1320. Corbett, P. T., Sanders, J. K., & Otto, S. Systems chemistry: pattern formation in random dynamic combinatorial libraries. *Angew Chem Int Ed Engl* **46**, 8858–8861 (2007).

1321. Otto, S., Furlan, R. L., & Sanders, J. K. Selection and amplification of hosts from dynamic combinatorial libraries of macrocyclic disulfides. *Science* **297**, 590–593 (2002).

1322. Lam, R. T., et al. Amplification of acetylcholine-binding catenanes from dynamic combinatorial libraries. *Science* **308**, 667–669 (2005).

1323. Huc, I. & Lehn, J. M. Virtual combinatorial libraries: dynamic generation of molecular and supramolecular diversity by self-assembly. *Proc Natl Acad Sci USA* **94**, 2106–2110 (1997).

1324. Cousins, G. R., Poulsen, S. A., & Sanders, J. K. Molecular evolution: dynamic combinatorial libraries, autocatalytic networks and the quest for molecular function. *Curr Opin Chem Biol* **4**, 270–279 (2000).

1325. Klekota, B. & Miller, B. L. Dynamic diversity and small-molecule evolution: a new paradigm for ligand identification. *Trends Biotechnol* **17**, 205–209 (1999).

1326. Corbett, P. T., Sanders, J. K., & Otto, S. Competition between receptors in dynamic combinatorial libraries: amplification of the fittest? *J Am Chem Soc* **127**, 9390–9392 (2005).

1327. Melkko, S., Scheuermann, J., Dumelin, C. E., & Neri, D. Encoded self-assembling chemical libraries. *Nat Biotechnol* **22**, 568–574 (2004).

1328. Rodda, S. J. Peptide libraries for T cell epitope screening and characterization. *J Immunol Methods* **267**, 71–77 (2002).

1329. Guiles, J. W., Lanter, C. L., & Rivero, R. A. A visual tagging process for mix and sort combinatorial chemistry. *Angew Chem Int Ed Engl* **37**, 926–928 (1998).

1330. Nicolaou, K. C. Radiofrequency encoded combinatorial chemistry. *Angew Chem Int Ed Engl* **34**, 2289–2291 (1995).

1331. Xiao, X. Y., et al. Solid-phase combinatorial synthesis using MicroKan reactors, Rf tagging, and directed sorting. *Biotechnol Bioeng* **71**, 44–50 (2000).

1332. Erb, E., Janda, K. D., & Brenner, S. Recursive deconvolution of combinatorial chemical libraries. *Proc Natl Acad Sci USA* **91**, 11422–11426 (1994).

1333. Lam, K. S., et al. Synthesis and screening of "one-bead one-compound" combinatorial peptide libraries. *Methods Enzymol* **369**, 298–322 (2003).

1334. Hiemstra, H. S., et al. The identification of CD4$^+$ T cell epitopes with dedicated synthetic peptide libraries. *Proc Natl Acad Sci USA* **94**, 10313–10318 (1997).

1335. Czarnik, A. W. Encoding strategies in combinatorial chemistry. *Proc Natl Acad Sci USA* **94**, 12738–12739 (1997).

1336. Muckenschnabel, I., Falchetto, R., Mayr, L. M., & Filipuzzi, I. SpeedScreen: label-free liquid chromatography–mass spectrometry-based high-throughput screening for the discovery of orphan protein ligands. *Anal Biochem* **324**, 241–249 (2004).

1337. Brown, N., et al. A chemoinformatics analysis of hit lists obtained from high-throughput affinity-selection screening. *J Biomol Screen* **11**, 123–130 (2006).

1338. Hajduk, P. J., et al. High-throughput nuclear magnetic resonance-based screening. *J Med Chem* **42**, 2315–2317 (1999).

1339. Mercier, K. A. & Powers, R. Determining the optimal size of small molecule mixtures for high throughput NMR screening. *J Biomol NMR* **31**, 243–258 (2005).

1340. Barnes, C. & Balasubramanian, S. Recent developments in the encoding and deconvolution of combinatorial libraries. *Curr Opin Chem Biol* **4**, 346–350 (2000).

1341. Liu, R., Marik, J., & Lam, K. S. A novel peptide-based encoding system for "one-bead one-compound" peptidomimetic and small molecule combinatorial libraries. *J Am Chem Soc* **124**, 7678–7680 (2002).

1342. Wang, X., Zhang, J., Song, A., Lebrilla, C. B., & Lam, K. S. Encoding method for OBOC small molecule libraries using a biphasic approach for ladder-synthesis of coding tags. *J Am Chem Soc* **126**, 5740–5749 (2004).

1343. Ohlmeyer, M. H., et al. Complex synthetic chemical libraries indexed with molecular tags. *Proc Natl Acad Sci USA* **90**, 10922–10926 (1993).

1344. Song, A., Zhang, J., Lebrilla, C. B., & Lam, K. S. A novel and rapid encoding method based on mass spectrometry for "one-bead-one-compound" small molecule combinatorial libraries. *J Am Chem Soc* **125**, 6180–6188 (2003).

1345. Kerr, J. M., Banville, S. C., & Zuckermann, R. N. Encoded combinatorial peptide libraries containing non-natural amino acids. *J Am Chem Soc* **115**, 2519–2531 (1993).

1346. Nikolaiev, V., et al. Peptide-encoding for structure determination of nonsequenceable polymers within libraries synthesized and tested on solid-phase supports. *Pept Res* **6**, 161–170 (1993).

1347. Stults, J. T. Matrix-assisted laser desorption/ionization mass spectrometry (MALDI-MS). *Curr Opin Struct Biol* **5**, 691–698 (1995).

1348. Bienvenut, W. V., et al. Matrix-assisted laser desorption/ionization-tandem mass spectrometry with high resolution and sensitivity for identification and characterization of proteins. *Proteomics* **2**, 868–876 (2002).

1349. Geysen, H. M., et al. Isotope or mass encoding of combinatorial libraries. *Chem Biol* **3**, 679–688 (1996).

1350. Brenner, S. & Lerner, R. A. Encoded combinatorial chemistry. *Proc Natl Acad Sci USA* **89**, 5381–5383 (1992).

1351. Needels, M. C., et al. Generation and screening of an oligonucleotide-encoded synthetic peptide library. *Proc Natl Acad Sci USA* **90**, 10700–10704 (1993).

1352. Fellouse, F. & Deshayes, K. Bacteriophage that display small molecules. *Chem Biol* **10**, 783–784 (2003).

1353. Woiwode, T. F., et al. Synthetic compound libraries displayed on the surface of encoded bacteriophage. *Chem Biol* **10**, 847–858 (2003).

1354. Odegrip, R., et al. CIS display: *in vitro* selection of peptides from libraries of protein–DNA complexes. *Proc Natl Acad Sci USA* **101**, 2806–2810 (2004).

1355. Halpin, D. R. & Harbury, P. B. DNA display. I. Sequence-encoded routing of DNA populations. *PLoS Biol* **2**, E173 (2004).

1356. Halpin, D. R. & Harbury, P. B. DNA display. II. Genetic manipulation of combinatorial chemistry libraries for small-molecule evolution. *PLoS Biol* **2**, E174 (2004).

1357. Wrenn, S. J., Weisinger, R. M., Halpin, D. R., & Harbury, P. B. Synthetic ligands discovered by *in vitro* selection. *J Am Chem Soc* **129**, 13137–13143 (2007).

1358. Halpin, D. R., Lee, J. A., Wrenn, S. J., & Harbury, P. B. DNA display. III. Solid-phase organic synthesis on unprotected DNA. *PLoS Biol* **2**, E175 (2004).

1359. Li, X. & Liu, D. R. DNA-templated organic synthesis: nature's strategy for controlling chemical reactivity applied to synthetic molecules. *Angew Chem Int Ed Engl* **43**, 4848–4870 (2004).

1360. Crabtree, G. R. & Schreiber, S. L. Three-part inventions: intracellular signaling and induced proximity. *Trends Biochem Sci* **21**, 418–422 (1996).

1361. Snyder, T. M., Tse, B. N., & Liu, D. R. Effects of template sequence and secondary structure on DNA-templated reactivity. *J Am Chem Soc* **130**, 1392–1401 (2008).

1362. Gartner, Z. J. & Liu, D. R. The generality of DNA-templated synthesis as a basis for evolving non-natural small molecules. *J Am Chem Soc* **123**, 6961–6963 (2001).

1363. Gartner, Z. J., Kanan, M. W., & Liu, D. R. Multistep small-molecule synthesis programmed by DNA templates. *J Am Chem Soc* **124**, 10304–10306 (2002).

1364. Kanan, M. W., Rozenman, M. M., Sakurai, K., Snyder, T. M., & Liu, D. R. Reaction discovery enabled by DNA-templated synthesis and *in vitro* selection. *Nature* **431**, 545–549 (2004).

1365. Calderone, C. T. & Liu, D. R. Nucleic-acid-templated synthesis as a model system for ancient translation. *Curr Opin Chem Biol* **8**, 645–653 (2004).

1366. Rupasinghe, C. N. & Spaller, M. R. The interplay between structure-based design and combinatorial chemistry. *Curr Opin Chem Biol* **10**, 188–193 (2006).

1367. Black, J. Drugs from emasculated hormones: the principle of syntopic antagonism. *Science* **245**, 486–493 (1989).

1368. Bradley, E. K., Miller, J. L., Saiah, E., & Grootenhuis, P. D. Informative library design as an efficient strategy to identify and optimize leads: application to cyclin-dependent kinase 2 antagonists. *J Med Chem* **46**, 4360–4364 (2003).

1369. Weaver, D. C. Applying data mining techniques to library design, lead generation and lead optimization. *Curr Opin Chem Biol* **8**, 264–270 (2004).

1370. Blower, P. E., et al. Comparison of methods for sequential screening of large compound sets. *Comb Chem High Throughput Screen* **9**, 115–122 (2006).

1371. Holton, J. & Alber, T. Automated protein crystal structure determination using ELVES. *Proc Natl Acad Sci USA* **101**, 1537–1542 (2004).

1372. Manjasetty, B. A., Turnbull, A. P., Panjikar, S., Bussow, K., & Chance, M. R. Automated technologies and novel techniques to accelerate protein crystallography for structural genomics. *Proteomics* **8**, 612–625 (2008).

1373. Thurmond, R. L., Gelfand, E. W., & Dunford, P. J. The role of histamine H1 and H4 receptors in allergic inflammation: the search for new antihistamines. *Nat Rev Drug Discov* **7**, 41–53 (2008).

1374. Gantz, I., et al. Molecular cloning of a gene encoding the histamine H2 receptor. *Proc Natl Acad Sci USA* **88**, 429–433 (1991).

1375. Gillard, M., Van Der Perren, C., Moguilevsky, N., Massingham, R., & Chatelain, P. Binding characteristics of cetirizine and levocetirizine to human H(1) histamine receptors: contribution of Lys(191) and Thr(194). *Mol Pharmacol* **61**, 391–399 (2002).

1376. Bryan, J. No longer a pain in the gut: how tagamet led peptic ulcer treatments. *The Pharmaceutical Journal* **279**, 656–657 (2007).

1377. Filizola, M. & Weinstein, H. The structure and dynamics of GPCR oligomers: a new focus in models of cell-signaling mechanisms and drug design. *Curr Opin Drug Discov Devel* **8**, 577–584 (2005).

1378. Achour, L., Labbe-Jullie, C., Scott, M. G. & Marullo, S. An escort for GPCRs: implications for regulation of receptor density at the cell surface. *Trends Pharmacol Sci* **29**, 528–535 (2008).

1379. Moniri, N. H., Covington-Strachan, D. & Booth, R. G. Ligand-directed functional heterogeneity of histamine H1 receptors: novel dual-function ligands selectively activate and block H1-mediated phospholipase C and adenylyl cyclase signaling. *J Pharmacol Exp Ther* **311**, 274–281 (2004).

1380. Welch, M. J., Meltzer, E. O., & Simons, F. E. H1-antihistamines and the central nervous system. *Clin Allergy Immunol* **17**, 337–388 (2002).

1381. Nguyen, T., et al. Discovery of a novel member of the histamine receptor family. *Mol Pharmacol* **59**, 427–433 (2001).

1382. Masaki, T. & Yoshimatsu, H. The hypothalamic H1 receptor: a novel therapeutic target for disrupting diurnal feeding rhythm and obesity. *Trends Pharmacol Sci* **27**, 279–284 (2006).

1383. Penrose, R. *Shadows of the Mind* (Oxford University Press, 1994).

1384. Klinman, J. P. The role of tunneling in enzyme catalysis of C–H activation. *Biochim Biophys Acta* **1757**, 981–987 (2006).

1385. Nagel, Z. D. & Klinman, J. P. Tunneling and dynamics in enzymatic hydride transfer. *Chem Rev* **106**, 3095–3118 (2006).

1386. Pu, J., Gao, J., & Truhlar, D. G. Multidimensional tunneling, recrossing, and the transmission coefficient for enzymatic reactions. *Chem Rev* **106**, 3140–3169 (2006).

1387. Ball, P. Enzymes: by chance, or by design? *Nature* **431**, 396–397 (2004).

1388. Aspuru-Guzik, A., Dutoi, A. D., Love, P. J., & Head-Gordon, M. Simulated quantum computation of molecular energies. *Science* **309**, 1704–1707 (2005).

1389. Friesner, R. A. *Ab initio* quantum chemistry: methodology and applications. *Proc Natl Acad Sci USA* **102**, 6648–6653 (2005).

1390. van Mourik, T. First-principles quantum chemistry in the life sciences. *Philos Transact A Math Phys Eng Sci* **362**, 2653–2670 (2004).

1391. Oldfield, E. Quantum chemical studies of protein structure. *Philos Trans R Soc Lond B Biol Sci* **360**, 1347–1361 (2005).

1392. Tour, J. M. & He, T. Electronics: the fourth element. *Nature* **453**, 42–43 (2008).

1393. Aaronson, S. The limits of quantum computers. *Sci Am* **298**, 50–57 (2008).

1394. Chohan, K. K., Paine, S. W., & Waters, N. J. Quantitative structure activity relationships in drug metabolism. *Curr Top Med Chem* **6**, 1569–1578 (2006).

1395. Vedani, A., Descloux, A. V., Spreafico, M., & Ernst, B. Predicting the toxic potential of drugs and chemicals *in silico*: a model for the peroxisome proliferator-activated receptor gamma (PPAR gamma). *Toxicol Lett* **173**, 17–23 (2007).

1396. Khedkar, S. A., Malde, A. K., Coutinho, E. C., & Srivastava, S. Pharmacophore modeling in drug discovery and development: an overview. *Med Chem* **3**, 187–197 (2007).

1397. Hartman, G. D., et al. Non-peptide fibrinogen receptor antagonists. 1. Discovery and design of exosite inhibitors. *J Med Chem* **35**, 4640–4642 (1992).

1398. Kapetanovic, I. M. Computer-aided drug discovery and development (CADDD): *in silico*-chemico-biological approach. *Chem Biol Interact* **171**, 165–176 (2008).

1399. Holtje, H.-D., Sippl, W., Rognan, D., & Folkers, G. *Molecular Modeling. Basic Principles and Applications*, 2nd edition (Wiley-VCH, 2003).

1400. Patrick, G. L. *An Introduction to Medicinal Chemistry*, 3rd edition (Oxford University Press, 2005).

1401. Walters, W. P. & Goldman, B. B. Feature selection in quantitative structure–activity relationships. *Curr Opin Drug Discov Devel* **8**, 329–333 (2005).

1402. Meek, P. J., et al. Shape signatures: speeding up computer aided drug discovery. *Drug Discov Today* **11**, 895–904 (2006).

1403. Lill, M. A. Multi-dimensional QSAR in drug discovery. *Drug Discov Today* **12**, 1013–1017 (2007).

1404. Teague, S. J. Implications of protein flexibility for drug discovery. *Nat Rev Drug Discov* **2**, 527–541 (2003).

1405. Gunasekaran, K. & Nussinov, R. How different are structurally flexible and rigid binding sites? Sequence and structural features discriminating proteins that do and do not undergo conformational change upon ligand binding. *J Mol Biol* **365**, 257–273 (2007).

1406. Watson, J. D. & Crick, F. H. Molecular structure of nucleic acids; a structure for deoxyribose nucleic acid. *Nature* **171**, 737–738 (1953).

1407. Richon, A. B. Current status and future direction of the molecular modeling industry. *Drug Discov Today* **13**, 665–669 (2008).

1408. Bostrom, J., Hogner, A., & Schmitt, S. Do structurally similar ligands bind in a similar fashion? *J Med Chem* **49**, 6716–6725 (2006).

1409. Keiser, M. J., et al. Relating protein pharmacology by ligand chemistry. *Nat Biotechnol* **25**, 197–206 (2007).

1410. Nettles, J. H., et al. The binding mode of epothilone A on alpha,beta-tubulin by electron crystallography. *Science* **305**, 866–869 (2004).

1411. Kubinyi, H. Drug research: myths, hype and reality. *Nat Rev Drug Discov* **2**, 665–668 (2003).

1412. Swaan, P. W. & Ekins, S. Reengineering the pharmaceutical industry by crash-testing molecules. *Drug Discov Today* **10**, 1191–1200 (2005).

1413. Ghosh, S., Nie, A., An, J., & Huang, Z. Structure-based virtual screening of chemical libraries for drug discovery. *Curr Opin Chem Biol* **10**, 194–202 (2006).

1414. Bajorath, J. Integration of virtual and high-throughput screening. *Nat Rev Drug Discov* **1**, 882–894 (2002).

1415. McInnes, C. Virtual screening strategies in drug discovery. *Curr Opin Chem Biol* **11**, 494–502 (2007).

1416. Goto, J., Kataoka, R., & Hirayama, N. Ph4Dock: pharmacophore-based protein–ligand docking. *J Med Chem* **47**, 6804–6811 (2004).

1417. Kitchen, D. B., Decornez, H., Furr, J. R., & Bajorath, J. Docking and scoring in virtual screening for drug discovery: methods and applications. *Nat Rev Drug Discov* **3**, 935–949 (2004).

1418. McGovern, S. L. & Shoichet, B. K. Information decay in molecular docking screens against holo, apo, and modeled conformations of enzymes. *J Med Chem* **46**, 2895–2907 (2003).

1419. Erickson, J. A., Jalaie, M., Robertson, D. H., Lewis, R. A., & Vieth, M. Lessons in molecular recognition: the effects of ligand and protein flexibility on molecular docking accuracy. *J Med Chem* **47**, 45–55 (2004).

1420. Perola, E. & Charifson, P. S. Conformational analysis of drug-like molecules bound to proteins: an extensive study of ligand reorganization upon binding. *J Med Chem* **47**, 2499–2510 (2004).

1421. Klebe, G. Virtual ligand screening: strategies, perspectives and limitations. *Drug Discov Today* **11**, 580–594 (2006).

1422. Sousa, S. F., Fernandes, P. A., & Ramos, M. J. Protein–ligand docking: current status and future challenges. *Proteins* **65**, 15–26 (2006).

1423. Davies, J. W., Glick, M., & Jenkins, J. L. Streamlining lead discovery by aligning *in silico* and high-throughput screening. *Curr Opin Chem Biol* **10**, 343–351 (2006).

1424. Joseph-McCarthy, D., Baber, J. C., Feyfant, E., Thompson, D. C., & Humblet, C. Lead optimization via high-throughput molecular docking. *Curr Opin Drug Discov Devel* **10**, 264–274 (2007).

1425. Schapira, M., et al. Discovery of diverse thyroid hormone receptor antagonists by high-throughput docking. *Proc Natl Acad Sci USA* **100**, 7354–7359 (2003).

1426. Schapira, M., Abagyan, R., & Totrov, M. Nuclear hormone receptor targeted virtual screening. *J Med Chem* **46**, 3045–3059 (2003).

1427. McInnes, C. Improved lead-finding for kinase targets using high-throughput docking. *Curr Opin Drug Discov Devel* **9**, 339–347 (2006).

1428. Ballesteros, J. & Palczewski, K. G protein-coupled receptor drug discovery: implications from the crystal structure of rhodopsin. *Curr Opin Drug Discov Devel* **4**, 561–574 (2001).

1429. Cherezov, V., et al. High-resolution crystal structure of an engineered human beta2-adrenergic G protein-coupled receptor. *Science* **318**, 1258–1265 (2007).

1430. Audet, M. & Bouvier, M. Insights into signaling from the beta2-adrenergic receptor structure. *Nat Chem Biol* **4**, 397–403 (2008).

1431. Costanzi, S. On the applicability of GPCR homology models to computer-aided drug discovery: a comparison between *in silico* and crystal structures of the beta2-adrenergic receptor. *J Med Chem* **51**, 2907–2914 (2008).

1432. Muller, P., et al. *In silico*-guided target identification of a scaffold-focused library: 1,3,5-triazepan-2,6-diones as novel phospholipase A2 inhibitors. *J Med Chem* **49**, 6768–6778 (2006).

1433. Do, Q. T., et al. Reverse pharmacognosy: application of selnergy, a new tool for lead discovery. The example of epsilon-viniferin. *Curr Drug Discov Technol* **2**, 161–167 (2005).

1434. Schneider, G. & Fechner, U. Computer-based *de novo* design of drug-like molecules. *Nat Rev Drug Discov* **4**, 649–663 (2005).

1435. Mauser, H. & Guba, W. Recent developments in *de novo* design and scaffold hopping. *Curr Opin Drug Discov Devel* **11**, 365–374 (2008).

1436. Pierce, A. C., Rao, G., & Bemis, G. W. BREED: generating novel inhibitors through hybridization of known ligands. Application to CDK2, p38, and HIV protease. *J Med Chem* **47**, 2768–2775 (2004).

1437. Boda, K. & Johnson, A. P. Molecular complexity analysis of *de novo* designed ligands. *J Med Chem* **49**, 5869–5879 (2006).

1438. Vagner, J., Qu, H., & Hruby, V. J. Peptidomimetics, a synthetic tool of drug discovery. *Curr Opin Chem Biol* **12**, 292–296 (2008).

1439. Ji, H., et al. Minimal pharmacophoric elements and fragment hopping, an approach directed at molecular diversity and isozyme selectivity. Design of selective neuronal nitric oxide synthase inhibitors. *J Am Chem Soc* **130**, 3900–3914 (2008).

1440. Brown, N. & Jacoby, E. On scaffolds and hopping in medicinal chemistry. *Mini Rev Med Chem* **6**, 1217–1229 (2006).

1441. Bains, W. & Tacke, R. Silicon chemistry as a novel source of chemical diversity in drug design. *Curr Opin Drug Discov Devel* **6**, 526–543 (2003).

1442. Showell, G. A. & Mills, J. S. Chemistry challenges in lead optimization: silicon isosteres in drug discovery. *Drug Discov Today* **8**, 551–556 (2003).

1443. Altieri, A. S. & Byrd, R. A. Automation of NMR structure determination of proteins. *Curr Opin Struct Biol* **14**, 547–553 (2004).

1444. Malawski, G. A., et al. Identifying protein construct variants with increased crystallization propensity: a case study. *Protein Sci* **15**, 2718–2728 (2006).

1445. Anderson, M. J., Hansen, C. L., & Quake, S. R. Phase knowledge enables rational screens for protein crystallization. *Proc Natl Acad Sci USA* **103**, 16746–16751 (2006).

1446. Kobilka, B. & Schertler, G. F. New G-protein-coupled receptor crystal structures: insights and limitations. *Trends Pharmacol Sci* **29**, 79–83 (2008).

1447. Columbus, L. & Hubbell, W. L. A new spin on protein dynamics. *Trends Biochem Sci* **27**, 288–295 (2002).

1448. Fanucci, G. E. & Cafiso, D. S. Recent advances and applications of site-directed spin labeling. *Curr Opin Struct Biol* **16**, 644–653 (2006).

1449. Xie, L. & Bourne, P. E. Functional coverage of the human genome by existing structures, structural genomics targets, and homology models. *PLoS Comput Biol* **1**, e31 (2005).

1450. Fogg, M. J. & Wilkinson, A. J. Higher-throughput approaches to crystallization and crystal structure determination. *Biochem Soc Trans* **36**, 771–775 (2008).

1451. Zwart, P. H., et al. Automated structure solution with the PHENIX suite. *Methods Mol Biol* **426**, 419–435 (2008).

1452. Kuhlman, B. & Baker, D. Exploring folding free energy landscapes using computational protein design. *Curr Opin Struct Biol* **14**, 89–95 (2004).

1453. Bradley, P., Misura, K. M., & Baker, D. Toward high-resolution *de novo* structure prediction for small proteins. *Science* **309**, 1868–1871 (2005).

1454. Dill, K. A. Polymer principles and protein folding. *Protein Sci* **8**, 1166–1180 (1999).

1455. Dill, K. A., Ozkan, S. B., Shell, M. S., & Weikl, T. R. The protein folding problem. *Annu Rev Biophys* **37**, 289–316 (2008).

1456. Tsai, C. J., Kumar, S., Ma, B., & Nussinov, R. Folding funnels, binding funnels, and protein function. *Protein Sci* **8**, 1181–1190 (1999).

1457. Lindorff-Larsen, K., Rogen, P., Paci, E., Vendruscolo, M., & Dobson, C. M. Protein folding and the organization of the protein topology universe. *Trends Biochem Sci* **30**, 13–19 (2005).

1458. Lindberg, M. O. & Oliveberg, M. Malleability of protein folding pathways: a simple reason for complex behaviour. *Curr Opin Struct Biol* **17**, 21–29 (2007).

1459. Bedard, S., Krishna, M. M., Mayne, L., & Englander, S. W. Protein folding: independent unrelated pathways or predetermined pathway with optional errors. *Proc Natl Acad Sci USA* **105**, 7182–7187 (2008).

1460. Baker, D. A surprising simplicity to protein folding. *Nature* **405**, 39–42 (2000).

1461. Plaxco, K. W., Simons, K. T., & Baker, D. Contact order, transition state placement and the refolding rates of single domain proteins. *J Mol Biol* **277**, 985–994 (1998).

1462. Zarrine-Afsar, A., Larson, S. M. & Davidson, A. R. The family feud: do proteins with similar structures fold via the same pathway? *Curr Opin Struct Biol* **15**, 42–49 (2005).

1463. Dill, K. A., Ozkan, S. B., Weikl, T. R., Chodera, J. D., & Voelz, V. A. The protein folding problem: when will it be solved? *Curr Opin Struct Biol* **17**, 342–346 (2007).

1464. Unger, R. & Moult, J. Finding the lowest free energy conformation of a protein is an NP-hard problem: proof and implications. *Bull Math Biol* **55**, 1183–1198 (1993).

1465. Hart, W. E. & Istrail, S. Robust proofs of NP-hardness for protein folding: general lattices and energy potentials. *J Comput Biol* **4**, 1–22 (1997).

1466. Zhang, Y. Progress and challenges in protein structure prediction. *Curr Opin Struct Biol* **18**, 342–348 (2008).

1467. Dunbrack, R. L., Jr. Rotamer libraries in the 21st century. *Curr Opin Struct Biol* **12**, 431–440 (2002).

1468. Liwo, A., Khalili, M., & Scheraga, H. A. *Ab initio* simulations of protein-folding pathways by molecular dynamics with the united-residue model of polypeptide chains. *Proc Natl Acad Sci USA* **102**, 2362–2367 (2005).

1469. Bujnicki, J. M. Protein-structure prediction by recombination of fragments. *Chembiochem* **7**, 19–27 (2006).

1470. Zhang, Y. & Skolnick, J. The protein structure prediction problem could be solved using the current PDB library. *Proc Natl Acad Sci USA* **102**, 1029–1034 (2005).

1471. Das, R. & Baker, D. Macromolecular modeling with rosetta. *Annu Rev Biochem* **77**, 363–382 (2008).

1472. Butterfoss, G. L. & Kuhlman, B. Computer-based design of novel protein structures. *Annu Rev Biophys Biomol Struct* **35**, 49–65 (2006).

1473. Lupas, A. N., Ponting, C. P., & Russell, R. B. On the evolution of protein folds: are similar motifs in different protein folds the result of convergence, insertion, or relics of an ancient peptide world? *J Struct Biol* **134**, 191–203 (2001).

1474. Alva, V., Ammelburg, M., Soding, J., & Lupas, A. N. On the origin of the histone fold. *BMC Struct Biol* **7**, 17 (2007).

1475. Aloy, P., Stark, A., Hadley, C., & Russell, R. B. Predictions without templates: new folds, secondary structure, and contacts in CASP5. *Proteins* **53** (Suppl 6), 436–456 (2003).

1476. Vincent, J. J., Tai, C. H., Sathyanarayana, B. K., & Lee, B. Assessment of CASP6 predictions for new and nearly new fold targets. *Proteins* **61** (Suppl 7), 67–83 (2005).

1477. Wodak, S. J. & Mendez, R. Prediction of protein–protein interactions: the CAPRI experiment, its evaluation and implications. *Curr Opin Struct Biol* **14**, 242–249 (2004).

1478. Moult, J., Pedersen, J. T., Judson, R., & Fidelis, K. A large-scale experiment to assess protein structure prediction methods. *Proteins* **23**, ii–v (1995).

1479. Moult, J. A decade of CASP: progress, bottlenecks and prognosis in protein structure prediction. *Curr Opin Struct Biol* **15**, 285–289 (2005).

1480. Shi, S. Y., et al. Analysis of CASP8 targets and predictions. (http://prodata.swmed.edu/CASP8, 2009).

1481. Moult, J., et al. Critical assessment of methods of protein structure prediction: round VII. *Proteins* **69** (Suppl 8), 3–9 (2007).

1482. Jauch, R., Yeo, H. C., Kolatkar, P. R., & Clarke, N. D. Assessment of CASP7 structure predictions for template free targets. *Proteins* **69** (Suppl 8), 57–67 (2007).

1483. Bonneau, R., Ruczinski, I., Tsai, J., & Baker, D. Contact order and *ab initio* protein structure prediction. *Protein Sci* **11**, 1937–1944 (2002).

1484. Richardson, J. S., et al. Looking at proteins: representations, folding, packing, and design. Biophysical Society National Lecture, 1992. *Biophys J* **63**, 1185–1209 (1992).

1485. Raes, J., Harrington, E. D., Singh, A. H., & Bork, P. Protein function space: viewing the limits or limited by our view? *Curr Opin Struct Biol* **17**, 362–369 (2007).

1486. Regan, L. & DeGrado, W. F. Characterization of a helical protein designed from first principles. *Science* **241**, 976–978 (1988).

1487. Hecht, M. H., Richardson, J. S., Richardson, D. C., & Ogden, R. C. *De novo* design, expression, and characterization of Felix: a four-helix bundle protein of native-like sequence. *Science* **249**, 884–891 (1990).

1488. Pierce, N. A. & Winfree, E. Protein design is NP-hard. *Protein Eng* **15**, 779–782 (2002).

1489. Voigt, C. A., Gordon, D. B., & Mayo, S. L. Trading accuracy for speed: a quantitative comparison of search algorithms in protein sequence design. *J Mol Biol* **299**, 789–803 (2000).

1490. Gordon, D. B., Hom, G. K., Mayo, S. L., & Pierce, N. A. Exact rotamer optimization for protein design. *J Comput Chem* **24**, 232–243 (2003).

1491. De Maeyer, M., Desmet, J., & Lasters, I. All in one: a highly detailed rotamer library improves both accuracy and speed in the modelling of sidechains by dead-end elimination. *Fold Des* **2**, 53–66 (1997).

1492. Park, S., Yang, X., & Saven, J. G. Advances in computational protein design. *Curr Opin Struct Biol* **14**, 487–494 (2004).

1493. Dahiyat, B. I. & Mayo, S. L. *De novo* protein design: fully automated sequence selection. *Science* **278**, 82–87 (1997).

1494. Offredi, F., et al. *De novo* backbone and sequence design of an idealized alpha/beta-barrel protein: evidence of stable tertiary structure. *J Mol Biol* **325**, 163–174 (2003).

1495. Kuhlman, B., et al. Design of a novel globular protein fold with atomic-level accuracy. *Science* **302**, 1364–1368 (2003).

1496. Nanda, V., Andrianarijaona, A., & Narayanan, C. The role of protein homochirality in shaping the energy landscape of folding. *Protein Sci* **16**, 1667–1675 (2007).

1497. Kim, D. E., Gu, H., & Baker, D. The sequences of small proteins are not extensively optimized for rapid folding by natural selection. *Proc Natl Acad Sci USA* **95**, 4982–4986 (1998).

1498. Watters, A. L., et al. The highly cooperative folding of small naturally occurring proteins is likely the result of natural selection. *Cell* **128**, 613–624 (2007).

1499. Dantas, G., et al. Mis-translation of a computationally designed protein yields an exceptionally stable homodimer: implications for protein engineering and evolution. *J Mol Biol* **362**, 1004–1024 (2006).

1500. Poole, A. M. & Ranganathan, R. Knowledge-based potentials in protein design. *Curr Opin Struct Biol* **16**, 508–513 (2006).

1501. Cooper, W. J. & Waters, M. L. Molecular recognition with designed peptides and proteins. *Curr Opin Chem Biol* **9**, 627–631 (2005).

1502. Jiang, L., et al. *De novo* computational design of retro-aldol enzymes. *Science* **319**, 1387–1391 (2008).

1503. Rothlisberger, D., et al. Kemp elimination catalysts by computational enzyme design. *Nature* **453**, 190–195 (2008).

1504. Sterner, R., Merkl, R., & Raushel, F. M. Computational design of enzymes. *Chem Biol* **15**, 421–423 (2008).

1505. Bolon, D. N., Voigt, C. A., & Mayo, S. L. *De novo* design of biocatalysts. *Curr Opin Chem Biol* **6**, 125–129 (2002).

1506. Rosenberg, M. & Goldblum, A. Computational protein design: a novel path to future protein drugs. *Curr Pharm Des* **12**, 3973–3997 (2006).

1507. Alvizo, O., Allen, B. D., & Mayo, S. L. Computational protein design promises to revolutionize protein engineering. *Biotechniques* **42**, 31, 33, 35 passim (2007).

1508. Tuchscherer, G., Grell, D., Mathieu, M., & Mutter, M. Extending the concept of template-assembled synthetic proteins. *J Pept Res* **54**, 185–194 (1999).

1509. Kent, S. Novel forms of chemical protein diversity: in nature and in the laboratory. *Curr Opin Biotechnol* **15**, 607–614 (2004).

1510. Zheng, Z. M., Tang, S., & Tao, M. Development of resistance to RNAi in mammalian cells. *Ann N Y Acad Sci* **1058**, 105–118 (2005).

1511. van Rij, R. P. & Andino, R. The silent treatment: RNAi as a defense against virus infection in mammals. *Trends Biotechnol* **24**, 186–193 (2006).

1512. Isaka, Y. DNAzymes as potential therapeutic molecules. *Curr Opin Mol Ther* **9**, 132–136 (2007).

1513. Liu, J. Control of protein synthesis and mRNA degradation by microRNAs. *Curr Opin Cell Biol* **20**, 214–221 (2008).

1514. Love, T. M., Moffett, H. F., & Novina, C. D. Not miR-ly small RNAs: big potential for microRNAs in therapy. *J Allergy Clin Immunol* **121**, 309–319 (2008).

1515. Sall, A., et al. MicroRNAs-based therapeutic strategy for virally induced diseases. *Curr Drug Discov Technol* **5**, 49–58 (2008).

1516. Birmingham, A., et al. A protocol for designing siRNAs with high functionality and specificity. *Nat Protoc* **2**, 2068–2078 (2007).

1517. Sullenger, B. A. & Gilboa, E. Emerging clinical applications of RNA. *Nature* **418**, 252–258 (2002).

1518. Puttaraju, M., Jamison, S. F., Mansfield, S. G., Garcia-Blanco, M. A. & Mitchell, L. G. Spliceosome-mediated RNA trans-splicing as a tool for gene therapy. *Nat Biotechnol* **17**, 246–252 (1999).

1519. Das, R. & Baker, D. Automated *de novo* prediction of native-like RNA tertiary structures. *Proc Natl Acad Sci USA* **104**, 14664–14669 (2007).

1520. Rossi, J. J. Partnering aptamer and RNAi technologies. *Mol Ther* **14**, 461–462 (2006).

1521. Vornlocher, H. P. Antibody-directed cell-type-specific delivery of siRNA. *Trends Mol Med* **12**, 1–3 (2006).

1522. McNamara, J. O., 2nd, et al. Cell type-specific delivery of siRNAs with aptamer–siRNA chimeras. *Nat Biotechnol* **24**, 1005–1015 (2006).

1523. An, C. I., Trinh, V. B., & Yokobayashi, Y. Artificial control of gene expression in mammalian cells by modulating RNA interference through aptamer–small molecule interaction. *RNA* **12**, 710–716 (2006).

1524. Noble, D. Will genomics revolutionise pharmaceutical R&D? *Trends Biotechnol* **21**, 333–337 (2003).

1525. Germain, R. N. The art of the probable: system control in the adaptive immune system. *Science* **293**, 240–245 (2001).

1526. Lederberg, J. & McCray, A. T. 'Ome sweet' omics: a genealogical treasury of words. *The Scientist* **15**, 8–9 (2001).

1527. Kirschner, M. W. The meaning of systems biology. *Cell* **121**, 503–504 (2005).

1528. Crameri, A., Dawes, G., Rodriguez, E., Jr., Silver, S., & Stemmer, W. P. Molecular evolution of an arsenate detoxification pathway by DNA shuffling. *Nat Biotechnol* **15**, 436–438 (1997).

1529. Chatterjee, R. & Yuan, L. Directed evolution of metabolic pathways. *Trends Biotechnol* **24**, 28–38 (2006).

1530. Barabasi, A. L. & Oltvai, Z. N. Network biology: understanding the cell's functional organization. *Nat Rev Genet* **5**, 101–113 (2004).

1531. Milo, R., et al. Network motifs: simple building blocks of complex networks. *Science* **298**, 824–827 (2002).

1532. Dobrin, R., Beg, Q. K., Barabasi, A. L., & Oltvai, Z. N. Aggregation of topological motifs in the *Escherichia coli* transcriptional regulatory network. *BMC Bioinformatics* **5**, 10 (2004).

1533. Krishnan, A., Giuliani, A., & Tomita, M. Indeterminacy of reverse engineering of gene regulatory networks: the curse of gene elasticity. *PLoS ONE* **2**, e562 (2007).

1534. Sams-Dodd, F. Target-based drug discovery: is something wrong? *Drug Discov Today* **10**, 139–147 (2005).

1535. Brown, D. Unfinished business: target-based drug discovery. *Drug Discov Today* **12**, 1007–1012 (2007).

1536. Bailey, J. E. Lessons from metabolic engineering for functional genomics and drug discovery. *Nat Biotechnol* **17**, 616–618 (1999).

1537. Hellerstein, M. K. A critique of the molecular target-based drug discovery paradigm based on principles of metabolic control: advantages of pathway-based discovery. *Metab Eng* **10**, 1–9 (2008).

1538. Fischer, E. & Sauer, U. A novel metabolic cycle catalyzes glucose oxidation and anaplerosis in hungry *Escherichia coli. J Biol Chem* **278**, 46446–46451 (2003).

1539. Goh, K. I., et al. The human disease network. *Proc Natl Acad Sci USA* **104**, 8685–8690 (2007).

1540. Yildirim, M. A., Goh, K.-I., Cusick, M. E., Barabasi, A.-L. & Vidal, M. Drug–target network. *Nat Biotechnol* **25**, 1119–1126 (2007).

1541. Haseltine, E. L. & Arnold, F. H. Synthetic gene circuits: design with directed evolution. *Annu Rev Biophys Biomol Struct* **36**, 1–19 (2007).

1542. Motter, A. E., Gulbahce, N., Almaas, E., & Barabasi, A. L. Predicting synthetic rescues in metabolic networks. *Mol Syst Biol* **4**, 168 (2008).

1543. Lee, I., et al. A single gene network accurately predicts phenotypic effects of gene perturbation in *Caenorhabditis elegans. Nat Genet* **40**, 181–188 (2008).

1544. Lai, C. S., Fisher, S. E., Hurst, J. A., Vargha-Khadem, F. & Monaco, A. P. A forkhead-domain gene is mutated in a severe speech and language disorder. *Nature* **413**, 519–523 (2001).

1545. Vargha-Khadem, F., Gadian, D. G., Copp, A., & Mishkin, M. FOXP2 and the neuroanatomy of speech and language. *Nat Rev Neurosci* **6**, 131–138 (2005).

1546. Ohlson, S. Designing transient binding drugs: a new concept for drug discovery. *Drug Discov Today* **13**, 433–439 (2008).

1547. Barbosa, M. D. & Celis, E. Immunogenicity of protein therapeutics and the interplay between tolerance and antibody responses. *Drug Discov Today* **12**, 674–681 (2007).

1548. Gibson, D. G., et al. Complete chemical synthesis, assembly, and cloning of a *Mycoplasma genitalium* genome. *Science* **319**, 1215–1220 (2008).

1549. Rousseau, F. & Schymkowitz, J. A systems biology perspective on protein structural dynamics and signal transduction. *Curr Opin Struct Biol* **15**, 23–30 (2005).

1550. Kenakin, T. Allosteric modulators: the new generation of receptor antagonist. *Mol Interv* **4**, 222–229 (2004).

1551. Snoep, J. L. The silicon cell initiative: working towards a detailed kinetic description at the cellular level. *Curr Opin Biotechnol* **16**, 336–343 (2005).

1552. Csete, M. E. & Doyle, J. C. Reverse engineering of biological complexity. *Science* **295**, 1664–1669 (2002).

1553. Chatterjee, S. & Templin, R. J. Biplane wing planform and flight performance of the feathered dinosaur *Microraptor gui. Proc Natl Acad Sci USA* **104**, 1576–1580 (2007).

1554. Endy, D. Genomics. Reconstruction of the genomes. *Science* **319**, 1196–1197 (2008).

1555. Koch, C. & Hepp, K. Quantum mechanics in the brain. *Nature* **440**, 611 (2006).

1556. Barrow, J. D. *Impossibility. The Limits of Science and the Science of Limits* (Oxford University Press, 1998).

1557. Hyatt, S. M., et al. On the intractability of estrogen-related receptor alpha as a target for activation by small molecules. *J Med Chem* **50**, 6722–6724 (2007).

1558. Cornforth, J. W. The trouble with synthesis. *Aust J Chem* **46**, 157–170 (1993).

1559. Eichler, R., et al. Chemical characterization of element 112. *Nature* **447**, 72–75 (2007).

1560. Gimona, M. Protein linguistics: a grammar for modular protein assembly? *Nat Rev Mol Cell Biol* **7**, 68–73 (2006).

1561. Cleland, C. E. & Chyba, C. F. Defining 'life'. *Orig Life Evol Biosph* **32**, 387–393 (2002).

1562. Oliver, J. D. & Perry, R. S. Definitely life but not definitively. *Orig Life Evol Biosph* **36**, 515–521 (2006).

1563. DeFrancis, J. *The Chinese Language. Fact and Fantasy* (University of Hawaii Press, Honolulu, HI 1984).

1564. Unger, J. M. *Ideogram. Chinese Characters and the Myth of Disembodied Meaning* (University of Hawaii Press, Honolulu, HI 2004).

1565. Crofts, A. R. Life, information, entropy, and time: vehicles for semantic inheritance. *Complexity* **13**, 14–50 (2007).

1566. Clement, M. J., et al. Toward a better understanding of the basis of the molecular mimicry of polysaccharide antigens by peptides: the example of *Shigella flexneri* 5a. *J Biol Chem* **281**, 2317–2332 (2006).

1567. Hofstadter, D. R. *Le Ton Beau de Marot* (Basic Books, 1997).

1568. Steiner, G. *After Babel: Aspects of Language and Translation* (Oxford University Press, 1975).

1569. Wagner, G. P., Pavlicev, M., & Cheverud, J. M. The road to modularity. *Nat Rev Genet* **8**, 921–931 (2007).

1570. Xiang, X. D., et al. A combinatorial approach to materials discovery. *Science* **268**, 1738–1740 (1995).

1571. Grimm, D. & Kay, M. A. Combinatorial RNAi: a winning strategy for the race against evolving targets? *Mol Ther* **15**, 878–888 (2007).

1572. Gardner, M. *Science. Good, Bad, and Bogus* (Prometheus Books, Buffalo, NY, 1981).

1573. Pinker, S. *How The Mind Works* (W. W. Norton, 1997).

1574. Swift, J. *Gulliver's Travels*, 1992 edition (Wordsworth Classics, 1726).

1575. Monod, J. *Chance and Necessity* (Vintage Books, New York, 1971).

1576. Fechner, U. & Schneider, G. Flux (2): comparison of molecular mutation and crossover operators for ligand-based *de novo* design. *J Chem Inf Model* **47**, 656–667 (2007).

1577. Whelan, K. E. & King, R. D. Intelligent software for laboratory automation. *Trends Biotechnol* **22**, 440–445 (2004).

1578. Soldatova, L. N., Clare, A., Sparkes, A., & King, R. D. An ontology for a robot scientist. *Bioinformatics* **22**, e464–e471 (2006).

1579. Zykov, V., Mytilinaios, E., Adams, B., & Lipson, H. Robotics: self-reproducing machines. *Nature* **435**, 163–164 (2005).

REFERENCES IN FIGURE LEGENDS

1. Vogel, C., Berzuini, C., Bashton, M., Gough, J., & Teichmann, S. A. Supra-domains: evolutionary units larger than single protein domains. *J Mol Biol* **336**, 809–823 (2004).

2. Berman, H., Henrick, K., & Nakamura, H. Announcing the worldwide Protein Data Bank. *Nat Struct Biol* **10**, 980 (2003).

3. Vitagliano, L., Masullo, M., Sica, F., Zagari, A., & Bocchini, V. The crystal structure of *Sulfolobus solfataricus* elongation factor 1alpha in complex with GDP reveals novel features in nucleotide binding and exchange. *EMBO J* **20**, 5305–5311 (2001).

4. al-Karadaghi, S., Aevarsson, A., Garber, M., Zheltonosova, J., & Liljas, A. The structure of elongation factor G in complex with GDP: conformational flexibility and nucleotide exchange. *Structure* **4**, 555–565 (1996).

5. Moreland, J. L., Gramada, A., Buzko, O. V., Zhang, Q., & Bourne, P. E. The Molecular Biology Toolkit (MBT): a modular platform for developing molecular visualization applications. *BMC Bioinformatics* **6**, 21 (2005).

6. Wagner, H. & Bauer, S. All is not Toll: new pathways in DNA recognition. *J Exp Med* **203**, 265–268 (2006).

7. Rudolph, M. G., Stanfield, R. L., & Wilson, I. A. How TCRs bind MHCs, peptides, and coreceptors. *Annu Rev Immunol* **24**, 419–466 (2006).

8. Ding, Y. H. et al. Two human T cell receptors bind in a similar diagonal mode to the HLA-A2/Tax peptide complex using different TCR amino acids. *Immunity* **8**, 403–411 (1998).

9. Borbulevych, O. Y. et al. Structures of MART-126/27-35 peptide/HLA-A2 complexes reveal a remarkable disconnect between antigen structural homology and T cell recognition. *J Mol Biol* **372**, 1123–1136 (2007).

10. Beltrami, A. et al. Citrullination-dependent differential presentation of a self-peptide by HLA-B27 subtypes. *J Biol Chem* **283**, 27189–27199 (2008).

11. Stemmer, W. P. Rapid evolution of a protein *in vitro* by DNA shuffling. *Nature* **370**, 389–391 (1994).

12. Stemmer, W. P. DNA shuffling by random fragmentation and reassembly: *in vitro* recombination for molecular evolution. *Proc Natl Acad Sci USA* **91**, 10747–10751 (1994).

13. Lubkowski, J., Hennecke, F., Pluckthun, A., & Wlodawer, A. The structural basis of phage display elucidated by the crystal structure of the N-terminal domains of g3p. *Nat Struct Biol* **5**, 140–147 (1998).

14. Marvin, D. A. Filamentous phage structure, infection and assembly. *Curr Opin Struct Biol* **8**, 150–158 (1998).

15. Deng, L. W. & Perham, R. N. Delineating the site of interaction on the pIII protein of filamentous bacteriophage fd with the F-pilus of *Escherichia coli*. *J Mol Biol* **319**, 603–614 (2002).

16. Zahnd, C., Amstutz, P., & Pluckthun, A. Ribosome display: selecting and evolving proteins *in vitro* that specifically bind to a target. *Nat Methods* **4**, 269–279 (2007).

17. Roberts, R. W. & Szostak, J. W. RNA–peptide fusions for the *in vitro* selection of peptides and proteins. *Proc Natl Acad Sci USA* **94**, 12297–12302 (1997).

18. Nemoto, N., Miyamoto-Sato, E., Husimi, Y., & Yanagawa, H. *In vitro* virus: bonding of mRNA bearing puromycin at the 3'-terminal end to the C-terminal end of its encoded protein on the ribosome *in vitro*. *FEBS Lett* **414**, 405–408 (1997).

19. Kurz, M., Gu, K., & Lohse, P. A. Psoralen photo-crosslinked mRNA–puromycin conjugates: a novel template for the rapid and facile preparation of mRNA–protein fusions. *Nucleic Acids Res* **28**, E83 (2000).

20. Mastrobattista, E. et al. High-throughput screening of enzyme libraries: *in vitro* evolution of a beta-galactosidase by fluorescence-activated sorting of double emulsions. *Chem Biol* **12**, 1291–1300 (2005).

21. van Pouderoyen, G., Eggert, T., Jaeger, K. E., & Dijkstra, B. W. The crystal structure of *Bacillus subtilis* lipase: a minimal alpha/beta hydrolase fold enzyme. *J Mol Biol* **309**, 215–226 (2001).

22. Acharya, P., Rajakumara, E., Sankaranarayanan, R., & Rao, N. M. Structural basis of selection and thermostability of laboratory evolved *Bacillus subtilis* lipase. *J Mol Biol* **341**, 1271–1281 (2004).

23. Kawasaki, K., Kondo, H., Suzuki, M., Ohgiya, S., & Tsuda, S. Alternate conformations observed in catalytic serine of *Bacillus subtilis* lipase determined at 1.3 Å resolution. *Acta Crystallogr D Biol Crystallogr* **58**, 1168–1174 (2002).

24. Guex, N. & Peitsch, M. C. SWISS-MODEL and the Swiss-PdbViewer: an environment for comparative protein modeling. *Electrophoresis* **18**, 2714–2723 (1997).

25. Lang, D., Thoma, R., Henn-Sax, M., Sterner, R., & Wilmanns, M. Structural evidence for evolution of the beta/alpha barrel scaffold by gene duplication and fusion. *Science* **289**, 1546–1550 (2000).

26. Wierenga, R. K. The TIM-barrel fold: a versatile framework for efficient enzymes. *FEBS Lett* **492**, 193–198 (2001).

27. Galperin, M. Y., Walker, D. R., & Koonin, E. V. Analogous enzymes: independent inventions in enzyme evolution. *Genome Res* **8**, 779–790 (1998).

28. Varghese, J. N. et al. Three-dimensional structures of two plant beta-glucan endohydrolases with distinct substrate specificities. *Proc Natl Acad Sci USA* **91**, 2785–2789 (1994).

29. Hahn, M., Olsen, O., Politz, O., Borriss, R., & Heinemann, U. Crystal structure and site-directed mutagenesis of *Bacillus macerans* endo-1,3-1,4-beta-glucanase. *J Biol Chem* **270**, 3081–3088 (1995).

30. Subramanya, H. S., Doherty, A. J., Ashford, S. R., & Wigley, D. B. Crystal structure of an ATP-dependent DNA ligase from bacteriophage T7. *Cell* **85**, 607–615 (1996).

31. Dalal, S., Balasubramanian, S. & Regan, L. Protein alchemy: changing beta-sheet into alpha-helix. *Nat Struct Biol* **4**, 548–552 (1997).

32. Alexander, P. A., Rozak, D. A., Orban, J., & Bryan, P. N. Directed evolution of highly homologous proteins with different folds by phage display: implications for the protein folding code. *Biochemistry* **44**, 14045–14054 (2005).

33. Shi, H. & Moore, P. B. The crystal structure of yeast phenylalanine tRNA at 1.93 Å resolution: a classic structure revisited. *RNA* **6**, 1091–1105 (2000).

34. Wang, L., Xie, J., & Schultz, P. G. Expanding the genetic code. *Annu Rev Biophys Biomol Struct* **35**, 225–249 (2006).

35. Xie, J. & Schultz, P. G. A chemical toolkit for proteins: an expanded genetic code. *Nat Rev Mol Cell Biol* **7**, 775–782 (2006).

36. Murray, J. B., Szoke, H., Szoke, A., & Scott, W. G. Capture and visualization of a catalytic RNA enzyme-product complex using crystal lattice trapping and X-ray holographic reconstruction. *Mol Cell* **5**, 279–287 (2000).

37. Tsukiji, S., Pattnaik, S. B., & Suga, H. An alcohol dehydrogenase ribozyme. *Nat Struct Biol* **10**, 713–717 (2003).

38. Flinders, J. et al. Recognition of planar and nonplanar ligands in the malachite green–RNA aptamer complex. *Chembiochem* **5**, 62–72 (2004).

39. Eriksson, M. & Nielsen, P. E. Solution structure of a peptide nucleic acid–DNA duplex. *Nat Struct Biol* **3**, 410–413 (1996).

40. Braden, B. C. et al. Three-dimensional structures of the free and the antigen-complexed Fab from monoclonal anti-lysozyme antibody D44.1. *J Mol Biol* **243**, 767–781 (1994).

41. Hu, S. et al. Epitope mapping and structural analysis of an anti-ErbB2 antibody A21: molecular basis for tumor inhibitory mechanism. *Proteins* **70**, 938–949 (2008).

42. Stanfield, R. L., Dooley, H., Verdino, P., Flajnik, M. F., & Wilson, I. A. Maturation of shark single-domain (IgNAR) antibodies: evidence for induced-fit binding. *J Mol Biol* **367**, 358–372 (2007).

43. De Genst, E. et al. Molecular basis for the preferential cleft recognition by dromedary heavy-chain antibodies. *Proc Natl Acad Sci USA* **103**, 4586–4591 (2006).

44. Gouverneur, V. E. et al. Control of the exo and endo pathways of the Diels–Alder reaction by antibody catalysis. *Science* **262**, 204–208 (1993).

45. Jacobsen, J. R. & Schultz, P. G. The scope of antibody catalysis. *Curr Opin Struct Biol* **5**, 818–824 (1995).

46. Binz, H. K. et al. High-affinity binders selected from designed ankyrin repeat protein libraries. *Nat Biotechnol* **22**, 575–582 (2004).

47. Wahlberg, E. et al. An affibody in complex with a target protein: structure and coupled folding. *Proc Natl Acad Sci USA* **100**, 3185–3190 (2003).

48. Chu, R. et al. Redesign of a four-helix bundle protein by phage display coupled with proteolysis and structural characterization by NMR and X-ray crystallography. *J Mol Biol* **323**, 253–262 (2002).

49. Ku, J. & Schultz, P. G. Alternate protein frameworks for molecular recognition. *Proc Natl Acad Sci USA* **92**, 6552–6556 (1995).

50. Dickinson, C. D. et al. Crystal structure of the tenth type III cell adhesion module of human fibronectin. *J Mol Biol* **236**, 1079–1092 (1994).

51. Xu, L. et al. Directed evolution of high-affinity antibody mimics using mRNA display. *Chem Biol* **9**, 933–942 (2002).

52. Skerra, A. Imitating the humoral immune response. *Curr Opin Chem Biol* **7**, 683–693 (2003).

53. Skerra, A. Alternative non-antibody scaffolds for molecular recognition. *Curr Opin Biotechnol* **18**, 295–304 (2007).

54. Murata, M. & Yasumoto, T. The structure elucidation and biological activities of high molecular weight algal toxins: maitotoxin, prymnesins and zooxanthellatoxins. *Nat Prod Rep* **17**, 293–314 (2000).

55. Wright, E. M. et al. 'Active' sugar transport in eukaryotes. *J Exp Biol* **196**, 197–212 (1994).

56. Wilson, J. X. Regulation of vitamin C transport. *Annu Rev Nutr* **25**, 105–125 (2005).

57. Huc, I. & Lehn, J. M. Virtual combinatorial libraries: dynamic generation of molecular and supramolecular diversity by self-assembly. *Proc Natl Acad Sci USA* **94**, 2106–2110 (1997).

58. Silverman, R. B. *The Organic Chemistry of Drug Design and Drug Action*, 2nd edition (Elsevier, 2004).

59. Dill, K. A., Ozkan, S. B., Shell, M. S., & Weikl, T. R. The protein folding problem. *Annu Rev Biophys* **37**, 289–316 (2008).

60. Tsai, C. J. et al. Synonymous mutations and ribosome stalling can lead to altered folding pathways and distinct minima. *J Mol Biol* (2008).

61. Kuhlman, B. et al. Design of a novel globular protein fold with atomic-level accuracy. *Science* **302**, 1364–1368 (2003).

62. Silverman, S. K. *In vitro* selection, characterization, and application of deoxyribozymes that cleave RNA. *Nucleic Acids Res* **33**, 6151–6163 (2005).

63. Breaker, R. R. Natural and engineered nucleic acids as tools to explore biology. *Nature* **432**, 838–845 (2004).

64. McNamara, J. O., 2nd et al. Cell type-specific delivery of siRNAs with aptamer–siRNA chimeras. *Nat Biotechnol* **24**, 1005–1015 (2006).

65. Amarzguioui, M., Rossi, J. J., & Kim, D. Approaches for chemically synthesized siRNA and vector-mediated RNAi. *FEBS Lett* **579**, 5974–5981 (2005).

INDEX
